Handbook of Technical Writing

Third Edition

Handbook of Technical Writing

Third Edition

Charles T. Brusaw
NCR Corporation

Gerald J. Alred
University of Wisconsin—Milwaukee

Walter E. Oliu
U.S. Nuclear Regulatory Commission

St. Martin's Press
New York

For our wives and families

Library of Congress Catalog Number: 86-60646

Manufactured in the United States of America.
ghijkl

For information, write:
St. Martin's Press, Inc.
175 Fifth Avenue
New York, NY 10010

cover design: Joel Weltman
text design: Arthur Ritter
art: G&H/Soho

ISBN: 0-312-35810-5

Preface

The *Handbook of Technical Writing*, Third Edition, is designed to be a comprehensive, thoroughly practical reference guide for students, as well as a desktop reference for people already working in industry or government. In the classroom, it can be used along with a standard textbook, although instructors wishing to avoid the constraints imposed by a textbook might easily use this handbook as the basic resource.

The *Handbook* offers coverage far beyond the scope of conventional English handbooks. In addition to a comprehensive treatment of grammar, usage, style, format, and writing procedures (planning, research, outlining, methods of development, and so forth), it provides information on all types of technical communication—reports of various kinds (including oral presentations), proposals, instructions, specifications, job descriptions, letters, and memorandums. It also covers the use of word processors. It gives abundant examples, all drawn from technical or industrial contexts to provide the greatest possible relevance for professionally oriented readers.

We believe that the unique four-way access system in the *Handbook of Technical Writing* makes it more flexible and easier to use than other writing handbooks we have seen. The main body of the book is organized alphabetically, so that readers can usually go immediately to the topic at hand. The index provides an exhaustive list of the topics, indexed not only by the terms actually used in the entries but also by various other terms that readers might think of instead. The Topical Key to the Alphabetical Entries groups the entries into categories, helping readers pull together information on broad subjects covered by several different entries and facilitating correlation of the *Handbook* with standard textbooks or writing guides. Five Steps to Successful Writing offers a brief, step-by-step guide through the writing process, which is then summarized in list fashion, with page references to pertinent entries, by the Checklist of the Writing Process.

In preparing this third edition, we have carefully reexamined every entry, clarified or updated a great many, deleted a few that proved to have been seldom used, and added a substantial number of new ones. Among the entries added are those on **word processing, nominalizations**, and **resignation letter or memorandum.** Other new entries discuss **concluding, notes, plurals, possessive case, whether or not**, and **who's/whose.** In addition to the new entries, major revisions have been made to the entries on **correspondence, executive summaries, job search, memorandums, oral presentations, refusal letters, résumés, sales proposals, style, technical manuals, voice**, and a great many others. We have also consolidated and expanded coverage of bibliographic documentation in **documenting sources.** Furthermore, some of the smaller grammatical entries have been deleted and the information has been incorporated into larger "parent" entries. A number of titles of entries have also been changed to make them more precise: for example, "house organ articles" has been changed to **newsletter articles**, and "oral reports" is now **oral presentations.**

As in earlier editions, the entries themselves are complete without being exhaustive; that is, they focus on solutions to specific writing problems. Therefore, the general tone is prescriptive. For example, questions about the history of a term such as *data* and the controversy surrounding its use may be interesting, but they are secondary to the advice the user seeks. So, although debate over divided usage is acknowledged, a guideline is offered to get the writer on with the business of writing. This principle holds true for the grammatical entries as well; they do not attempt to settle the debate between traditional and linguistically oriented grammarians, although current linguistic thinking about grammatical problems is presented where it contributes to the explanation of a concept. The guidance offered in the entries will solve most of the problems users of this book will encounter. However, students should understand that some companies and government agencies have their own standards governing certain subjects dealt with in this book and their employees must conform to those standards.

The *Handbook of Technical Writing* shares not only its format but a majority of the topics in the alphabetical entries with *The Business Writer's Handbook*, also published in a third edition by St. Martin's Press. Within the alphabetical entries common to both books, however, the examples given, as well as portions of the text, are often different—illustrating the same writing principles but in language that reflects a technical context in one handbook and a business context in the other. In a course combining both technical and business writing, either book might be used, the choice depending upon the emphasis the instructor thinks most appropriate.

We are deeply grateful to the many instructors, students, technical writers, and others who have helped us prepare the *Handbook of Technical Writing*. For his contributions to the first edition, which continue to lend strength to the second and third, we are indebted to Conrad R. Winterhalter, who reviewed the entire manuscript in several drafts and wrote the original entries on the **résumé, letter of application**, and **letter of acceptance**, and to Keith Palmer, who wrote the entry on **government proposals**. Others whose suggestions for the first edition continue to be reflected here, and to whom we are happy to renew our thanks, are James H. Chaffee, Paul Falon, Wayne Losano, Lee Newcomer, Thomas Pearsall, John Taylor, and Arthur Young.

For assistance with the second edition, we renew our thanks to Thomas Warren and his staff at Oklahoma State University, whose detailed advice was as valuable as it was generous. We are also deeply indebted to Edmund Dandridge, North Carolina State University; James Farrelly, University of Dayton; Patrick Kelley, New Mexico State University; Carol Mablekos, Temple University; Roger Masse, New Mexico State University; Judy McInish, Hornbake Library, University of Maryland; Wendy Osborne, Society of American Foresters; Diana Reep, University of Akron; Charles Stratton, University of Idaho; Mary Thompson, Krannert Graduate School of Management, Purdue Univeristy, and Merrill Whitburn, Rensselaer Polytechnic Institute.

For their assistance with the third edition, we wish to recognize the contributions of William Van Pelt of the University of Wisconsin–Milwaukee, who wrote the entry on **word processing**; Eleanor Berry of the University of Wisconsin–Milwaukee, who contributed to the entry on **documenting sources**; Mary Mullins and Sandra Balzo of First Wisconsin National Bank who provided the sample newsletter article; and Ann Thomas, Rayleona Sanders, and Mimi Mejac—all of the Nuclear Regulatory Commission—for their helpful comments. We would also like to express our heartfelt appreciation to our academic reviewers for the third edition, who did a truly superb job: Deborah Barret, Department of Languages, Houston

Baptist University, Houston; Mary Coney, STC Program, University of Washington, Seattle; Susan Hitchcock, University of Virginia, Covesville; Lois Rew, English Department, San Jose State University, San Jose, California.

We also express our appreciation to the following organizations for permission to use examples of their technical writing: Biospherics, Inc., *Chemical Engineering*, First Wisconsin National Bank, Harnischfeger Corporation, Johnson Service Company, NCR Corporation, and Professional Secretaries International.

We would also like to acknowledge the editorial guidance we received from St. Martin's Press, especially from production supervisor Christine Pearson, project editor Julie Nord, and acquiring editor Susan Anker.

And finally, we want to extend a special thanks to Barbara Brusaw for her truly exceptional secretarial work in the preparation of the third edition.

C.T.B.
G.J.A.
W.E.O.

Contents

Five Steps to Successful Writing

Many technical people have difficulty writing because they do not know how to begin. They sit down at a typewriter or with pen in hand and hope to fill a blank page without realizing that skill in writing, like the technical skill in which they are proficient, requires a disciplined and systematic approach.

Successful writing on the job is not the product of inspiration, nor is it merely the spoken word transferred to paper—it is primarily the result of knowing how to structure ideas on paper. The best way to ensure that a writing task will be successful—whether it is a letter, a proposal, or a formal report—is to divide the writing process into the following five steps:

- Preparation
- Research
- Organization
- Writing the draft
- Revision

At first, these five steps must be consciously—even self-consciously—followed. But so, at first, must the steps in operating a computer, designing a circuit, troubleshooting a motor, interviewing a candidate for a job, or chairing a meeting of a board of directors. With practice, the steps in each of these processes become nearly automatic. This is not to suggest that writing becomes easy; it does not because writing well requires that you exercise judgment during each step. But the easiest and most efficient way to do it is systematically.

As you master the five steps of the writing process, keep in mind that the steps are interrelated and overlap at points. For example, your reader's needs and your objective could affect decisions in each of the other steps. You may also find that you need to retrace steps. As you organize the information you gathered, for example, you may discover a gap that will require further research.

Be aware also that the time required for each step varies with different writing tasks. For example, when writing a brief informal memo, you might follow the first three steps (preparation, research, and organization) by simply listing the points you want to cover in the order you want to cover them. In other words, the information gathering and organizing occurs mentally as you consider your purpose in writing the memo. For a formal report, on the other hand, these three steps require well-organized research, careful note taking, and detailed outlining. In short, the five steps expand, contract, and at times must be reviewed or even repeated to fit the complexity of the writing task.

Use the summary of the following five steps of the writing process to guide you. The Checklist of the Writing Process that follows this summary can also help you detect and solve problems by referring to the specific entry that explains the cause of the problem and shows you how to solve it. In the following discussion, as elsewhere throughout this book, words and phrases printed in **boldface type** refer to specific alphabetical entries.

Step 1. Preparation

Writing, like most technical tasks, requires solid **preparation.** In fact, adequate preparation is as important as writing the draft. Preparation for writing consists of (1) establishing your objective, (2) identifying your reader, and (3) determining the scope of your coverage.

Establishing your **objective** is simply determining what you want your readers to know or be able to do when they have finished reading your report or paper. But you must be precise; it is all too common for a writer to state an objective in terms so broad that it is almost useless. An objective such as "To report on possible locations for a new manufacturing facility" is too general to be of real help. But "To compare the relative advantages of Chicago, Minneapolis, and Salt Lake City as possible locations for a new manufacturing facility so that top management can choose the best location'' gives you an objective that can guide you throughout the writing process.

The next task is to identify your **reader.** Again, be precise. What are your reader's needs in relation to your subject? What does your reader already know about your subject? You need to know, for example, whether you must define basic terminology or whether such definitions will merely bore, or even insult, your reader. Is your reader actually several readers with different interests and levels of technical knowledge? For the objective stated in the previous paragraph, the reader was described as ''top management.'' But *who* is included in this category? Will one of the people evaluating the report be the personnel manager? If so, that person is likely to have a special interest in such things as the availability of qualified personnel in each city, the presence of colleges where special training would be available for employees, housing conditions, and perhaps even recreational facilities. The purchasing manager will be concerned about available sources for the materials needed by the plant. The marketing manager will give priority to such things as the plant's closeness to its primary markets and the available transport facilities for distribution of the plant's products. The financial vice-president will want to know about land and building costs and the local tax structure. The president may be interested in all these things and more; for example, he or she might be concerned about the convenience of personal travel between corporate headquarters and the new plant.

In addition to knowing the needs and interests of these readers, you should know as much as you can about their backgrounds. For example, have they visited all three cities? Have they already seen other studies of the three cities? Is this the first new plant, or have they chosen locations for new plants before? If you have many readers, one way to accommodate their needs is to combine all your readers into one composite reader and write for *that* reader. Another way is to aim parts of a document to different readers: an **executive summary** to the top manager and an **appendix** with detailed data to technical specialists, and the body of the report to middle managers.

Determining your objective and identifying your readers will help you decide what to include and not to include in your writing. Then you will have established the **scope** of your writing project. If you do not clearly define the scope before beginning your research (the next step), you will spend needless extra hours on research because you will not be sure what kind of information you need, or even how much. For example, given the objective and reader established in the preceding two paragraphs, the scope of your report on plant locations would include such things as land and building costs, available labor force, transportation facilities, proximity to sources of supply, and so forth; however, it probably would not include the history of the cities being considered or the altitude and geological features (unless these were directly related to your particular business).

Step 2. Research

The purpose of most technical writing is to explain something—usually something that is complex. This kind of writing cannot be done by someone who does not understand the subject he or she is writing about. The only way to be sure that you can deal adequately with a complex subject is to compile a complete set of notes during your **research** and then to create a working outline from the notes. Three sources of information are available to you: the library (see **library research**), personal interviews (see **interviewing for information** and **questionnaires**), and your own knowledge (see research). Consider them all when you begin your research, and use those that fit your needs. Of course, the amount of research you will need to do depends on your project; for a simple memo or letter, your "research" may amount to nothing more than jotting down the pertinent ideas before you begin to organize them.

Step 3. Organization

Without **organization**, the material gathered during your research would be incomprehensible to your reader. Thus you must determine the best sequence in which your ideas should be presented—that is, you must choose a **method of development.**

An appropriate method of development is the writer's tool for keeping things under control and the reader's means of following the writer's presentation. Your subject may readily lend itself to a particular method of development. For example, if you were giving instructions for starting an engine, you would naturally present the steps of the process in the order of their occurrence. For this task, the obvious method of development would be sequential. If you were writing about the history of the computer, your account would go from the beginning to the present. This is the chronological method of development. If your subject, like those mentioned, naturally lends itself to a certain method of development, use it—don't attempt to impose another method on it. This book includes all the methods of development that are likely to be used by technical people: chronological, sequential, spatial, increasing or decreasing order of importance, comparison, division and classification, analysis, general to specific, specific to general, and cause and effect. As the writer, you must choose the one that best suits your subject, your reader, and your objective. You are then ready to prepare your outline.

Outlining makes large or complex subjects easier for you to organize, by breaking them into manageable parts, and it ensures that your finished writing will move logically from idea to idea without omitting anything important. It also enables you to emphasize your key points by placing them in the positions of greatest importance. Finally, by forcing you to structure your thinking at an early stage, a good outline releases you to concentrate exclusively on writing when you begin the rough draft. Even if the task is only a letter or short memo, successful writing needs the logic and structure that a method of development and an outline provide, although for such simple projects the method of development and outline may be in your head rather than on paper.

If you intend to include **illustrations** with your writing, a good time to think about them is when you have completed your outline—especially if the illustrations need to be prepared by someone else while you are writing and revising the draft. If your outline is reasonably detailed, you should be able to determine which ideas will benefit from graphic support.

Make sure at this point that you use a **format** that is both helpful to your reader and appropriate to your subject and objective.

Step 4. Writing the Draft

When you have established your objective, reader's needs, and scope and have completed your research and your outline, you will be well prepared to write your first draft. To do so, simply expand the notes from your outline into **paragraphs**, without worrying about **grammar**, refinements of language, or such mechanical aspects of writing as **spelling**. Refinements will come with revision.

Write the rough draft quickly, concentrating entirely on converting your outline to sentences and paragraphs. To facilitate this step, write as though you were explaining your subject to someone sitting across the desk from you. Don't worry about a good opening. Just start. There is no need in the rough draft to be concerned about an **introduction** or **transition** unless they come quickly and easily—concentrate instead on *ideas*. Don't attempt to polish or revise. Writing and revising are different activities. Keep writing quickly so as to achieve **unity** and proportion.

Even with good preparation, writing the draft remains a chore and an obstacle for many writers. The entry on **writing the draft** describes tactics used by experienced writers to get started and keep moving. Discover which ones are the most helpful and appropriate for you.

However, the most effective way to start and keep going is to use a good outline as a map for your first draft. Above all, don't wait for inspiration to write the rough draft—you need to treat writing a draft in technical writing as you would any on-the-job task.

Step 5. Revision

Chances are that the clearer a piece of writing seems to the reader, the more effort the writer has put into **revision**. If you have followed the steps of the writing process to this point, you will have a very rough draft that could hardly be considered a finished product. Revision, the obvious final step, requires a different frame of mind than writing the draft does. Read and evaluate the draft from the reader's point of view. Be eager to find and correct faults, and be honest. Be hard on yourself for the reader's convenience; never be easy on yourself at the reader's expense.

Do not try to do all your revising at once. Read your rough draft several times, each time looking for and correcting a different set of problem or errors.

Check your draft for accuracy and completeness. Your draft should give readers exactly what they need, but it should not burden them with unnecessary information or sidetrack them into loosely related subjects.

If you have not yet written an **opening** or **introduction**, now is the time to do it. Your introduction should serve as a frame into which readers can fit the information that follows in the body of your draft; check your introduction to see that it does, in fact, provide readers with such assistance. Your opening should both announce the subject and give the reader essential background information about it.

Check your draft for unity, coherence, and transition. If it has unity, all of the sentences in each paragraph will contribute to the development of that paragraph's central idea (expressed in the topic sentence), and all of the paragraphs will contribute to the development of the main topic. If the draft has **coherence**, the sentences and the paragraphs will flow smoothly from one to the next, and the relationship of each sentence or paragraph to the one before it will be clear. Coherence can be achieved in many ways, but especially by the careful use of **transition** devices and by the maintenance of a consistent **point of view.** Also check your draft for the proper **emphasis** and **subordination** of your ideas. Now is also a good time to adjust the **pace**: if you find places where too many ideas are jammed together, space out the ideas and slow the pace; conversely, if you find a series of simple ideas expressed in short, choppy sentences, combine some of the sentences to quicken the pace.

Check your draft for **clarity**. Although much of your revising will already have improved the clarity, check now for any terms that must be defined or explained (see **defining terms**). In addition, check for **ambiguity**. Is your writing free of **affectation**? Is it free of **jargon** that your readers may not understand? Are there **abstract words** that could be replaced by concrete words? Check your entire draft for appropriate **word choice**.

Check your draft for **style**. By now you will have already done much to improve the style, but there is more you can do. Check for **conciseness**: can you eliminate useless words or phrases—what some writers call "deadwood"? It is almost always possible to do so, and your reader will benefit greatly. Get rid of **clichés** and other **trite language**. Is your writing active? Especially in technical writing there is a great temptation to use the passive **voice** when the active voice would be far stronger and more concise. Replace negative writing with **positive writing**. Check, too, for **parallel structure**. Examine your **sentence construction** and look for ways to create **sentence variety**.

Check your draft for **awkwardness** and for departures from the appropriate **tone**. Awkwardness and tone are hard to define because they are the result of a great many different things, most of which you will have already dealt with in your revision. Read your draft aloud, or have someone read it while you listen. You will recognize awkwardness because the writing sounds forced or clumsy—like something you would never say if you were *talking* to your reader because it would sound unnatural. You will recognize departures from the proper tone because they are words, phrases, or statements that are inappropriate to the relationship between you and your reader. For example, in a memo to a superior in your organization you would want to avoid phrases or statements that sound like lecturing. If your readers are serious about your subject (and it is best to assume that they are), don't try to be witty. Correcting the tone is frequently a matter of replacing a word with another that has more appropriate **connotations**. Consider the difference in tone between, "I'm puzzled by your *stubbornness* in opposing the plan," and "I'm puzzled by the *firmness* of your opposition to the plan." Finally, check slowly and carefully for problems of **grammar, punctuation**, and mechanics (**spelling, abbreviations, capital letters, acronyms and initialisms**, and the like).

The following Checklist of the Writing Process is designed to help you follow the five steps to successful writing. It can help you diagnose any writing problem you may have, and it refers you to the entry that explains the causes of the problem and shows you how to correct it.

Checklist of the Writing Process

This checklist arranges the key entries of the *Handbook* according to a recommended sequence of steps useful for any writing project. The exact titles of the entries are in **boldface type** for quick reference, followed by the page numbers. When you turn to the entries themselves, you will find boldface cross-references to other entries that may be helpful. (You may also wish to refer to the Topical Key to the Alphabetical Entries and to the Index at the back of the book.)

5. Revision 583

Topical Key to the Alphabetical Entries

This tabulation arranges the alphabetical entries into subject categories. Within each category the entries are listed mainly, but not strictly, in alphabetical order. Excluded from this tabulation are the numerous entries concerned with the usage of particular words or phrases (*imply/infer, already/all ready*, and the like). For a list of the key entries in a recommended sequence of steps for any writing project, see the Checklist of the Writing Process.

Types of Technical Writing

Organization, Writing, and Revision

Format and Illustrations

Language, Style, and Usage

Paragraphs, Sentences, Clauses, and Phrases

Parts of Speech, Inflection, and Agreement

Punctuation and Mechanics

A

a/an

A and *an* are indefinite **articles**; "indefinite" implies that the **noun** designated by the article is not a specific person, place, or thing but is one of a group.

EXAMPLE The computer operator ran *a* program. (Not a specific program, but an unnamed program.)

Use *a* before words beginning with a consonant or consonant sound (including *y* or *w* sounds).

EXAMPLES The year's activities are summarized in *a* one-page report. (Although the *o* in *one* is a vowel, the first *sound* is *w*, a consonant sound. Hence the article *a* precedes the word.)
A manual has been written on that subject.
It was *a* historic event for the laboratory.
The project manager felt that it was *a* unique situation.
A sonar operator first detected the wreck.

Use *an* before words beginning with a vowel or vowel sound.

EXAMPLES The report is *an* overview of the year's activities.
He seems *an* unlikely candidate for the job.
The interviewer arrived *an* hour early. (Although the *h* in *hour* is a consonant, the word begins on a vowel *sound*; hence it is preceded by *an*.)
He bought *an* SLR camera.

Be careful not to use unnecessary indefinite articles in a sentence.

CHANGE I will meet you in *a* half *an* hour.
TO I will meet you in *a* half hour.
OR I will meet you in half *an* hour.

a lot/alot

A lot is often incorrectly written as one word (*alot*). Write the **phrase** as two words: *a lot*. The phrase *a lot* is very informal, however, and should not normally be used in technical writing.

CHANGE The peer review group had *a lot* of objections.
TO The peer review group had many objections.

abbreviations

Abbreviations may be shortened versions of words, or they may be formed by the first letters of words. (See also **acronyms and initialisms**.)

EXAMPLES company/co.
boulevard/blvd.
electromotive force/emf
horsepower/hp
National Retail Dry Goods Association/NRDGA
cash on delivery/c.o.d.

Abbreviations, like **symbols**, can be important space savers in technical writing, as it is often necessary to provide the maximum amount of information in a limited amount of space. Use abbreviations, however, only if you are certain that your **reader** will understand them as readily as he or she would the terms for which they stand. Remember also that a **memorandum** or **report** addressed to a specific person may be read by others and that you must consider those readers as well. Take into account your reader's level of knowledge of your subject when deciding whether to use abbreviations. Do not use them if they might become an inconvenience to the reader. A good rule of thumb: When in doubt, spell it out.

In general, use abbreviations only in charts, **tables**, **graphs**, footnotes, **bibliographies**, and other places where space is at a premium. Normally you should not make up your own abbreviations, for they will probably confuse your reader. Except for commonly used abbreviations (U.S.A., a.m.), a term to be abbreviated should be spelled out the first time it is used, with the abbreviation enclosed in **parentheses** following the term. Thereafter, the abbreviation may be used alone.

EXAMPLE The National Aeronautics and Space Administration (NASA) has accomplished much in its short history. NASA will now develop a permanent orbiting space laboratory.

Do not add an additional period at the end of a sentence that ends with an abbreviation.

EXAMPLE The official name of the company is Data Base, Inc.

For abbreviations specific to your profession, use the style guide provided by your company or professional organization.

FORMING ABBREVIATIONS

Measurements. When you abbreviate terms that refer to measurement, be sure your reader is familiar with the abbreviated form. The following list contains some common abbreviations used with units of measurement. Notice that except for *in.* (inch), *tan.* (tangent), and others that form words, abbreviations of measurements do not require periods.

amp, ampere
atm, atmosphere
bbl, barrel
bhp, brake horsepower
Btu, British thermal unit

bu, bushel
cal, calorie
cd, candela
cm, centimeter
cos, cosine

ctn, cotangent
doz or dz, dozen
emf or EMF, electromotive force
F, Fahrenheit
fig., figure (illustration)

ft, foot (or feet)
gal., gallon
hp, horsepower
hr, hour

Hz, hertz
in., inch
kc, kilocycle
kg, kilogram
km, kilometer

lb, pound
lm, lumen
min, minute
oz, ounce
pa, pascal

ppm, parts per million
pt, pint
qt, quart
rad,radian
rev, revolution

sec, second or secant
tan., tangent
yd, yard
yr, year

Abbreviations of units of measure are identical in the singular and plural: one *cm* and three *cm* (not three *cms*).

Personal Names and Titles. Personal names should generally not be abbreviated.

CHANGE Chas., Thos., Wm., Geo.
TO Charles, Thomas, William, George

An academic, civil, religious, or military title should be spelled out when it does not precede a name.

EXAMPLE The doctor asked for the patient's chart.

When they precede names, some titles are customarily abbreviated. (See also **Ms./Miss/Mrs.**)

EXAMPLES Dr. Smith, Mr. Mills, Mrs. Katz

An abbreviation of a title may follow the name; however, be certain that it does not duplicate a title before the name.

CHANGE Dr. William Smith, Ph.D.
TO Dr. William Smith
OR William Smith, Ph.D.

When addressing **correspondence** and including names in other documents, you should normally spell out titles.

EXAMPLES The Honorable Mary J. Holt
Professor Charles Matlin
Captain Juan Ramirez
the Reverend James MacIntosh, D.D.

The following is a list of common abbreviations for personal and professional titles:

Atty.	Attorney
B.A.	Bachelor of Arts
B.S.	Bachelor of Science
B.S.E.E.	Bachelor of Science in Electrical Engineering
D.D.	Doctor of Divinity
D.D.S.	Doctor of Dental Science (or Surgery)
Dr.	Doctor (used with any doctor's degree)
Drs.	Plural of Dr.
Ed.D.	Doctor of Education
Hon.	Honorable
Jr.	Junior (used when a father with the same name is living)
LL.B.	Bachelor of Law
LL.D.	Doctor of Law
M.A.	Master of Arts
M.B.A.	Master of Business Administration
M.D.	Doctor of Medicine
Messrs.	Plural of Mr.
Mr.	Mister (spelled out only in the most formal contexts)

Mrs.	Married woman
Ms.	Woman of unspecified marital status
M.S.	Master of Science
Ph.D.	Doctor of Philosophy
Rev.	Reverend
Sr.	Senior (used when a son with the same name is living)

Names of Organizations. Many companies include in their names such terms as *Brothers, Incorporated, Corporation,* and *Company*. If these terms appear as abbreviations in the official company names, use them in their abbreviated forms: *Bros., Inc., Corp.*, and *Co.* If not abbreviated in the official names, such terms should be spelled out in most writing other than addresses, footnotes, **bibliographies**, and **lists**, where abbreviations may be used. A similar guideline applies for use of an **ampersand** (&); this symbol should be used only if it appears in an official company name. Titles of divisions within organizations, such as Department (Dept.) and Division (Div.) should be abbreviated only when space is limited.

Geographical Locations. It is generally better to spell out geographical locations, including the names of streets, cities, states, regions, and countries (except for long multiword names, such as United States of America, Union of Soviet Socialist Republics, and United Kingdom). Spelling out such names avoids possible misunderstanding and makes reading much easier. The following list includes some common geographical abbreviations:

N.E.	Northeast	Ave.	Avenue
N.H.	New Hampshire	Blvd.	Boulevard
N.Y.	New York	Ct.	Court
U.K.	United Kingdom	Dr.	Drive
U.S.A.	United States of America	St.	Street

The U.S. Postal Service recommends the following abbreviations for states and protectorates in business **correspondence**. Notice that they are written in capital letters without punctuation.

Alabama	AL	Connecticut	CT
Alaska	AK	Delaware	DE
Arizona	AZ	District of Columbia	DC
Arkansas	AR	Florida	FL
California	CA	Georgia	GA
Colorado	CO	Guam	GU

Hawaii	HI	New York	NY
Idaho	ID	North Carolina	NC
Illinois	IL	North Dakota	ND
Indiana	IN	Ohio	OH
Iowa	IA	Oklahoma	OK
Kansas	KS	Oregon	OR
Kentucky	KY	Pennsylvania	PA
Louisiana	LA	Puerto Rico	PR
Maine	ME	Rhode Island	RI
Maryland	MD	South Carolina	SC
Massachusetts	MA	South Dakota	SD
Michigan	MI	Tennessee	TN
Minnesota	MN	Texas	TX
Mississippi	MS	Utah	UT
Missouri	MO	Vermont	VT
Montana	MT	Virgin Islands	VI
Nebraska	NE	Virginia	VA
Nevada	NV	Washington	WA
New Hampshire	NH	West Virginia	WV
New Jersey	NJ	Wisconsin	WI
New Mexico	NM	Wyoming	WY

The names of the Canadian provinces are abbreviated as follows:

British Columbia	B.C.	New Brunswick	N.B.
Alberta	ALTA.	Nova Scotia	N.S.
Saskatchewan	SASK.	Prince Edward Island	P.E.I.
Manitoba	MAN.	Newfoundland	NFLD.
Ontario	ONT.	Northwest Territories	N.W.T.
Quebec	P.Q.	Yukon	YKN.

Dates and Time. The following are common abbreviations.

A.D.	*anno Domini* (beginning of calendar time—A.D. 1790)	Jan.	January
		Feb.	February
B.C.	before Christ (before the beginning of calendar time—647 B.C.)	Mar.	March
		Apr.	April
		Aug.	August
a.m.	*ante meridiem* (before noon)	Sept.	September
p.m.	*post meridiem* (after noon)	Dec.	December

Months should always be spelled out when only the month and year are given. The standard and alternative forms of abbreviations appear in current dictionaries, either in regular alphabetical order (by the letters of the abbreviation) or in a separate index. (See **dictionaries**.)

above

Avoid using *above* to refer to a preceding passage, **illustration**, or **table** in your writing. Its reference is often vague, and if the passage referred to is far from the current one, the use of *above* is distracting because your **reader** must go back to the earlier passage to understand your reference. The same is true of *aforesaid, aforementioned, above mentioned, the former*, and *the latter*. In addition to distracting the reader, these words also contribute to a heavy, wooden **style**. If you must refer to somcthing previously mentioned, either repeat the **noun** or **pronoun** or construct your **paragraph** so that your reference is obvious (see also **former/latter** and **aforesaid**).

CHANGE Please fill out and submit the above by March 1.
TO Please fill out and submit the vacation schedule on the previous page by March 1.

absolute words

Absolute words (such as *round, unique, exact*, and *perfect*) are not logically subject to **comparison** (*rounder, roundest*); nevertheless, these words are sometimes used comparatively.

CHANGE Phase-locked loop circuits make the FM tuner performance *more exact* by decreasing tuner distortion.
TO Phase-locked loop circuits make the FM tuner performance *more accurate* by decreasing tuner distortion.

Absolute words should be used comparatively only with the greatest caution in technical writing, where accuracy and precision are crucial.

absolutely

Absolutely means "definitely," "entirely," "completely," or "unquestionably." It is not an **intensifier** and should not be used to mean "very" or "much." When used as an intensifier, it can be deleted.

CHANGE A new analysis is *absolutely* impossible.
TO A new analysis is impossible.

abstracts

An *abstract* summarizes and highlights the major points of a longer piece of writing. Abstracts are written for many **formal reports**, **journal articles**, and most dissertations, as well as for many other long works. Their primary purpose is to enable readers to decide whether to read the work in full. (For a discussion of how summaries differ from abstracts, see **executive summaries**.)

Although abstracts are published with the longer works they condense, they are also often published independently in computer-retrievable periodical indexes. These periodicals, such as the *American Statistics Index, Chemical Abstracts*, and *Metal Abstracts*, are devoted exclusively to abstracting information in specific fields of study. They enable researchers to review a great deal of **literature** in a short time. (For a discussion of computer-retrievable indexes, see **library research**, and for a listing of typical periodicals that publish abstracts, see **reference books**.)

The abstracts for **reports** and articles are also frequently the source of terms (called *keywords*) used to **index** the original document by subject for computerized information-retrieval systems. Because they are the source of keywords, abstracts must accurately but concisely describe the original work so that researchers in the field will not miss valuable information. Accordingly, they should contain no information not discussed in the original. Abstracts do vary, however, in the amount of information they provide about the original work. Depending on their **scope**, abstracts are usually classified as either *descriptive* or *informative*.

DESCRIPTIVE ABSTRACTS

A descriptive abstract includes information about the purpose, scope, and methods used to arrive at the findings contained in the original document. It is almost an expanded **table of contents** in sentence form. A descriptive abstract need not be longer than several sentences if it adequately summarizes the information.

The following descriptive abstract comes from a 30-page report that describes how a select group of foreign countries provide engineering expertise to their control-room operators for round-the-clock shift work at nuclear power plants:

ABSTRACT

Purpose and scope This report describes the practices of selected foreign countries for providing engineering expertise on shift in nuclear power plants. The report discusses the extent to which engineering expertise is made available and the alternative models of providing such expertise. The implications of foreign practices for U.S. consideration are discussed, with particular reference to the shift technical adviser (STA) position and to a proposed shift engineer position. The relevant information for Methods used this study came from the open literature, interviews with utility staff and officials, and governmental and nuclear utility reports.

INFORMATIVE ABSTRACTS

The informative abstract is an expanded version of the descriptive abstract. In addition to information about the purpose, scope, and methods of the original document, the informative abstract includes the results, **conclusions**, and recommendations, if any. The informative abstract retains the **tone** and essential scope of the original work while omitting its details.

The following informative abstract expands the scope of the sample descriptive abstract by including the report's findings, conclusions, and recommendation:

ABSTRACT

Purpose and scope This report describes the practices of selected foreign countries for providing engineering expertise on shift in nuclear power plants. The report discusses the extent to which engineering expertise is made available and the alternative models of providing such expertise. The implications of foreign practices for U.S. consideration are discussed, with particular reference to the shift technical adviser (STA) position and to a proposed shift engineer position. The relevant information for Methods used this study came from the open literature, interviews with utility staff and officials, and governmental and nuclear utility reports.

> Finding / General conclusions / Recommendation

The countries studied used two approaches to provide engineering expertise on shift: (1) employing a graduate engineer in a line management operations position and (2) creating a specific engineering position to provide expertise to the operations staff. The comparison of these two models did not indicate that one system inherently functions more effectively than does the other for safe operations. However, the alternative modes are likely to affect crew relationships and performance; labor supply, recruitment, and retention; and system implementation. Of the two systems, the nonsupervisory engineering position seems more advantageous within the context of current recruitment and career-path practices.

Which of the two types should you write? The answer depends on the organization or publication for which you are writing. If it has a policy, comply with it. Otherwise, aim at the needs of the principal readers of your document. To satisfy the widest possible readership, including those who must index the original document based on the abstract, write informative abstracts. Descriptive abstracts, on the other hand, are indispensable for information surveys, conference proceedings, **progress reports** that combine information from more than one project, and other publications that compile a variety of information. For these types of publications, conclusions and recommendations either do not exist in the original or are too numerous to include in an abstract.

LENGTH OF ABSTRACTS

A long abstract defeats the purpose of an abstract. For this reason, abstracts are usually no longer than 200 to 250 words. Descriptive abstracts may be considerably shorter, of course.

SCOPE

Include in your abstract the following kinds of information, bearing in mind that your readers know nothing, except what your title announces, about your document:

- the subject of the study
- the scope of the study
- the purpose of the study
- the methods used
- the results obtained (informative abstract only)
- the recommendations made, if any (informative abstract only)

Do not include the following kinds of information:

- the background of the study
- a detailed discussion or explanation of the methods used
- administrative details about how the study was undertaken, who funded it, who worked on it, and the like
- figures, **tables**, charts, **maps**, and bibliographic references
- any information that does not appear in the original

Also, abstracts published independently of the main document will make no sense if they refer to tables or **illustrations** in the document; therefore, the abstract must make no reference to such tables or illustrations.

WRITING STYLE

Write the abstract after completing the report, article, or other work. Otherwise, your abstract may not accurately reflect the original document. Use the major and minor heads of your outline to help distinguish primary from secondary ideas.

Begin with a topic sentence that announces at least the subject and scope of the original document. Then combine the other relevant material for conciseness and clarity, eliminating unnecessary words and ideas. Write in complete sentences, but instead of stringing a group of short sentences end to end, combine ideas by using **subordination** and **parallel structure**. The degree of conciseness possible for **abbreviations, acronyms**, and technical **jargon** will vary depending on your readers. As a rule, spell out all **acronyms and initialisms** and all but the most common abbreviations (C°, F°, mph). In your attempts to write concisely, do not slip into a **telegraphic style** in which you omit **articles** (*a, an, the*) and important transitional words (*however, therefore, but, in summary*). Finally, as you summarize, keep the **tone** and **emphasis** consistent with those of the original document. (For guidance about the placement of abstracts in **reports**, see **formal report**.)

abstract nouns (see **abstract words/concrete words**)

abstract words/concrete words

The difference between abstract and concrete words is the difference between *durability* (abstract) and *stone* (concrete).

Abstract nouns and other words refer to general ideas, qualities, conditions, acts, or relationships. Abstract words refer to something that is intangible, something that cannot be discerned by the five senses.

EXAMPLES work, courage, crime, kindness, idealism, love, hate, fantasy, sportsmanship

Abstract words must frequently be qualified by other words.

CHANGE What the members of the Research and Development Department need is *freedom*.
TO What the members of the Research and Development Department need is *freedom to explore the problem further*.

Concrete nouns and other words refer to specific persons, places, objects, and acts that can be perceived by the senses.

EXAMPLES wrench, book, house, scissors, gold, water
Skiing (concrete) is a strenuous *sport* (abstract).

Concrete words are easier to understand, for they create images in the mind of your **reader**. Still, you cannot express ideas without using some abstract words. Actually, the two kinds of words are usually used best together, in support of each other. For example, the abstract idea of *transportation* is made clearer with the use of specific concrete words, such as *subways, jets,* or *automobiles*.

In fact, a word that is concrete in one context can be less concrete in another. The same word may even be abstract in one context and concrete in another. The choice between abstract and concrete terms always depends on the context. Just how concrete a particular context might require you to be is shown in the accompanying **illustration**, which goes from the most abstract, on the left, to the most concrete, on the right.

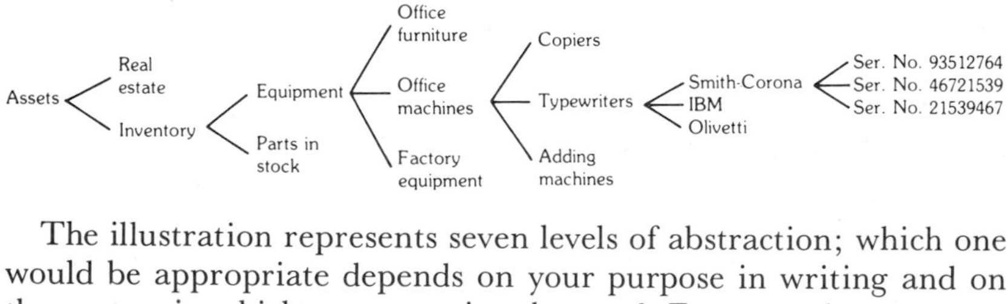

The illustration represents seven levels of abstraction; which one would be appropriate depends on your purpose in writing and on the context in which you are using the word. For example, a company's **annual report** might logically use the most abstract term, *assets*, to refer to all the property and goods the firm owns; shareholders would probably not require a less abstract term. Interoffice **memorandums** between the company's accounting and legal departments would appropriately call the firm's holdings *real estate* and *inventory*. To the company's inventory control department, however, the word *inventory* is much too abstract to be useful, and a report on inventory might contain the more concrete terms *equipment* and *parts in stock*. But to the assistant inventory control manager in charge of equipment, that term is still too abstract; he or she would speak of several particular kinds of equipment: *office furniture, factory equipment*, and *office machines*. The breakdown of the types of office machines for which the inventory control assistant is responsible might include *copiers, adding machines*, and *typewriters*. But even this classification would not be concrete enough to enable the company's purchasing department to obtain service contracts for the normal maintenance of its typewriters. Because the department must deal with different typewriter manufacturers, *typewriters* would have to be listed by brand name: *Olivetti, IBM,* and *Smith-Corona*. And the Smith-Corona technician who performs the maintenance must go one step further and identify each Smith-Corona typewriter by serial number. As the illustration shows, then, a term that is sufficiently concrete in one context may be too abstract in another. (See also **word choice**.)

accept/except

Accept is a **verb** meaning "consent to," "agree to take," or "admit willingly."

> EXAMPLE I *accept* the responsibility that goes with the appointment.

Except is normally a **preposition** meaning "other than" or "excluding."

> EXAMPLE We agreed on everything *except* the schedule.

acceptance letters

When you have received an offer of a job that you want to accept, reply as soon as possible—certainly within a week. The **format** for such a letter is simple: Begin by accepting, with pleasure, the job you have been offered. Identify the job you are accepting, and state

State the job and salary you are accepting. →

Specify the date on which you will report to work. →

State your pleasure at joining the firm. →

2647 Patterson Road
Beechwood, OH 45432
March 6, 19--

Mr. F. E. Cummins
Personnel Manager
Calcutex Industries, Inc.
3275 Commercial Park Drive
Bintonville, MI 49474

Dear Mr. Cummins:

I am pleased to accept your offer of a position as assistant personnel manager at a salary of $X,XXX per month.

Since graduation is August 30, I plan to leave Dayton on Tuesday, September 2. I should be able to locate suitable living accommodations within a few days and be ready to report for work on the following Monday, September 8. Please let me know if this date is satisfactory to you.

I look forward to a rewarding future with Calcutex.

Sincerely,

Craig Adderly

Craig Adderly

Letter 1 Acceptance Letter

the salary so that there is no confusion on these two important points.

The second **paragraph** might go into detail about moving dates and reporting for work. The details will vary depending on what occurred during your job interview. Complete the letter with a statement that you are looking forward to working for your new employer. Letter 1 is an example of an acceptance letter. (See also **correspondence** and **refusal letters**.)

accumulative/cumulative

Accumulative and *cumulative* are **synonyms** that mean "amassed" or "added up over a period of time." *Accumulative* is rarely used except with reference to accumulated property or wealth.

EXAMPLES Last month's X-ray exposures were acceptable, but we must be careful so that *cumulative* doses do not exceed our guidelines.

The corporation's *accumulative* holdings include 19 real-estate properties.

acknowledgment letters

It is sometimes necessary (and always considerate), as well as good public relations, to let someone know that you have received something sent to you. An acknowledgment letter serves such a function. It should usually be a short, polite note that mentions when the item arrived and that expresses thanks. See the **correspondence** entry for general advice on letter writing; study the example (Letter 1 on page 16) for an idea of how to phrase an acknowledgment letter.

acronyms and initialisms

An *acronym* is an **abbreviation** that is formed by combining the first letter or letters of several words. Acronyms are pronounced as words and are written without periods.

EXAMPLES *ra*dio *d*etecting *a*nd *r*anging/radar
*a*lternating *g*radient *s*ynchrotron (AGS)
*r*andom-*a*ccess *m*emory (RAM)
*r*ecirculation *a*ctuation *s*ymbol (RAS)
*r*ead-*o*nly *m*emory (ROM)

An *initialism* is an abbreviation that is formed by combining the initial letter of each word in a multiword term. Initialisms are pronounced as separate letters.

EXAMPLES *e*nd *o*f *m*onth/e.o.m.
*S*ociety for *T*echnical *C*ommunication/STC
*c*ash *o*n *d*elivery/c.o.d.
*p*ost *m*eridiem/p.m.

Energy Savings Systems
501 North Springfield
Phoenix, AZ 85302

November 8, 19--

Ms. Wanda Evans, Consultant
936 East Avenue
Phoenix, AZ 85301

Dear Ms. Evans:

I received your report today; it appears to be complete and well done. Thank you for sending it so promptly.

When I finish studying the report, I will send you our cost estimate for the installation of the Mark II Energy-Saving System.

Sincerely,

Robert A. Martinez

Robert A. Martinez
Administrative Assistant

RAM/mo

Letter 1 Acknowledgment Letter

In business and industry, acronyms and initialisms are often used by people working together on particular projects or having the same specialties—as, for example, mechanical engineers or organic chemists. As long as such people are communicating with one another, the abbreviations are easily recognized and understood. If the same acronyms or initialisms were used in **correspondence** to someone outside the group, however, they might be incomprehensible to that **reader** and should be explained.

Acronyms and initialisms can be convenient—for the reader and the writer alike—if they are used appropriately. Technical people, however, often overuse them, either as an **affectation** or in a misguided attempt to make their writing concise.

WHEN TO USE ACRONYMS AND INITIALISMS

The following are two sample guidelines to apply in deciding whether to use acronyms and initialisms:

1. If you must use a multiword term as much as once each **paragraph**, you should instead use its acronym or initialism. For example, a **phrase** such as "primary software overlay area" can become tiresome if repeated again and again in one piece of writing; it would be better, therefore, to use *PSOA*.
2. If something is better known by its acronym or initialism than by its formal term, you should use the abbreviated form. The initialism *a.m.*, for example, is much more common than the formal *ante meridiem*.

If these conditions do not exist, however, always spell out the full term.

HOW TO USE ACRONYMS AND INITIALISMS

The first time an acronym or initialism appears in a written work, write the complete term, followed by the abbreviated form in **parentheses**.

> EXAMPLE The transaction processing monitor (TPM) controls all operations in the processor.

Thereafter, you may use the acronym or initialism alone. In a long document, however, you will help your reader greatly by repeating the full term in parentheses after the abbreviation at regular intervals so that he or she does not have to search back to the first time the acronym or initialism was used to find its meaning.

EXAMPLE Remember that the TPM (transaction processing monitor) controls all operations in the processor.

Write acronyms in **capital letters** without **periods**. The only exceptions are those acronyms that have become accepted as **common nouns**, which are written in lowercase letters.

EXAMPLES NASA, HUD, laser, scuba

Initialisms may be written either uppercase or lowercase. Generally, do not use periods when they are uppercase, but use periods when they are lowercase. Two exceptions are geographic names and academic degrees.

EXAMPLES EDP/e.d.p., EOM/e.o.m., OD/o.d.
U.S.A., U.S.S.R.
B.A., M.B.A.

Form the plural of an acronym or initialism by adding an *s*. Do not use an *apostrophe*.

EXAMPLES MIRVs, CRTs

activate/actuate

Even linguists disagree on the distinction between these two words, although both mean "make active." *Actuate* is usually applied only to mechanical processes.

EXAMPLES The governor *activated* the National Guard.
The relay *actuates* the trip hammer. (mechanical process)
The electrolyte *activates* the battery. (chemical process)

actually

Actually is an **adverb** meaning "really" or "in fact." Although it is often used for **emphasis** in speech, such use of the word should be avoided in writing. (See also **intensifiers**.)

CHANGE Did he *actually* finish the report on time?
TO Did he finish the report on time?

adapt/adept/adopt

Adapt is a **verb** meaning "adjust to a new situation." *Adept* is an **adjective** meaning "highly skilled." *Adopt* is a verb meaning "take or use as one's own."

EXAMPLE The company will *adopt* a policy of finding engineers who are *adept* administrators and who can *adapt* to new situations.

ad hoc

Ad hoc is Latin for "for this" or "for this particular occasion." An ad hoc committee is one set up to consider a particular issue, as opposed to a permanent committee. This term has been fully assimilated into English and thus does not have to be italicized (or underlined). (See also **foreign words in English**.)

adjectives

An adjective makes the meaning of a **noun** or **pronoun** more specific by pointing out one of its qualities (descriptive adjective) or by imposing boundaries on it (limiting adjective).

EXAMPLES a *hot* iron (descriptive)
He is *cold*. (descriptive)
ten automobiles (limiting)
his desk (limiting)

Limiting adjectives include these common and important categories:

Articles (a, an, the)
Demonstrative adjectives (this, that, these, those)
Possessive adjectives (my, his, her, your, our, their)
Interrogative and relative adjectives (whose, which, what)
Numeral adjectives (two, first)
Indefinite adjectives (all, none, some, any)

The demonstrative adjectives are *this, these, that,* and *those*. They are adjectives that "point to" the thing they modify, specifying its position in space or time. *This* and *these* specify a closer position; *that* and *those* specify a more remote position.

EXAMPLES *This* proposal is the one we accepted.
That proposal would have been impracticable.
These problems remain to be solved.
Those problems are not insurmountable.

Because possessive adjectives (*my, your, our, his, her, its, their*) directly modify **nouns**, they function as adjectives, even though they are **pronoun** forms. Anything that modifies a noun functions as an adjective.

EXAMPLE The *proposal's* virtues outweighed *its* defects.

The possessives *mine, yours, hers, ours,* and *theirs* normally function as pronouns rather than as adjectives.

Like **relative pronouns**, the relative adjectives (*whose, which,* and *what*) are **pronoun** forms that link **dependent clauses** to main **clauses**. The difference is that relative pronouns perform this linking function while taking the place of **nouns**, and relative adjectives perform this linking function while *modifying* (describing) nouns. Examples of the uses of relative adjectives follow:

EXAMPLES The president commended the managers *whose* divisions were the most profitable. (*Whose* modifies "divisions.")
He also told us *which* divisions needed to show the greatest improvement. (*Which* modifies "divisions.")
Finally, he told us *what* steps we should take to improve our profits. (*What* modifies "steps.")

COMPARISON OF ADJECTIVES

Most adjectives add the **suffix** *-er* to show **comparison** with one other item and the suffix *-est* to show comparison with two or more other items. The three degrees of comparison are called the *positive, comparative,* and *superlative*.

EXAMPLES The first ingot is *bright*. (positive form)
The second ingot is *brighter*. (comparative form)
The third ingot is *brightest*. (superlative form)

However, many two-syllable adjectives and most three-syllable adjectives are preceded by the words *more* or *most* to form the comparative or the superlative.

EXAMPLES The new facility is *more impressive* than the old one.
The new facility is the *most impressive* in the city.

A few adjectives have irregular forms of comparison (*much, more, most; little, less, least*).

Absolute words (round, unique) are not logically subject to comparison (rounder, roundest); nevertheless, these words are sometimes used comparatively.

CHANGE Phase-locked loop circuits make FM tuner performance *more exact* by decreasing tuner distortion.

TO Phase-locked loop circuits make FM tuner performance *more accurate* by decreasing tuner distortion.

PLACEMENT OF ADJECTIVES

When limiting and descriptive adjectives appear together, the limiting adjectives precede the descriptive adjectives, with the **articles** usually in the first position.

EXAMPLE *The ten gray* cars were parked in a row. (article, limiting adjective, descriptive adjective)

Within a sentence, adjectives may appear before the nouns they modify (the attributive position) or after the nouns they modify (the predicative position).

EXAMPLES The *small* jobs are given priority. (attributive position)
Tests are taken even when exposure is *brief*. (predicative position)

An adjective in an attributive position may shift to a predicative position in a larger, more complex construction.

EXAMPLES We passed a *big* budget. (attributive position)
We negotiated a contract *bigger by far than theirs*. (predicative position for adjective phrase)

When an adjective follows a **linking verb**, it is called a predicate **adjective**.

EXAMPLES The warehouse is *full*.
The lens is *convex*.
His department is *efficient*.

A predicate adjective is one kind of **subjective complement**. By completing the meaning of a linking verb, a predicate adjective describes or limits the subject of the **verb**.

EXAMPLES The job is *easy*.
The manager was very *demanding*.

An adjective that follows a transitive verb and modifies its direct object is one kind of **objective complement**. (See also **complements**.)

EXAMPLE The lack of lubricant rendered the bearing *useless*.

USE OF ADJECTIVES

Because of the need for precise qualification in technical writing, it is often necessary to use nouns as adjectives.

EXAMPLE The *test* conclusions led to a redesign of the system.

When adjectives modifying the same noun can be reversed and still make sense or when they can be separated by *and* or *or*, they should be separated by commas.

EXAMPLE The company is seeking a *young, energetic, creative* engineering team.

When an adjective modifies a phrase, no comma is required.

EXAMPLE He was wearing his *old cotton tennis hat* (*old* modifies the phrase *cotton tennis hat*).

Never separate a final adjective from its noun.

CHANGE He is a conscientious, honest, *reliable, worker*.
TO He is a conscientious, honest, *reliable worker*.

Technical people frequently string together a series of nouns to form a unit modifier, thereby creating **jammed modifiers**. Be aware of this problem when using nouns as adjectives.

CHANGE The test control group meeting was held on Wednesday.
TO The meeting of the test control group was held on Wednesday.

As a rule of thumb, it is better to avoid general (*nice, fine, good*) and trite (a *fond* farewell) adjectives; in fact, it is good practice to question the need for most adjectives in your writing. Often, your writing will not only read as well without an adjective, but it may be even better without it. If an adjective is needed, try to find one that is as precise as possible for your meaning.

adjective clauses (see **clauses**)

adjustment letters

An adjustment letter is written in response to a **complaint letter** and tells the customer what your firm intends to do about his or her complaint.

Although it is sent in response to a problem, an adjustment letter actually provides an excellent opportunity to build goodwill for the firm. An effective adjustment letter both repairs the damage that has been done and restores the customer's confidence in your firm.

You should settle claims quickly and courteously, trying always to satisfy the customer at a reasonable cost to your company.

Grant adjustments graciously, for a settlement made grudgingly will do more harm than good. **Tone** is critical. No matter how unpleasant or unreasonable the complaint letter, your response must remain both respectful and positive. Avoid emphasizing the unfortunate situation at hand; put your **emphasis** instead on what you are doing to correct it. Not only must you be gracious, but also you must admit your error in such a way that the buyer will not lose confidence in your firm.

Before granting an adjustment to a claim for which your company is at fault, you must investigate what happened and decide what you can do to satisfy the customer. Be certain that you know your company's policy regarding adjustments before you attempt to write an adjustment letter. Also, be careful about how you put certain words together; for example, "we have just received your letter of May 7 about our *defective product*" could be ruled in a court of law as an admission that the product is in fact defective. Treat every claim individually, and lean toward giving the customer the benefit of the doubt. The following guidelines might help you write adjustment letters:

1. Open with whatever you believe the **reader** will consider good news:
   Grant the adjustment for uncomplicated situations ("Enclosed is a replacement for the damaged part").
   Reveal that you intend to grant the adjustment by admitting that the customer was in the right ("Yes, you were incorrectly billed for

ELECTRONIC Parts Inc.
One South Park Avenue
Columbia, AL 36319

January 17, 19--

Mr. Paul E. Denlinger
Denlinger Television Services, Inc.
4873 Wenton Way
Birmingham, AL 35214

Dear Mr. Denlinger:

Thank you for your letter regarding your order for nine TR-5771-3 tuners. We have shipped the correct tuners by United Parcel; you should receive them shortly after you receive this letter. I have also instructed our Billing Department to charge you our preferred-customer rate--normally reserved for orders of more than $2000. Please accept our apologies for not sending the proper tuners and for incorrectly billing you.

Evidently, when your package arrived at our loading docks, a dock worker failed to see your letter in the container. He set the box aside with several boxes of parts destined for our Rebuilt Parts Department; therefore, your note did not come to the attention of our parts manager.

If I can be of further help, please let me know.

Sincerely,

Jonathan L. Pennington

Jonathan L. Pennington
Office Manager

JLP/jq

Letter 1 Adjustment Letter

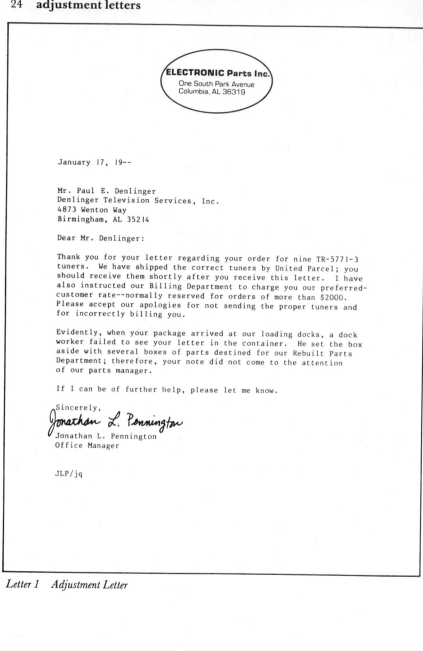

SWELCO Coffee Maker, Inc.
9025 North Main Street
Butte, MT 59702

August 26, 19--

Mr. Carlos Ortiz
638 McSwaney Drive
Butte, MT 59702

Dear Mr. Ortiz:

Enclosed is your SWELCO Coffee Maker, which you sent to us on August 17.

In various parts of the country, tap water may contain a high mineral content. If you fill your SWELCO Coffee Maker with water for breakfast coffee before going to bed, a mineral scale will build up on the inner wall of the water tube--as explained on page 2 of your SWELCO Instruction Booklet.

We have removed the mineral scale from the water tube of your coffee maker and thoroughly cleaned the entire unit. To ensure the best service from your coffee maker in the future, clean it once a month by operating it with four ounces of white vinegar and eight cups of water. To rinse out the vinegar taste, operate the unit twice with clear water.

With proper care, your SWELCO Coffee Maker will serve you faithfully and well for many years to come.

Sincerely,

Helen Upham

Helen Upham
Customer Services

HU/mo
Enclosure

Letter 2 Adjustment Letter

the oil delivery''). Then later explain the specific details of the adjustment. This method is good for adjustments that require detailed explanations.

Apologize for the error (''Please accept our apologies for not acting sooner to correct your account''). This method is effective when the customer's inconvenience is as much an issue as money.

Use a combination of these techniques. Often, situations requiring an adjustment are unique and do not fit a single pattern.

2. Explain what caused the problem—if such an explanation will help restore your reader's confidence or goodwill.
3. Explain specifically how you intend to make the adjustment—if it is not obvious in your opening.
4. Close with an offer to be of further service.

Express appreciation to the customer for calling your attention to the situation, explaining that this helps your firm keep the quality of its product or service high. Point out any steps you may be taking to prevent a recurrence of whatever went wrong, giving the customer as much credit as the facts allow.

Close pleasantly—looking forward, not back. Avoid recalling the problem in your closing (''Again, we apologize . . .''). Letter 1 is a typical adjustment letter.

You may sometimes need to educate your reader about the use of your product or service. Customers sometimes submit claims that are not justified, even though they honestly believe them to be. The customer may actually be at fault—for not following maintenance instructions properly for example. Such a claim is granted only to build goodwill. When you write a letter of adjustment in such a situation, it is wise to give the explanation before granting the claim—otherwise, your reader may never get to the explanation. If your explanation establishes customer responsibility, be sure that it is by implication rather than by outright statement. Look at the example of such an adjustment letter (see Letter 2).

Sometimes you may decide to grant a partial adjustment, even though the claim is not really justified, as in Letter 3.

adverbs

An adverb modifies the action or condition expressed by a **verb**.

EXAMPLE The recording head hit the surface of the disc *hard.* (The adverb tells *how* the recording head hit the disc.)

General Television, Inc.
5521 West 23rd Street
New York, NY 10062

Customer Relations
(212) 574-3894

September 28, 19--

Mr. Fred J. Swesky
7811 Ranchero Drive
Tucson, AZ 85761

Dear Mr. Swesky:

Thank you for your letter regarding the replacement of your KL-71 television set.

You stated in your letter that you used the set on an uncovered patio. As our local service representative pointed out, this model is not designed to operate in extreme heat conditions. As the instruction manual accompanying your new set stated, such exposure can produce irreparable damage to this model. Since your set was used in such extreme heat conditions, therefore, we cannot honor the two-year replacement warranty.

However, we are enclosing a certificate entitling you to a trade-in allowance equal to your local GTI dealer's markup for the set. This certificate will enable you to purchase a new set from the dealer at the wholesale price when you return the original set to your local dealer

Sincerely yours,

Susan Siegel

Susan Siegel
Assistant Director

SS/mr
Enclosure

Letter 3 Adjustment Letter

An adverb may also modify an **adjective**, another adverb, or a **clause**.

EXAMPLES The graphics department used *extremely* bright colors. (modifying an adjective)
The redesigned brake pad lasted *much* longer. (modifying another adverb)
Surprisingly, the machine failed. (modifying a clause)

An adverb answers one of the following questions:

WHERE? (adverb of place)

EXAMPLE Move the throttle *forward* slightly.

WHEN? (adverb of time)

EXAMPLE Replace the thermostat *immediately*.

HOW? (adverb of manner)

EXAMPLE Add the reagent *cautiously*.

HOW MUCH? (adverb of degree)

EXAMPLE The *nearly* completed report was lost in the move.

TYPES OF ADVERBS

An adverb may be a common, a conjunctive, an interrogative, or a numeric **modifier**.

Typical common adverbs are *almost, seldom, down, also, now, ever,* and *always*.

EXAMPLE I *rarely* work on the weekend.

Typical **conjunctive adverbs** are *however, therefore, nonetheless, nevertheless, consequently, accordingly,* and *then*.

EXAMPLE I rarely work on the weekend; *however*, this weekend will be an exception.

Interrogative adverbs ask questions. Common interrogative adverbs include *where, when, why*, and *how*.

EXAMPLES *How* many hours did you work last week?
Where are we going when the new project is finished?
Why did it take so long to complete the job?

Typical numeric adverbs are *once* and *twice*.

EXAMPLE I have worked overtime *twice* this week.

COMPARISON OF ADVERBS

Adverbs are normally compared by adding *-er* or *-est* to them or by inserting *more* or *most* in front of them. One-syllable adverbs use the comparative ending *-er* and the superlative ending *-est*.

EXAMPLES This copier is *faster* than the old one.
This copier is the *fastest* of the three tested.

Most adverbs with two or more **syllables** end in *-ly*, and most adverbs ending in *-ly* are compared by inserting the comparative *more* or *less* or the superlative *most* or *least* in front of them.

EXAMPLES He moved *more quickly* than the other company's salesman.
Most surprisingly, the engine failed during the test phase.
This filter can be changed *less quickly* than the previous filter.
Least surprisingly, the engine failed during the test phase.

There are a few irregular adverbs that require a change in form to indicate **comparison**.

EXAMPLES The training program functions *well*.
Our training program functions *better* than most others in the industry.
Many consider our training program the *best* in the industry.

PLACEMENT OF ADVERBS

An adverb may appear almost anywhere in a sentence, but its position may affect the meaning of the sentence. Avoid placing an adverb between two verb forms where it can be read ambiguously as modifying either.

CHANGE The operator tried *belatedly* to close the relief valve.
TO The operator *belatedly* tried to close the relief valve.

The adverb is commonly placed in front of the verb it modifies.

EXAMPLE The pilot *meticulously* performed the preflight check.

An adverb may, however, follow the verb (or the verb and its **object**) that it modifies.

EXAMPLES The gauge dipped *suddenly*.
They repaired the computer *quickly*.

The adverb may be placed between a **helping verb** and a main verb.

EXAMPLE In this temperature range, the pressure will *quickly* drop.

If an adverb modifies only the main verb, and not any accompanying helping verbs, place the adverb immediately before the main verb.

EXAMPLE The alternative proposal has been *effectively* presented.

An adverb **phrase** should not be inserted between the words of a compound verb.

CHANGE This suggestion has *time and time again* been rejected.
TO This suggestion has been rejected *time and time again*.

To place **emphasis** on an adverb, put it before the **subject** of the sentence.

EXAMPLES *Clearly*, he was ready for the promotion when it came.
Unfortunately, fuel rationing has been unavoidable.

In writing, such adverbs as *nearly, only, almost, just,* and *hardly* are placed immediately before the words they limit. A speaker can place these words earlier and avoid **ambiguity** by stressing the word to be limited; in writing, however, only the appropriate placement of the adverb will ensure **clarity**.

CHANGE The punch press with the auxiliary equipment *only* costs $47,000.
TO The punch press with the auxiliary equipment costs *only* $47,000.

The first sentence is ambiguous because it might be understood to mean that *only* the punch press with auxiliary equipment costs $47,000. (See also **misplaced modifiers.**)

adverb clauses (see **clauses**)

advice/advise

Advice is a **noun** that means "counsel" or "suggestion."

EXAMPLE My *advice* is to sign the contract immediately.

Advise is a **verb** that means "give advice."

EXAMPLE I *advise* you to sign the contract immediately.

affect/effect

Affect is a **verb** that means "influence."

EXAMPLE The public utility commission's decisions *affect* all state utilities.

Effect can function either as a verb that means "bring about" or "cause" or as a **noun** that means "result." It is best, however, to avoid using *effect* as a verb. A less formal word, such as *made*, is usually preferable.

CHANGE The new manager *effected* several changes that had a good *effect* on morale.

TO The new manager *made* several changes that had a good *effect* on morale.

affectation

Affectation is the use of language that is more technical, formal, or showy than is necessary to communicate information to the **reader**. A writer who is unnecessarily ornate, pompous, or pretentious is usually attempting to impress the reader by showing off a repertoire of fancy or flashy words and phrases. Using unnecessarily formal words (such as *herewith*) and outdated phrases (such as *please find enclosed*) is another cause of affectation.

EXAMPLES *pursuant to* (instead of *about* or *regarding*)
in view of the foregoing (instead of *therefore*)
in view of the fact that (instead of *because*)
it is interesting to note that (omit)
it may be said that (omit)

Affected writing forces the reader to work harder to understand the writer's meaning. Affected writing typically contains abstract, highly technical, pseudotechnical, pseudolegal, or foreign words and is often liberally sprinkled with **vogue words**. **Jargon** can become affectation if it is misused. **Euphemisms** can contribute to affected writing if their purpose is to hide the facts of a situation rather than treat them with dignity or restraint.

The easiest kind of affectation to be lured into is the use of **long variants**: words created by adding **prefixes** and **suffixes** to simpler words (*analyzation* for analysis, *telephonic communication* for telephone call). The practice of **elegant variation**—attempting to avoid repeating the same word in the same **paragraph** by substituting pretentious **synonyms**—is also a form of affectation. Another contributor to affectation is **gobbledygook**, which is wordy, roundabout writing that has many pseudolegal and psuedoscientific terms sprinkled throughout.

Attempts to make the trivial seem important can also cause affectation. This attempt is apparent in the first version of the following example, taken from a specification soliciting bids from local merchants for the operation of a television repair shop in an Air Force Post Exchange.

CHANGE In addition to performing interior housekeeping services, the concessionaire shall perform custodial maintenance on the exterior of the facility and grounds. Where a concessionaire shares a facility with one or more other concessionaires, exterior custodial maintenance responsibilities will be assigned by Post Exchange management on a fair and equitable basis. In those instances where the concessionaire's activity is located in a Post Exchange complex wherein predominant tenancy is by Post Exchange-operated activities, then Post Exchange management shall be responsible for exterior custodial maintenance except for those described in 1, 2, 3, and 4 below. The necessary equipment and labor to perform exterior custodial maintenance, when such a responsibility has been assigned to the concessionaire, shall be furnished by the concessionaire. Exterior custodial maintenance shall include the following tasks:

1. Clean entrance door and exterior of storefront windows daily.
2. Sweep and clean the entrance and customer walks daily.
3. Empty and clean waste and smoking receptacles daily.
4. Check exterior lighting and report failures to the contracting officer's representative daily.

TO The merchant will, with his or her own equipment, perform the following duties daily:

1. Maintain a clean and neat appearance inside the store.

> 2. Clean the entrance door and the outsides of the store windows.
> 3. Sweep the entrance and the sidewalk.
> 4. Empty and clean wastebaskets and ashtrays.
> 5. Check the exterior lighting, and report failures to the representative of the contracting officer.
>
> The merchant will also maintain the grounds surrounding the store. Where two or more merchants share the same building, Post Exchange management will assign responsibility for the grounds. Where the merchant's store is in a building that is occupied predominantly by Post Exchange operations, Post Exchange management will be responsible for the grounds.

Affectation is a widespread problem in technical writing because many people apparently feel that affectation lends a degree of formality, and hence authority, to their writing. Nothing could be further from the truth. Affectation puts up a smoke screen that the reader must penetrate to get to the writer's meaning. It can alienate the customer or client, making him or her feel that the writer is difficult to communicate with. (See also **conciseness/wordiness** and **clichés**.)

affinity

Affinity refers to the attraction of two persons or things to each other.

EXAMPLE The *affinity* between these two elements can be explained in terms of their valence electrons.

Affinity should never be used to mean "ability" or "aptitude."

CHANGE She has an *affinity* for chemistry.
TO She has an *aptitude* for chemistry.
OR She has a *talent* for chemistry.

aforesaid

Aforesaid means "stated previously," but it is legal **jargon** and should be avoided in writing. (See also **above**.)

CHANGE The *aforesaid* problems can be solved in six months.
TO The three problems explained in the preceding paragraphs can be solved in six months.

agree to/agree with

When you *agree to* something, you are "giving consent."

EXAMPLE I *agree to* a road test of the new model by August 1.

When you *agree with* something, you are "in accord" with it.

EXAMPLE I *agree with* the recommendations of the advisory board.

agreement

Agreement, grammatically, means the correspondence in form between different elements of a sentence to indicate **person, number, gender,** and **case**. A **pronoun** must agree with its antecedent, and a **verb** must agree with its **subject**.

A subject and its verb must agree in number and in person.

EXAMPLES The *design* is an acceptable one. (The first person singular subject, *design*, requires the first person singular verb, *is*.)
The new *products are* going into production soon. (The third person plural subject, *products*, requires the third person plural form of the verb *are*.)

A pronoun and its antecedent must agree in person, number, and gender.

EXAMPLES The *employees* report that *they* become less efficient as the humidity rises. (The third person plural subject, *employees*, requires the third person plural pronoun, *they*.)
Mr. Joiner said that *he* would serve as a negotiator. (The third person singular, masculine subject, *Mr. Joiner*, requires the third person singular, masculine form of the pronoun, *he*.)

(See also **agreement of pronouns and antecedents** and **agreement of subjects and verbs**.)

agreement of pronouns and antecedents

Every **pronoun** must have an antecedent, or a **noun** to which it refers. In the following sentence, the pronoun *it* has no noun to which it could logically refer. The solution is to use a noun instead of the pronoun or to provide a noun to which it could logically refer.

CHANGE Electronics technicians must constantly struggle to keep up with the professional literature because *it* is a dynamic science.

TO Electronics technicians must constantly struggle to keep up with the professional literature because *electronics* is a dynamic science.

OR Electronics technicians must continue to study *electronics* because *it* is a dynamic science.

Using the **relative pronoun** *which* to refer to an idea instead of a specific noun can be confusing.

CHANGE He acted independently on the advice of his consultant, *which* the others thought unjust. (Was it the fact that he acted independently or was it the advice that the others thought unjust?)

TO The others thought it unjust for him to act independently.

OR The others thought it unjust for him to act on the advice of his consultant.

Using the pronouns *it, they, these, those, that*, and *this* can also lead to vague or uncertain references.

CHANGE Studs and thick treads make snow tires effective. *They* are implanted with an air gun. (Which are implanted with an air gun: studs or thick treads?)

TO Studs and thick treads make snow tires effective. *The studs* are implanted with an air gun.

CHANGE The inadequate quality-control procedure has resulted in an equipment failure. *This* is our most serious problem at present. (Which is our most serious problem at present: the procedure or the failure?)

TO The inadequate quality-control procedure has resulted in an equipment failure. Quality control is our most serious problem at present.

OR The inadequate quality-control procedure has resulted in an equipment failure. The equipment failure is our most serious problem at present.

GENDER

A pronoun must agree in **gender** with its antecedent.

EXAMPLE Mr. Swivet in the Accounting Department acknowledges *his* share of the responsibility for the misunderstanding, just as

Mrs. Barkley in the Research Division must acknowledge *hers*.

Traditionally, a masculine, singular pronoun has been used to agree with such indefinite antecedents as *anyone* and *person*.

EXAMPLE *Each* may stay or go as *he* chooses.

It is now recognized that there is an implied sexual bias in such usage. When alternatives are available, use them. One solution is to use the plural. Another is to use both feminine and masculine pronouns, although this combination is clumsy when used too often.

CHANGE Every *employee* must sign *his* time card.
TO All *employees* must sign *their* time cards.
OR Every *employee* must sign *his or her* time card.

Do not, however, attempt to avoid expressing gender by resorting to a plural pronoun when the antecedent is singular.

CHANGE A *technician* can expect to advance on *their* merit.
TO *Technicians* can expect to advance on *their* merit.
OR A *technician* can expect to advance on *his or her* merit.

NUMBER

A pronoun must agree with its antecedent in **number**. Many problems of agreement are caused by expressions that are not clear in number.

CHANGE Although the typical *engine* runs well in moderate temperatures, *they* often stall in extreme cold.
TO Although the typical *engine* runs well in moderate temperatures, *it* often stalls in extreme cold.

Use singular pronouns with the antecedents *everybody* and *everyone* unless to do so would be illogical because the meaning is obviously plural. (See also **everybody/everyone**.)

EXAMPLES *Everyone* pulled *his* share of the load.
Everyone laughed at my sales slogan, and I really couldn't blame *them*.

The demonstrative adjectives sometimes cause problems with agreement of number. *This* and *that* are used with singular **nouns**, and *these* and *those* are used with plural nouns. Demonstrative adjectives often cause problems when they modify the nouns *kind, type*, and *sort*. Demonstrative adjectives used with these nouns should agree with them in **number**.

EXAMPLES this kind / these kinds
that type / those types
this sort / these sorts

Confusion often develops when the **preposition** *of* is added ("this kind *of*," "these kinds *of*") and the **object** of the preposition is not made to conform in number to the demonstrative adjective and its noun.

CHANGE *This kind* of hydraulic *cranes* is best.
TO *This kind* of hydraulic *crane* is best.
CHANGE *These kinds* of hydraulic *crane* are best.
TO *These kinds* of hydraulic *cranes* are best.

The error can be avoided by remembering to make the demonstrative adjective, the noun, and the object of the preposition—all three—agree in number. This **agreement** makes the sentence not only correct but also more precise.

Using demonstrative adjectives with words like *kind, type*, and *sort* can easily lead to vagueness. It is better to be more specific.

CHANGE *These kinds* of hydraulic cranes are best.
TO *Computer-controlled* hydraulic cranes are best.

A compound antecedent joined by *or* or *nor* is singular if both elements are singular, and plural if both elements are plural.

EXAMPLES Neither the *engineer* nor the *draftsman* could do *his* job until *he* understood the new concept.
Neither the *executives* nor the *directors* were pleased at the performance of *their* company.

When one of the antecedents connected by *or* or *nor* is singular and the other plural, the pronoun agrees with the nearer antecedent.

EXAMPLES Either the *supervisor* or the *operators* will have *their* licenses suspended.
Either the *operators* or the *supervisor* will have *his* license suspended.

A compound antecedent with its elements joined by *and* requires a plural pronoun.

EXAMPLE Martha and Joan took *their* layout drawings with *them*.

If both elements refer to the same person, however, use the singular pronoun.

EXAMPLE The respected *economist and author* departed from *her* prepared speech.

Collective nouns may be singular or plural, depending on meaning.

EXAMPLES The *committee* arrived at the recommended solutions only after *it* had deliberated for days.

The *committee* quit for the day and went to *their* respective homes.

agreement of subjects and verbs

A **verb** must agree with its **subject**. Do not let intervening **phrases** and **clauses** mislead you.

CHANGE The use of insecticides, fertilizers, and weed killers, although they offer unquestionable benefits, often result in unfortunate side effects.

TO The *use* of insecticides, fertilizers, and weed killers, although they offer unquestionable benefits, often *results* in unfortunate side effects. (The verb *results* must agrcc with thc subject of the sentence, *use*, rather than with the subject of the preceding clause, *they*.)

Be careful to avoid making the verb agree with the **noun** immediately in front of it if that noun is not its subject. This problem is especially likely to occur when a plural noun falls between a singular subject and its verb.

EXAMPLES Only *one* of the emergency lights *was* functioning when the accident occurred. (The subject of the verb is *one*, not *lights*.)

Each of the switches *controls* a separate circuit. (The subject of the verb is *each*, not *switches*.)

Each of the managers *supervises* a very large region. (The subject of the verb is *each*, not *managers*.)

Be careful not to let modifying phrases obscure a simple subject.

EXAMPLE The *advice* of two engineers, one lawyer, and three executives *was* obtained prior to making a commitment. (The subject of the verb is *advice*, not *executives*.)

Sentences with inverted word order can cause problems with agreement between subject and verb.

EXAMPLE From this work *have come* several important *improvements*. (The subject of the verb is *improvements*, not *work*.)

Such words as *type, part, series*, and *portion* take singular verbs even when they precede a phrase containing a plural noun.

EXAMPLES A *series* of meetings *was* held about the best way to market the new product.

A large *portion* of most industrial annual reports *is* devoted to promoting the corporate image.

Subjects expressing measurement, weight, mass, or total often take singular verbs even when the subject word is plural in form. Such subjects are treated as a unit.

EXAMPLES *Four years is* the normal duration of the training program.

Twenty dollars is the wholesale price of each unit.

Indefinite pronouns such as *some, none, all, more,* and *most* may be singular or plural depending upon whether they are used with a mass noun (*oil* in the following examples) or with a count noun (*drivers* in the following examples). Mass nouns are singular and count nouns are plural.

EXAMPLES *None* of the oil *is* to be used.

None of the truck drivers *are* to go.

Most of the oil *has* been used.

Most of the drivers *know* why they are here.

Some of the oil *has* leaked.

Some of the drivers *have* gone.

One and *each* are normally singular.

EXAMPLES *One* of the brake drums *is* still scored.

Each of the original founders *is* scheduled to speak at the dedication ceremony.

A verb following the **relative pronouns** *who* and *that* agrees in number with the noun to which the **pronoun** refers (its antecedent).

EXAMPLES Steel is one of those *industries* that *are* hardest hit by high energy costs. (*That* refers to *industries*.)

This is one of those engineering *problems* that *require* careful analysis. (*That* refers to *problems*.)
She is one of those *employees* who *are* rarely absent. (*Who* refers to *employees*.)

The word *number* sometimes causes confusion. When used to mean a specific number, it is singular.

EXAMPLE *The number* of committee members *was* six.

When used to mean an approximate number, it is plural.

EXAMPLE *A number* of people *were* waiting for the announcement.

Relative pronouns (*who, which, that*) may take either singular or plural verbs, depending upon whether the antecedent is singular or plural. (See also **who/whom**.)

EXAMPLES He is a chemist *who takes* work home at night.
He is one of those chemists *who take* work home at night.

Some abstract nouns are singular in meaning though plural in form: *mathematics, news, physics,* and *economics*.

EXAMPLES *News* of the merger *is* on page 4 of the *Chronicle*.
Textiles is an industry in need of import quotas.

Some words are always plural, such as *trousers* and *scissors*.

EXAMPLE His *trousers were* torn by the machine.
BUT A *pair* of trousers *is* on order.

Collective subjects take singular verbs when the group is thought of as a unit, plural verbs when the individuals are thought of separately. (See also **collective nouns**.)

A singular subject that is followed by a phrase or clause containing a plural noun still requires a singular verb.

CHANGE One in twenty transistors we receive from our suppliers are faulty.
TO *One* in twenty transistors we receive from our suppliers *is* faulty. (The subject is *one*, not *transistors*.)

The number of a **subjective complement** does not affect the number of the verb—the verb must always agree with the subject.

EXAMPLE The *topic* of his report *was* rivers. (The subject of the sentence is *topic*, not *rivers*.)

A book with a plural title requires a single verb.

EXAMPLE Romig's *Binomial Tables is* a useful source.

COMPOUND SUBJECTS

A compound subject is one that is composed of two or more elements joined by a **conjunction** such as *and, or, nor, either . . . or*, or *neither . . . nor*. Usually, when the elements are connected by *and*, the subject is plural and requires a plural verb.

EXAMPLE *Chemistry and accounting are* both prerequisites for this position.

One exception occurs when the elements connected by *and* form a unit or refer to the same thing. In this case, the subject is regarded as singular and takes a singular verb.

EXAMPLES *Bacon and eggs is* a high-cholesterol meal.
The *red, white, and blue flutters* from the top of the capitol.
Our greatest *technical challenge and business opportunity is* the Model MX Calculator.

A compound subject with a singular and a plural element joined by *or* or *nor* requires that the verb agree with the element nearest to it.

EXAMPLES Neither the office manager nor the *secretaries were* there.
Neither the secretaries nor the office *manager was* there.
Either they or *I am* going to write the report.
Either I or *they are* going to write the report.

If *each* or *every* modifies the elements of a compound subject, use the singular verb.

EXAMPLES *Each* manager and supervisor *has* a production goal to meet.
Every manager and supervisor *has* a production goal to meet.

all around/all-around/all-round

All-round and *all-around* both mean "comprehensive" or "versatile."

EXAMPLE The company started an *all-round* training program.

Do not confuse these words with the two-word **phrase** *all around*, as in "The fence was installed *all around* the building."

all right/all-right/alright

All right means "all correct," as in "The answers were all right." In formal writing it should not be used to mean "good" or "acceptable." It is always written as two words, with no **hyphen**; *all-right* and *alright* are incorrect.

CHANGE The decision that the committee reached was *all right*.
TO The decision that the committee reached was acceptable.

all together/altogether

All together means "all acting together," or "all in one place."

EXAMPLE The necessary instruments were *all together* on the tray.

Altogether means "entirely" or "completely."

EXAMPLE The trip was *altogether* unnecessary.

allude/elude

Allude means to make an indirect reference to something not specifically mentioned.

EXAMPLE The report simply *alluded* to the problem, rather than stating it clearly.

Elude means to escape notice or detection.

EXAMPLES The discrepancy in the account *eluded* the auditor.
The leak *eluded* the inspectors.

allude/refer

Allude means to make an indirect reference to something not specifically mentioned.

EXAMPLES In his speech he *alluded* to the rumors of widespread safety infractions.
The memo *alluded* to past equipment failures.

Refer is used to indicate a direct reference to something.

EXAMPLE He *referred* to the chart three times during his speech.

allusion

The use of allusion (implied or indirect reference) promotes economical writing because it is a shorthand way of referring to a body of material in a few words, or of helping to explain a new and unfamiliar process in terms of one that is familiar. Be sure, however, that your **reader** is familiar with the material to which you allude. In the following **paragraph**, the writer sums up the argument he has been developing with an allusion to the Bible. The biblical story is well known, and the allusion, with its implicit reference to "right standing up to might," concisely emphasizes the writer's point.

EXAMPLE As it presently exists, the review process involves the consumer's attorney sitting alone, usually without adequate technical assistance, faced by two or three government attorneys, two or three attorneys from the XYZ Corporation, and large teams of experts who support the government and corporation's attorneys. The entire proceeding is reminiscent of David versus Goliath.

Allusions should be used with restraint. If overdone, they can lead to **affectation**. (See also **analogy**.)

allusion/illusion

An *allusion* is an indirect reference to something not specifically mentioned.

EXAMPLE The report made an *allusion* to metal fatigue in support structures.

An *illusion* is a mistaken perception or a false image.

EXAMPLE County officials are under the *illusion* that the landfill will last indefinitely.

almost/most

Do not use *most* as a colloquial substitute for *almost* in your writing.

CHANGE New shipments arrive *most* every day.
TO New shipments arrive *almost* every day.

If you can substitute *almost* for *most* in a sentence, *almost* is the word you need.

already/all ready

Already is an **adverb** expressing time.

EXAMPLE We had *already* shipped the transistors when the stop order arrived.

All ready is a two-word **phrase** meaning "completely prepared."

EXAMPLE He was *all ready* to start work on the project when it was suddenly canceled.

also

Also is an **adverb** that means "additionally."

EXAMPLE Two 500,000-gallon tanks have recently been constructed on site. Several 10,000-gallon tanks are *also* available, if needed.

It should not be used as a **connective** in the sense of "and."

CHANGE He brought the reports, letters, *also* the section supervisor's recommendations.
TO He brought the reports, letters, *and* the section supervisor's recommendations.

Avoid opening sentences with *also*. It is a weak transitional word that suggests an afterthought rather than planned writing.

CHANGE *Also* he brought statistical data to support his proposal.
TO *In addition*, he brought statistical data to support his proposal.
OR He *also* brought statistical data to support his proposal.

ambiguity

A word or passage is ambiguous when it is susceptible to two or more interpretations, yet provides the **reader** with no certain basis for choosing among the alternatives.

EXAMPLE Mathematics is more valuable to an engineer than a computer. (Does this mean that an engineer is more in need of mathematics than a computer is? Or does it mean that mathematics is more valuable to an engineer than a computer is?)

Ambiguity can take many forms: ambiguous **pronoun reference, misplaced modifiers, dangling modifiers**, ambiguous coordination, ambiguous juxtaposition, incomplete **comparison**, incomplete **idiom**, ambiguous **word choice**, and so on.

CHANGE Inadequate quality-control procedures have resulted in more equipment failures. *This is* our most serious problem at present. (ambiguous pronoun reference; does *this* refer to "quality-control procedures" or "equipment failures"?)

TO Inadequate quality-control procedures have resulted in more equipment failures. These failures are our most serious problem at present.

OR Inadequate quality-control procedures have resulted in more equipment failures. Quality control is our most serious problem at present.

CHANGE Ms. Jones values rigid quality-control standards more than Mr. Rosenblum. (incomplete comparison)

TO Ms. Jones values rigid quality-control standards more than Mr. Rosenblum *does*.

CHANGE His hobby was cooking. He was especially fond of cocker spaniels. (missing modifier)

TO His hobby was cooking. He was *also* especially fond of cocker spaniels.

CHANGE All navigators are *not* talented in mathematics. (misplaced modifier; the implication is that *no* navigator is talented in mathematics)

TO *Not* all navigators are talented in mathematics.

Ambiguity is also often caused by thoughtless word choice.

CHANGE The general manager has denied reports that the plant's recent fuel allocation cut will be *restored*. (inappropriate word choice)

TO The general manager has denied reports that the plant's recent fuel allocation cut will be *rescinded*.

amount/number

Amount is used with things thought of in bulk (mass **nouns**).

EXAMPLES The *amount* of electricity available for industrial use is limited.
The *amount* of oxygen was insufficient for combustion.

Number is used with things that can be counted as individual items (count nouns).

EXAMPLES A large *number* of stockholders attended the meeting.
The *number* of employees who are qualified for early retirement has increased in recent years.

Avoid using *amount* when referring to countable items.

CHANGE The *amount* of people in the room gradually increased.
TO The *number* of people in the room gradually increased.
CHANGE I was surprised at the *amount* of errors in the report.
TO I was surprised at the *number* of errors in the report.
CHANGE Because the *amount* of thefts has increased, the doors will be locked in the evening.
TO Because the *number* of thefts has increased, the doors will be locked in the evening.

ampersands

The ampersand (&) is a **symbol** sometimes used to represent the word *and*, especially in the names of organizations.

EXAMPLES Chicago & Northwestern Railway
Watkins & Watkins, Inc.

The ampersand is appropriate for footnotes, **bibliographies**, **lists**, and references if the ampersand appears in the name being listed. When writing the name of an organization in sentences or in an address, however, spell out the word *and* unless the ampersand appears in the official name of the company. (See also **abbreviations**.)

An ampersand must always be set off by normal word spacing but should never be preceded by a **comma**.

CHANGE Carlton, Dillon, & Manchester, Inc.
TO Carlton, Dillon & Manchester, Inc.

analogy

Analogy is a **comparison** between two objects or concepts to show ways in which they are similar. In effect, analogies say, "A is to B as C is to D."

EXAMPLE Pollution is to the environment as cancer is to the body.

The resemblance between the concepts represented in an analogy must be close enough to illuminate the relationship the writer wants to establish. Analogies may be brief or extended, depending on the writer's purpose. Analogy often helps writers explain unfamiliar things by comparing them to things with which the reader is familiar.

Analogy can be a particularly useful tool to the technical person writing for an intelligent and educated but nontechnical audience, such as top management, because of its effectiveness in **defining terms** and explaining processes. Like all figurative language, analogy can provide a shortcut means of communication if it is used with care and restraint.

The following example explains a computer search technique for finding information in a data file by comparing it to the method used by most people to find a word in a **dictionary**.

> The search technique used in this kind of file processing is similar to the search technique used to look up a word in a dictionary. To locate a specific word, you scan the key words located at the top of each dictionary page (these key words identify the first and last words on the page) until you find the key words that confine your word to a specific page. Assume that all the key words that reference the last word on each page of a dictionary were placed in a file, along with their corresponding page numbers, and that all the words in the dictionary were placed in another file. To locate any word, you would simply scan the first file until you found a key word greater in alphabetical sequence than the desired word and go to the place in the second file indicated by the key word to find the desired word. This search technique is called indexed sequential processing.
>
> —*The NCR File Management Manual* (Dayton: NCR Corporation, 1974), p. 76.

and/or

And/or means that either both circumstances are possible or only one of two circumstances is possible; however, it is clumsy and awkward because it makes the **reader** stop to puzzle over your distinction. Also, writers more often than not use the term inexactly. As Wilson Follett said, "English speakers and writers have managed to express this simple relationship without *and/or* for over six centuries." Avoid it.

CHANGE Use A and/or B.
TO Use A or B or both.

anecdotes (see **openings**)

annual reports

The corporate annual report is a legally required document that is, in effect, a "state of the company" message. It is written primarily for stockholders, although many other audiences must be kept in mind when the report is prepared (bankers, financial press, labor unions, employees, and the like). An annual report usually covers the high points of the previous year's operations and finances and attempts to forecast the coming year's operations. It may also explain the company's present directions and highlight its present strengths. If weaknesses have developed or failures have occurred, the annual report may analyze them and explain the efforts being made to overcome them.

A wide variety of annual reports is published every year. Some are lavish in presentation and paint the company and its operations in glowing terms; others are spartan financial recitations that merely meet the legal requirements for annual financial reporting. Most annual reports are combinations of legally required financial reporting and narrative articles presenting the company with its best foot forward.

Although they vary greatly, annual reports typically have five major divisions: (1) financial highlights, (2) a statement to the stockholders or letter from the president, (3) a narrative section containing articles on the company's operations, (4) a financial statement, and (5) a listing of the company's board of directors and officers.

Financial Highlights. The financial highlights section is a quick review of the company's sales and earnings that usually precedes the president's message to the stockholders, sometimes even appearing on the inside of the front cover. The most-read part of the annual report, this section often compares sales and earnings for three years, and it often includes the percentage of change from year to year.

Statement to Stockholders. This section of the annual report is a direct statement to stockholders from the company's president or chairman of the board of directors. This statement (which is sometimes presented in the form of a letter) is second in readership only to the financial highlights. It is the place to set the stage for the rest of the report. It should avoid repeating financial facts already cited in the financial highlights; instead, it can be used to (1) interpret the entire year's performance, (2) touch upon plans and future directions, and (3) give the company's explanations for any failures.

The statement to stockholders may be an in-depth review of the company's operations during the past year, or it may be a brief summary of the entire report. A brief summary of the annual report is sometimes followed by an article in "interview" (question and answer) **format** that reviews the past year's operations as though an actual interview were taking place between the writer and the president of the company. Avoid stereotyped language, **clichés**, and technical terms when you write this statement. Use simple sentences and short **paragraphs**, and use a straightforward and informal **style**.

Narrative Feature Articles. This section of the annual report is normally used to present company operations and new products or developments in a positive light. Select the **topic** for these articles very carefully, making certain that they are timely and meaningful and that they contribute to the primary objective of the annual report. The following list includes topics that are often dealt with in annual reports:

Major profit factors in last year's performance
Sales trends
The company's performance compared with that of its competition
International operations
World problems that affect the company
The dependability of the company's overseas markets
Research and development efforts
Significant organizational changes
The company's labor relations outlook
Productivity
Acquisitions and their significance
Prospects for increasing stock dividends
Outlook for next year (for the company as a whole or by divisions)
Significant new products or services
Current market performance of existing products or services
Service and support operations
Social responsibilities—on environmental protection, energy conservation, safety, hiring minorities and the handicapped, and so on. (Be certain to document the company's efforts in this area because overstated claims by a few companies in the past have generated suspicion of all such claims.)

Most **readers** merely scan an annual report, but they do stop to look at **photographs** and **illustrations** and to read their captions. Therefore, use photographs and illustrations liberally in this section of the annual report, and make your captions informative and well written.

Financial Statement. The financial statement should not be forbidding in appearance; it should be uncluttered, inviting, and consistent with the other sections of the annual report. The financial statement may appear at the beginning, middle, or end of the annual report; however, it most often appears at the end. It is frequently prepared as a separate booklet, printed on colored paper, and then stitched into the report.

Minimum normal contents of the financial statement include (1) balance sheet, (2) statement of income, (3) changes in financial position, (4) the auditor's statement, and (5) footnotes. Most annual reports also include a fairly comprehensive **comparison** of financial results of the past ten to fifteen years.

Footnotes in the financial section should be written in simple, direct language rather than in accounting terms. The auditor's statement should be limited to no more than one-third of a page.

Board of Directors and Company Officers. The final part of the annual report lists the company's board of directors, along with their corporate affiliations. Many annual reports also include a portrait photograph of each director, and the company's officers are listed by name and title (sometimes with photographs). These generally include the chairman of the board of directors, president, vice-presidents, secretary, treasurer, and legal counsel. To avoid hard feelings about some being shown in photographs and others not, establish a logical cutoff point. For example, if there are too many officers to include photographs of them all, show only the executive committee. Do not, however, underestimate the importance of these photographs; all your reading audiences are interested in seeing what your company's top executives look like.

PREPARING THE REPORT

Assuming you are to prepare the annual report, the first thing you must do is interview the president of the company or the chairman of the board of directors to determine the general directions the report is to take. Next, you must interview the various vice-presidents, or division heads, to learn the highlights of each division's operations during the year. Determine from these executives the proper **emphasis** to place on their divisions' performances—but stay within the general direction established by the president or chairman of the board.

The primary objective of the annual report often depends on which of several audiences (shareholders, bankers, business and financial analysts, the financial press, employees, labor unions, the communities in which your company operates) you are most concerned about. Since you cannot satisfy all these groups with one report, you must establish priorities among them. Having done that, you can compose a list of "must" topics, then a list of secondary topics (such as diversification, expansion of markets, and product development). You can also select the main topics for brief coverage in the statement to stockholders, decide whether to show sales and earnings by division, determine the amount of space to allot to various divisions and subsidiaries, determine the contents of charts and graphs, decide whether to include photographs of the executives of branches, divisions, and subsidiaries, decide whether to include a frank discussion of company problems and solutions, and decide how lavish or spartan the report should be in appearance and costs.

Tone is a critical factor in writing and designing an annual report. Companies are as different as people; some are conservative, and some are aggressively youthful. The annual report should convey the image your company has or wants to establish. A static, formal page layout with traditional typography will suggest a conservative, dignified company, for example; and bold graphics, a dynamic design, and unusual typography will suggest an aggressive company. Don't mix the two, however; photographs of executives in dark business suits would not be compatible with bold graphics.

Be selective in your choice of photographs and illustrations, choosing only those that will make the maximum contribution to the report's theme. Remember that most readers merely scan an annual report, looking carefully only at photographs and captions. So both photographs and their captions are critical to the success of your annual report. A photograph of a machine can be dull. But put an operator in it, and the caption can tell what the operator is doing and why it is meaningful.

Study your company's annual reports of the past several years for content, style, and format. Within reason, try to emulate these. Then review the writing process outlined in the Checklist of the Writing Process at the beginning of this book, and use all the steps listed there to the best of your ability—your company's annual report could well be the most important writing you will ever be asked to do.

The following information is normally printed on the inside front and back covers of the annual report: (1) notice of the annual meeting, (2) corporate address, (3) names of transfer agents, (4) registrar, and (5) stock exchange. Some annual reports use the inside front cover for the announcement of the annual meeting alone.

Use of Charts and Graphs. **Graphs** and charts enable readers to grasp statistical material quickly and easily—provided the charts and graphs are not so complex that they defeat that purpose. Some companies scatter graphs throughout the annual report at strategic locations near the financial highlights, in the statement to stockholders, throughout the feature articles, and so forth. Others group them in the financial statement.

The subjects that most easily lend themselves to graphs and charts—usually shown in a five-year comparison arrangement—are assets, capital expenditures, dividends, earnings (by product groups or by divisions), industry growth, inventories, liabilities, net worth, prices (trend of), reserves, sales (by product groups or by divisions), and source and disposition of funds (taxes, wages, working capital).

ante-/anti-

Ante- means "before" or "in front of."

EXAMPLES *Ante*room, *ante*date, *ante*diluvian (before the flood)

Anti- means "against" or "opposed to."

EXAMPLES *anti*body, *anti*clerical, *anti*social

Anti- is hyphenated when joined to proper nouns or to words beginning with the letter *i*.

EXAMPLES *anti*-American, *anti*-intellectual

When in doubt, consult your **dictionary**. (See also **prefixes**.)

antonyms

An antonym is a word that is nearly the opposite, in meaning, of another word.

EXAMPLES good/bad, well/ill, fresh/stale

Many pairs of words that look as if they are antonyms are not. Be careful not to use these words incorrectly.

EXAMPLES famous/infamous, flammable/inflammable, limit/delimit

When in doubt about the pronunciation, correct use, or exact meaning of a word, use your **dictionary**.

apostrophes

The apostrophe (') is used to show possession, to mark the omission of letters, and sometimes to indicate the plural of arabic numbers, letters, and **acronyms**. Do not confuse the apostrophe used to show the plural with the apostrophe used to show possession.

EXAMPLES The entry required five 7's in the appropriate columns. (The apostrophe is used here to indicate the plural, not possession.)

The letter's purpose was clearly evident in its opening paragraph. (The apostrophe here is used to show possession, not the plural.)

TO SHOW POSSESSION

An apostrophe is used with an *s* to form the possessive **case** of some **nouns**.

EXAMPLE A recent scientific analysis of *New York City's* atmosphere concluded that a New Yorker on the street took into his or her lungs the equivalent in toxic materials of 38 cigarettes a day.

With coordinate nouns, the last noun takes the possessive form to show joint possession.

EXAMPLE Michelson and *Morley's* famous experiment on the velocity of light was made in 1887.

To show individual possession with coordinate nouns, each noun should take the possessive form.

EXAMPLE The difference between *Tom's* and *Mary's* test results is statistically insignificant.

Singular nouns ending in *s* may form the possessive either by an apostrophe alone or by *'s*. Whichever way you do it, however, be consistent.

EXAMPLES a waitress' uniform, an actress' career
a waitress's uniform, an actress's career.

Singular nouns of one syllable form the possessive by adding *'s*.

EXAMPLE The boss's desk was cluttered.

Use only an apostrophe with plural nouns ending in *s*.

EXAMPLES a managers' meeting, the technicians' handbook, the waitresses' lounge

When a noun ends in multiple consecutive *s* sounds, form the possessive by adding only an apostrophe.

EXAMPLES Jesus' disciples, Moses' sojourn.

The apostrophe is not used with possessive **pronouns.**

EXAMPLES yours, its, his, ours, whose, theirs

It's is a **contraction** of *it is; its* is the possessive form of the pronoun. Be careful not to confuse the two words. (See **its/it's**.)

EXAMPLE *It's* important that the sales force meet *its* quota.

In names of places and institutions, the apostrophe is usually omitted.

EXAMPLES Harpers Ferry, Writers Book Club

TO SHOW OMISSION

An apostrophe is used to mark the omission of letters in a word or date.

EXAMPLES can't, I'm, I'll
the class of '61

TO FORM PLURALS

An apostrophe and an *s* may be added to show the plural of a word as a word. (The word itself is underlined, or italicized, to call attention to its use.)

EXAMPLE There were five *and's* in his first sentence.

If the term is in all **capital letters** or ends with a capital letter, however, the apostrophe is not required to form the plural. (See also **acronyms and initialisms**.)

EXAMPLES The university awarded seven *Ph.D.s* in engineering last year.
He had included 43 *ADDs* in his computer program.

Do not use apostrophes to indicate the plural of numbers and letters unless confusion would result without one.

EXAMPLES 5s, 30s, two 100s, seven I's

appendix/appendixes/appendices

An appendix contains material at the end of a **formal report** or book that supplements or clarifies. (The plural form of the word may be either **appendixes** or **appendices**.)

Although not a mandatory part of a **report**, an appendix can be useful for explanations that are too long for notes but that could be helpful to the reader seeking further assistance or a clarification of points made in the report. Information placed in an appendix is too detailed or voluminous to appear in text without impeding the orderly presentation of ideas. This information typically includes passages from documents and laws that reinforce or illustrate the text, long lists of charts and **tables**, letters and other supporting documents, calculations (in full), computer printouts of raw data, and case histories. An appendix, however, should not be used for miscellaneous bits and pieces of information you were unable to work into the text.

Generally, each appendix contains only one type of information. The contents of each appendix should be identifiable without the reader having to refer to the body of the report. An introductory paragraph describing the contents of the appendix, therefore, is necessary for some appendixes, especially those containing computer printouts of data, long tables, or similar information.

When the report contains more than one appendix, arrange them in the order in which they are referred to in the text. Thus, a reference in the text to Appendix A should precede the first text reference to Appendix B.

Each appendix begins on a new page. (For guidance on where to locate appendixes in reports, see **formal reports**.) Identify each with a **title** and a **head**:

Appendix A
Sample Questionnaire

Appendixes are ordinarily labeled Appendix A, Appendix B, and so on. If your report has only one appendix, label it Appendix, followed by the title. To call it Appendix A implies that an Appendix B will follow.

The titles and beginning page numbers of the appendixes are listed in the **table of contents** of the report in which they appear.

application letters

The letter of application is essentially a **sales letter**. You are marketing your skills, abilities, and knowledge. Remember that you may be competing with many other applicants. The immediate **objective** of an application letter is to get the attention of the person who screens and hires job applicants. Your ultimate goal is to obtain a job interview (see **interviewing for a job**).

The successful application letter does three things: catches the **reader's** favorable attention, convinces the reader that you are qualified for consideration, and requests an interview. It should be concisely written.

A letter of application should provide the following information:

1. Identify an employment area or state a specific job title.
2. Point out your source of information about the job.
3. Summarize your qualifications for the job, specifically education, work experience, and activities showing leadership skills.
4. Refer the reader to your **résumé**.
5. Ask for an interview, stating where you can be reached and when you will be available for an interview.

If you are applying for a specific job, include information pertinent to the position—details not included on the more general résumé.

Personnel directors review many letters each week. To save them time, you should state your job objective directly at the beginning of the letter.

EXAMPLE I am seeking a position in an engineering department where I can use my computer science training to solve engineering problems.

If you have been referred to a prospective employer by one if its employees, a placement counselor, a professor, or someone else, however, you might say so before stating your job objective.

EXAMPLE During the recent NOMAD convention in Washington, D.C., a member of your sales staff, Mr. Dale Jarrett, informed me of a possible opening for a manager in your Dealer Sales Division. My extensive background in the office machine

6819 Locustview Drive
Topeka, Kansas 66614
June 14, 19--

Loudons, Inc.
4619 Drove Lane
Kansas City, Kansas 63511

Dear Personnel Manager:

The Kansas Dispatch recently reported that Loudons is building a new data processing center just north of Kansas City. I would like to apply for a position as an entry-level programmer at the center.

I am a recent graduate of Fairview Community College in Topeka, with an Associate Degree in Computer Science. In addition to taking a broad range of courses, I have served as a computer consultant at the college's computer center, where I helped train novice computer users. Since I understand Loudons produces both in-house and customer documentation, my technical writing skills (as described in the enclosed resume) may be particularly useful.

I will be happy to meet with you at your convenience and provide any additional information you may need. You can reach me either at my home address or at (913) 233-1552.

Sincerely,

David B. Edwards

Enclosure: Resume

Letter 1 Sample Application Letter

2701 Wyoming Street
Atlanta, Georgia 30307
May 29, 19--

Ms. Laura Goldman
Chief Engineer
Acton, Inc.
80 Roseville Road
St. Louis, Missouri 63130

Dear Ms. Goldman:

I am seeking a responsible position in an engineering department in which I may use my training in computer sciences to solve engineering problems. I would be interested in exploring the possibility of obtaining such a position within your firm.

I expect to receive a Bachelor of Science degree in Engineering from Georgia Institute of Technology in June, when I will have completed the Computer Systems Engineering Program of the Engineering Department. Since September 19-- I have been participating, through the university, in the Professional Training Program at Computer Systems International, in Atlanta. In the program I was assigned, on a rotating basis, to several staff sections in apprentice positions. Most recently I have been a programmer trainee in the Engineering Department and have gained a great deal of experience in computer applications. Details of the academic courses I have taken are contained in the enclosed resume.

I look forward to meeting you soon in an interview. I can be contacted at my office (415-866-7000, ext. 312) or at home (415-256-6320).

Sincerely yours,

Victoria T. Fromme

Victoria T. Fromme

Enclosure: Resume

Letter 2 Sample Application Letter

522 Beethoven Drive
Roanoke, Virginia 24017
November 15, 19--

Ms. Cecilia Smathers
Vice President, Dealer Sales
Hamilton Office Machines, Inc.
6194 Main Street
Hampton, Virginia 23661

Dear Ms. Smathers:

During the recent NOMAD convention in Washington, a member of your sales staff, Mr. Dale Jarrett, informed me of a possible opening for a Manager in your Dealer Sales Division. I believe that my extensive background in the office machine industry qualifies me for the position.

I was with the Technology, Inc., Dealer Division from its formation in 19-- to its phase-out last year. During this period, I was involved in all areas of dealer sales, both within Technology, Inc., and through personal contact with a number of independent dealers. Between 19-- and 19-- I served as Assistant to the Dealer Sales Manager as a Special Representative. My education and work experience are detailed in the enclosed resume.

I would like to discuss my qualifications in an interview at your convenience. Please write to me or telephone me at 703-449-6743 any weekday.

Sincerely,

Gregory Mindukakis

Enclosure: Resume

Letter 3 Sample Application Letter

industry, I believe, makes me highly qualified for the position.

In succeeding **paragraphs** expand upon the qualifications you mentioned in your **opening**. Add any appropriate details, highlighting the experience listed on your résumé that is especially pertinent to the job you are seeking. Close your letter with a request for an interview.

Type your letter, proofread it carefully for errors, and *keep a copy* for future reference. (See also **proofreading**.)

See the accompanying three sample letters of application. The first one is written by a recent college graduate, the second by a college student about to graduate, and the third by a writer with many years of work experience.

Letter 1 is in response to a local newspaper article about a company's plan to build a new plant. The writer is not applying for a specific job opening but describes the position he is looking for. In Letter 2 the writer does not specify where she learned of the opening because she does not know whether a position is actually available. Letter 3 opens with an indication of where the writer learned of the job vacancy. (See also **job search** and **reference letters**.)

appositives

An appositive is a **noun** or noun **phrase** that follows and amplifies another noun or noun phrase. It has the same grammatical function as the noun it complements.

EXAMPLES Dennis Gabor, *a British scientist*, experimented with coherent light in the 1940s.

The British scientist *Dennis Gabor* experimented with coherent light in the 1940s.

George Thomas, *head of the Economic and Planning Branch of PRC*, summarized the president's speech in a confidential memo to the advertising staff.

For detailed information on the use of **commas** with appositives, see **restrictive and nonrestrictive elements**.

When in doubt about the **case** of an appositive, you can check it by substituting the appositive for the noun it modifies.

EXAMPLE My boss gave the two of us, Jim and *me*, the day off. (You wouldn't say, "My boss gave *I* the day off.")

argumentation (see **persuasion**)

articles

As a **part of speech**, articles are considered to be **adjectives** because they modify the items they designate by either limiting them or making them more precise. There are two kinds of articles, indefinite and definite.

Indefinite: *a* and *an* (denotes an unspecified item)

EXAMPLE *A* program was run on our new computer. (Not a specific program, but an unspecified program. Therefore, the article is indefinite.)

Definite: *the* (denotes a particular item)

EXAMPLE *The* program was run on the computer. (Not just any program, but *the* specific program. Therefore, the article is definite.)

The choice between *a* and *an* depends on the sound rather than the letter following the article. Use *a* before words beginning with a consonant sound (*a* person, *a* happy person, *a* historical event). Use *an* before words beginning with a vowel sound (*an* uncle, *an* hour). With **abbreviations**, use *a* before initial letters having a consonant sound: *a* TWA flight. Use *an* before initial letters having a vowel sound: *an* SLN report (note that the first sound is "ess").

Do not omit all articles from your writing. This is, unfortunately, an easy habit to develop. To include the articles costs nothing; to eliminate them makes the reading more difficult. (See also **telegraphic style**.)

CHANGE Pass card through punch area for any debris.
TO Pass *a* card through *the* punch area *to clear away* any debris.
CHANGE There has been decline in domestic output of crude oil.
TO There has been *a* decline in *the* domestic output of crude oil.

On the other hand, don't overdo it. An article can be superfluous.

EXAMPLES I'll meet you in *a* half *an* hour. (Choose one and eliminate the other.)
Fill with *a* half *a* pint of fluid. (Choose one and eliminate the other.)

Do not capitalize articles when they appear in titles except as the first word of the title. (See also **capital letters**.)

EXAMPLE *Time* magazine reviewed *The Nuclear Option in Europe for the Eighties.*

as (see also **like/as**)

Since *as* can mean so many things (*since, because, for, that, at that time, when, while*, and so on) and can be at least four **parts of speech (conjunction, preposition, adverb, pronoun)**, it is often overused and misused, especially in speech. In writing, *as* is often weak or even ambiguous. Notice that the following sentence has two possible meanings:

EXAMPLE *As* we were together, he revealed his plans.
MEANS *Because* we were together, he revealed his plans.
OR *While* we were together, he revealed his plans.

The word *as* can also contribute to **awkwardness** by appearing too many times in a sentence.

CHANGE *As* we realized *as* soon *as* we began the project, the problem needed a solution.
TO We realized the moment we began the project that the problem needed a solution.

(See also **because**.)

as much as/more than

These two **phrases** are sometimes incorrectly run together, especially when intervening phrases delay the completion of the phrase.

CHANGE The engineers had *as much*, if not *more*, influence in planning the program *than* the accountants.
TO The engineers had *as much* influence in planning the program *as* the accountants, if not *more*.
OR The engineers had *as much* influence *as* the accountants, if not *more*, in planning the program.

as regards/with regard to/in regard to/regarding

With regards to and *in regards to* are incorrect **idioms** for *with regard to* and *in regard to*. *As regards* and *regarding* both are acceptable variants.

CHANGE *With regards to* the building contract, this question is pertinent.
TO *With regard to* the building contract, this question is pertinent.
OR *In regard to* the building contract, this question is pertinent.
OR *As regards* the building contract, this question is pertinent.
OR *Regarding* the building contract, this question is pertinent.

Avoid using *with regard to* where *about* would be more specific.

CHANGE I am writing you *with regard to* your design for a new product.
TO I am writing you *about* your design for a new product.

as such

The **phrase** *as such* is seldom useful and should be omitted.

CHANGE The drafting department, *as such*, worked overtime to meet the deadline.
TO The drafting department worked overtime to meet the deadline.
CHANGE This program is poor. *As such*, it should be eliminated.
TO This program is poor and should be eliminated.

as to whether

The **phrase** *as to whether* (*as to when* or *as to where*) is clumsy and redundant. Either omit it altogether or use only *whether*.

CHANGE *As to whether* we will redesign the fuel-injection system, we are still undecided.
TO *Whether* we will redesign the fuel-injection system is still undecided.

Be wary of all phrases starting with *as to*; they are often redundant, vague, or indirect.

CHANGE *As to* his policy, I am in full agreement.
TO I am in full agreement with his policy.

as well as

Do not use *as well as* together with *both*. The two expressions have similar meanings; use one or the other.

CHANGE *Both* General Motors, *as well as* Ford, are developing electric cars.
TO *Both* General Motors *and* Ford are developing electric cars.
OR General Motors, *as well as* Ford, is developing an electric car.

attribute/contribute

Attribute (with the accent on the second syllable) is a **verb** that means "point to a cause or source."

EXAMPLE He *attributes* the plant's improved safety record to the new training program.

Attribute (with the first syllable accented) is a **noun** meaning a quality or characteristic belonging to someone or something.

EXAMPLE His mathematical skill is his most valuable *attribute*.

Contribute means "give."

EXAMPLE His mathematical skills will *contribute* much to the project.

audience (see **reader**)

augment/supplement

Augment means to increase or magnify in size, degree, or effect.

EXAMPLE Many employees *augment* their incomes by working overtime.

Supplement means to add something to make up for a deficiency.

EXAMPLE The physician told him to *supplement* his diet with vitamins.

average/mean/median

The *average*, or arithmetical *mean*, is determined by dividing a sum of two or more quantities by the number of quantities. For example, if one report is 10 pages, another is 30 pages, and a third is 20 pages, their *average* (or *mean*) length is 20 pages. It is incorrect, therefore, to say that "*each* averages 20 pages" because *each* report is a specific length.

CHANGE Each report *averages* 20 pages.
TO The three reports *average* 20 pages.

A *median* is the middle number in a sequence of numbers.

EXAMPLE The *median* of the series, 1, 3, 4, 7, 8 is 4.

awhile/a while

Awhile is an **adverb** meaning "for a short time." It is not preceded by *for* because the meaning of *for* is inherent in the meaning of *awhile*. *A while* is a **noun** phrase that means "a period of time."

CHANGE Wait for *awhile* before investing more heavily.
TO Wait for *a while* before investing more heavily.
OR Wait *awhile* before investing more heavily.

awkwardness

Any writing that strikes the **reader** as awkward—that is, as forced or unnatural—impedes the reader's understanding. Awkwardness has many causes, including overloaded sentences, overlapping **subordination**, grammatical errors, ambiguous statements, overuse of **expletives** or of the passive **voice**, faulty **logic**, unintentional **repetition, garbled sentences**, and **jammed modifiers**.

To avoid awkwardness, make your writing as direct and concise as possible. The following three guidelines will help you smooth out most awkward passages: (1) In general, you should keep your sentences uncomplicated. (2) Use the active voice unless you have a particular reason to use the passive voice. (3) Tighten up your writing by eliminating excess words.

B

bad/badly

Bad is the **adjective** form that follows such **linking verbs** as *feel* and *look*.

EXAMPLES With the flu, you will feel *bad* for three days.
We don't want our department to look *bad* at the meeting.

Badly is an **adverb**.

EXAMPLE The test model performed *badly* during the trial run.

To say "I feel badly" would mean, literally, that your sense of touch was impaired. (See also **good/well**.)

balance/remainder

One meaning of *balance* is "a state of equilibrium"; another meaning is "the amount remaining in a bank account after balancing deposits and withdrawals." *Remainder*, in all applications, is "what is left over." *Remainder* is the more accurate word, therefore, to mean "that which is left over."

EXAMPLES The accounting department must attempt to maintain a *balance* between looking after the company's best financial interests and being sensitive to the company's research and development work.
The *balance* in the corporate account after the payroll has been met is a matter for concern.
Round the fraction off to its nearest whole number and drop the *remainder*.

When using these words figuratively (outside of banking and mathematical contexts), use *remainder* to mean "what is left over."

CHANGE Four of the speakers at the conference were from West Germany, and the *balance* were from the United States.
TO Four of the speakers at the conference were from West Germany, and the *remainder* were from the United States.

be sure and/be sure to

The **phrase** *be sure and* is colloquial and unidiomatic when used for *be sure to*. (See also **idioms**.)

CHANGE When the claxon sounds, *be sure and* phone the guard desk immediately.
TO When the claxon sounds, *be sure to* phone the guard desk immediately.

because

To express cause, *because* is the strongest and most specific connective (others are *for, since, as, inasmuch as, insofar as*). *Because* is unequivocal in stating causal relationship.

EXAMPLE We didn't complete the project *because* the raw materials became too costly. (The use of *because* emphasizes the cause-and-effect relationship.)

For can express causal relationships, but it is weaker than *because*, and it allows the **clause** that follows to be a separate **independent clause**.

EXAMPLE We didn't complete the project, *for* raw materials became too costly. (The use of *for* expresses but does not emphasize the cause-and-effect relationship.)

As a connective to express cause, *since* is also a weak substitute for *because*.

EXAMPLE *Since* the computer is broken, paychecks will be delayed.

However, *since* is an appropriate connective when the **emphasis** is on circumstances, conditions, or time rather than on cause and effect.

EXAMPLES *Since* I was in town anyway, I decided to visit the test site.
Since 1944, the company has earned a profit every year.

As is the least definite connective to indicate cause; its use for this purpose is better avoided.

CHANGE I left the office early, *as* I had finished my work.
TO I left the office early *because* I had finished my work.
OR *Since* I had finished my work, I left the office early.

Inasmuch as implies concession; that is, it suggests that a statement is true "in view of the circumstances."

EXAMPLE You may as well complete the project, *inasmuch as* you have already begun it.

(See also **reason is because**.)

being as/being that

These **phrases** are nonstandard English and should not be used in writing. Use *because* or *since*.

beside/besides

Besides, meaning "in addition to" or "other than," should be carefully distinguished from *beside*, meaning "next to" or "apart from."

EXAMPLE *Besides* the two of us from the Systems Department, three people from Production were standing *beside* the president when he presented the award.

between/among

Between is normally used to relate two items or persons.

EXAMPLES The roll pin is located *between* the grommet and the knob.
Preferred stock offers a buyer a middle ground *between* bonds and common stock.

Among is used to relate more than two.

EXAMPLE The subcontracting was distributed *among* the three firms.

between you and me

People sometimes use the incorrect expression *between you and I*. Because the **pronouns** are **objects** of the **preposition** *between*, the objective form of the **personal pronoun** (*me*) must be used. (See also **case [grammar]**.)

CHANGE *Between you and I*, John should be taken off the job.
TO *Between you and me*, John should be taken off the job.

bi/semi

When used with periods of time, *bi* means "two" or "every two." *Bimonthly* means "once in two months"; *biweekly* means "once in two weeks."

When used with periods of time, *semi* means "half of" or "occurring twice within a period of time." Semimonthly means "twice a month"; semiweekly means "twice a week."

Both *bi* and *semi* are normally joined with the following element without space or **hyphen**.

biannual/biennial

By conventional usage, *biannual* means "twice during the year," and *biennial* means "every other year." (See also **bi/semi**.)

bibliography

A bibliography is a list of the books, articles, and other source materials consulted in the preparation of a paper, **report**, or article. It provides a convenient alphabetical listing of these sources in a standardized form for **readers** interested in getting further information on the topic or in assessing the scope of the **research.** A bibliography is normally placed at the end of a document. (See **formal reports** for guidance on the placement of a bibliography in a report.)

Those works consulted for background information, in addition to those actually cited in the text, should be included in a bibliography. A bibliography is often appropriate as a supplement to **lists** of "Works Cited" or "Reference" sections, since such lists include only works actually referred to in the text. If a list of "References" or "Works Cited" is used, the bibliography should follow the same format. (See **documenting sources** for detailed format guidelines.)

The entries in a bibliography are listed alphabetically by the author's last name. If the author is unknown, the entry is alphabetized by the first word in the title (following *a*, *an*, or *the*). Entries may also be arranged into subject categories and then by alphabetical order within these categories.

An annotated bibliography is one that includes complete bibliographic information about a work (author, title, publisher) followed by a brief description or evaluation of what the work contains. (See also **literature reviews**.) The following annotation concisely evaluates a book:

> Kingslake, Rudolf. *Lens Design Fundamentals*. New York: Academic Press, 1978.
>
> > Intended for the designer, the book presents equations prepared for use with calculators and computers rather than log and trig tables. It contains valuable sections on calculating achromats, apochromats, catadioptric systems, and eyepieces.

The format of this sample entry follows the guidelines in the Modern Language Association's *MLA Handbook for Writers of Research Papers*. For explanations and further samples of this and other formats, see **documenting sources**.

blend words

A blend word is formed by combining part of one word with part of another.

EXAMPLES motor + hotel = motel
breakfast + lunch = brunch
smoke + fog = smog
electric + execute = electrocute
chuckle + snort = chortle

Although blend words (sometimes called "portmanteau words") may occasionally be created by a specialist to meet a specific need—such as *stagflation* (a stagnant economy coupled with inflation)—resist creating blend words in your writing unless an obvious need arises. If you must create a blend word, be sure to define it clearly for your **reader**. Otherwise, creating blend words is at best merely "cute" and could be confusing. (See also **new words**.)

both . . . and

Statements using the *both . . . and* construction should always be balanced both grammatically and logically.

EXAMPLE A successful photograph must be *both* clearly focused *and* adequately lighted.

Notice that *both* and *and* are followed logically by ideas of equal weight and grammatically by identical constructions.

CHANGE For success in engineering, it is necessary both *to develop* writing skills and *mastering* calculus.
TO For success in engineering, it is necessary both *to develop* writing skills and *to master* calculus.

Do not substitute *as well as* for *and* in this construction.

CHANGE For success in engineering, it is necessary *both* to master calculus *as well as* to develop writing skills.
TO For success in engineering, it is necessary *both* to master calculus *and* to develop writing skills.

(See also **parallel structure** and **correlative conjunctions**.)

brackets

The primary use of brackets is to enclose a word or words inserted by an editor or writer into a **quotation** from another source.

EXAMPLE The text stated, "Fissile and fertile nuclei spontaneously emit characteristic nuclear radiations [such as neutrons and gamma rays] that are sufficiently energetic to penetrate the container or cladding."

Brackets are also used to set off a parenthetical item within **parentheses**.

EXAMPLE We should be sure to give Emanuel Foose (and his brother Emilio [1812–1882] as well) credit for his role in founding the institute.

Brackets are also used in academic writing to insert the Latin word **sic**, which indicates that the writer has quoted material exactly as it appears in the original, even though it contains an obvious error.

EXAMPLE Dr. Smith pointed out that "The earth does not revolve around the son [*sic*] at a constant rate."

(See also **mathematical equations**.)

bunch

Bunch refers to like things that grow or are fastened together. Do not use the word *bunch* to refer to people.

CHANGE A *bunch* of trainees toured the site.
TO A *group* of trainees toured the site.

C

can/may

In writing, *can* refers to capability, and *may* refers to possibility or permission.

EXAMPLES I *can* have the project finished by January 1. (capability)
I *may* be in Boston on Thursday. (possibility)
I *can* be in Boston on Thursday. (capability)
May I have an extra week to finish the project? (permission)

cannot/can not

Cannot is one word.

CHANGE We *can not* meet the deadline specified in the contract.
TO We *cannot* meet the deadline specified in the contract.

cannot help but

Avoid the **phrase** *cannot help but* in writing. (See also **double negatives**.)

CHANGE We *cannot help but* cut our staff.
TO We cannot avoid cutting our staff.

canvas/canvass

Canvas is a **noun** meaning "heavy, coarse, closely woven cotton or hemp fabric." *Canvass* is a **verb** meaning "to solicit votes or opinions."

EXAMPLES The maintenance crew spread the *canvas* over the equipment.
The executive committee decided to *canvass* the employees.

capital/capitol

Capital may refer either to financial assets or to the city that hosts the government of a state or a nation. *Capitol* refers to the building in which the state or national legislature meets. *Capitol* is often written with a small *c* when it refers to a state building, but it is always capitalized when it refers to the home of the United States Congress in Washington, D.C.

capital letters

The use of capital letters (or uppercase letters) is determined by custom and tradition. Capital letters are used to call attention to certain words, such as **proper nouns** and the first word of a sentence. Care must be exercised in using capital letters because they can affect the meaning of words (march/March, china/China, turkey/Turkey). Thus, capital letters can help eliminate **ambiguity**.

PROPER NOUNS

Proper nouns name a specific person, place, thing, concept, or quality and therefore are capitalized.

EXAMPLES Physics 101, General Electric, John Doe

COMMON NOUNS

Common nouns name a general class or category of persons, places, things, concepts, or qualities rather than specific ones and therefore are not capitalized.

EXAMPLES a physics class, a company, a person

FIRST WORDS

The first letter of the first word in a sentence is always capitalized.

EXAMPLE Of all the plans you mentioned, the first one seems the best.

The first word after a **colon** may be capitalized if the statement following is a complete sentence or if it introduces a formal resolution or question.

EXAMPLE Today's meeting will deal with only one issue: What is the firm's role in environmental protection?

If a subordinate element follows the colon, however, or if the thought is closely related, use a lowercase letter following the colon.

EXAMPLE We had to keep working for one reason: our deadline was upon us.

The first word of a complete sentence in **quotation marks** is capitalized.

EXAMPLE Dr. Vesely stated, "It is possible to postulate an imaginary world in which no decisions are made until all the relevant information is assembled."

Complete sentences contained as numbered items within a sentence may also be capitalized.

EXAMPLE To make correct decisions, you must do three things: (1) Identify the information that would be pertinent to the decision anticipated, (2) Establish a systematic program for acquiring this pertinent information, and (3) Rationally assess the information so acquired.

The first word in the salutation and complimentary close of a letter is capitalized. (See also **correspondence**.)

EXAMPLES Dear Mr. Smith:
Sincerely yours,
Best regards,

SPECIFIC PEOPLE AND GROUPS

Capitalize all personal names.

EXAMPLES Walter Bunch, Mary Fortunato, Bill Krebs

Capitalize names of ethnic groups and nationalities.

EXAMPLES American Indian, Italian, Jew, Chicano
Thus Italian immigrants contributed much to the industrialization of the United States.

Do not capitalize names of social and economic groups.

EXAMPLES middle class, working class, ghetto dwellers

SPECIFIC PLACES

Capitalize the names of all political divisions.

EXAMPLES Chicago, Cook County, Illinois, Ontario, Iran, Ward Six.

Capitalize the names of geographical divisions.

EXAMPLES Europe, Asia, North America, the Middle East, the Orient.

Do not capitalize geographic features unless they are part of a proper name.

EXAMPLE The mountains in some areas, such as the Great Smoky Mountains, make television transmission difficult.

The words *north, south, east*, and *west* are capitalized when they refer to sections of the country. They are not capitalized when they refer to directions.

EXAMPLES I may travel south when I relocate to Delaware.
We may build a new plant in the South next year.
State Street runs east and west.

Capitalize the names of stars, constellations, and planets.

EXAMPLES Saturn, Andromeda, Jupiter, Milky Way

Do not capitalize *earth, sun,* and *moon*, however, except when they are used with the names of other planets.

EXAMPLES Although the sun rises in the east and sets in the west, the moon may appear in any part of the evening sky when darkness settles over the earth.
Mars, Pluto, and Earth were discussed at the symposium.

SPECIFIC INSTITUTIONS, EVENTS, AND CONCEPTS

Capitalize the names of institutions, organizations, and associations.

EXAMPLE The American Society of Mechanical Engineers and the Department of Housing and Urban Development are cooperating in the project.

An organization usually capitalizes the names of its internal divisions and departments.

EXAMPLES Faculty, Board of Directors, Engineering Department

Types of organizations are not capitalized unless they are part of an official name.

EXAMPLES Our group decided to form a writers' association; we called it the American Association of Writers.
I attended Post High School. What high school did you attend?

Capitalize historical events.

EXAMPLE Dr. Jellison discussed the Boston Tea Party at the last class.

Capitalize words that designate specific periods of time.

EXAMPLES Labor Day, the Renaissance, the Enlightenment, January, Monday, the Great Depression, Lent

Do not, however, capitalize seasons of the year.

EXAMPLES spring, autumn, winter, summer

Capitalize scientific names of classes, families, and orders, but do not capitalize species or English derivatives of scientific names.

EXAMPLES Mammalia, Carnivora/mammal, carnivorous

TITLES OF BOOKS, ARTICLES, PLAYS, FILMS, REPORTS, AND MEMOS (SUBJECT LINE)

Capitalize the initial letters of the first and last words of a title of a book, article, play, or film, as well as all major words in the title. Do not capitalize **articles** (*a, an, the*), **conjunctions** (*and, but, if*), or short **prepositions** (*at, in, on, of*) unless they begin the title. Capitalize prepositions that contain more than four letters (*between, because, until, after*).

EXAMPLES The microbiologist greatly admired the book *The Lives of a Cell*.

Her favorite article is still "On the Universe Around Us."

The book *Year After Year* recounts the life story of a great scientist.

The report entitled "Alternative Sites for Plant Relocation" was submitted in February.

The memo "Analysis of Quality Assurance Procedures," dated November 13, explains the problem.

Some reference systems for scientific and technical publications do not follow these guidelines. See **documenting sources**.

PERSONAL, PROFESSIONAL, AND JOB TITLES

Titles preceding proper names are capitalized.

EXAMPLES Miss March, Professor Galbraith, Senator Church

Appositives following proper names are not normally capitalized. (The word *President* is usually capitalized when it refers to the chief executive of a national government.)

EXAMPLE Frank Jones, senator from New Mexico (but Senator Jones)

The only exception is an epithet, which actually renames the person.

EXAMPLES Alexander the Great, Solomon the Wise

Job titles used with personal names are capitalized.

EXAMPLE John Reems, Division Manager, will meet with us on Wednesday.

Job titles used without personal names are not capitalized.

EXAMPLE The division manager will meet with us on Wednesday.

Use capital letters to designate family relationships only when they occur before a name or substitute for a name.

EXAMPLES One of my favorite people is Uncle Fred.
Jim and Mother went along.
Jim and my mother went along.

ABBREVIATIONS

Capitalize **abbreviations** if the words they stand for would be capitalized.

EXAMPLES UCLA (University of California at Los Angeles)
p. (page)
Ph.D. (Doctor of Philosophy)

LETTERS

Certain single letters are always capitalized. Capitalize the **pronoun** *I* and the **interjection** *O* (but do not capitalize *oh* unless it is the first word in a sentence).

EXAMPLES When I say writing, O believe me, I mean rewriting.
When I say writing, oh believe me, I mean rewriting.

Capitalize letters that serve as names or indicate shapes.

EXAMPLES X-ray, vitamin B, T-square, U-turn, I-beam

MISCELLANEOUS CAPITALIZATIONS

The word *Bible* is capitalized when it refers to the Christian Scriptures; otherwise, it is not capitalized.

EXAMPLE He quoted a verse from the Bible, then read from Blackstone, the lawyer's bible.

All references to deities (Allah, God, Jehovah, Yahweh) are capitalized.

EXAMPLE God is the One who sustains us.

A complete sentence enclosed in **dashes, brackets**, or **parentheses** is not capitalized when it appears as part of another sentence.

EXAMPLES We must make an extra effort in safety this year (accidents last year were up 10 percent).
Extra effort in safety should be made this year. (Accidents were up 10 percent.)

Certain units, such as parts and chapters of books and rooms in buildings, when specifically identified by number, are normally capitalized.

EXAMPLES Chapter 5, Ch. 5; Room 72, Rm. 72

Minor divisions within such units are not capitalized unless they begin a sentence.

EXAMPLES page 11, verse 14, seat 12

When in doubt about whether or not to capitalize, check a **dictionary** or a secretarial manual, such as the current edition of *The Gregg Reference Manual* by William A. Sabin (McGraw-Hill).

case (grammar)

Grammatically, *case* indicates the functional relationship of a **noun** or a **pronoun** to the other words in a sentence. Nouns change form only in the possessive case; pronouns may show change for the subjective, the objective, or the possessive case. The case of a noun or pronoun is always determined by its function in its **phrase, clause**, or sentence. If it is the **subject** of its phrase, clause, or sentence, it is in the subjective case; if it is an **object** within its phrase, clause, or sentence, it is in the objective case; if it reflects possession or ownership and modifies a noun, it is in the possessive case. The subjective case indicates the person or thing acting (*he* sued the vendor); the objective case indicates the thing acted upon (The vendor sued *him*); and the possessive case indicates the person or thing owning or possessing something (it was *his* company).

The different forms of a noun or pronoun indicate whether it is functioning as a subject (subjective case), as a **complement** (usually objective case), or as a **modifier** (possessive case).

Subjective Case	*Objective Case*	*Possessive Case*
I	me	my, mine
we	us	our, ours
he	him	his, his
she	her	her, hers
they	them	their, theirs
you	you	your, yours
who	whom	whose
it	it	its

SUBJECTIVE CASE

A pronoun is in the subjective case (also called the nominative case) when it represents the person or thing acting.

EXAMPLE I wrote a letter to that company before I graduated.

A **linking verb** links a pronoun to its antecedent to show that they identify the same thing. Because they represent the same thing, the pronoun is in the subjective case even when it follows the **verb**, which makes it a **subjective complement**.

EXAMPLES *He* is the head of the Quality Control Group. (subject)
The head of the Quality Control Group is *he*. (subjective complement)

Whether a pronoun is a subject or a subjective complement, it is in the subjective case.

The subjective case is used after the words *than* and *as* because of the understood (although unstated) portion of the clauses in which these words appear.

EXAMPLES George is as good a designer as *I* [am].
Our subsidiary can do the job better than *we* [can].

OBJECTIVE CASE

A pronoun is in the objective case when it indicates the person or thing receiving the action expressed by the verb. (The objective case is also called the accusative case.)

EXAMPLE They informed *me* by letter that they had received my résumé.

A pronoun is in the objective case when it is the object of a verb, **gerund**, or **preposition** and when it is the subject of an **infinitive**. Pronouns that follow prepositions must be in the objective case.

EXAMPLES Between you and *me*, his facts are questionable.
Many of *us* attended the conference.

Pronouns that follow action verbs (which excludes all forms of the verb *be*) must be in the objective case. Don't be confused by an additional name.

EXAMPLES The company promoted *me* in June.
The company promoted John and *me* in June.

Pronouns that follow gerunds must be in the objective case.

EXAMPLE Training *him* was the best thing I could have done.

Subjects of infinitives must be in the objective case.

EXAMPLE We asked *them* to recalibrate the instruments.

For determining the case of an object, English does not differentiate between direct objects and indirect objects; both require the objective form of the pronoun. (See also **complements**.)

EXAMPLES The interviewer seemed to like *me*. (direct object)
They wrote *me* a letter. (indirect object)

POSSESSIVE CASE

A noun or a pronoun is in the possessive case when it represents a person or thing owning or possessing something.

EXAMPLE Dr. Peterson's risk calculations appear in Appendix A of *his* report.

Nouns. Although exceptions are relatively common, it is a good rule of thumb to use the *'s* form of the possessive case with nouns referring to persons and living things and to use an *of* phrase for the possessive case of nouns referring to inanimate objects.

EXAMPLES The *chairman's* address was well received.
The leaves *of the tree* look healthy.

If this rule leads to awkwardness or wordiness, however, be flexible.

EXAMPLES The *company's* pilot plants are doing well.
The *plane's* landing gear failed.

Only the possessive form of a noun or pronoun should precede a **gerund**.

EXAMPLES John's working has not affected his grades.
His working has not affected his grades.

The established **idiom** calls for the possessive case in many stock phrases.

EXAMPLES a day's journey, a day's work, a moment's notice, at his wit's end, the law's delay

In a few cases, the idiom even calls for a double possessive employing both the *of* and *'s* forms.

EXAMPLE That colleague *of* George*'s* was at the conference.

Plural words ending in *s* need only add an **apostrophe** to form the possessive case.

EXAMPLE the *laborers'* union

When several words compose a single term, add the *'s* to the last word only.

EXAMPLES The *Chairman of the Board's* statement was brief.
The *Department of Energy's* fiscal 19— budget shows increased revenues of $192 million for uranium enrichment.

To show individual possession with coordinate nouns, make both nouns possessive.

EXAMPLE The *Senate's and House's* chambers were packed.

To show joint possession with coordinate nouns, make only the last possessive.

EXAMPLE The *Senate and House's* joint declaration was read to the press.

Pronouns. The use of possessive pronouns does not normally cause problems except with gerunds and **indefinite pronouns**. Several indefinite pronouns (*all, any, each, few, most, none,* and *some*) require *of* phrases to form the possessive case.

EXAMPLE Both dies were stored in the warehouse, but rust had ruined the surface *of each*.

Others, however, use the apostrophe.

EXAMPLE *Anyone's* contribution is welcome.

Only the possessive form of a pronoun should be used with a gerund.

EXAMPLES The safety officer insisted on *my* wearing a respirator.
Our monitoring was not affected by changing weather conditions.

Pronouns in compound constructions should be in the same case.

EXAMPLES This is just between *them* and *us*.
Both *they* and *we* must agree to the arrangement.

APPOSITIVES

An appositive is a noun or noun phrase that follows and amplifies another noun or noun phrase. Because it has the same grammatical function as the noun it complements, an appositive should be in the same case as the noun with which it is in apposition.

EXAMPLES Two design engineers, Jim Knight and *I*, were asked to review the drawings. (subjective case)
The group leader selected two members to represent the department—Jim Knight and *me*. (objective case)
We all came—Jim and Carol and *I*. (subjective case)
He gave us both, Rod and *me*, a week to make up our minds. (objective case)

TIPS ON DETERMINING THE CASE OF PRONOUNS

One test to determine the proper case of a pronoun is to try it with some transitive verb such as *resembled* or *hit*. If the pronoun would logically precede the verb, use the subjective case; if it would logically follow the verb, use the objective case.

EXAMPLES *She* (he, they) resembled her father. (subjective case)
Angela resembled *him* (her, them). (objective case)

In the following type of sentence, try omitting the noun to determine the case of the pronoun.

EXAMPLES (*We/Us*) pilots fly our own airplanes.
We [pilots] fly our own airplanes. (This correct usage sounds right.)
Us [pilots] fly our own airplanes. (This incorrect usage is obviously wrong.)

To determine the case of a pronoun that follows *as* or *than*, try mentally adding the words that are normally omitted.

EXAMPLES The other operator is not paid as well *as she* [is paid]. (You would not write, "*Her* is paid.")
His partner was better informed that *he* [was informed]. (You would not write, "*Him* was informed.")

If compound pronouns cause problems, try using them singly to determine the proper case.

EXAMPLES (*We/Us*) and the Johnsons are going to the Grand Canyon.
We are going to the Grand Canyon. (You would not write, "*Us* are going to the Grand Canyon.")

WHO/WHOM

Who and *whom* cause much trouble in determining case. *Who* is the subjective case form, whereas *whom* is the objective case form. When in doubt about which form to use, try substituting a personal pronoun to see which one fits. If *he* or *they* fits, use *who*.

EXAMPLES *Who* is the congressman from the 45th district?
He is the congressman from the 45th district.

If *him* or *them* fits, use *whom*.

EXAMPLES It depended on *them*.
It depended on *whom*?
It was they on *whom* it depended.

It is becoming common to use *who* for the objective case when it begins a clause or sentence, although some readers still object to such an "ungrammatical" construction, especially in formal contexts.

EXAMPLE *Who* should I call to report a fire?

The best advice is to know your **reader**. (See also **who/whom**.)

case (usage)

The word *case* is often merely filler. Be critical of the word, and eliminate it if it contributes nothing.

CHANGE An exception was made in the *case* of those closely connected with the project.

TO An exception was made for those closely connected with the project.

cause-and-effect method of development

When your purpose is to explain why something happened or why you think something will happen, the cause-and-effect **method of development** is a useful writing strategy.

The goal of the cause-and-effect method of development is to make as plausible as possible the relationship between a situation and either its cause or its effect. The conclusions you draw about the relationships should be based on the evidence you have gathered. Because not all evidence will be of equal value to you, keep some guidelines in mind for evaluating evidence.

EVALUATING EVIDENCE

The facts and arguments you gather should be pertinent to your **topic**. Be careful not to draw a conclusion that your evidence does not lead to or support. You may have researched some statistics, for example, which show that an increasing number of Americans are licensed to fly small airplanes. But you cannot use this information as evidence that there is a slowdown in interstate highway construction in the United States—the evidence does not lead to that conclusion. Statistics on the increase in small-plane licensing may be relevant to other conclusions, however. You could argue that the upswing has occurred because small planes save travel time, provide easy access to remote areas, and, once they are purchased, are economical to operate.

Your evidence should be adequate. Incomplete evidence can lead to false conclusions.

> EXAMPLE Driver training classes do not help prevent auto accidents. Two people I know who completed driver training classes were involved in accidents.

Although the evidence cited to support the conclusion may be accurate, there is not enough of it. A thorough investigation of the usefulness of driver training classes in keeping down the accident rate would require many more than two examples. And it would require a comparison of the driving records of those who had completed driver training with drivers who had not.

Your evidence should be representative. If you conduct a survey to obtain your evidence, be sure that you do not solicit responses only from individuals or groups whose views are identical to yours; that is, be sure you obtain a representative sampling.

Your evidence should also be plausible. Two events that occur close to each other in time or place may or may not be causally related. Thunder and black clouds do not always signal rain, but they do so often enough that if we are outdoors and the sky darkens and we hear thunder, we seek shelter. If you sprain your ankle after walking under a ladder, however, you cannot conclude that a ladder brings bad luck. Merely to say that *X* caused *Y* (or will cause *Y*) is inadequate. You must demonstrate the relationship with pertinent facts and arguments.

LINKING CAUSES TO EFFECTS

To show a true relationship between a cause and an effect, you must demonstrate that the existence of the one *requires* the existence of the other. It is often difficult to establish beyond any doubt that one event was *the* cause of another event. More often, a result will have more than one cause. As you research your subject, your task is to determine which cause or causes are most plausible.

When several probable causes are equally valid, report your findings accordingly, as in the following **paragraph** on the use of an energy-saving device called a furnace-vent damper. The damper is a metal plate fitted inside the flue or vent pipe of natural-gas or fuel-oil furnaces. When the furnace is on, the damper opens to allow the gases to escape up the flue. When the furnace shuts off, the damper closes, thus preventing warm air from escaping up the flue stack.

MEMORANDUM

To: James K. Arburg, Safety Officer
From: Lawrence T. Baker, Foreman of Section A-40
Date: November 30, 19--

Subject: Personal-Injury Accident in Section A-40
October 10, 19--

On October 10, 19--, at 10:15 p.m., Jim Hollander, operating punch press #16, accidentally brushed the knee switch of his punch press with his right knee as he swung a metal sheet over the punching surface. The switch actuated the punching unit, which severed Hollander's left thumb between the first and second joints as his hand passed through the punch station. While an ambulance was being summoned, Margaret Wilson, R.N., administered first aid at the plant dispensary. There were no witnesses to the accident.

The ambulance arrived from Mercy Hospital at 10:45 p.m., and Hollander was admitted to the emergency room at the hospital at 11:00 p.m. He was treated and kept overnight for observation, then released the next morning.

Hollander returned to work one week later, on October 17. He has been given temporary duties in the tool room until his injury heals.

Conclusions About the Cause of the Accident

The Maxwell punch press on which Hollander was working has two switches, a hand switch and a knee switch, and both must be pressed to actuate the punch mechanism. The hand switch must be pressed first, and then the knee switch, to trip the punch mechanism. The purpose of the knee switch is to leave the operator's hands free to hold the panel being punched. The hand switch, in contrast, is a safety feature. Because the knee switch cannot activate the press until the hand switch has been pressed, the operator cannot trip the punching mechanism by touching the knee switch accidentally.

Inspection of the punch press that Hollander was operating at the time of the accident made it clear that Hollander had taped the hand switch of his machine in the ON position, effectively eliminating its safety function. He could then pick up a panel, swing it onto the machine's punching surface, press the knee switch, stack the newly punched panel, and grab the next unpunched panel, all in one continuous motion--eliminating the need to let go to the panel, after placing it on the punching surface, in order to press the hand switch.

Preventive Training

To prevent a recurrence of this accident, I have conducted a brief safety session with all punch press operators, at which I described Hollander's experience and cautioned them against tampering with the safety feature of their machines.

mo

Figure 1 Cause-and-Effect Method of Development

The dampers are potentially dangerous, however. If the dampers fail to open at the proper time, they could allow poisonous furnace gases to back into the house and asphyxiate anyone in a matter of minutes. Tests run on several dampers showed a number of probable causes for their malfunctioning.

EXAMPLE One damper was sold without proper installation instructions, and another was wired incorrectly. Two of the units had slow-opening dampers (15 seconds) that prevented the [furnace] burner from firing. And one damper jammed when exposed to a simulated fuel temperature of more than 700 degrees.

—Don DeBat, "Save Energy but Save Your Life, Too," *Family Safety* (Fall 1978), 27.

The investigator located all the causes of damper malfunctions and reported on them. Without such a thorough account, recommendations to prevent similar malfunctions would be based on incomplete evidence.

By substituting *problem* for "cause" and *solution* for "effect," you can use the same approach to develop a **report** dealing with a solution to a problem.

The report shown in Figure 1 is developed from effect to cause.

center around

Be careful always to substitute *on* or *in* for *around* in this redundant and illogical expression.

CHANGE The experiments *center around* the new discovery.
TO The experiments *center on* the new discovery.

Usually the idea intended by *center around* is best expressed by *revolve around*.

EXAMPLE The subcommittee hearings on computer security *revolved around* access codes.

chair/chairperson/chairman/chairwoman

The terms *chair, chairperson, chairman*, and *chairwoman* all are used to refer to a presiding officer. The titles *chair* and *chairperson*, however, avoid any sexual bias that might be implied by the other titles.

EXAMPLES Mary Roberts preceded John Stevens as *chair* (or *chairperson*) of the executive committee.

Mary Roberts was *chairwoman* of the executive committee before John Stevens became *chairman*.

character

Character, used in the sense of "nature" or "quality," is often an unnecessary and inexact filler that should be omitted from your writing. Choose instead a word that conveys your meaning more specifically.

CHANGE The modifications changed the whole *character* of the engine.
TO The modifications changed the performance of the engine.

chronological method of development

The chronological **method of development** arranges the events under discussion in sequential order, beginning with the first event and continuing chronologically to the last event. **Trip reports, laboratory reports,** work schedules, some **minutes of meetings**, and certain **trouble reports** are among the types of writing in which information is organized chronologically.

In the **report** shown in Figure 1, a fire fighter describes a fire that took place at a lumber mill. After providing important background information, the writer presents the events as they occurred chronologically.

cite/site/sight

Cite means "acknowledge" or "quote an authority"; *site* is the place or plot of land where something is located; *sight* is the ability to see.

EXAMPLES The speaker *cited* several famous economists to support his prediction about the stock market.

The *site* for the new factory is three miles from the middle of town.

After the accident, his vision was blurred, and he feared that he might lose his *sight*.

clarity

No element is more essential to writing than clarity. You should strive to make all of your writing direct, orderly, and precise. Many

Woodworking Plant Fire*

Exposed Building Destroyed — July 24, 19—
Notification Delayed — Burney, California

Setting

Wood bark, sawdust, and wood chips were stored in three piles about 100 feet south of one building at this lumber mill and 150 feet west of a second building. The second building, called the "panel plant," consisted of one story and a partial attic and was used in part as an electric shop, and in part for the storage of finished lumber. About six pallet loads of Class I flammable liquids in 55-gallon drums were stored in the western section. The building contained a sprinkler system.

Cause of fire

Fire, caused by spontaneous ignition of the piled bark, spread to sawdust and chip piles, then to the chip-loading facilities, and finally to the panel plant. A 40-mph wind was blowing in the direction of the panel plant.

Fire first noticed

A watchman first noticed the fire in the bark pile about 6 a.m. He notified the plant superintendent, who arrived more than an hour later, hosed down the smoldering bark pile, and set up several irrigation sprinklers to wet the area.

Fire department called

At about 1:20 p.m., smoke was seen at the farther end of the bark pile. The hose was not long enough to reach this area and the local fire department was called, nearly eight hours after the fire was originally discovered.

Start of pumping equipment delayed

The fire burned up into the hollow joisted roof of the panel plant. The sprinklers were on a dry system and, from accounts of witnesses, it is estimated that the fire pump was not started until after the fire had been burning in the panel plant for 15 to 30 minutes.

The plant was a $350,000 loss. . . .

*"Bimonthly Fire Record," *Fire Journal* 72 (March 1977), 24.

Figure 1 Chronological Method of Development

factors contribute to clarity just as many other elements can defeat it. Logical development, unity, coherence, emphasis, subordination, pace, transition, an established point of view, conciseness, and word choice contribute to clarity. **Ambiguity, awkwardness**, vagueness, poor use of **idiom, clichés**, and inappropriate level of usage detract from clarity.

It is surely evident that a logical **method of development** and a good outline are essential to clarity. Without a logical method of development and an outline, you may communicate only isolated thoughts to your **reader**, and your **objective** cannot be achieved by a jumble of isolated thoughts. You must use a method of development that puts your thoughts together in a logical, meaningful sequence. Only then will your writing achieve the **unity** and **coherence** so vital to clarity.

Proper **emphasis** and **subordination** are mandatory if you wish to achieve clarity. If you do not use these two complementary techniques wisely, your **clauses** and sentences all may appear to be of equal importance. Your reader will be forced to guess which are most important, which are least important, and which fall between the two extremes. At the very least, this will puzzle and annoy your reader; at worst, it will render your writing incoherent.

The **pace** at which you present your ideas is important to clarity because if the pace is not carefully adjusted to both the **topic** and the reader, your writing will appear cluttered and unclear.

Point of view establishes through whose eyes, or from what vantage point, the reader views the subject. A consistent point of view is essential to clarity; if you switch from the first person to the third person in midsentence, you are certain to confuse your reader.

Clear **transition** contributes to clarity by providing the smooth flow that enables the reader to connect your thoughts with one another without conscious effort. This enables the reader to concentrate solely on absorbing your ideas.

That **conciseness** is a requirement of clearly written communication should be evident to anyone who has ever attempted to decipher an insurance policy or legal contract. Although words are our chief means of communication, too many of them can impede communication just as effectively as too many cars on a highway can impede traffic. For the sake of clarity, prune excess verbiage from your writing.

The selection of precise words over **vague words** is **word choice**. Thoughtful choice of the right word advances clarity by defeating ambiguity and awkwardness.

Another important contributor to clarity is a careful and methodical approach to your writing project, including proper application of all the steps of the writing process—**preparation, research, organization, writing the draft**, and **revision**.

clauses

A clause is a syntactical construction, or group of words, that contains a **subject** and a **predicate** and functions as part of a sentence. A clause that could stand alone as a **simple sentence** is an **independent clause**.

EXAMPLE *The scaffolding fell* when the rope broke.

A clause that could not stand alone if the rest of the sentence were deleted is a **dependent clause**.

EXAMPLE I was at the St. Louis branch *when the decision was made*.

Every subject-predicate word group in a sentence is a clause. Unlike a **phrase**, a clause can make a complete statement because it contains a finite **verb** (as opposed to **verbals**) as well as a subject. Every sentence must contain at least one independent clause, with the obvious exception of minor sentences—**sentence fragments** that are acceptable because the missing part is clearly understood, such as "At last." or "So much for that."

A clause may function as a **noun**, an **adjective**, or an **adverb** in a larger sentence, or it may be modified by one or more other clauses that are subordinate to it.

EXAMPLE While I was in college, I studied differential equations.

While I was in college is an **adverb clause** modifying the verb of the independent clause *I studied differential equations*.

A clause may be connected with the rest of its sentence by a **coordinating conjunction**, a **subordinating conjunction**, a **relative pronoun**, or a **conjunctive adverb**.

EXAMPLES It was 500 miles to the facility, *so* we made arrangements to fly. (coordinating conjunction)

Mission control will have to be alert *because* at launch the space laboratory will contain a highly flammable fuel. (subordinating conjunction)
It was Robert M. Fano *who* designed and developed the earliest "Multiple Access Computer" system at M.I.T. (relative pronoun)
It was dark when we arrived; *nevertheless*, we began the tour of the factory. (conjunctive adverb)

INDEPENDENT CLAUSES

Unlike a dependent clause, which is part of a larger construction, an independent clause is complete in itself; it could stand alone as a separate sentence if taken out of its larger sentence.

EXAMPLE *We abandoned the project* because the cost was excessive.

DEPENDENT CLAUSES

A dependent, or subordinate, clause is a group of words that has a subject and a verb but must nonetheless depend on a main clause to complete its meaning. Within the sentence as a whole, a dependent clause can function as a noun, an adjective, or an adverb.

A **noun clause** is a subordinate clause that functions as a noun. It can be a subject, an **object**, or a **complement**.

EXAMPLES *Why the report took three months to prepare* is a mystery to me. (subject)
Mr. Yen told me *that you are a key-punch operator.* (direct object)
That is not *what I meant.* (subjective complement)

Noun clauses are frequently introduced by interrogative and **relative pronouns** and adverbs (*which, who, what, when, where,* and *why*). The **conjunction** *that* (not the relative pronoun) is also commonly used.

A noun clause can fit any sentence position that can be occupied by a noun.

EXAMPLES *That we had succeeded* pleased us. (subject)
The treasurer admitted *that he had made a mistake.* (direct object of verb)
Upon retirement, she's going back to *where she came from.* (object of preposition)
Help out by doing *what you can.* (direct object of gerund)
Give *whoever comes* a ticket. (indirect object)

The main thing we should remember, *that we are in business to make a profit*, is the thing we seem to be forgetting. (appositive)

He made himself *what he wanted to be*. (objective complement)

Noun clauses appear often in definitions and explanations as **subjective complements**.

EXAMPLE A common theory is *that competition keeps prices low.*

An adjective clause is a subordinate **clause** that functions as an adjective by modifying a noun or pronoun in another clause.

EXAMPLE The designer *we commissioned last year* [modifying "designer"] has delivered the drawings, *which we have approved* [modifying "drawings"].

Adjective clauses are often introduced by **relative pronouns** (*who, that,* and *which*).

EXAMPLE Our culture is beginning to produce a breed of people *who are very mobile.*

Adjective clauses are also sometimes introduced by relative adverbs (*when, where*, and *why*).

EXAMPLE The fourth quarter is the period *when cost control will be crucial.*

Adjective clauses may be **restrictive** or nonrestrictive. If the clause is essential to limiting the meaning of the noun, it is restrictive and not set off by **commas**.

EXAMPLE The statistics *that accompany this report* were compiled from a series of questionnaires.

If the clause is not intended to limit the meaning of the noun but merely provides further information about it, the clause is nonrestrictive and set off by commas.

EXAMPLE New statistics, *which we will submit soon*, suggest a different conclusion.

If adjective clauses are not placed carefully, they may appear to modify the wrong noun or pronoun.

CHANGE The factory in the suburbs *that we bought* has increased in value. (This sentence implies that we bought the suburbs rather than the factory.)

TO The factory *that we bought* in the suburbs has increased in value.

OR The suburban factory *that we bought* has increased in value.

(See also **that/which/who**.)

An adverb clause is a subordinate clause used to modify a verb, **adjective,** or **adverb** in the main clause.

EXAMPLE The property is located *where the railroad track crosses the road.*

Like adverbs, adverb clauses normally express such ideas as those of time, place, condition, cause, manner, or **comparison**.

EXAMPLES It stopped *after we installed the new bearings*. (time)
It stopped *where the previous model stopped*. (place)
If we install the new bearings again, it will stop again. (condition)
It stopped *because we installed the bearings*. (cause)
It stopped *as though it had run into a wall*. (manner)
The new bearings perform no better *than the old bearings did.* (comparison)

Adverb clauses are often introduced by **subordinating conjunctions** (*because, when, where, since, though*, and the like). If the adverb clause follows the main clause, a comma should not normally separate the two clauses.

EXAMPLE The sonic boom problem caused by the hypersonic transport is greatly alleviated *because the craft makes a steep ascent*.

If the adverb clause precedes the main clause, the adverb clause should be set off by a comma.

EXAMPLE *Because the craft makes a steep ascent*, the sonic boom problem is greatly alleviated.

Placement of an adverb clause can change the **emphasis** of a sentence. Do not allow this to happen inadvertently.

EXAMPLES Ruins and artifacts are mere objects of curiosity *unless they are used to reconstruct the daily activities of prehistoric peoples*.
Unless they are used to reconstruct the daily activities of prehistoric peoples, ruins and artifacts are mere objects of curiosity.
Ruins and artifacts, *unless they are used to reconstruct the daily activities of prehistoric peoples*, are mere objects of curiosity.

An adverb clause cannot act as the **subject** of a sentence.

CHANGE *Because medical expenses have increased* is the reason health insurance rates have gone up.

TO Health insurance rates have gone up *because medical expenses have increased.*

OR *Because medical expenses have increased,* health insurance rates have gone up.

cliché

A cliché is an expression that has been used for so long that it is no longer fresh (although some clichés were, at one time, fresh **figures of speech**). Because they have been used continually over a long period of time, clichés come to mind easily. In addition to being stale, clichés are usually wordy and often vague. Each of the following clichés is followed by better, more direct words, or expressions:

EXAMPLES quick as a flash/quickly, in five minutes
straight from thc shoulder/frank
last but not least/last, finally
as plain as day/clear, obvious
abreast of the times/up to date, current
the modern business world/business today

Clichés are often used in an attempt to make writing elegant or impressive (see also **affectation**). Because they are wordy and vague, however, they slow communication and can even irritate your **reader**. So, although clichés come to mind easily while you are **writing the draft**, they normally should be eliminated during the **revision** phase of the writing process.

On rare occasions, certain clichés, because they are so much a part of the language, may provide (much like **jargon**) a time-saving and efficient means of relating an idea. For example, "guardedly optimistic" or "if I can help further, please let me know" may, in appropriate contexts, best express your idea—provided you are certain that the expression is meaningful and acceptable to your reader. Be on guard, however, because clichés can too easily bccome the pattern of your **word choice**. The best advice is to avoid clichés if *any* other choice of words will work. Considcr the following **paragraphs**, first with clichés and then rewritten without them:

CHANGE Our new computer system will have a positive impact on the company *as a whole.* It will keep us *abreast of the times* and make our competition *green with envy*. The committee deserves a *pat on the back* for its *herculean efforts* in convincing management that it was *the thing to do*. I'm sure that their *untiring efforts* will *not go unrewarded*.

TO Our new computer system will have a positive impact throughout the company. It will keep our operations up-to-date and make our competition envious. The committee deserves credit for their efforts in convincing management of the need for the computer. I'm sure that the value of their efforts will be recognized.

clipped forms of words

When the beginning or end of a word is cut off to create a shorter word, the result is called a *clipped form*.

EXAMPLES dorm, lab, demo, phone, memo

The work *specification*, for example, is often shortened to *spec*. Although acceptable in conversation, most clipped forms should not appear in writing unless they are commonly accepted as part of the special vocabulary of an occupational group.

Apostrophes are not normally used with clipped forms of words (not *'phone*, but *phone*). Since they are not strictly **abbreviations**, clipped forms are not followed by **periods** (not *lab.*, but *lab*).

Do not use clipped forms of **spelling** (*thru, nite*, and the like).

coherence

Writing is coherent when the relationships among ideas are made clear to the reader. Coherent writing moves logically and consistently from point to point. Each idea should relate clearly to the others, with one idea flowing smoothly to the next. Many elements contribute to smooth and coherent writing; however, the major components are (1) a logical sequence of ideas and (2) clear **transitions** between ideas.

A logical sequence of presentation is the most important single requirement in achieving coherence, and the key to achieving the most logical sequence of presentation is the use of a good **outline.** The outline forces you to establish a beginning (**introduction**), a middle (body), and an end (**conclusion**), and this alone contributes greatly to coherence. The outline also enables you to lay out the most direct route to your **objective**—without digressing into interesting but only loosely related side issues, a habit that inevitably defeats coherence. Drawing up an outline permits you to experiment with different sequences and choose the best one.

Thoughtful transition is also essential to coherence, for without it your writing cannot achieve the smooth flow from sentence to sentence and from **paragraph** to paragraph that is required for coherence. Notice the difference between the following two paragraphs; the first has no transition and the second has transition added.

CHANGE The moon has always been an object of interest to human beings. Until the 1960s, getting there was only a dream. Some thought that we were not meant to go to the moon. In 1969 Neil Armstrong stepped onto the lunar surface. Moon landings became routine to the general public.

TO The moon has always been an object of interest to human beings, *but* until the 1960s, getting there was only a dream. *In fact,* some thought that we were not meant to go to the moon. *However,* in 1969 Neil Armstrong stepped onto the lunar surface. *After that* moon landings became routine to the general public.

The transitional words and expressions of the second paragraph fit the ideas snugly together, making that paragraph read more smoothly than the first. Attention to transition in longer works is essential if your reader is to move smoothly from point to point in your writing.

Providing your readers with **sentence variety** also contributes greatly to making your writing coherent. Provide variety in (1) **sentence construction** (for example, don't begin every sentence with an **article** and a **noun**); (2) in sentence types (for example, use complex sentences in addition to **simple sentences** and **compound sentences**); and (3) in sentence length (if you write only long sentences, you'll put your readers to sleep; if you write only short sentences, you'll have a jackhammer effect on them).

Check your draft carefully for coherence during **revision**; if your writing is not coherent, you are not really communicating with your **reader**.

collective nouns

Collective nouns name a group or collection of persons, places, things, concepts, actions, or qualities.

EXAMPLES army, committee, crowd, team, public, class, jury, humanity

When a collective noun refers to a group as a whole, it takes a singular **verb** and **pronoun**.

EXAMPLE The staff *was* divided on the issue and could not reach *its* decision until May 15.

When a collective noun refers to individuals within a group, it takes a plural verb and pronoun.

EXAMPLE The staff returned to *their* offices after the conference.

A better way to emphasize the individuals on the staff would be to use the phrase *members of the staff.*

EXAMPLE The members of the staff returned to *their* offices after the conference.

Organizational titles, such as **company** and **department**, can be either singular or plural.

EXAMPLES LRM Company has grown 200 percent in the last three years; *it* will move to a new facility in January.
LRM Company will submit an environmental report in November. *They* will present it at the next County Commission meeting.

Where any **pronoun reference** confusion could occur, substitute an appropriate noun.

EXAMPLE LRM Company submitted several environmental impact statements. LRM *scientists* [rather than *they*] are meeting with government officials to revise these statements.

Some collective nouns regularly take singular verbs (*crowd*); others do not (*people*).

EXAMPLES *The crowd was* growing impatient.
Many *people were* able to watch the first space shuttle land safely.

Some collective nouns have regular plural forms (*team, teams*); others do not (*sheep*). (For additional information about the **case, number**, and function of collective nouns, see **nouns**.)

colons

The colon is a mark of anticipation and introduction that alerts the **reader** to the close connection between the first statement and the one that follows it.

A colon may be used to connect a list or series to the **clause**, word, or **phrase** with which it is in apposition.

EXAMPLE Three decontamination methods are under consideration: a zeolite-resin system, an evaporation and resin system, and a filtration and storage system.

Do not, however, place a colon between a **verb** and its **objects**.

CHANGE The three fluids for cleaning pipettes are: water, alcohol, and acetone.

TO The three fluids for cleaning pipettes are water, alcohol, and acetone.

One common exception is made when a verb is followed by a stacked list.

EXAMPLE The corporations that manufacture computers include:

NCR	CDC	Apple
Burroughs	IBM	DEC
Honeywell		

Do not use a colon between a **preposition** and its object.

CHANGE I would like to be transferred to: Tucson, Boston, or Miami.

TO I would like to be transferred to Tucson, Boston, or Miami.

A colon may be used to link one statement to another that develops, explains, amplifies, or illustrates the first. A colon may be used in this way to link two **independent clauses**.

EXAMPLE Any large organization is confronted with two separate, though related, information problems: it must maintain an effective internal communication system, and it must see that an effective overall communication system is maintained.

A colon may be used to link an **appositive** phrase to its related statement if greater **emphasis** is needed.

EXAMPLE There is only one thing that will satisfy Mr. Sturgess: our finished report.

Colons are used to link numbers signifying different identifying **nouns**.

EXAMPLES Matthew 14:1 (chapter 14, verse 1)
9:30 a.m. (9 hours, 30 minutes)

In proportions, the colon indicates the ratio of one amount to another.

EXAMPLE The cement is mixed with the water and sand at 7:5:14. (In this case, the colon replaces *to*.)

Colons are often used in mathematical ratios.

EXAMPLE $7:3 = 14:x$

In **bibliography**, footnote, and reference citations, colons may link the place of publication with the publisher and perform other specialized functions.

EXAMPLE Wine, R. L. *Statistics for Scientists and Engineers*. Englewood Cliffs, NJ: Prentice-Hall, 1964.

A colon follows the salutation in business letters, even when the salutation refers to a person by name.

EXAMPLES Dear Ms. Jeffers:
Dear Sir:
Dear George:

The initial **capital letter** of a **quotation** is retained following a colon if the quoted material originally began with a capital letter.

EXAMPLE The senator issued the following statement: "We are not concerned about the present. We are worried about the future."

A colon always goes outside **quotation marks**.

EXAMPLE This was the real meaning of his "suggestion": the division must show a profit by the end of the year.

When quoting material that ends in a colon, drop the colon and replace it with **ellipses**.

CHANGE "Any large corporation is confronted with two separate, though related, information problems:"
TO "Any large corporation is confronted with two separate, though related, information problems . . ."

The first word after a colon may be capitalized if (1) the statement following is a complete sentence or (2) it introduces a formal resolution or question.

EXAMPLE The members attending this year's conference passed a single resolution: Voting will be open to associate members next year.

If a subordinate element follows the colon, however, use a lowercase letter following the colon.

EXAMPLE There is only one way to stay within our present budget: to reduce expenditures for research and development.

comma splice

Do not attempt to join two **independent clauses** with only a **comma**; this is called a *comma splice.*

EXAMPLE It was five hundred miles to the facility, we made arrangements to fly.

Such a comma splice could be corrected in several ways.

1. Substitute a **semicolon**, or a semicolon and a **conjunctive adverb**.

CHANGE It was five hundred miles to the facility, we made arrangements to fly.
TO It was five hundred miles to the facility; we made arrangements to fly.
OR It was five hundred miles to the facility; *therefore*, we made arrangements to fly.

2. Add a **coordinating conjunction** following the comma.

EXAMPLE It was five hundred miles to the facility, *so* we made arrangements to fly.

3. Create two sentences. (Be aware, however, that putting a **period** between two closely related and brief statements may result in two weak sentences.)

EXAMPLE It was five hundred miles to the facility. We made arrangements to fly.

4. Subordinate one **clause** to the other.

EXAMPLE *Because* it was five hundred miles to the facility, we made arrangements to fly.

When a conjunctive adverb connects two independent clauses, the conjunctive adverb *must* be preceded by a semicolon and followed by a comma.

CHANGE It was five hundred miles to the facility, therefore, we made arrangements to fly.
TO It was five hundred miles to the facility; therefore, we made arrangements to fly.

commas

Like all **punctuation**, the comma helps **readers** understand the writer's meaning and prevents **ambiguity**. Notice how the comma helps make the meaning clear in the following examples:

CHANGE To be successful managers with MBAs must continue to learn. (At first reading, the **sentence** seems to be about "successful managers with MBAs.")
TO To be successful, managers with MBAs must continue to learn. (The comma makes clear where the main part of the sentence begins.)
CHANGE When you see an airport fly over it at an altitude of 1,500 feet. (At first glance, this sentence seems to have airports flying.)
TO When you see an airport, fly over it at an altitude of 1,500 feet. (The comma makes clear where the main part of the sentence begins.)

As these examples illustrate, effective use of the comma depends on your understanding of **sentence construction**.

To help you find the advice you need, the following entry is divided into the following uses of the comma:

Linking Independent Clauses
Enclosing Elements
Introducing Elements
  Clauses and Phrases
  Words and Quotations
Separating Items in a Series
Clarifying and Contrasting
Showing Omissions
Using with Other Punctuation
Using with Numbers and Names
Avoiding Unnecessary Commas

LINKING INDEPENDENT CLAUSES

Use a comma between **independent clauses** that are linked by a **coordinating conjunction** (*and, but, or, nor*, and sometimes *so, yet,* and *for*). The comma precedes the **conjunction**.

EXAMPLE Human beings have always prided themselves on their unique capacity to create and manipulate symbols, but today computers are manipulating symbols.

Although many writers omit the comma when the **clauses** are short and closely related, the comma can never be wrong.

EXAMPLES The cable snapped and the power failed.
The cable snapped, and the power failed.

ENCLOSING ELEMENTS

Commas are used to enclose nonrestrictive clauses and parenthetical elements. (For other means of punctuating parenthetical elements, see **dashes** and **parentheses**.)

EXAMPLES Our new Detroit factory, *which began operations last month*, should add 25 percent to total output. (nonrestrictive clause)
We can, *of course*, expect their lawyer to call us. (parenthetical element)

(See also **restrictive and nonrestrictive elements.**)

Yes and *no* are set off by commas in such uses as the following:

EXAMPLES I agree with you, *yes*.
No, I do not think we can finish as soon as we would like.

A **direct address** should be enclosed in commas.

EXAMPLE You will note, *Mark*, that the surface of the brake shoe complies with the specifications.

Phrases in apposition (which identify another expression) are enclosed in commas.

EXAMPLE Our company, *The Blaylok Precision Company*, did well this year.

Commas enclose nonrestrictive **participial phrases**.

EXAMPLE The lathe operator, *working quickly and efficiently*, finished early.

Interrupting transitional words or phrases are usually set off with commas.

EXAMPLE We must wait for the written authorization to arrive, *however*, before we can begin work on the project.

Commas are omitted when the word or phrase does not interrupt the continuity of thought.

EXAMPLE I *therefore* suggest that we begin construction.

INTRODUCING ELEMENTS

Clauses and Phrases. It is generally a good rule of thumb to put a comma after an introductory clause or phrase. Identifying where the introductory element ends helps indicate where the main part of the sentence begins.

EXAMPLE *Since many rare fossils seem never to occur free from their matrix*, it is wise to scan every slab with a hand lens.

In all cases, use a comma following a long introductory dependent clause.

EXAMPLES *When they artificially stimulated the electrochemical action of the brain,* scientists learned more about the brain.
Because the vertical wing extensions produce more lift, the engines can be throttled back and fuel can be saved.

When long modifying phrases precede the main clause, they should always be followed by a comma.

EXAMPLE *During the first series of field-performance tests last year at our Colorado proving ground*, the new motor failed to meet our expectations.

When an introductory phrase is short and closely related to the main clause, the comma may be omitted.

EXAMPLE *In two seconds* a 20°F temperature is created in the test tube.

A comma should always follow an introductory absolute phrase.

EXAMPLE *The tests completed*, we organized the data for the final report.

Words and Quotations. Certain types of introductory words are followed by a comma. One such is a **noun** used in direct address.

EXAMPLE *Bill*, enclosed is the article you asked me to review.

An introductory **interjection** (such as *oh, well, why, indeed, yes*, and *no*) is followed by a comma.

EXAMPLES *Yes*, I will make sure your request is approved.
Indeed, I will be glad to send you further information.

A transitional word or phrase like *moreover* or *furthermore* is usually followed by a comma.

EXAMPLE *Moreover*, steel can withstand a humidity of 99 percent, provided that there is no chloride or sulphur dioxide in the atmosphere.

Introductory **adverbs** and adverb phrases are set off with a comma when they serve the dual function of modifying the idea that follows and connecting it to the idea in the previous sentence. (See also **transition**.)

EXAMPLE We can expect a better balance of payments in the coming year. *In addition*, we should look for a better world market as a result. *However*, we should expect some shortages due to the overall economic climate.

When adverbs closely modify the **verb** or the entire sentence, however, they should not be followed by a comma.

EXAMPLE *Perhaps* we can still solve the environmental problem. *Certainly* we should try.

Use a comma to separate a direct quotation from its introduction.

EXAMPLE Morton and Lucia White said, "Men live in cities but dream of the countryside."

Do not use a comma, however, when giving an indirect quotation.

EXAMPLE Morton and Lucia White said that men dream of the countryside, even though they live in cities.

SEPARATING ITEMS IN SERIES

Although the comma before the last word in a series is sometimes omitted, it is generally clearer to include it. The confusion that may result from omitting the comma is illustrated in the following sentence:

CHANGE Random House, Irwin, Doubleday and Dell are publishing companies. (Is "Doubleday and Dell" one company or two?)

TO Random House, Irwin, Doubleday, and Dell are publishing companies.

The presence of the comma removes the ambiguity.

Phrases and clauses in coordinate series, like words, are punctuated with commas.

EXAMPLE It is well known that plants absorb noxious gases, act as receptors of dirt particles, and cleanse the air of other impurities.

When **adjectives** modifying the same noun can be reversed and make sense, or when they can be separated by *and* or *or*, they should be separated by commas.

EXAMPLE The drawing was of a *modern, sleek, swept-wing* airplane.

When an adjective modifies a phrase, no comma is required.

EXAMPLE He was investigating his *damaged radar beacon system.* (*damaged* modifies the phrase *radar beacon system*)

Never separate a final adjective from its noun.

CHANGE He is a conscientious, honest, reliable, worker.

TO He is a conscientious, honest, reliable worker.

CLARIFYING AND CONTRASTING

If you find you need a comma to separate the consecutive use of the same word to prevent misreading, rewrite the sentence.

CHANGE The assets we had, had surprised us.

TO We were surprised at the assets we had.

Use a comma to separate two contrasting thoughts or ideas.

EXAMPLES The project was finished on time, but not within the budget.
The specifications call for 100-ohm resistors, not 1,000-ohm resistors.
The project was finished on time, wasn't it?
It was Bill, not Matt, who decided to change the design.
I generally agreed with him, but not completely.

Use a comma following an independent clause that is only loosely related to the **dependent clause** that follows it.

EXAMPLE The plan should be finished by July, even though I lost time because of illness.

SHOWING OMISSIONS

A comma sometimes replaces a verb in certain elliptical constructions.

EXAMPLE Some were punctual; *others, late.* (replaces *were*)

However, it is better to avoid such constructions in on-the-job writing.

USING WITH OTHER PUNCTUATION

Conjunctive adverbs (*however, nevertheless, consequently, for example, on the other hand*) joining independent clauses are preceded by a **semicolon** and followed by a comma. Such adverbs function as both **modifiers** and connectives.

EXAMPLES Your idea is good; *however*, your format is poor.
He has held the project together; *moreover*, he has helped everyone's morale.

Use a semicolon to separate phrases or clauses in a series when one or more of the phrases or clauses contains commas.

EXAMPLE Among those present were John Howard, president of the Howard Paper Company; Thomas Martin, president of Copco Corporation; and Larry Stanley, president of Stanley Papers.

A comma always goes inside **quotation marks**.

EXAMPLE The operator placed the discharge bypass switch at "normal," which triggered a second discharge.

When an introductory phrase or clause ends with a parenthesis, the comma separating the introductory phrase or clause from the rest of the sentence always appears outside the parenthesis.

EXAMPLE Although we left late (at 7:30 p.m.), we arrived in time for the keynote address.

Except with **abbreviations**, a comma should not be used with a **period**, **question mark**, **exclamation mark**, or **dash**.

CHANGE "I have finished the project.," he said.
TO "I have finished the project," he said. (omit the first period)
CHANGE "Have you finished the project?," I asked.
TO "Have you finished the project?" I asked. (omit the comma)

USING WITH NUMBERS AND NAMES

Commas are conventionally used to separate distinct items. Use commas between the elements of an address written on the same line.

EXAMPLE Walter James, 4119 Mill Road, Dayton, Ohio 45401

A **date** can be written with or without a comma following the year if the date is in the month-day-year format.

EXAMPLES October 26, 19--, was the date the project began.
October 26, 19-- was the date the project began.

If the date is in the day-month-year format, do not set off the date with commas.

EXAMPLE The date was 26 October 19-- when the project began.

Use commas to separate the elements of arabic **numbers**.

EXAMPLE 1,528,200

Use a space rather than a comma in metric values, because many countries use the comma as the decimal marker.

EXAMPLE 1 528 200

A comma may be substituted for the **colon** in a personal letter. Do not use a comma in a business letter, however, even if you use the person's first name.

EXAMPLES Dear John, (personal letter)
Dear John: (business letter)

Use commas to separate the elements of geographical names.

EXAMPLE Toronto, Ontario, Canada

Use a comma to separate names that are reversed.

EXAMPLE Smith, Alvin

Use commas to separate certain elements of footnote, **reference,** and **bibliography** entries.

EXAMPLES Bibliography—Fowles, Jib, ed. *Handbook of Futures Research.* Westport, Conn.: Greenwood Press, 1978.

Footnote—[1]Jib Fowles, ed., *Handbook of Futures Research* (Westport, Conn.: Greenwood Press, 1978), p. 30.

Reference—1. Fowles, Jib, ed. *Handbook of Futures Research,* Westport, Conn.: Greenwood Press, 1978.

(See also **documenting sources.**)

AVOIDING UNNECESSARY COMMAS

A number of common writing errors involve placing commas where they do not belong. These errors often occur because writers assume that a pause in a sentence should be indicated by a comma. It is true that commas usually signal pauses, but it is not true that pauses *necessarily* call for commas.

Be careful not to place a comma between a **subject** and verb or between a verb and its **object**.

CHANGE The cold conditions at the test site in the Arctic, made accurate readings difficult.

TO The cold conditions at the test site in the Arctic made accurate readings difficult.

CHANGE He has often said, that one company's failure is another's opportunity.

TO He has often said that one company's failure is another's opportunity.

Do not use a comma between the elements of a compound subject or a compound **predicate** consisting of only two elements.

CHANGE The director of the engineering department, and the supervisor of the quality-control section both were opposed to the new schedules.

TO The director of the engineering department and the supervisor of the quality-control section both were opposed to the new schedules.

CHANGE The director of the engineering department listed five major objections, and asked that the new schedule be reconsidered.

TO The director of the engineering department listed five major objections and asked that the new schedule be reconsidered.

An especially common error is the placing of a comma after a coordinating conjunction such as *and* or *but* (especially *but*).

CHANGE The chairman formally adjourned the meeting, but, the members of the committee continued to argue.
TO The chairman formally adjourned the meeting, but the members of the committee continued to argue.
CHANGE I argued against the proposal. And, I gave good reasons for my position.
TO I argued against the proposal. And I gave good reasons for my position.

Do not place a comma before the first item or after the last item of a series.

CHANGE We are considering a number of new products, such as, calculators, typewriters, and cameras.
TO We are considering a number of new products, such as calculators, typewriters, and cameras.
CHANGE It was a fast, simple, inexpensive, process.
TO It was a fast, simple, inexpensive process.

Do not unnecessarily separate a prepositional phrase from the rest of the sentence with a comma.

CHANGE He met me, in the conference room, down the hall.
TO He met me in the conference room down the hall.
CHANGE We discussed the final report, on the new project.
TO We discussed the final report on the new project.

committee

Committee is a **collective noun** that takes a singular **verb**.

EXAMPLE The *committee is* to meet at 3:30 p.m.

If you wish to emphasize the individuals on the committee, use *the members of the committee* with the plural verb form.

EXAMPLE The *members of the committee were* all in agreement.

common nouns

A common noun names general classes or categories of persons, places, things, concepts, actions, or qualities. Common nouns can be abstract (*justice*), concrete (*jail*), or **collective** (*jury*).

Common Nouns	*Proper Nouns*
boy	Toby Wilson
city	Chicago
company	General Electric
day	Tuesday
document	Declaration of Independence

Common nouns are not capitalized unless they begin a sentence. (For information about the **case, number**, and function of common nouns, see **nouns**.)

comparative degree

Most **adjectives** and **adverbs** can be compared. Their common forms of **comparison** are as follows:

EXAMPLES The new copier is *fast*. (positive)
The new copier is *faster* than the old one. (comparative)
The new copier is the *fastest* I have ever used. (superlative)

Most one-syllable words use the comparative ending *-er* and the superlative ending *-est*.

EXAMPLES Ours was the *faster* of the two cars tested.
Ours was the *fastest* car on the track.

Most words with more than one syllable form the comparative degree by using the word *more* and the superlative degree by using the word *most*.

EXAMPLES She is *more* talented than the rest of the writers on the staff.
She is the *most* talented writer on the staff.

Some words may use either means of expressing degree.

EXAMPLES He was the *most* able technician at the meeting.
He was the *ablest* technician at the meeting.

A few words have irregular forms of comparison.

EXAMPLES much, more, most
good, better, best

Writers often use the comparative degree when comparing two persons or things and the superlative degree when comparing three or more.

EXAMPLES This is the *newer* of the two policies.
This is the *newest* of the three policies.

The superlative form is sometimes also used in expressions that do not really express comparison (*best* wishes, *deepest* sympathy, *highest* praise, and *most* sincerely).

compare/contrast

When you *compare* things, you point out similarities or both similarities and differences. When you *contrast* things you point out only the differences. In either case, you compare or contrast only things that are part of a common category.

EXAMPLES He *compared* all the features of the two brands before buying.
Their styles of selling *contrasted* sharply.

When *compared* is used to establish a general similarity, it is followed by *to*.

EXAMPLE *Compared to* the computer, the abacus is a primitive device.

When *compare* is used to indicate a close examination of similarities or differences, it is followed by *with* in formal **usage**.

EXAMPLE We *compared* the features of the new capacitor very carefully *with* those of the old one.

Contrast is normally followed by *with*.

EXAMPLE The new policy *contrasts* sharply *with* the earlier one in requiring that sealed bids be submitted.

When the **noun** form of *contrast* is used, one speaks of the *contrast between* two things or of one thing being *in contrast to* the other.

EXAMPLES There is a sharp *contrast between* the old and new policies.
The new policy is *in* sharp *contrast to* the earlier one.

comparison

When making a comparision, make certain that both or all the elements being compared are clearly evident to your **reader**.

CHANGE The third-generation computer is *better*.
TO The third-generation computer is *better than the second-generation computer.*

The things being compared must be of the same kind.

CHANGE *Imitation alligator hide* is almost as tough as a *real alligator*.
TO *Imitation alligator hide* is almost as tough as *real alligator hide*.

Be sure to point out the parallels or differences between the things being compared. Don't assume your reader will know what you mean.

CHANGE Washington is farther from Boston *than Philadelphia*.
TO Washington is farther from Boston *than it is from* Philadelphia.
OR Washington is farther from Boston *than Philadelphia is*.

A double comparison in the same sentence requires that the first be completed before the second is stated.

CHANGE The discovery of electricity was *one of the great if not the greatest* scientific discoveries in history.
TO The discovery of electricity was *one of the great* scientific discoveries in history, *if not the greatest*.

Do not attempt to compare things that are not comparable.

CHANGE Farmers say that storage space is reduced by 40 percent compared with baled hay. (*Storage space* is not comparable to *baled hay*.)
TO Farmers say that baled hay requires 40 percent less storage space than loose hay requires.

comparison method of development

As a method of development, comparison points out similarities and differences between the elements of your subject. The comparison **method of development** can be especially effective because it can explain a difficult or unfamiliar subject by relating it to a simpler or more familiar one.

You must first determine the basis for the **comparison**. For example, if you were responsible for the purchase of chain saws for a logging company, you would have a number of factors to take into account in order to establish your bases of comparison. Because loggers use the equipment daily, you would have to select durable saws with the appropriate size engines, chain thicknesses, and bar lengths for the type of wood most frequently cut. Since chain saws produce noise and vibration, you would want to compare the quality and cost of the various silencers on the market. You would not include in your comparison such irrelevant factors as color or place of manufacture. Taking into account all of the important elements, however, you would establish a number of bases for choosing from among the available chain saws—engine size, chain thickness, bar length, and noise mufflers.

Once you have determined the basis (or bases) for comparison, you must decide how to present it. In the whole-by-whole method, all the relevant characteristics of one item are discussed before those of the next item are considered. In the part-by-part method, the relevant features of each item are compared one by one. The following discussion of typical woodworking glues, organized according to the whole-by-whole method, describes each type of glue and its characteristics before going on to the next type:

> *White glue* is the most useful all-purpose adhesive for light construction, but it cannot be used on projects that will be exposed to moisture, high temperature, or great stress. Wood that is being joined with white glue must remain in a clamp until the glue dries, which will take about 30 minutes.
>
> *Aliphatic resin glue* has a stronger and more moisture-resistant bond than white glue. It must be used at temperatures above 50°F. The wood should be clamped for about 30 minutes. . . .
>
> *Plastic resin glue* is the strongest of the common wood adhesives. It is highly moisture resistant—though not completely waterproof. Sold in powdered form, this glue must be mixed with water and used at temperatures above 70°F. It is slow setting and the joint should be clamped for four to six hours. . . .
>
> *Contact cement* is a very strong adhesive that bonds so quickly it must be used with great care. It is ideal for mounting sheets of plastic laminate on wood. It is also useful for attaching strips of veneer to the edges of plywood. Since this adhesive bonds immediately when two pieces are pressed together, clamping is not necessary, but the parts to be joined must be very carefully aligned before being placed together. Most brands are quite flammable and the fumes can be harmful if inhaled. To meet current safety standards, this type of glue must be used in a well-ventilated area, away from flames or heat.
>
> —*Space and Storage* (Alexandria, Va.: Time-Life Books, 1977), p. 61.

As is often the case when the whole-by-whole method is used, the purpose of this comparison is to weigh the advantages and disadvantages of each glue for certain kinds of woodworking. The comparison could be expanded, of course, by the addition of other types of glue. If, on the other hand, your purpose were to consider, one at a time, the various characteristics of all the glues, the information might be arranged according to the part-by-part method, as in the following example:

> Woodworking adhesives are rated primarily according to their bonding strength, moisture resistance, and setting times.
>
> *Bonding strengths* are categorized as very strong, moderately strong, or adequate for use with little stress. Contact cement and plastic resin glue bond very strongly, while aliphatic resin glue bonds moderately strongly. White glue provides a bond least resistant to stress.
>
> The *moisture resistance* of woodworking glues is rated as high, moderate, and low. Plastic resin glues are highly moisture resistant. Aliphatic resin glues are moderately moisture resistant; white glue is least moisture resistant.
>
> *Setting times* for these glues vary from an immediate bond to a four-to-six-hour bond. Contact cement bonds immediately and requires no clamping. Because the bond is immediate, surfaces being joined must be carefully aligned before being placed together. White glue and aliphatic resin glue set in thirty minutes; both require clamping to secure the bond. Plastic resin, the strongest wood glue, sets in four to six hours and also requires clamping.

The part-by-part method could accommodate further comparison. Comparisons might be made according to temperature ranges, special warnings, common uses, and so on.

complaint letters

Businesses sometimes err when providing goods and services to customers, and so customers write complaint letters (or claim letters) asking that such situations be corrected. The **tone** of such a letter is important; the most effective complaint letters do not sound angry. Do not use a complaint letter to vent your anger; remember that the **reader** of your letter probably had nothing to do with whatever went wrong, and berating that person is not likely to achieve anything positive. In most cases, you need only state your claim, support it with all the pertinent facts, and then ask for the desired adjustment. Most companies are very willing to correct whatever went wrong.

BAKER MEMORIAL HOSPITAL

Television Services
501 Main Street
Springfield, OH 45321
(513) 683-8100

September 23, 19--

General Television, Inc.
5521 West 23rd Street
New York, NY 10062

Attention: Customer Relations Manager

On July 9th I ordered nine TV tuners for your model MX-15 color receiver. The tuner part number is TR-5771-3.

On August 2nd I received from your Newark, New Jersey, parts warehouse seven tuners, labeled TR-413-7. I immediately returned these tuners with a note indicating the mistake that had been made. However, not only have I failed to receive the tuners I ordered, but I have also been billed repeatedly.

Please either send me the tuners I ordered or cancel my order. I have enclosed a copy of my original order and the most recent bill.

Sincerely,

Paul Denlinger
Manager

PD:sj
Enclosure

Letter 1 Complaint Letter

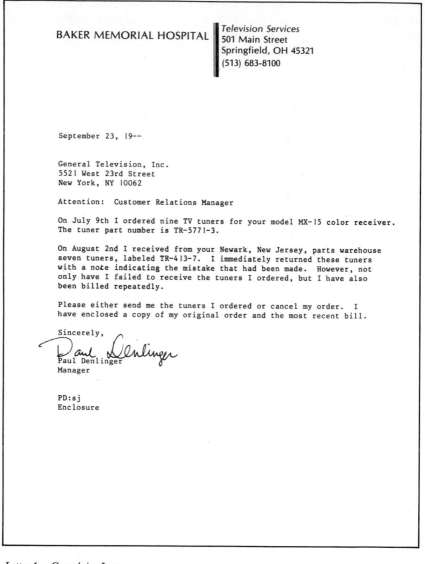

The **opening** of your complaint letter should include all identifying data concerning the transaction: item, date of purchase, place of purchase if pertinent, cost, invoice number, and so on.

The body of your letter should explain logically and clearly what happened. You should present any facts that prove the validity of your claim. Be sure of your facts, and present them concisely and objectively, carefully avoiding any overtones of accusation or threat. You may, however, wish to state any inconvenience or loss created by the problem, such as a broken machine stopping an entire assembly line.

Your **conclusion** should be friendly, and it should request action. State what you would like your reader to do to solve your problem.

Large organizations often have special departments to handle complaints. If you address your letter to one of these departments—for example, to Customer Relations or Consumer Affairs—it should reach someone who can respond to your claim. In smaller organizations you might write to a vice-president in charge of sales or service. For a very small business, write directly to the owner.

Letter 1 is an example of a typical complaint letter. (See also **adjustment letters** and **refusal letters**.)

complement/compliment

Complement means "anything that completes a whole." It is used as either a **noun** or a **verb**.

EXAMPLES A *complement* of four employees would bring our staff up to its normal strength. (noun)
The two programs *complement* one another perfectly. (verb)

Compliment means "praise." It too is used as either a noun or a verb.

EXAMPLES The manager *complimented* the staff on its efficient job. (verb)
The manager's *compliment* boosted staff morale. (noun)

complements

A complement is a word, **phrase**, or **clause** used in the **predicate** of a sentence to complete the meaning of the sentence.

EXAMPLES Pilots fly *airplanes*. (word)
To live is *to risk death*. (phrase)
John knew *that he would be late*. (clause)

Four kinds of complements are generally recognized: direct object (which completes the sense of a transitive **verb**); indirect object (which completes the meaning of a transitive verb and the verb's direct object); **objective complement** (which completes the meaning of a verb's object); and **subjective complement** (which completes the meaning of the subject).

A direct object is a **noun** or noun equivalent that receives the action of a transitive verb; it answers the question *what* or *whom* after the verb.

EXAMPLES John built *an antenna*. (noun)
I like *to work*. (verbal)
I like *it*. (pronoun)
I like *what I saw*. (noun clause)

An indirect object is a noun or noun equivalent that occurs with a direct object after certain kinds of transitive verbs such as *give, wish, cause*, and *tell*. It answers the question *to whom* or *for whom* (or *to what* or *for what*).

EXAMPLES Give *John* a wrench. (*wrench* is the direct object)
We should buy *the Milwaukee office* a computer. (*computer* is the direct object)

An objective complement completes the meaning of a sentence by revealing something about the object of its transitive verb. An objective complement may be either a noun or an **adjective**.

EXAMPLES They call him *a genius*. (noun)
We painted the building *white*. (adjective)

A subjective complement, which follows a **linking verb** rather than a transitive verb, describes the **subject**. A subjective complement may be either a noun or an adjective.

EXAMPLES His sister is *an engineer*. (noun)
His brother is *ill*. (adjective)

complex sentences

The complex sentence provides a means of subordinating one thought to another (or, put another way, of emphasizing one thought over another) because it contains one **independent clause** and at least one **dependent clause** that expresses a subordinate idea.

EXAMPLE We lost some of our efficiency (independent clause) when we moved (dependent clause).

Normally, the independent clause carries a main point, and the dependent clause carries a related subordinate point. A dependent clause may occur before, after, or within the independent clause. The dependent clause can serve within the sentence as its **subject**, as an **object**, or as a **modifier**.

EXAMPLES *What he proposed* is irrelevant. (subject)
We know *where it is supposed to be*. (object)
The computer, *which can make more calculations in one hour than thousands of scientists could make in a lifetime*, is perhaps one of the greatest challenges people have ever had to face. (modifier)

Complex sentences offer more variety than **simple sentences**. And frequently the meaning of a **compound sentence** can be made more precise by subordinating one of the two independent clauses to the other to create a complex sentence (thereby establishing the relationship of the two parts more clearly). (See also **subordination**.)

EXAMPLES We moved *and* we lost some of our efficiency. (compound sentence with **coordinating conjunction**)
When we moved, we lost some of our efficiency. (complex sentence with **subordinating conjunction**)

Learn to handle the complex sentence well; it can be a useful tool with which to express your thoughts clearly and exactly. However, be on guard against placing the main idea of a complex sentence in a dependent clause; this is one of the most common errors made with the complex sentence. For example, do not write the first sentence below if what you mean is expressed by the second sentence.

EXAMPLES Production was not as great as we expected, although we met our quota. (The main idea expressed here is that production fell short of expectations.)
Although production was not as great as we expected, we met our quota. (The main idea expressed here is that the production quota was met.)

compose/comprise (see **comprise/compose**.)

compound sentences

A compound sentence combines two or more related **independent clauses** that are of equal importance.

EXAMPLE Drilling a well is the only sure way to determine the presence of oil, *but* it is a very costly endeavor.

The independent clauses of a compound sentence may be joined by a **comma** and a **coordinating conjunction**, by a **semicolon**, or by a **conjunctive adverb** preceded by a semicolon and followed by a comma.

EXAMPLES People deplore violence, *but* they have an insatiable appetite for it on television and in films. (coordinating conjunction)

There is little similarity between the chemical composition of sea water and river water; the various elements are present in entirely different proportions. (semicolon)

People deplore violence; *however*, they have an insatiable appetite for it on television and in films. (conjunctive adverb)

A compound sentence is balanced when its clauses are of similar length and construction.

EXAMPLES The plan was sound, and the staff was eager to begin.

Fingerprints were used for personal identification in 200 B.C., but they were not used for criminal identification until about A.D. 1880.

compound-complex sentences

A compound-complex sentence consists of two or more **independent clauses** and at least one **dependent clause.**

EXAMPLE At the same time *that it cools and lubricates the bit and brings the cuttings to the surface*, the "drilling mud" deposits a sheath of mud cake on the wall of the hole to prevent cave-ins; it reduces the friction *which is created by the drill string's rubbing against the wall of the hole;* and *since the weight of the mud column bears against the wall of the hole*, it helps to contain formation pressures and prevent a blowout.

Here, the three parallel independent clauses are joined in a series linked by **semicolons** and the **coordinating conjunction** *and*; each independent clause contains a dependent clause (in **italics**). The first two dependent clauses are adjective clauses modifying *time* and *friction*, and the third is an adverb clause that modifies its independent clause as a whole.

The compound-complex sentence offers a vehicle for a more elaborate grouping of ideas than other sentence types permit. It is not really recommended for inexperienced writers because it is diffic·ılt to handle skillfully and often leads them into sentences that become so complicated that they lose all **logic** and **coherence**. The more technically complex the subject, the more difficult the compound-complex sentence is to control effectively.

compound words

A compound word is made from two or more words that are either hyphenated or written as one word. (If you are not certain whether a compound word should be hyphenated, check a **dictionary**.)

EXAMPLES nevertheless, mother-in-law, courthouse, run-of-the-mill, low-level, high-energy

Be careful to distinguish between compound words and words that frequently appear together but do not constitute compound words, such as *high school* and *post office*. Also be careful to distinguish between compound words and word pairs that mean different things, such as *greenhouse* and *green house*.

Plurals of compound words are usually formed by adding an *s* to the last letter.

EXAMPLES bedrooms, masterminds, overcoats, cupfuls

When the first word of the compound is more important to its meaning than the last, however, the first word takes the *s* (when in doubt, check your dictionary).

EXAMPLES editors-in-chief, fathers-in-law

Possessives are formed by adding *'s* to the end of the compound word.

EXAMPLES the *vice-president's* speech, his *brother-in-law's* car, the *pipeline's* diameter, the *antibody's* action

comprise/compose

Comprise means "include," contain," or "consist of." The whole *comprises* the parts.

EXAMPLE The mechanism *comprises* 13 moving parts.

Compose means "create" or "make up the whole." The parts *compose* the whole.

EXAMPLE The 13 moving parts *compose* the mechanism.
OR The mechanism *is composed of* 13 moving parts.

Do not use *comprise* in place of *compose.*

CHANGE The engineering design team *is comprised of* 12 members.
TO The engineering design team *is composed of* 12 members.
OR The engineering design team *comprises* 12 members.

concept/conception

A *concept* is a thought or an idea. A *conception* is the sum of a person's ideas, or concepts, on a subject.

EXAMPLES This final *conception* of the whole process evolved from many smaller *concepts.*
From the *concept* of combustion evolved the *conception* of the internal combustion engine.

conciseness/wordiness

Effective writers make all words, sentences, and **paragraphs** count by eliminating unnecessary words and **phrases**. Wordiness results from needless **repetition** of the same idea in different words.

CHANGE Modern students *of today* are more sophisticated than their parents. (The phrase *of today* repeats the thought already expressed by the adjective *modern.*)
TO Modern students are more sophisticated than their parents.
CHANGE The walls were sky-blue *in color.* (The phrase *in color* is redundant.)
TO The walls were sky-blue.

Careful writers remove every word, **phrase**, **clause**, or **sentence** they can without sacrificing **clarity**. In doing so, they are striving to be as concise as **clarity** permits—but note that conciseness is not a **synonym** for brevity. Brevity may or may not be desirable in a given passage depending on the writer's **objective**, but conciseness is always desirable. The writer must distinguish between language that is used for effect and mere wordiness that stems from lack of care or judgment. The following **anecdote** was related by Benjamin Franklin to Thomas Jefferson as members of the Continental Congress undertook to trim excess words and phrases from the Declaration of Independence:

> The wise old Pennsylvanian told Jefferson, in an aside, about a hatter who was opening a shop and wanted a signboard for it. What the new proprietor had in mind was a message reading, "John Thompson, Hatter, makes and sells hats for ready money," but one of his friends suggested that the word "hatter" was superfluous. Another told him that no buyer would care who made the hats, so the word "makes" was omitted. Someone else advised him to leave out "ready money," since nobody expected to buy on credit, and yet another man said that since Thompson did not propose to give the hats away the word "sells" should go. When the sign was finally erected, Franklin smiled, all that remained was the name "John Thompson" and the picture of a hat.
>
> —Richard M. Ketchum, *The Winter Soldiers* (New York: Doubleday, 1973), p. 19.

A concise sentence is not guaranteed to be effective, but a wordy sentence always loses some of its readability and **coherence** because of the extra load it must carry. Wordiness is to be expected in a first draft, but it should never survive **revision**. "I would have written a shorter letter if I'd had more time" is a truism.

CAUSES OF WORDINESS

Modifiers that repeat an idea already implicit or present in the word being modified contribute to wordiness by being redundant.

EXAMPLES
active consideration
final outcome
personal opinion
completely finished
round circles
basic essentials
hot boiling water
advance planning
small *in size*
balance *of equilibrium*
realization of a dream *come true*
visible *to the eye*
present status
true facts
past history
isolated *by himself*
tall skyscrapers
descended *down*
circle *around*
square *in shape*
worthy *of merit*
the reason *is because*
cooperate *together*

Another cause of wordiness is the "it . . . that . . ." construction. Not only is this construction wordy, but often it forces you to use the passive **voice** (a major cause of wordiness in its own right).

> CHANGE *It* is agreed *that* our new design will strive for simplicity.
> TO We agree that our new design will strive for simplicity.

The **gobbledygook** addict is fond of such words and phrases as *factor, case, basis, elements, field, phenomenon, in terms of, in the nature of, with reference to.* Key words that seem to invite redundancy are *situation, angle, line, factor, aspect, element, consideration, considering.* Although all these words have legitimate uses, technical people seem to have a particular weakness for their overuse and inexact use, under the mistaken impression that they add formality to writing.

Coordinating synonyms that merely repeat one another contribute to wordiness.

> EXAMPLES *any and all*
> *finally and for good*
> *first and foremost*
> *each and every*
> *basic and fundamental*

Wordiness can be caused by unnecessary or redundant phrases and clauses.

> EXAMPLE Our control-room simulators, *which are manufactured in this country*, are more sophisticated than Russian simulators. (unnecessary clause)

Excess qualification also contributes to wordiness, as the following examples demonstrate:

> EXAMPLES utterly rejected—rejected
> perfectly clear—clear
> completely compatible—compatible
> completely accurate—accurate
> radically new—new

The use of **expletives, relative pronouns**, and relative adjectives, although they have legitimate purposes, often results in wordiness.

> CHANGE *There are* (expletive) many supervisors in the area *who* (relative pronoun) are planning to attend the workshop *which* (relative adjective) is scheduled for Friday.
> TO Many supervisors in the area plan to attend the workshop scheduled for Friday.

Circumlocution (a long, roundabout way of expressing things) is a leading cause of wordiness.

CHANGE We urge you to submit any suggestions you may have on the subject, and you may be certain that your suggestions will be given our most careful attention.

TO Please give us your suggestions. We will consider them carefully.

Conciseness can be overdone. If you respond to a written request that you cannot understand with "Your request was unclear" or "I don't understand," you will probably offend your reader. Instead of attacking the writer's ability to phrase a request, consider that what you are really doing is asking for more information. Say so.

EXAMPLE I will need more information before I can answer your request. Specifically, can you give me the title and the date of the report you are looking for?

This version is a little longer than the others, but it is both more polite and more helpful.

Although conciseness and clarity *usually* reinforce each other, there are times when clarity legitimately demands more words. See the example of "rendering" in **descriptions** as an example.

HOW TO ACHIEVE CONCISENESS

Conciseness can be achieved by effective use of **subordination**. This is, in fact, the best means of tightening wordy writing.

CHANGE The chemist's report was carefully illustrated, and it covered five pages.

TO The chemist's five-page report was carefully illustrated.

Conciseness can be achieved by using simple words and phrases.

CHANGE It is the policy of the company to provide the proper equipment to enable each employee to conduct the telephonic communication necessary to discharge his responsibilities; such should not be utilized for personal communications.

TO Your telephone is provided for company business; do not use it for personal calls.

Conciseness can be achieved by eliminating undesirable **repetition**.

CHANGE Postinstallation testing, which is offered to all our customers at no further cost to them whatsoever, is available with each Line Scan System One purchased from this company.

TO Free postinstallation testing is offered with each Line Scan System One.

Conciseness can be achieved by making sentences positive. (See also **positive writing**.)

CHANGE If the error does not involve data transmission, the special function is not used.

TO The special function is used only when the error involves data transmission.

Conciseness can sometimes be achieved by changing a sentence from the passive to the active voice or from the indicative to the imperative **mood**. The following example does both:

CHANGE Card codes are normally used when it is known that the cards are to be processed by a computer, and control punches are normally used when it is known that the cards are designed to be processed at a tab installation.

TO Use card codes when you process the cards on a computer, and use control punches when you process them at a tab installation.

Eliminate wordy introductory phrases or pretentious words and phrases of any kind.

EXAMPLES
In view of the foregoing
It appears that
In the case of
It may be said that
It is interesting to note that
As you may recall
As you know
Needless to say
In view of the fact that

CHANGE *In connection with* your recent accident, we must *discontinue* your insurance *due to the fact that* this is your third claim within the past year. *Inasmuch* as you were insured with us *at the time of the accident, we are prepared* to honor your claim *along the lines* of your policy.

TO Concerning your recent accident, we are canceling your insurance because you have filed three claims in the past year. Since you were insured with us at the time of the accident, however, we will honor your claim.

CHANGE *in order to, so as to, so as to be able to, with a view to*
TO *to*
CHANGE *due to the fact that, for the reason that, owing to the fact that, the reason for*
TO *because*
CHANGE *by means of, by using, utilizing, through the use of*
TO *by* or *with*
CHANGE *at this time, at this point in time, at present, at the present*
TO *now*
CHANGE *at that time, at that point, at that point in time, as of that date, during that period*
TO *then*

Overuse of **intensifiers** (such as *very, more, most, best, quite*) contributes to wordiness; conciseness can be achieved by eliminating them. The same is true of excessive use of **adjectives** and **adverbs**.

concluding

Concluding a piece of writing not only ties together all the main ideas but can do so emphatically by making a final significant point. This final point may be to recommend a course of action, to make a prediction, to offer a judgment, to speculate on the implications of your ideas, or merely to summarize your main points.

The way you conclude depends on both the purpose of your writing and the needs of your **reader**. For example, a committee **report** about possible locations for a new manufacturing plant could end with a recommendation. A report on a company's annual sales figures might conclude with a judgment about why sales are up or down. A letter about consumer trends could end by speculating on the implications of these trends. A document that is particularly lengthy will often end with a summary of its main points. Study the following examples:

recommendation — These results indicate that you may need to alter your testing procedure to eliminate the impurities we found in specimens A through E.

prediction — Although my original estimate on equipment ($20,000) has been exceeded by $2,300, my original labor estimate ($60,000) has been reduced by $3,500; therefore, I will easily stay within the limits of my original bid. In addition, I see no difficulty in having the arena finished for the December 23 Christmas program.

judgment — Although our estimate calls for a substantially higher budget than in the three previous years, we believe that it is justified by our planned expansion.

summary — As this letter has indicated, we would attract more recent graduates by (1) increasing our advertising in local student newspapers, (2) resuming our co-op program, (3) sending a representative to career day programs at local colleges, (4) inviting local college instructors to teach in-house courses here at the plant, and (5) encouraging our employees to attend evening classes at local colleges.

The concluding statement may merely present ideas for consideration, but also it may call for action or deliberately provoke thought.

ideas for consideration — The new prices become effective the first of the year. Price adjustments are routine for the company, but some of your customers will not consider them so. Please consider the needs of both your customers and the company as you implement these new prices.

call for action — Send us a check for $250 now if you wish to keep your account active. If you have not responded to our previous letters because of some special hardship, I will be glad personally to work out a solution with you.

thought-provoking statement — Can we continue to accept the losses incurred by careless workmanship? Must we accept it as inevitable? Or should we consider steps to control it firmly now?

Because this statement appears in a position of great **emphasis**, be careful to avoid closing with a cliché. Some such expressions do say what you want to say so precisely that they are considered acceptable closings, however.

EXAMPLES It has been a pleasure to be of assistance.
May we thank you again for your cooperation.

Be especially careful not to introduce a new **topic** when you conclude. A concluding thought should always relate to and reinforce the ideas presented in your writing. (See also **conclusions**.)

conclusions

The conclusion section of a **report** or article pulls together the results or findings and interprets them in the light of the study's purpose and the methods by which it was conducted. Consequently, the

CONCLUSIONS

Brief statement of purpose

This study was undertaken to test the hypotheses that nuclear power plants have an adverse effect on (1) community growth and (2) residential property values. Implicit in the second hypothesis is that prices paid for residential properties reflect perceptions of house and lot quality, community services and attributes, and environmental characteristics. If people in general perceive nuclear plants as a possible threat to health and safety, they will tend to avoid living close to them. Such actions should be reflected in the real estate market for residential properties, and everything else being equal, the closer to the plant, the lower the price for housing.

Findings based on methods specified in introduction

The following conclusions are apparent from an analysis of time series data on total assessed real property values from 1960 to 1976 and from a regression analysis of 540 single-family house sales in 1975, 1976, and 1977 in the vicinity of four nuclear power generating plants in the Northeast.

1. Assuming that the real (deflated) increase in total assessed property values is a good indicator of growth, the presence of the nuclear plants has had no adverse effects on the growth rates of communities in close proximity to such plants.
2. Because (a) growth rates were inversely related to distance from the plant and (b) annual growth rates for the years following plant construction were higher than for the years before plant construction, with the increase in the growth rate for the host communities being higher than the increase for the region as a whole, the presence of the nuclear plant may actually stimulate growth. The likely explanation for this is the lower tax rates in the host municipalities, which are made possible by the large tax assessments levied on the plants.
3. The presence of the nuclear power plants, at least in 1975, 1976, and 1977, exerted no influence on the price of single-family housing within 20 miles of these plants. Therefore, our hypothesis that residential property values are directly related to distance from the nuclear power plants must be rejected and the null hypothesis accepted.
4. For most people in these study areas, the proximity of a nuclear power plant does not appear as a factor in residential location choice. The fears for health and safety expressed by some individuals and groups in society are not reflected in the housing decisions of residents in communities near the nuclear plants studied.

Restatement of scope and its implications

These conclusions apply only to the four areas studied. Since the plants were preselected and do not represent a random sample of all nuclear plants, the findings cannot be used to predict influences on growth and property values at other plant sites. Moreover, society's perceptions and values change over time, and what may have been true in the mid-1970s may not hold true in the future.

conclusion is the focal point of the work, the goal toward which the study was aiming. The evidence for these findings makes up the discussions in the body of the report or article; the conclusions must grow out of the information discussed there. Moreover, the conclusions must be consistent with what the **introduction** stated that the report would examine (its purpose) and how it would do so (its method). If the introduction stated that the report had as its **objective** to learn the economic costs of relocating a plant from one city to another, then the conclusion should not discuss the social or aesthetic impacts the new plant could have on the new location. Of course, if "costs" are defined in the introduction to include social and aesthetic concerns, then these also must be accounted for in the report and discussed in the conclusions.

The sample conclusion (on page 129) comes from a report about how community growth and real estate values in four specific areas were affected by nearby nuclear power plants. Note that the authors briefly restate the purpose of the study: to test two hypotheses. They then clarify some implicit assumptions in their original purpose before indicating that the findings are based on the methods specified in the introduction to the report (that is, the analysis of time series data and a regression analysis). After listing the findings, they go on to qualify these, noting that the results are valid for the scope of this study but that they may not apply to areas at other plant sites.

Not all reports or articles require a separate conclusions section. Periodic **progress reports** for a long-term project, for example, may simply indicate what was done during the last month or quarter. The final report for such a project, however, would require a conclusions section in which the findings and their implications were discussed. Nor do many other types of writing require a fully developed conclusion. For examples of less formal conclusions, see **concluding.**

Finally, because the conclusions are frequently read independently of the rest of the work, they should be written accordingly. The conclusions should restate the purpose of the document and specify the methodology (as was done in the power plant sample) and should define **symbols** and **abbreviations** as though they were being used for the first time. (For guidance about the location of the conclusions section in a report, see **formal reports**.)

concrete nouns (see **abstract words/concrete words**)

conjunctions

A conjunction connects words, **phrases**, or **clauses**. A conjunction can also indicate the relationship between the two elements it connects (*and* joins together, *or* selects and separates).

Occasionally, a conjunction may begin a sentence and even a paragraph, as in the following example:

EXAMPLE The executive is impressed by the marvels of computer technology before him; he has difficulty in understanding the programming endeavor that makes the computer run. He walks away feeling that the annual report will look better because of these machines—and the data processing manager dreams of a larger paycheck.

But the balloon bursts the first time the system gets into trouble and requires a modification, and the lead programmer is no longer with the company or cannot recall the details of the program due to the passage of time. . . .

—William L. Harper, *Data Processing Documentation* (Englewood Cliffs, N.J.: Prentice-Hall, 1980), p. 145.

TYPES OF CONJUNCTIONS

A **coordinating conjunction** is a word that joins two sentence elements that have identical functions. The coordinating conjunctions are *and, but, or, for, nor, yet*, and *so*.

EXAMPLES Nature *and* technology are only two conditions that affect petroleum operations around the world. (joining two nouns)

To hear *and* to obey are two different things. (joining two phrases)

He would like to include the test results, *but* that would make the report too long. (joining two clauses)

Correlative conjunctions are coordinating conjunctions that are used in pairs. The correlative conjunctions are *either . . . or, neither . . . nor, not only . . . but also, both . . . and,* and *whether . . . or*.

EXAMPLE The inspector will arrive *either* on Wednesday *or* on Thursday.

A **subordinating conjunction** connects sentence elements of different weights, normally **independent clauses** and **dependent clauses**. The most frequently used are *so, although, after, because, if, where, than, since, as, unless, before, that, though, when*, and *whereas*.

EXAMPLE I left the office *after* I had finished writing the report.

A **conjunctive adverb** is an **adverb** that has the force of a conjunction because it is used to join two independent clauses. The most common conjunctive adverbs are *however, moreover, therefore, further, then, consequently, besides, accordingly, also, thus*.

EXAMPLE The engine performed well in the laboratory; *however*, it failed under road conditions.

PUNCTUATING CONJUNCTIONS

Two independent clauses separated by a coordinating conjunction should have a **comma** immediately preceding the coordinating conjunction if the clauses are relatively long.

EXAMPLE The Alpha project was her third assignment, *and* it was the one that made her reputation as a project leader.

If two independent clauses that are joined by a coordinating conjunction have commas within them, a **semicolon** may precede the conjunction.

EXAMPLE Even though the schedule is tight, we must meet our deadline; *and* all the staff, including programmers, will have to work overtime this week.

CONJUNCTIONS IN TITLES

Conjunctions in the titles of books, articles, plays, movies, and so on should not be capitalized unless they are the first or last word in the title. (See also **capital letters**.)

EXAMPLE The book *Technical and Professional Writing* was edited by Herman Estrin.

conjunctive adverbs

A conjunctive adverb is an **adverb** because it modifies the **clause** that it introduces; it operates as a **conjunction** because it joins two **independent clauses**. The most common conjunctive adverbs are *however, nevertheless, moreover, therefore, further, then, consequently, besides, accordingly, also,* and *thus*.

EXAMPLE The engine performed well in the laboratory; *however*, it failed under road conditions.

Two independent clauses that are joined by a conjunctive adverb require a **semicolon** before and a **comma** after the conjunctive adverb because the conjunctive adverb is introducing the independent clause that follows it and is indicating its relationship with the independent clause preceding it. The conjunctive adverb connects ideas but does not grammatically connect the clause stating the ideas; the semicolon must do that. The conjunctive adverb both connects and modifies. As a **modifier**, it is part of one of the two clauses that it connects. The use of the semicolon makes it a part of the clause it modifies.

EXAMPLES The building was not finished on the scheduled date; *nevertheless*, Rogers moved in and began to conduct the business of the department from the unfinished offices.

The new project is almost completed and is already over cost estimates; *however*, we must emphasize the importance of adequate drainage.

connected with/in connection with

Connected with and *in connection with* are wordy **phrases** that can usually be replaced by *in* or *with*.

CHANGE He is *connected with* the TFT Corporation.
TO He is with the TFT Corporation.
OR He is employed by the TFT Corporation.

CHANGE The fringe benefits *in connection with* (or *connected with*) the job are quite good.
TO The fringe benefits with the job are quite good.

connectives (see **conjunctions**)

connotation/denotation

The terms *connotation* and *denotation* refer to two different ways of interpreting a word's meaning. The *denotation* of a word is its literal and objective meaning. Defined this way, *communism* is a socioeconomic system that aims at a classless society, and a *bureaucrat* is an unelected public official who administers government policy. Of course both words bring to mind a host of secondary or implied meanings that arise from our emotional reactions to them. Many Americans react negatively to the word *communism*, associating it with a totalitarian form of government that is antireligious and committed to world conquest. Likewise, the word *bureaucrat* frequently conjures up visions of someone who insists on rigid adherence to arbitrary rules and regulations. These secondary meanings are a word's connotations.

In your writing you must be sure to consider the connotations of your words. You know what you mean to say, but you must be careful to say it in words that will not be misinterpreted by your **reader**. For example, you might not want to refer to an item as *cheap*, meaning inexpensive, for the word *cheap* also implies shoddy workmanship or poor materials. The **usage** notes in a reputable dictionary can help you make these distinctions. Technical writing in particular requires an objective and impartial presentation of information, and so you should always be as careful as possible to use words without misleading or unwanted connotations.

On the other hand, a word may have several equally valid denotative meanings. Context will make the meaning clear, as in the following varied uses of the word *check*:

EXAMPLES The bank verified the *check*.
The driver was instructed to *check* the oil before every trip.
The chess champion was in *check* three times, but he eventually won the match.
The hockey player executed a bruising *check*.

(See also **defining terms**.)

consensus of opinion

Since *consensus* normally means "harmony of opinion," the **phrase** *consensus of opinion* is redundant. The word *consensus* can be used only to refer to a group, never one or two people.

CHANGE The *consensus of opinion* of the members was that the committee should change its name.
TO The *consensus* of the members was that the committee should change its name.

continual/continuous

Continual means "happening over and over" or "frequently repeated."

EXAMPLE Writing well requires *continual* practice.

Continuous means "occurring without interruption" or "unbroken."

EXAMPLE The *continuous* roar of the machines was deafening.

contractions

A contraction is a shortened **spelling** of a word or **phrase** with an **apostrophe** substituting for the missing letters.

EXAMPLES cannot/can't
will not/won't
have not/haven't
it is/it's

Contractions are often used in speech but should be used discriminatingly in **reports**, formal letters, and most technical writing. (See also **style**.)

coordinating conjunctions

A coordinating conjunction is a word that joins two sentence elements that have identical functions. The coordinating conjunctions include *and, but, or, for, nor, yet*, and *so*.

EXAMPLE The first impact of the cutback was on the purchasing department, *but* it is now making itself felt in our department as well.

Coordinating conjunctions can join words, **phrases**, or **clauses** of equal rank.

EXAMPLES Only crabs *and* sponges managed to escape the red tide. (words)
Our twin objectives are to increase power *and* to decrease noise levels. (phrases)
The average city dweller in the United States now has 0.17 parts per million of lead in his blood, *and* the amount is increasing yearly. (clauses)

Under normal circumstances, use coordinating conjunctions only to connect **parallel structures**.

copyright

Copyright is the right of exclusive ownership by an author of the benefits resulting from the creation of his or her work. This right gives authors, or others to whom they transfer ownership of copyright, control over the reproduction and dissemination of their works. Once copyrighted, a work cannot be indiscriminately reproduced unless the copyright owner is compensated by royalties or other arrangements.

This right is protected for the life of the author plus 50 years by a federal law that became effective on January 1, 1978. This law protects any work from the moment of its creation, regardless of whether it is ever published and whether it contains a notice of copyright. At the end of the copyright period, the work becomes "public domain" and may be published or reproduced by anyone without permission or compensation to the copyright owner's estate.

The copyright law provides a "fair use" provision that allows teachers, librarians, reviewers, and others to reproduce copyrighted materials for educational and illustrative purposes without compensation to the copyright owners. The law refers to such purposes as "criticism, comment, news reporting, teaching (including multiple copies for classroom use), scholarship, or research." Although the law does not exactly define "fair use," it sets out four criteria for determining whether a given use is fair: (1) the purpose and nature of the use, (2) the amount of material used in relation to the copyrighted work as a whole, (3) the nature of the copyrighted work, and (4) the effect of the use on the potential market for or value of the copyrighted work. Publications that explain the copyright law in detail are available from the Copyright Office, Library of Congress, Washington, D.C. 20559.

If you use copyrighted materials in your written work, be aware of the following guidelines:

1. A small amount of material from a copyrighted source may be used in your written work without permission or payment as long as the use is reasonable and not harmful to the rights of the copyright owner.
2. A work created after January 1, 1978, receives copyright protection whether or not it bears a notice of copyright, although all published materials contain such a notice, usually on the back of the title page. (For an example of a copyright notice, turn to the back of the title page of this book.)
3. All publications of U.S. government agencies are in the public domain—that is, they are *not* copyrighted.
4. One way to determine whether a work from which you wish to quote is in the public domain is to write to the Copyright Office

at the address cited in this entry. There is a search fee for this service.

(See also **plagiarism**.)

correlative conjunctions

Correlative conjunctions are **coordinating conjunctions** that are used in pairs. The correlative conjunctions are *either . . . or, neither . . . nor, not only . . . but also, both . . . and,* and *whether . . . or.*

EXAMPLES The shipment will arrive *either* on Wednesday *or* on Thursday.
The shipment contains the parts *not only* for the seven machines scheduled for delivery this month *but also* for those scheduled for delivery next month.

To add not only symmetry but also logic to your writing, follow correlative conjunctions with parallel sentence elements (such as *on Wednesday* and *on Thursday* in the previous example).

A problem can develop with correlative conjunctions joining a singular **noun** to a plural noun. The problem is with the **verb**. If you use this construction, make the verb agree with the nearest noun.

EXAMPLE Ordinarily, neither the pressure switch nor the relay *coils fail* the test. ("Ordinarily, neither the pressure switch nor the relay coils fails the test" would be grammatically incorrect.)

A common solution is to change the construction of the sentence.

EXAMPLE Ordinarily, the test is failed by neither the pressure switch nor the relay coils.

(See also **parallel structure**.)

correspondence

The process of writing letters involves many of the same steps that go into most other on-the-job tasks. The following list summarizes these steps:

1. Establish your **objective**, and determine your **reader's** attitude and needs.
2. Prepare an **outline**, even if it is only a list of points to be covered in the order you wish to cover them.
3. Write the first draft (see also **writing the draft**).

4. Allow a "cooling" period (time for weaknesses to become obvious).
5. Revise the draft (see also **revision**).
6. Use **proofreading** techniques.

These guidelines will help you write a clear, well-organized letter. (See also **memorandums**.) You might also find it valuable to look at the "Five Steps to Successful Writing" at the front of this book. Keep in mind, however, that one very important element in business letters is the impression they leave on the reader. To convey the right impression—of yourself as well as of your company or organization—you must take particular care with both the **tone** and the **style** of your writing.

TONE

Letters are generally written directly to another person who is identified by name. You may or may not know the person, but never forget that you are writing to another human being. For this reason, letters are always more personal than are **reports** or other forms of business writing. Successful writers find that it helps to imagine their reader sitting across the desk from them as they write; they then write to the reader as if they were talking to him or her in person. This technique helps them keep their language natural. Of course, they adhere to all the guidelines of good writing, including grammar and punctuation. As a letter writer addressing yourself directly to your reader, you are in a good position to take into account your reader's needs. If you ask yourself, "How might I feel if I were the recipient of such a letter?" you can gain some insight into the likely needs and feelings of your reader—and then tailor your message to fit those needs and feelings. Furthermore, you have a chance to build goodwill for your business or organization. Many companies spend millions of dollars to create a favorable public image. A letter to a customer that sounds impersonal and unfriendly can quickly tarnish that image, but a thoughtful letter that communicates sincerity can greatly enhance it.

Suppose, for example, you are a department-store manager who receives a request for a refund from a customer who forgot to enclose the receipt with the request. In a letter to that customer, you might write the following:

EXAMPLE The sales receipt must be enclosed with the merchandise before we can process a refund.

But if you consider how you might keep the customer's goodwill, you might word the request this way:

EXAMPLE Please enclose the sales receipt with the merchandise so that we can send you your refund promptly.

However, you can go one step further as a business-letter writer. You can put the reader's needs and interests at the center of the letter by writing from the reader's perspective. Often, although not always, doing so means using the words *you* and *your* rather than *we, our, I,* and *mine*. That is why the technique has been referred to as the **"you" viewpoint**. Consider the following revision, which is written with the "you" viewpoint:

EXAMPLE So that you can receive your refund promptly, please enclose the sales receipt with the merchandise.

In this example, the reader's benefit and interest is stressed: to get a refund as quickly as possible. By emphasizing the reader's needs, the writer will be more likely to accomplish his or her objective, to get the reader to act.

Obviously, both goodwill and the "you" viewpoint can be overdone. Used thoughtlessly, both techniques can produce a fawning, insincere tone—what might be called "plastic goodwill," as in the following:

CHANGE You're the sort of forward-thinking person whose outstanding good judgment is obvious from your selection of the Model K-50 word processor.

TO Congratulations on selecting the Model K-50 word processor. We believe the K-50 is one of the finest on the market.

Language that is full of false praise will seem insincere and thus be counterproductive.

DIRECT AND INDIRECT PATTERN: GOOD NEWS AND BAD NEWS LETTERS

It is generally more effective to present good news directly and bad news indirectly, because readers form their impressions and attitudes very early in letters. In fact, some readers do not finish reading a letter when bad news is presented first. Other readers may finish a letter that presents bad news at the outset, but they tend to read what follows with a predetermined opinion. They may be skeptical about an explanation, or they may reject a reasonable alternative presented by the writer. Furthermore, even if you must refuse a person's request or say no to someone, you still may wish to work with him or her in the future, and an abruptly phrased rejection early in the letter may prevent you from reestablishing an amicable relationship in the future.

Consider the thoughtlessness of the rejection in Letter 1. Although the letter is direct and uses the pronouns *you* and *your*, the writer has apparently not considered how the reader will feel as she reads the letter. There is no expression of regret that Mrs. Mauer is being rejected for the position nor any appreciation of her efforts in applying for the job. The letter is, in short, rude. The pattern of this letter is (1) bad news, (2) explanation, (3) close.

A better general pattern for "bad news" letters is as follows:

1. Buffer
2. Bad news
3. Goodwill

The "buffer" may be either neutral information or an explanation that makes the bad news *understandable*. Bad news is never pleasant; however, information that either puts the bad news in perspective or makes the bad news seem reasonable maintains goodwill between the writer and the reader. Consider, for example, a revision of the rejection letter, as shown in Letter 2. This letter carries the same disappointing news as the first one does, but the writer is careful to thank the reader for her time and effort, to explain why she was not accepted for the job, and to offer her encouragement in finding a position in another office. Presenting good news is, of course, easier. Present good news early—at the outset, if at all possible. The pattern for "good news" letters should be as follows:

1. Good news
2. Explanation or facts
3. Goodwill

By presenting the good news first, you will increase the likelihood that the reader will pay careful attention to details, and you will achieve goodwill from the start. Letter 3 is a good example of a "good news" letter.

Southtown Dental Center
3221 Ryan Road San Diego, CA 92217
(714) 321-1579

November 11, 19--

Mrs. Barbara L. Mauer
157 Beach Drive
San Diego, CA 92113

Dear Mrs. Mauer:

Your application for the position of dental receptionist at Southtown Dental Center has been rejected. We have found someone more qualified than you.

Sincerely,

Mary Hernandez

Mary Hernandez
Office Manager

MH/bt

Letter 1 Poor "Bad News" Letter

Southtown Dental Center

3221 Ryan Road San Diego, CA 92217
(714) 321-1579

November 11, 19--

Mrs. Barbara L. Mauer
157 Beach Drive
San Diego, CA 92113

Dear Mrs. Mauer:

Buffer Thank you for your time and effort in applying for the position of dental receptionist at Southtown Dental Center.

Bad news Since we need someone who can assume the duties here with a minimum of training, we have selected an applicant with over ten years of experience.

Goodwill I am sure that with your excellent college record you will find a position in another office.

Sincerely,

Mary Hernandez
Office Manager

MH/bt

Letter 2 Courteous and Effective "Bad News" Letter

Southtown Dental Center

3221 Ryan Road San Diego, CA 92217
(714) 321-1579

November 11, 19--

Mrs. Barbara L. Mauer
157 Beach Drive
San Diego, CA 92113

Dear Mrs. Mauer:

Good News

Please accept our offer of the position of dental receptionist at Southtown Dental Center.

Explanation

If the terms we discussed in the interview are acceptable to you, please come in at 9:30 a.m. on November 15. At that time we will ask you to complete our personnel form, in addition to. . . .

Goodwill

Everyone here at Southtown is looking forward to working with you. We all were very favorably impressed with you during your interviews.

Sincerely,

Mary Hernandez

Mary Hernandez
Office Manager
MH/bt

Letter 3 Effective "Good News" Letter

The following tips will help you achieve a tone that builds goodwill with the reader:

1. *Be respectful, not demanding.*

CHANGE Submit your answer in one week.
TO I would appreciate your answer within one week.

2. *Be modest, not arrogant.*

CHANGE My report is thorough, and I'm sure that you won't be able to continue without it.
TO I have tried to be as thorough as possible in my report, and I hope you find it useful.

3. *Be polite, not sarcastic.*

CHANGE I just received the shipment we ordered six months ago. I'm sending it back—we can't use it now. Thanks!
TO I am returning the shipment we ordered on March 12, 19--. Unfortunately, it arrived too late for us to be able to use it.

4. *Be positive and tactful, not negative and condescending.*

CHANGE Your complaint about our prices is way off target. Our prices are definitely not any higher than those of our competitors.
TO Thank you for your suggestion concerning our prices. We believe, however, that our prices are competitive with, and in some cases are below, those of our competitors.

WRITING STYLE

Letter-writing style may vary from informal, in a letter to a close business associate, to formal, or restrained, in a letter to someone you do not know. (Even if you are writing a business letter to a close associate, you should always follow the rules of standard grammar, spelling, and punctuation.)

INFORMAL It worked! The new process is better than we had dreamed.
RESTRAINED You will be pleased to know that the new process is more effective than we had expected.

You will probably find yourself relying on the restrained style more frequently than on the informal one, since an obvious attempt to sound casual, like overdone goodwill, may strike the reader as insincere. Do not adopt such a formal style, however, that your letters read like legal contracts. Using legalistic-sounding words in an effort to impress your reader will make your writing seem stuffy and pompous—and may well irritate your reader.

CHANGE In response to your query, I wish to state that we no longer have an original copy of the brochure requested. Be advised that a photographic copy is enclosed herewith. Address further correspondence to this office for assistance as required.

TO Because we are currently out of original copies of our brochure, I am sending you a photocopy of it. If I can help further, please let me know.

The excessively formal writing style in the original version is full of largely out-of-date business language; expressions like *query, I wish to state, be advised that,* and *herewith* are both old-fashioned and pretentious. Good business letters today have a more personal, down-to-earth style, as the revision illustrates.

The revised version not only is less stuffy but also slightly more concise. Being concise in writing is important, but don't be so concise that you become blunt (or lapse into **telegraphic style**). If you respond to a written request that you cannot understand with "Your request was unclear" or "I don't understand," you will probably offend your reader. Instead of attacking the writer's ability to phrase a request, consider that what you are really doing is asking for more information. Say so.

EXAMPLE I will need more information before I can answer your request. Specifically, can you give me the title and the date of the report you are looking for?

This version is a bit longer, but it is both more polite and more helpful.

ACCURACY

Since a letter is a written record, it must be accurate. Facts, figures, dates, and explanations that are incorrect or misleading may cost time, money, and goodwill. Remember that when you sign a letter, you are responsible for its contents. Always allow yourself time to review a letter before mailing it. When time permits, ask someone who is familiar with its contents to review an important letter. Listen with an open mind to the criticism of others about what you have said, and make any changes you believe necessary.

A second kind of accuracy to check for is the mechanics of writing—**punctuation, grammar**, and **spelling**. Regardless of who types the letter, if you sign it, *you* are responsible for its contents and form.

APPEARANCE

The appearance of a business letter may be crucial in influencing a recipient who has never seen you. The rules for preparing a neat, attractive letter are not difficult to master, and they are important—particularly if you type your own letters. Type on unruled white bond paper of standard size, and use envelopes of the same quality. Center the letter on the page vertically and horizontally—a "picture frame" of white space should surround the contents. When you use your company's letterhead stationery, consider the bottom of the letterhead as the top edge of the paper.

PARTS OF A BUSINESS LETTER

If your employer recommends or requires a particular **format** and typing style, use it. You may also wish to consult a secretarial guide, such as *The Gregg Reference Manual* by William A. Sabin (McGraw-Hill). Otherwise, follow the guidelines provided here, and review the illustrations for placement with full- and modified-block style (pages 153-154).

Heading. The writer's full address (street, city, state, and ZIP code) and the date are given in the heading. Because the writer's name appears at the end of the letter, it need not be included in the heading. Words like *street, avenue, first,* or *west* should be spelled out rather than abbreviated. You may either spell out the name of the state in full or use the U.S. Postal Service abbreviations given in **abbreviations**. The date usually goes directly beneath the last line of the address. Do not abbreviate the name of the month.

EXAMPLE 1638 Parkhill Drive East
Great Falls, MT 59407
April 8, 19—

Align the heading on the page at the center line. If you are using company letterhead that gives the address, type in only the date, two spaces below the last line of printed copy.

The Inside Address. The recipient's full name and address are given in the inside address. You can begin the inside address two spaces below the date if the letter is long, or four spaces below if the letter is quite short. The inside address should be aligned with the left margin—and the left margin should be at least one inch wide. Include the reader's full name and title (if you know them) and his or her full address, including ZIP code.

EXAMPLE Ms. Gail Silver
Production Manager
Quicksilver Printing Company
14 President Street
Sarasota, FL 33546

The Salutation. Place the salutation, or greeting, two spaces below the inside address, also aligned with the left margin. In most business letters, the salutation contains the recipient's title (*Mr., Ms., Dr.*, and so on) and last name, followed by a **colon**. If you are on a first-name basis with the recipient, include his or her title and full name on the inside address, but use only the first name in the salutation. Notice that the titles *Mr., Ms., Mrs.*, and *Dr.* may be abbreviated but that other titles, such as *Captain* or *Professor*, should always be spelled out.

EXAMPLES Dear Ms. Silver:
Dear Dr. Lee:
Dear Captain Ortiz:
Dear Professor Murphy:
Dear George:

Address women without a professional title as *Ms.*, whether they are married or unmarried. If a woman has expressed a preference for *Miss* or *Mrs.*, however, honor her preference. If you do not know whether the recipient is a man or a woman, you may use a title appropriate to the context of the letter. The following are examples of the kinds of titles you may find suitable:

EXAMPLES Dear Customer: (Letter from a department store)
Dear Homeowner: (Letter from an insurance agent soliciting business)
Dear Parts Manager: (Letter to an auto-parts dealer)

In the past, correspondence to large companies or organizations was customarily addressed to "Gentlemen." Today, however, writers who do not know the name or title of the recipient often address the letter to an appropriate department or identify the subject in a "subject line" and use no salutation.

EXAMPLE National Business Systems
501 West National Avenue
Minneapolis, MN 55407

Attention: Customer Relations Department

I am returning three calculators that failed to operate. . . .

EXAMPLE National Business Systems
501 West National Avenue
Minneapolis, MN 55407

Subject: Defective Parts for SL-100 Calculators

I am returning three calculators that failed to operate. . . .

When a person's first name could be either feminine or masculine, one solution is to use both the first and the last name in the salutation.

EXAMPLE Dear Pat Smith:

The Body. The body of the letter should begin two spaces below the salutation (or directly below the heading if there is no salutation). Single-space within paragraphs, and double-space between paragraphs. If a letter is very short and you want to suggest a fuller appearance, you may instead double-space throughout and indicate paragraphs by indenting the first line of each paragraph five spaces from the left. The right margin should be approximately as wide as the left margin. (In very short letters, you may increase both margins to about an inch and a half.)

Two very important elements in the body are the **opening** and **closing**. One effective way to arrange your letter is to open with a short **paragraph**, followed by one or more longer paragraphs for the message and another short paragraph for concluding. This is called a diamond arrangement, and it can help focus your letter whether you are using the direct or indirect pattern. Never underestimate the importance of the opening and closing; in fact, their positions of **emphasis** make them particularly significant.

In your opening you should identify your subject so as to focus its relevance for the reader. Remember that your reader may not immediately recognize or see the importance of your topic—indeed, he or she may be preoccupied with some other business. Therefore, it is important to focus his or her attention on the subject at hand. Be particularly careful to get directly to the point; leave out any less important details.

EXAMPLE Yesterday, I received your letter and the defective tuner, number AJ 50172, that you described. I sent the tuner to our laboratory for tests.

Carol Moore, our lead technician, reports that preliminary tests indicate. . . .

Your closing should let the reader know what he or she should do next or establish goodwill—or often both.

EXAMPLE Thanks again for the report, and let me know if you want me to send you a printout of the tests.

Because a closing is in a position of emphasis, be especially careful to avoid **clichés**. Of course, some very commonly used closings are so precise that they are hard to replace.

EXAMPLES Thank you for your advice.

If you have further questions, please let me know.

The Complimentary Closing. Type the complimentary closing two spaces below the body. Use a standard expression like *Yours truly, Sincerely,* or *Sincerely yours*. (If the recipient is a friend as well as a business associate, you can use a less formal closing, such as *Best wishes* or *Best regards*.) Capitalize only the initial letter of the first word, and follow the expression with a **comma**. Type your full name four spaces below, aligned with the closing at the left. On the next line you may type in your business title, if it is appropriate to do so. Write your signature in the space between the complimentary closing and your typed name. If you are writing to someone with whom you are on a first-name basis, it is acceptable to sign only your given name; otherwise, sign your full name.

EXAMPLE Sincerely,

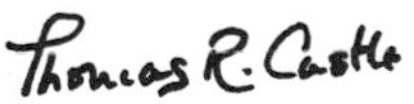

Thomas R. Castle
Treasurer

Sometimes the complimentary closing is followed by the name of the firm. Then comes the actual signature, between the name of the firm and the typed name.

EXAMPLE Sincerely yours,
VIKING SUPPLY COMPANY, INC.

Laura A. Newland
Controller

A Second Page. If a letter requires a second page, always carry at least two lines of the body text over to that page; do not use a continuation page to type only the letter's closing. The second page should be typed on plain paper of quality equivalent to that of the letterhead stationery. It should have a heading with the recipient's name, the

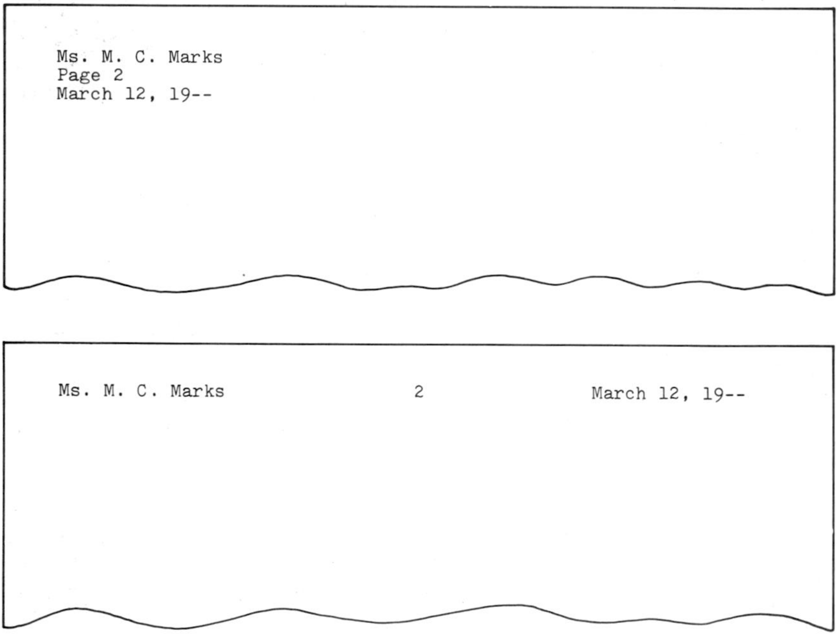

Ms. M. C. Marks
Page 2
March 12, 19--

Ms. M. C. Marks 2 March 12, 19--

Figure 1 Headings for the Second Page of a Letter

page number, and the date. The heading may go in the upper left-hand corner or across the page, as shown in Figure 1.

Additional Information. Business letters sometimes require the typist's initials, an enclosure notation, or a notation that a copy of the letter is being sent to one or more people. Place any such information at the left margin, two spaces below the last line of the complimentary closing in a long letter, four spaces below in a short letter.

The *typist's initials* should follow the letter writer's initials, and the two sets of initials should be separated by either a colon or a slash. The writer's initials should be in capital letters, and the typist's initials should be in lowercase letters. (When the writer is also the typist, no initials are needed.)

EXAMPLES CBG:pbg
APM/sjl

Enclosure notations indicate that the writer is sending material along with the letter (an invoice, an article, and so on). They may take several forms, as illustrated below; choose the form that seems most helpful to your readers.

EXAMPLES Enclosure: Preliminary report invoice
Enclosures (2)
Enc. (Encs.)

Enclosure notations are included in long, formal letters or in any letters where the enclosed items would not be obvious to the reader. Remember, though, that an enclosure notation cannot stand alone. You must mention the enclosed material in the body of the letter.

Copy notations tell the reader that a copy of the letter is being sent to one or more named individuals. You should add a copy notation for both carbon copies and photocopies.

EXAMPLE cc: Ms. Marlene Brier
Mr. David Williams
Ms. Bonnie Ng
Ms. Robin Horton

A business letter may, of course, contain all three items of additional information.

EXAMPLE Sincerely yours,

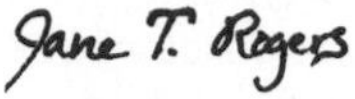

Jane T. Rogers

JTR/pst
Enclosure: Preliminary report invoice
cc: Ms. Marlene Brier
Mr. David Williams
Ms. Bonnie Ng
Ms. Robin Horton

Letter 4 shows a typical business letter using most of the standard elements.

LETTER FORMATS

The two most common formats of business letters are the full block style (shown in Letter 3) and the modified block style. The full block style, though easier to type because every line begins at the left margin, is suitable only with business letterhead stationery. In the modified block style, the return address, the date, and the complimentary closing all begin at the center of the page, and the other elements are aligned at the left margin. All other letter styles are variations of these two styles. Again, if your employer recommends or requires a particular style, follow it carefully. Otherwise, choose a style and follow it consistently. Letter 5 is an example of the modified block style.

For information on specific types of correspondence, see **acceptance letters, acknowledgment letters, adjustment letters, application letters, complaint letters, inquiry letters, reference letters, refusal letters, resignation letter or memorandum, technical information letters**, and **memorandums**.

count nouns (See **nouns**.)

cover letters (See **transmittal letters** and **application letters**.)

Letterhead	EVANS & ASSOCIATES 520 Niagara Street Lexington, KY 40502 (502) 787-1175 TELEX 5-72118
Date	May 15, 19--
Inside address	Mr. George W. Nagel Director of Operations Boston Transit Authority 57 West City Avenue Boston, MA 02110
Salutation	Dear Mr. Nagel:
Body	Enclosed is our final report evaluating the safety measures for the Boston Intercity Transit System. We believe the report covers the issues you raised and that it is self-explanatory. However, if you have any further questions, we would be happy to meet with you at your convenience. We would also like to express our appreciation to Mr. L. K. Sullivan of your committee for his generous help during our trips to Boston.
Complimentary close	Sincerely,
Signature	Carolyn Brown
Typed name	Carolyn Brown, Ph.D.
Title	Director of Research
Additional information	CB/ls Enclosure: Final Safety Report cc: ITS Safety Committee Members

Letter 4 Business Letter (Full Block) with Standard Elements

Center

Heading just right of center

3814 Oak Lane
Lexington, KY 40514
December 8, 19--

Inside address

Dr. Carolyn Brown
Director of Research
Evans & Associates
520 Niagara Street
Lexington, KY 40502

Salutation

Dear Dr. Brown:

Body

Thank you very much for allowing me to tour your testing facilities. The information I gained from the tour will be of great help to me in preparing the report for my class at Marshall Institute. The tour has also given me some insight into the work I may eventually do as a laboratory technician.

I especially appreciated the time and effort Steve Dement spent in showing me your facilities. His comments and advice were most helpful.

Again, thank you.

Complimentary close aligned with heading

Sincerely,

Signature

Leslie Warden

Typed name

Leslie Warden

Letter 5 Modified Block Style

credible/creditable

Something is *credible* if it is believable.

EXAMPLE The statistics in this report are *credible.*

Something is *creditable* if it is worthy of praise or credit.

EXAMPLE The chief engineer did a *creditable* job.

criterion/criteria/criterions

Criterion means "an established standard for judging or testing." *Criteria* and *criterions* both are acceptable plural forms of *criterion*, but *criteria* is generally preferred.

EXAMPLE In evaluating this job, we must use three *criteria.* The most important *criterion* is quality of workmanship.

critique

A *critique* (**noun**) is a written or oral evaluation of something. Avoid using *critique* as a **verb** meaning "criticize."

CHANGE Please *critique* his job description.
TO Please prepare a *critique* of his job description.

D

dangling modifiers

Verbal phrases **(gerund, participle, infinitive)** that do not clearly and logically refer to the proper **noun** or **pronoun** are called *dangling modifiers*. Dangling modifiers usually appear at the beginning of a sentence as an introductory phrase.

CHANGE While eating lunch in the cafeteria, the computer malfunctioned. (The problem, of course, is that the sentence neglects to mention *who* was eating lunch in the cafeteria.)

TO While *the operator* was eating lunch in the cafeteria, the computer malfunctioned.

They can, however, appear at the end of the sentence as well. Probably the clearest example of this type of dangling modifier is the sentence "The mouse was caught *using the trap*." For a more technically oriented example, consider the following sentence:

CHANGE The program gains in efficiency by eliminating the superfluous instructions. (*Who* is to eliminate the instructions? The program can hardly do it! An action is stated, but no one is identified who could perform the stated action.)

TO The program gains in efficiency *when you* eliminate the superfluous instructions.

OR *When you* eliminate the superfluous instructions, the program gains in efficiency.

Edmund Weiss, in *The Writing System for Engineers and Scientists*, offers this example: "The car was designed by our new safety engineer with an impact-absorbing rear end."

Dangling modifiers are often caused by overuse of the passive **voice**. In an active-voice clause, the doer of the action appears in front of the verb; since the doer of the action is normally the intended subject of the introductory phrase, the phrase does not dangle. If the clause is in the passive voice, however, the doer of the action appears after the verb or is not identified, thereby causing the phrase to dangle.

CHANGE To evaluate the feasibility of the project, the centralized plan will be compared with the present system of dispersing facility sites. (The main clause is in the passive voice, and the doer of the action expressed by the verb *compared* is not identified; therefore, the introductory phrase "To evaluate the feasibility of the project" has nothing that it can logically modify because there is no one in the main clause who could possibly evaluate anything.)

TO To evaluate the feasibility of the project, *the committee* will compare the centralized plan with the present system of dispersing facility sites.

DANGLING GERUND PHRASES

A dangling gerund phrase can be corrected by adding a noun or a pronoun for it to modify.

CHANGE After finishing the research, the job was easy.
TO After finishing the research, *we* found the job to be easy.

Another way to correct a dangling gerund phrase is to make the **phrase** into a **clause.**

CHANGE After finishing the research, the job was easy.
TO After we finished the research, the job was easy.
OR The job was easy after we finished the research.

DANGLING PARTICIPIAL PHRASES

A dangling **participial phrase** can be corrected by providing a subject that stands near the phrase.

CHANGE Keeping busy, the afternoon passed swiftly.
TO Keeping busy, *I* felt that the afternoon passed swiftly.

A dangling participial phrase may also be corrected by making the phrase into a clause.

CHANGE Keeping busy, the afternoon passed swiftly.
TO Because I kept busy, the afternoon passed swiftly.

CHANGE Entering the gate, the administration building is visible.
TO As you enter the gate, the administration building is visible.

DANGLING INFINITIVE PHRASES

Correct a dangling **infinitive phrase** by supplying the missing noun or pronoun that provides the subject of the infinitive phrase.

CHANGE To improve typing skills, practice is needed.
TO To improve typing skills, *you* must practice.

CHANGE To evaluate the feasibility of the project, the centralized plan will be compared with the present system of dispersing facility sites.
TO To evaluate the feasibility of the project, the *committee* will compare the centralized plan with the present system of dispersing facility sites.

DANGLING SUBORDINATE CLAUSES

Occasionally a **subject** and **verb** are omitted from a **dependent clause,** forming what is known as an *elliptical clause.* If the omitted subject of the elliptical clause is not the same as the subject of the main clause, the construction will dangle. Simply adding the subject and verb to the elliptical clause solves the problem.

CHANGE When ten years old, his father started the company.
TO When *Bill* was ten years old, his father started the company.
OR Bill was ten years old when his father started the company.

(See also **misplaced modifiers** and **verbals**.)

dashes

The dash is a versatile, yet limited, mark of **punctuation.** It is versatile because it can perform all the duties of punctuation (to link, to separate, and to enclose). It is limited because it is an especially emphatic mark that is easily overused. Use the dash cautiously, therefore, to indicate more informality or **emphasis** (a dash gives an impression of abruptness) than would be achieved by the conventional punctuation marks. In some situations, a dash is required; in others, a dash is a forceful substitute for other marks.

A dash can emphasize a sharp turn in thought.

EXAMPLE The project will end January 15—unless the company provides additional funds.

A dash can indicate an emphatic pause.

EXAMPLES Consider the potential danger of a household item that contains mercury—a very toxic substance.
The job will be done—after we are under contract.

Sometimes, to emphasize contrast, a dash is also used with *but.*

EXAMPLE We may have produced work more quickly—but the result was not as good.

A dash can be used before a final summarizing statement or before **repetition** that has the effect of an afterthought.

EXAMPLE It was hot near the ovens—steaming hot.

Such a thought may also complete the meaning.

EXAMPLE We try to speak as we write—or so we believe.

A dash can be used to set off an explanatory or **appositive** series.

EXAMPLE Three of the applicants—John Evans, Mary Fontana, and Thomas Lopez—seem well qualified for the job.

Dashes set off parenthetical elements more sharply and emphatically than do **commas.** Unlike dashes, **parentheses** tend to reduce the importance of what they enclose. Compare the following sentences:

EXAMPLES Only one person—the president—can authorize such activity.
Only one person, the president, can authorize such activity.
Only one person (the president) can authorize such activity.

Use dashes for **clarity** when commas appear within a parenthetical element; this avoids the confusion of too many commas.

EXAMPLE Retinal images are patterns in the eye—made up of light and dark shapes in addition to areas of color—but we do not see patterns; we see objects.

The first word after a dash is never capitalized unless it is a proper noun. In typing, use two consecutive **hyphens** (--) to indicate a dash, with no spaces before or after the hyphens.

data/datum

The debate over whether *data* should be treated as a plural or a collective singular noun continues. In much technical writing, **data** is considered a collective singular. In formal scientific and scholarly writing, however, *data* is generally used as a plural, with *datum* as the singular form.

EXAMPLES The *data are* voluminous in support of a link between smoking and lung cancer. (formal)
The *data* is now ready to be evaluated. (less formal)

Base your decision on whether your **reader** should consider the data as a single collection or as a group of individual facts. Whatever you decide, be sure that your **pronouns** and **verbs** agree in **number** with the selected **usage.**

EXAMPLES The *data are* voluminous. *They indicate* a link between smoking and lung cancer.
The *data is* now ready for evaluation. *It is* in the mail.

dates

In business and industry, dates have traditionally been indicated by the month, day, and year, with a **comma** separating the figures.

EXAMPLE October 26, 19--

The day-month-year system used by the military does not require commas.

EXAMPLE 26 October 19--

A date can be written with or without a comma following the year if the date is in the month-day-year format.

EXAMPLES October 26, 19--, was the date the project began.
October 26, 19-- was the date the project began.

If the date is in the day-month-year format, do not set off the date with commas.

EXAMPLE 26 October 19-- was the date the project began.

The strictly numerical form for dates (10/26/87) should be used sparingly, and never on business letters or formal documents, since it is less immediately clear. When this form is used, the order in American usage is always month/day/year. For example, 5/7/87 is May 7, 1987.

CENTURIES

Confusion often occurs because the spelled-out names of centuries do not correspond to the numbers of the years. (See also **numbers.**)

EXAMPLES The twentieth century is the 1900s (1900–1999).
The nineteenth century is the 1800s (1800–1899).
The fifth century is the 400s (400–499).

When the century is written as a noun, do not use a hyphen.

EXAMPLE The sixteenth century produced great literature.

When the century is written as an adjective, however, use a hyphen.

EXAMPLE Twentieth-century technology must rely on clear technical communications.

decided/decisive

When something is *decided,* it is "clear-cut," "without doubt," "unmistakable." When something is *decisive,* it is "conclusive" or "has the power to settle a dispute."

EXAMPLE The executive committee's *decisive* action gave our firm a *decided* advantage over the competition.

decreasing-order-of-importance method of development

Decreasing order of importance is a **method of development** in which the writer's major points are arranged in descending order of importance. It begins with the most important fact or point, then goes to the next most important, and so on, ending with the least important. It is an especially appropriate method of development for a **report** addressed to a busy decision maker, who may be able to reach a decision after considering only the most important points, or for a report addressed to various **readers,** some of whom may be interested only in the major points and others in all of the points.

The advantages of this method of development are that (1) it gets the reader's attention immediately by presenting the most important point first, (2) it makes a strong initial impression, and (3) it ensures that the hurried reader will not miss the most important point.

The example shown in Figure 1 uses decreasing order of importance.

To: Mary Vincenti, Chief, Personnel Department
From: Frank W. Russo, Chief, Claims Department FWR
Date: November 13, 19--
Subject: Selection of Chief of the Claims Processing Section

The most qualified candidate for Chief of the Claims Processing Section is Mildred Bryand, who is at present Acting Chief of the Claims Processing Section. In her twelve years in the Claims Department, Ms. Bryand has gained wide experience in all facets of the department's operations. She has maintained a consistently high production record and has demonstrated the skills and knowledge that are required for the supervisory duties she is now handling in an acting capacity. A number of additional qualifications also make her the best candidate: her valuable contributions to and cooperation during the automation of the department's claims records; her performance on several occasions as acting chief in other sections of the Claims Department; the continual rating of "outstanding" she has received in all categories of her job-performance appraisals.

Michael Bastik, Claims Coordinator, our second choice, also has strong potential for the position. An able administrator, he has been with the company for seven years. For the past year he has been enrolled in several management training courses at the university. He is ranked second, however, because he lacks supervisory experience and because his most recent work has been with the department's maintenance and supply components. He would be the best person to fill many of Mildred Bryand's responsibilities if she should be made full-time Chief of the Claims Processing Section.

Jane Fine, our third-ranking candidate, has shown herself a skilled administrator in her three years with the Claims Consideration Section. Despite her obvious potential, my main objection to her is that compared with the other top candidates, she lacks the breadth of experience in claims processing that would be required of someone responsible for managing the Claims Processing Section. Jane Fine also lacks on-the-job supervisory experience.

mo

Figure 1 Decreasing Order of Importance

de facto/de jure

De facto means that something "exists" or is "a fact" and therefore is accepted for practical purposes. *De jure* means that something "legally" exists.

EXAMPLE The law states that no signs should be erected along Highway 127. Store owners have disregarded this law, and, therefore, many signs exist along Highway 127. The presence of the signs along Highway 127 is *de facto* but not *de jure*—their presence is "a fact," but it is not "lawful."

defective/deficient

If something is *defective*, it is faulty.

EXAMPLE The wiring was *defective.*

If something is *deficient*, it lacks a necessary ingredient.

EXAMPLE The compound was found to be *deficient* in calcium.

defining terms

One of the most basic rules of good writing is to be sure that your **readers** understand and follow what they are reading. You must always keep in mind their familiarity with the topic and be careful to define any terms that they may not understand.

Terms can be defined either formally or informally, depending on your purpose and on your readers. (For techniques on expanding a definition to explain the meaning of a term, see **definition method of development.**)

A *formal definition* is a form of classification. A term is defined by placing it in a category and then identifying what features distinguish it from other members of the same category. The following are several examples of formal definitions.

Term	*Category*	*Distinguishing Features*
A spoon is	an eating utensil	that consists of a small shallow bowl on the end of a handle.
An auction is	a public sale	in which property passes to the highest bidder through successively increased offers.

Term	*Category*	*Distinguishing Features*
An annual is	a plant	that completes its life cycle, from seed to natural death, in one growing season.
A contract is	a binding agreement	between two or more persons or parties.
A lease is	a contract	that conveys real estate for a term of years at a specified rent.

An *informal definition* explains a term by giving a familiar word or **phrase** as a **synonym.**

EXAMPLES An *invoice* is a bill.
Many states have set up wildlife *habitats* (or living spaces).
Plants have a *symbiotic,* or mutually beneficial, relationship with certain kinds of bacteria.

Definitions should normally be stated positively; focus on what the term is rather than on what it is not. For example, "In a legal transaction, real property is not personal property," although a true statement, does not tell the reader exactly what real property *is*. The definition could just as easily be stated positively: "Real property is legal terminology for the right or interest a person has in land and the permanent structures on that land." (For a discussion of when negative definitions are appropriate, see **definition method of development.**)

When writing a definition, keep in mind the following pitfalls that may result in confusing, inaccurate, or incomplete definitions.

Avoid circular definitions, which merely restate the term to be defined and therefore fail to clarify it.

CIRCULAR *Spontaneous combustion* is fire that begins spontaneously.
REVISED *Spontaneous combustion* is the self-ignition of a flammable material through a chemical reaction like oxidation and temperature buildup.

Avoid "is when" and "is where" definitions. Such definitions fail to include the category and are too indirect: They say when or where rather than what.

"IS WHEN" A *contract* is when two or more people agree to something.

REVISED A *contract* is a binding agreement between two or more people.

"IS WHERE" A *day-care center* is where working parents can leave their preschool children during the day.

REVISED A *day-care center* is a facility at which working parents can leave their preschool children during the day.

Avoid definitions that include terms that may be unfamiliar to your readers. Even informal writing will occasionally require terms used in a special sense that your readers may not understand; such terms should always be defined.

EXAMPLE In these specifications the term *safety can* refers to an approved container of not more than five-gallon capacity having a spring-closing spout cover designed to relieve internal pressure when the can is exposed to fire.

definite/definitive

Definite and *definitive* both apply to what is "precisely defined," but *definitive* more often refers to what is complete and authoritative.

EXAMPLE When the committee took a *definite* stand, the president made it the *definitive* company policy.

definition method of development

To define something is basically to identify precisely its fundamental qualities. In technical writing, as in all writing, clear and accurate definitions are critical. Sometimes simple, dictionary-like definitions suffice (see **defining terms**), but at other times definitions need to be expanded with additional details, examples, comparisons, or other explanatory devices.

When more than a **phrase** or a sentence or two is needed to explain an idea, use an *extended definition,* which explores a number of qualities of the item being defined. How an extended definition is developed depends on your reader's needs and on the complexity of the subject. A **reader** familiar with a **topic** or an area might be able to handle a long, fairly technical definition, whereas one less familiar with a topic would require simpler language and more basic information.

Probably the easiest way to give an extended definition is with specific examples. Listing examples gives your reader easy-to-picture details that help him or her to see and thus to understand the term being defined.

EXAMPLE Form, which is the shape of landscape features, can best be represented by both small-scale features, such as *trees and shrubs,* and by large-scale elements, such as *mountains and mountain ranges.*

Another useful way to define a difficult concept, especially when you are writing for nonspecialists, is to use an **analogy** to link the unfamiliar concept with a simpler or more familiar one. An analogy is a resemblance in certain aspects between things otherwise unalike, and it can help the reader understand an unfamiliar term by showing its similarities with a more familiar one. Defining radio waves in terms of the length (long) and frequency (low), a writer develops an analogy to show why a low frequency is advantageous.

EXAMPLE The low frequency makes it relatively easy to produce a wave having virtually all its power concentrated at one frequency. Think, for example, of a group of people lost in a forest. If they hear sounds of a search party in the distance, they all will begin to shout for help in different directions. Not a very efficient process, is it? But suppose that all the energy which went into the production of this noise could be concentrated into a single shout or whistle. Clearly the chances that the group will be found would be much greater.

Some terms are best defined by an explanation of their causes. Writing in a professional journal, a nurse describes an apparatus used to monitor blood pressure in severely ill patients. Called an indwelling catheter, the device displays blood-pressure readings on an oscilloscope and on a numbered scale. Users of the device, the writer explains, must understand what a *dampened wave form* is.

EXAMPLE The *dampened wave form*, the smoothing out or flattening of the pressure wave form on the oscilloscope, is *usually caused by an obstruction* that prevents blood pressure from being freely transmitted to the monitor. The obstruction may be a *small clot or bit of fibrin* at the catheter tip. More likely, *the catheter tip* has become *positioned against the artery wall* and is preventing the blood from flowing freely.

Sometimes a formal definition of a concept can be made simpler by breaking the concept into its component parts. In the following example, the formal definition of *fire* is given in the first paragraph, and the component parts are given in the second:

EXAMPLE Fire is the visible heat energy released from the rapid oxidation of a fuel. A substance is "on fire" when the release of heat energy from the oxidation process reaches visible light levels.

The classic fire triangle illustrates the elements necessary to create fire: *oxygen, heat,* and *burnable material* or *fuel.* Air provides sufficient oxygen for combustion; the intensity of the heat needed to start a fire depends on the characteristics of the burnable material or fuel. A burnable substance is one that will sustain combustion after an initial application of heat to start it.

Under certain circumstances, the meaning of a term can be clarified and made easier to remember by an exploration of its origin. Scientific and medical terms, because of their sometimes unfamiliar Greek and Latin roots, benefit especially from an explanation of this type. Tracing the derivation of a word can also be useful when you want to explain why a word has favorable or unfavorable associations—particularly if your goal is to influence your reader's attitude toward an idea or an activity.

EXAMPLE Efforts to influence legislation generally fall under the head of *lobbying,* a term that once referred to people who prowl the lobbies of houses of government, buttonholing lawmakers and trying to get them to take certain positions. Lobbying today is all of this, and much more, too. It is a respected—and necessary—activity. It tells the legislator which way the winds of public opinion are blowing, and it helps inform him of the implications of certain bills, debates, and resolutions he must contend with.

—Bill Vogt, *How to Build a Better Outdoors* (New York: David McKay Company, Inc., 1978), p. 93.

Sometimes it is useful to point out what something is *not* in order to clarify what it is. A negative definition is effective only when the reader is familiar with the item with which the defined item is contrasted. If you say *x* is not *y*, your readers must understand the meaning of *y* for the explanation to make sense. In a crane operators' manual, for instance, a negative definition is used to show that for safety reasons, a hydraulic crane cannot be operated in the same manner as a lattice boom crane.

EXAMPLE A hydraulic crane is *not* like a lattice boom crane in one very important way. In most cases, the safe lifting capacity of a lattice boom crane is based on the *weight needed to tip the machine.* Therefore, operators of friction machines sometimes depend on signs that the machine might tip to warn them of impending danger.

This practice is very dangerous with a hydraulic crane. . . .

—*Operator's Manual* (Model W-180), Harnishfeger Corporation.

demonstrative adjectives

The demonstrative adjectives are *this, these, that,* and *those*. They are **adjectives** that "point to" the thing they modify, specifying its position in space or time. *This* and *these* specify closer position; *that* and *those* specify a more remote position.

EXAMPLES *This* proposal is the one we accepted.
That proposal would have been impracticable.
These problems remain to be solved.
Those problems are not insurmountable.

Notice that *this* and *that* are used with singular **nouns** and that *these* and *those* are used with plural nouns. Demonstrative adjectives often cause problems when they modify the nouns *kind, type,* and *sort.* Demonstrative adjectives used with these nouns should agree with them in **number.**

EXAMPLES this kind/these kinds
that type/those types
this sort/these sorts

Confusion often develops when the **preposition** *of* is added ("this kind *of*," "these kinds *of*") and the **object** of the preposition is not made to conform in number to the demonstrative adjective and its noun.

CHANGE *This kind* of hydraulic *cranes* is best.
TO *This kind* of hydraulic *crane* is best.

CHANGE *These kinds* of hydraulic *crane* are best.
TO *These kinds* of hydraulic *cranes* are best.

The error can be avoided by remembering to make the demonstrative adjective, the noun, and the object of the preposition—all three—agree in number. This **agreement** not only makes the sentence correct but also more precise.

Using demonstrative adjectives with words like *kind, type,* and *sort* can easily lead to vagueness. It is better to be more specific.

CHANGE It was *kind of* a bad year for the firm.
TO The firm's profits were down 10 percent from last year.

Do not use the pronoun *this* and *that* to make broad or ambiguous references to a whole sentence or to an abstract thought. (See also **pronoun reference.** And, for recommendations on the placement and usage of demonstrative adjectives, see **adjectives.**)

demonstrative pronouns

Demonstrative pronouns are actually the same words as **demonstrative adjectives**, but they differ in **usage** (a demonstrative adjective modifies a **noun,** a demonstrative pronoun substitutes for a noun). The antecedent of a demonstrative pronoun must either be the last noun of the preceding sentence—not the idea of the sentence—or be clearly identified within the same sentence, as in the following examples:

EXAMPLES *This* is the *specification* as we wrote it.
These are the *statistics* we needed to begin the project.
That was a *proposal* for drilling a water well.
Those were *salesmen.*

Notice that as with demonstrative adjectives, *this* and *that* are used for singular nouns and *these* and *those* for plural nouns. Do not use the pronoun *this* and *that* to make broad or ambiguous references to a whole sentence or to an abstract thought. (See also **pronoun reference.**) One way to avoid the problem is to insert the noun immediately after the pronoun.

CHANGE The inadequate quality-control procedure has resulted in an equipment failure. *This* is our most serious problem at present.
TO The inadequate quality-control procedure has resulted in an equipment failure. *This failure* is our most serious problem at present.

dependent clauses

A dependent (or subordinate) clause is a group of words that has a **subject** and a **predicate** but nonetheless depends on a main **clause** to complete its meaning. A dependent clause can function in a sentence as a **noun,** an **adjective,** or an **adverb.**

As nouns, dependent clauses may function in the sentence as subjects, **objects,** or **complements.**

EXAMPLES *That human beings can learn to control their glands and internal organs by direct or indirect means* is an established fact. (subject)

I learned *that drugs ordered by brand name can cost several times as much as drugs ordered by generic name.* (direct object)

The trouble is *that we cannot finish the project by May.* (subjective complement)

As adjectives, dependent clauses can modify nouns or **pronouns.** Dependent clauses are often introduced by **relative pronouns** and **relative adjectives** (*who, which, that*).

EXAMPLE The man *who called earlier* is here. (modifying man)

As adverbs, dependent clauses may express relationships of time, cause, result, or degree.

EXAMPLES You are making an investment *when you buy a house.* (time)

A title search was necessary *because the bank would not issue a loan without one.* (cause)

The cost of financing a home will be higher *if discount points are charged.* (result)

Monthly mortgage payments should not be much more *than a person earns in one week.* (degree)

Dependent clauses are useful in making the relationship between thoughts more precise and succinct than if the ideas were presented in a series of **simple sentences** or **compound sentences.**

CHANGE The sewage plant is located between Millville and Darrtown. Both villages use it. (two thoughts of approximately equal importance)

TO The sewage plant, *which is located between Millville and Darrtown,* is used by both villages. (one thought subordinated to the other)

CHANGE Title insurance is a policy issued by a title insurance company, *and* it protects against any title defects, such as outside claimants. (two thoughts of approximately equal importance)

TO Title insurance, *which* protects against any title defects such as outside claimants, is a policy issued by a title insurance company. (one thought subordinated to the other)

Subordinate clauses are especially effective, therefore, for expressing thoughts that describe or explain another statement. Too much **subordination,** or a string of dependent clauses, however, may be worse than none at all.

CHANGE Since interest rates on conventional loans tend to follow general conditions in the money market, *if* money is plentiful for lending and the demand for loans is low, interest rates will be forced down, *whereas* if money is scarce and the demand for loans is high, interest rates will be forced up.

TO Interest rates on conventional loans tend to follow general conditions in the money market: if money is plentiful for lending and the demand for loans is low, interest rates will be forced down; if money is scarce and the demand for loans is high, interest rates will be forced up.

CHANGE He had selected teachers *who* taught classes *that* had a slant *that* was specifically directed toward students *who* intended to go into business.

TO He had selected teachers *who* taught classes *that* were specifically directed to business students.

description

Effective description uses words to transfer a mental image from the writer's mind to the reader's. The key to effective description is the accurate presentation of details. To select appropriate details, determine what use your **reader** will make of the description. Will it be used to identify an object? Will it be used to repair or assemble the device you are describing?

The following **paragraph** describes the assembly of an electronic typewriter that feeds coded tape or punch cards into the machine. Intended for the typewriter mechanic, this description includes an **illustration** of the assembly mechanism (Figure 1).

Number of features and relative size differences noted

Use of analogy

The Die Block Assembly consists of two machined block sections, eight Code Punch Pins, and a Feed Punch Pin. The larger section, called the Die Block, is fashioned of a hard, noncorrosive beryllium-copper alloy. It houses the eight Code Punch Pins and the smaller Feed Punch Pin in nine finely machined guide holes. The guide holes at the upper part of the Die Block are made smaller to conform to the thinner tips of the Punch Pins. Extending over the top of the Die Block and secured to it at one end is a smaller, arm-like block called the Stripper Block. The Stripper Block is made from hardened tool steel, and it also has been drilled through with nine finely machined guide holes. It is carefully fitted to the Die Block at the factory so that its holes will be precisely above those in the Die Block and so that the space left between the blocks will measure .015" (plus or minus .003"). This space should be barely wide enough to allow the passage of a tape or edge-card. It is here, as the Punch Pins are driven up through the media and into the Stripper Block, that the actual cutting, or Definition punching, of the code and feed holes takes place. The residue (chad) from the hole punching operation is pushed out through the top of the Stripper Block and guided out of the assembly by means of a plastic Chad Collector and Chad Collector Extender.

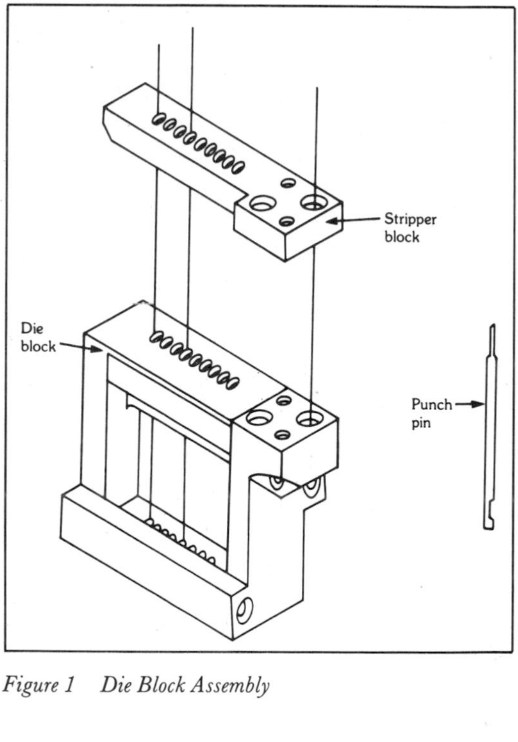

Figure 1 Die Block Assembly

Note that this description concentrates on the number of pieces—their sizes, shapes, dimensions—and on their relationship to one another to perform their function. It also specifies the materials of which the hardware is made. Because the description is illustrated (with identifying captions) and is intended for technicians who have been trained on the equipment, it does not require the use of "bridging devices" to explain the unfamiliar in relation to the familiar. Even so, an important term is defined (chad), and crucial alignment dimensions are specified (plus or minus .003″).

In technical writing you may need to describe a mechanism. In describing a mechanical device, for example, you might want to give your reader a general description of the whole device. Such a description should include the primary functions and the main parts of the device. This provides a frame of reference for understanding the more specific details that follow—the physical description of the various parts and each part's location and function in relation to the whole. The description should conclude with an explanation of the way the parts work together to get their particular jobs done, as in the following example:

General description of the whole device

> An internal combustion engine is one in which the fuel is burned directly inside the machine which converts the heat energy into work. Practically all motor cars use this kind of engine, the fuel being petrol (gasoline). The engine consists of several cylinders in which pistons are fitted. The top of each cylinder is covered over with a head which contains a hollowed-out . . . combustion chamber. A rod connects each piston with a . . . crankshaft. The burning of fuel and air in the combustion chamber produces heat energy which causes expansion of the gases formed by the process of combustion. The gases push the pistons down in the cylinder and thus cause the crankshaft to rotate. A flywheel on the end of the crankshaft keeps the latter rotating smoothly between the power impulses from each cylinder; it also transmits the power developed by the engine to the clutch.

Detailed description of the principles of the engine's operation

> The easiest way to understand how an engine works is to keep in mind only one cylinder. For the engine to run and do work, certain things must take place over and over again in a regular order. The fuel must be sucked in, it must be compressed and burned; the resulting gases expand, and thus do work, and after expansion they must be removed. These steps make up what is known as a cycle; in most motor car engines it takes four strokes of the piston to make up one cycle. During this time the crankshaft will have made two complete revolutions. This type of engine is said to operate on a four-stroke cycle. The four strokes of the piston are named according to what they accomplish. The first is the suction, the second the compression, the third the power, and the fourth the exhaust stroke.

At the start of the suction stroke, the piston is at the top of the cylinder. It moves downward and the inlet valve opens. This sucks in the charge of gases from the carburetor. The compression stroke starts when the inlet valve closes and the piston is moving upward. The mixture is ignited while under pressure by a high-tension electric spark, which is produced across the points of the sparking plug. . . . The power stroke begins with a downward thrust of the piston. The burning of the gases gives off heat, which causes expansion. Pressure is thus exerted on the piston, forcing it downward as far as it will go. This completes the power stroke. The last movement of the piston is the exhaust stroke. The exhaust valve opens toward the end of the power stroke and most of the burned gases rush out. The piston moves upward, pushing the remaining gases out of the cylinder. The four-stroke cycle is now complete.

Summation of how the parts work together to perform the engine's function

—*Reprinted by permission* © *Encyclopaedia Britannica*, 14th edition (1956).

For descriptions intended for readers unfamiliar with a topic, more details and bridging devices are needed. This type of description is sometimes called *rendering*, which is the act of showing or demonstrating, as opposed to telling, primarily through the use of images and details. Rendering makes writing vivid by providing the reader with more concrete details. This technique is especially useful to technical people because it can help clarify and make more concrete the complex, abstract, and sometimes subtle subjects with which they must often deal. It also uses **analogy** to explain unfamiliar concepts in terms of familiar concepts. For example, the plastic device that holds the letters and symbols for some typewriters and word-processing printers is a wheel-shaped piece of plastic, with each letter and symbol positioned at the circumference of the wheel and attached to the hub at the center of the wheel by thin plastic strips. Using analogy, you can think of the strips as petals radiating from the center of a flower. Because the strips are uniform in length and width, similar to the petals of a daisy, the print elements are commonly referred to, in an extension of the flower analogy, as daisy wheel print elements.

The following sentence, typical of much technical writing, contains a number of complex notions unfamiliar to the nonexpert:

EXAMPLE Computers can scan and read punched cards.

The complex thought contained in this sentence can be rendered clearly by the skillful use of images and details from everyday life (the familiar) used as a bridge to the unfamiliar.

> Familiar (print/page) When information appears in print, as on this page, people like you and me are able to read and understand it. However, if information is to be processed by a machine, like a computer, then other ways must be found to put these same letters Unfamiliar (machine) and words into the machine. Computers, of course, do not have eyes like humans—but they do have electrical sensing equipment that in certain ways does almost the same thing.
>
> For example, most of us have walked into supermarkets Familiar (doors at market) through doors that open by themselves. These doors are controlled by electrical sensing equipment known as photoelectric cells, which act like eyes. Each door is controlled by two photoelectric cells that shine a beam of light to each other. As you Unfamiliar (photoelectic cell) walk through the door, the light beam is broken causing the photoelectric cells or "eyes" to sense that someone is beginning to enter. The electric eye reacts in a split second by sending a burst of electrical energy to the door's mechanical hinge, and this automatically swings the door open.
>
> Transitional ideas (beam of light/hole in card) The photoelectric sensing idea can also be used to detect the presence of a hole in a card or piece of paper.
>
> Thus, if holes are punched in paper to represent certain let- New concept (machine "reading" text) ters of the alphabet or words, it is possible for a machine to electrically sense or read this information.
>
> —Joseph Becker, *The First Book of Information Science* (Washington, D.C.: The U.S. Atomic Energy Commission, 1973), p. 20.

Illustrations can be powerful aids in descriptive writing, especially when they show details too cumbersome to explain in writing.

despite/in spite of

Despite and *in spite of* (both meaning "notwithstanding") should not be blended into *despite of*.

CHANGE *Despite of* our best efforts, the plan failed.
TO *In spite of* (or *despite*) our best efforts, the plan failed.

Although there is no literal difference between *despite* and *in spite of*, *despite* suggests an effort to avoid blame.

EXAMPLES *Despite* our best efforts, the plan failed. (We are not to blame for the failure.)

In spite of our best efforts, the plan failed. (We did everything possible, but failure overcame us.)

diacritical marks

Diacritical marks are **symbols** added to letters to indicate their specific sounds. They include the phonetic symbols used by **dictionaries** as well as the marks used with foreign words. Some dictionaries place a reference list of common marks and sounds at the bottom of each page. The following is a list of common diacritical marks with their equivalent sound values. (See also **foreign words in English**.)

Name	*Symbol*	*Example*	*Meaning*
macron	¯	cāke	a "long" sound that signifies the standard pronunciation of the letter.
breve	˘	brăcket	a "short" sound, in contrast to the standard pronunciation of the letter.
dieresis	¨	coöperate	indicates that the second of two consecutive vowels is to be pronounced separately.
acute accent	´	cliché	indicates a primary vocal stress on the indicated letter.
grave accent	`	crèche	indicates a deep sound articulated toward the back of the mouth.
circumflex	^	crêpe	indicates a very soft sound.
tilde	~	cañon	a Spanish diacritical mark identifying the palatal nasal "ny" sound.
cedilla	ç	garçon	a mark placed beneath the letter *c* in French, Portuguese, and Spanish to indicate an *s* sound.

diagnosis/prognosis

Because they sound somewhat alike, these words are often confused with each other. *Diagnosis* means "an analysis of the nature of something" or "the conclusions reached by such analysis."

EXAMPLE The chairman of the board *diagnosed* the problem as the allocation of too few dollars for research.

Prognosis means "a forecast or prediction."

EXAMPLE He offered his *prognosis* that the problem would be solved next year.

dictation

If you write many short letters or **memorandums**, dictation may save you numerous hours each week. (See also **correspondence**.)

You should be aware, however, that dictation has certain limitations. First, although some people can compose sentences easily while speaking into a microphone or telephone, others find dictating a chore and can compose faster on paper. Only experience will tell whether dictation is for you. Second, certain kinds of material do not lend themselves to dictation. **Reports**, for example, or any material with **mathematical equations**, **tables**, **graphs**, or scientific terminology pose special problems for the person dictating as well as for the person transcribing. In general, dictation proves less effective with materials that have a complicated **format**. Finally, dictation sometimes causes a choppy and unconversational **style**, and so you may need to allow time to edit and polish your draft after you receive the transcription from the typist.

If you determine that dictation will save you time and effort, you may find the following tips helpful:

1. Once you learn how to use your firm's dictating equipment, you might want to make a few practice runs in order to become familiar with switches and buttons or the response time of a centralized dictating service.
2. Outline before you dictate. The length and complexity of the letter will determine how detailed your outline should be. However, even if you jot down only several key words in the order you wish to present your ideas, don't skip the outline—it will help prevent you from "freezing" when the machine is running.
3. Begin your dictation with any special **instructions** for the person transcribing: "This is a rough draft—please double-space" or "Prepare four copies."
4. Speak in a normal voice and enunciate clearly. Avoid vocal pauses ("uh," "er," "hm"); simply stop speaking or turn off the machine when you cannot think of what to say next.
5. Spell out words that sound like other words (*ensure*, *insure*); personal or company names when used for the first time ("capital

M, small *a*, *c*, capital *G*, small *i*, *l*, *l*, *i*, *s*''); **abbreviations**, and unusual terms.

6. Slow down for **numbers**, addresses, and difficult phrases (''*send six showroom* display cases'').
7. Dictate as much **punctuation** as your transcriber needs; however, *always* specify ''paragraph,'' ''quote,'' ''unquote,'' ''open parenthesis,'' ''close parenthesis,'' and ''underline.''
8. Give a good secretary or transcriber the latitude to make corrections in punctuation, grammar, or **usage**. You, however, are responsible for the content and **tone**.
9. Proofread the transcribed dictation carefully, and make any necessary revisions—especially in important correspondence. Never skip this important stage, no matter how much you trust the person transcribing the dictation. (See also **proofreading**.)
10. Set a time each day for dictation, preferably early enough to allow time to review the correspondence before mailing it.

diction

The term *diction* is often misunderstood because it means both ''the choice of words used in writing and speech'' *and* ''the degree of distinctness of enunciation of speech.'' In discussions of writing, *diction* applies to choice of words. (See also **word choice**.)

dictionaries

A dictionary lists, in alphabetical order, a selection of the words in a language. It defines them, gives their **spelling** and pronunciation, and indicates their function as a **part of speech**. In addition, it provides information about a word's origin and historical development. Often it provides a list of **synonyms** for a word and, where pertinent, an **illustration** to help clarify the meaning of the word.

The explanation of a word's meaning makes up the bulk of a dictionary entry. A dictionary gives primarily denotative rather than connotative meanings. It also labels the field of knowledge to which a specific meaning applies (grid, *electricity*; merger, *law*). Some specialized dictionaries list such technical terms exclusively. (See **library research** for examples.) The order in which a word's meanings are given varies. Some dictionaries give the most widely accepted current meaning first; others list the meaning in historical order, with the oldest meaning first and the current meaning last. If a word has two or more fundamental meanings, they are listed in separate paragraphs, with secondary meanings numbered consecutively within each paragraph. A dictionary's preface normally indicates whether current or historical meanings are listed in an entry.

A dictionary entry also includes a word's spelling. It lists any variant spellings (align/aline, catalog/catalogue) and indicates which is most commonly used. Entries show if and where a word ought to be hyphenated—information especially helpful for **compound words**—and how the word can be abbreviated.

A word's pronunciation, indicated by phonetic symbols, is also given. The **symbols** are explained in a key in the front pages of the dictionary, and an abbreviated symbol key is often found at the bottom of each page as well. (See also **diacritical marks**.)

Information about the origin and history of a word (its etymology) is also given, usually in **brackets** at the end of the entry. This information can help clarify a word's correct meaning.

Entries also include information about a word's part of speech and its inflected forms (how it forms plurals or how it changes form to express comparative and superlative degree). Many entries include a list of synonyms in a separate paragraph following the main paragraph. Some dictionaries label words according to **usage** levels to indicate whether a word is considered standard English, nonstandard English, slang, and so on. Many dictionaries supplement a word's definition by providing illustrations in the form of **maps**, **photographs**, **tables**, **graphs**, and the like.

TYPES OF DICTIONARIES

Desk dictionaries are often abridged versions of larger dictionaries. There is no single "best" dictionary, but there are several guidelines for selecting a good desk dictionary. Choose a recent edition. The older the dictionary, the less likely it is to have the up-to-date information you need. Select a dictionary with upward of 125,000 entries. Pocket dictionaries are convenient for checking spelling, but for detailed information the larger range of a desk dictionary is necessary. The following are considered good desk dictionaries:

1. *The Random House College Dictionary*. Rev. ed. New York: Random House.
2. *The American Heritage Dictionary of the English Language*. Boston: Houghton Mifflin.

3. *Funk & Wagnalls Standard College Dictionary*. New, updated ed. New York: Funk & Wagnalls.
4. *Webster's New World Dictionary of the American Language*. Cleveland: World Publishing Co.
5. *Webster's Ninth New Collegiate Dictionary*. Springfield, Mass.: G. & C. Merriam Company.

Unabridged dictionaries provide complete and authoritative linguistic information, but they are impractical for desk use because of their size and expense.

1. *Webster's Third New International Dictionary of the English Language, Unabridged* (Springfield, Mass.: G. & C. Merriam Company, 1981) contains over 450,000 entries. Since word meanings are listed in historical order, the current meaning is given last. This dictionary does not list personal and geographical names, nor does it include usage information.
2. *The Random House Dictionary of the English Language* (New York: Random House, 1973) contains 260,000 entries and is encyclopedic in its use of examples. It gives a word's most widely used current meaning first and includes personal and geographical names.

The following unabridged dictionary attempts to be exhaustive:

The Oxford English Dictionary (Oxford, England: Clarendon Press, 1933) is the standard historical dictionary of the English language. It follows the chronological developments of over 240,000 words, from about the year 1000 to the early part of this century, providing numerous examples of uses and sources. *A Supplement to the Oxford English Dictionary*, Vol. 1, A–G, and Vol. 2, H–N (New York: Oxford University Press, 1972 and 1976), containing 17,000 and 13,000 entries respectively, is now available. Volumes 3 and 4 were published in 1986.

(See also **reference books**.)

differ from/differ with

Differ from is used to suggest that two things are not alike.

EXAMPLE The vice-president's background *differs from* the general manager's background.

Differ with is used to indicate disagreement between persons.

EXAMPLE Our attorney *differed with* theirs over the selection of the last jury member.

different from/different than

In formal writing, the **preposition** *from* is used with *different*.

EXAMPLE The fourth-generation computer is *different from* the third-generation computer.

Different than is acceptable when it is followed by a **clause**.

EXAMPLE The job cost was *different than* we had estimated it.

direct address

Direct address refers to a sentence or **phrase** in which the person being spoken or written to is explicity named.

EXAMPLE *John*, call me as soon as you arrive at the airport.

The person's name in a direct address is set off by **commas**.

direct objects (see **objects**)

discreet/discrete

Discreet means "having or showing prudent or careful behavior."

EXAMPLE Since the matter was personal, he asked Bob to be *discreet* about it at the office.

Discrete means something is "separate, distinct, or individual."

EXAMPLE The publications department was a *discrete* unit of the marketing division.

disinterested/uninterested

Disinterested means "impartial, objective, unbiased."

EXAMPLE Like good judges, scientists should be passionately interested in the problems they tackle but completely *disinterested* when they seek to solve these problems.

Uninterested means simply "not interested."

> EXAMPLE Despite Jane's enthusiasm, her manager remained *uninterested* in the project.

division-and-classification method of development

One effective way to develop a complex subject is to divide it into logical parts and then go on to explain each part separately. This technique is especially well suited to subjects that can be readily broken down into units, and it can be used as the **method of development** in much technical writing. You might use this approach to describe a physical object, like a machine; to examine an idea, like the terms of a new labor–management contract; to explain a process, like the stages of an illness; or to give **instructions**, like the steps necessary to prepare a rusty metal surface for painting.

To explain the different types of printing processes currently in use, for example, you could divide the field into its major components, and where a fuller explanation is required, subdivide those (Figure 1). The emphasis in division is on breaking down a complex whole into a number of like units, because it is easier to consider smaller units and to examine the relationship of each to the other.

The process by which a subject is divided is similar to the process by which a subject is classified. Division involves the separation of a whole into its component parts; classification is the grouping of a

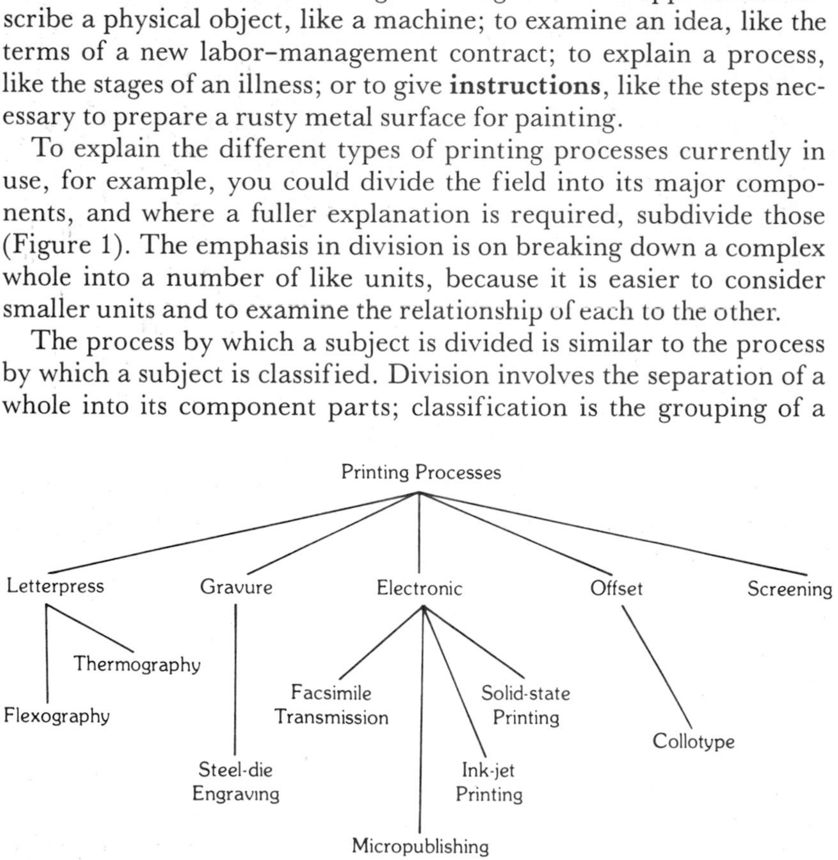

Figure 1 Printing Processes

number of units (people, objects, ideas, and so on) into related categories. Consider the following list:

Triangular file	Steel tape ruler
Needle-nose pliers	Vise
Pipe wrench	Keyhole saw
Mallet	Tin snips
C clamps	Rasp
Hacksaw	Plane
Glass cutter	Ball-peen hammer
Steel square	Spring clamp
Claw hammer	Utility knife
Crescent wrench	Folding extension ruler
Slip-joint pliers	Crosscut saw
Tack hammer	Utility scissors

To group the items in the list, you would first determine what they have in common. The most obvious characteristic they share is that they all belong in a carpenter's tool chest. With that observation as a starting point, you can begin to group the tools into related categories. Pipe wrenches belong with slip-joint pliers because both tools grip objects. The rasp and the plane belong with the triangular file because all three tools smooth rough surfaces. By applying this kind of thinking to all the items in the list, you can group each tool according to function (Figure 2).

To divide or classify a subject, you must first divide the subject into its largest number of equal units. The basis for division depends, of course, on your subject and your purpose. For explaining the functions of tools, the classifications in Figure 2 (smoothing, hammering, measuring, gripping, and cutting) are excellent. For recommending which tools a new homeowner should buy first, however, the classifications are not helpful—they *all* contain tools that a new homeowner might want to purchase right away. To give homeowners advice on purchasing tools, you should clarify the types of repairs they are most likely to have to do. That classification can serve as a guide to tool purchase.

Once you have established the basis for the division or classification, apply it consistently, putting each item in only one category. For example, it might seem logical to classify needle-nose pliers both as "tools that cut" and as "tools that grip" because most needle-nose pliers have a small section for cutting wires. However, the *pri-*

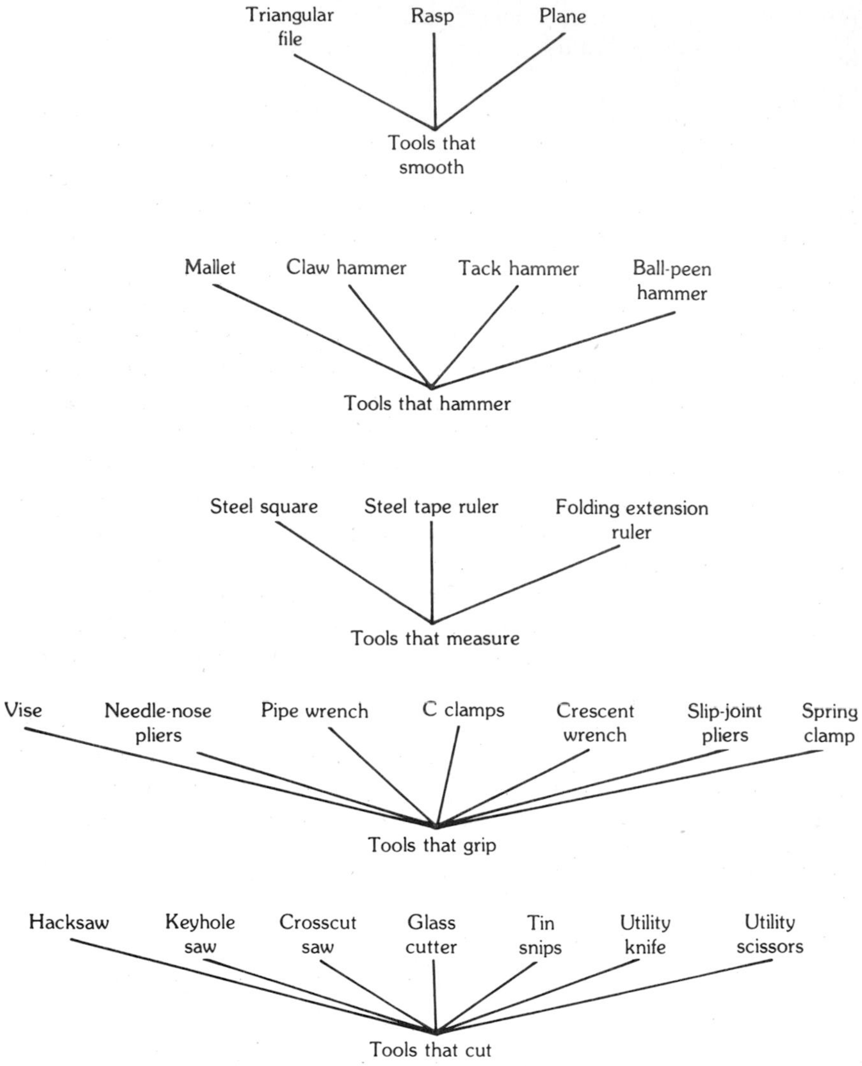

Figure 2 Function of Tools

mary function of needle-nose pliers is to grip. So listing them only under "tools that grip" would be most consistent with the basis used for listing the other tools.

In the example given in Figure 3, two Canadian park rangers classify typical park users according to four categories; the rangers then

Dealing With Groups

First, recognize the various types of campers. They can be broken down as follows:

1. Family for weekend stay
2. Small groups ("a few of the boys")
3. Large groups or conventions
4. Hostile gangs

Persons in groups A and B can probably be dealt with on a one-to-one basis. For example, suppose a member of the group is picking wild flowers, which is an offense in most park areas. Two courses of action are open. You could either issue a warning or charge the person with the offense. In this situation, a warning is preferable to a charge. First, advise the person that this action is an offense but, more importantly, explain why. Point out that the flowers are for all to enjoy and that most wild flowers are delicate and die quickly when picked.

For large groups, other approaches may be necessary. Every group has a leader. The leader may be official or, in informal or hostile groups, unofficial. If the group is organized, seek the official leader and hold this person responsible for the group's behavior. For informal groups, seek the person who assumes command and try to deal with this person.

—A.W. Moore and J. Mitchell, "Vandalism and Law Enforcement in Wildland Areas," in *Proceedings of the Wildland Recreation Conference,* Feb. 28–March 3, 1977, Banff Centre, Alberta, Canada.

Figure 3 Division and Classification

go on to discuss how to deal with potential rule breaking by members of each group. The rangers could have classified the visitors in a variety of other ways, of course: as city and country residents, backpackers and drivers of recreational vehicles, United States and Canadian citizens and so on. But for law enforcement agents in public parklands, the size of a group and the relationships among its members were the most significant factors.

documentation

The word *documentation* in the academic world refers to the giving of formal credit to sources used or quoted in a research paper or **report**—that is, the form used in a **bibliography**, footnote or a reference. In the business and industrial world, *documentation* may refer also to information that is recorded, or "documented," on paper. The **technical manuals** provided to customers by manufacturers are examples of such documentation. Even **flowcharts** and engineering **drawings** are considered documentation.

documenting sources

By documenting sources, writers identify where they obtained the facts, ideas, quotations, and paraphrases used in preparing a report, journal article, book, or other document. The information can come from books, manuals, correspondence, interviews, software documentation, reference works, and many other kinds of sources. Full and accurate documentation allows readers to locate and consult the information given and to find further information on the subject. It also gives proper credit to others so that the writer avoids **plagiarism**. The sources of all facts and ideas that are not common knowledge to the intended audience should be documented, as well as the sources of all direct quotations. The documentation must be complete, accurate, and consistent in format.

This entry describes and provides examples of the three principal methods of documenting sources: (1) *parenthetical documentation*—putting brief citations in parentheses in the text and providing full information in a list of "Works Cited"; (2) *reference documentation*—referring to sources with numbers in parentheses or by superscripts in the text and providing full information in a "References" section where the entries are listed numerically in the order of their first citation in the text; and (3) *notational documentation*—using superscript numbers in the text to refer to notes either at the bottom of the page (footnotes) or at the end of the paper, article, or chapter (endnotes).

Many professional organizations and journals publish style manuals that describe their own formats for documenting sources. A list of such style manuals appears at the end of this entry. If you are writing for publication in a professional field, consult the manual for the particular field or the style sheet for the journal to which you are submitting your article, and follow it exactly. Whatever format you choose, be sure to follow it consistently, in every detail of order, **punctuation**, and capitalization.

Use the following contents guide to this entry to locate information quickly. Samples of each format appear at the end of each discussion. Parenthetical samples begin on page 191; reference samples begin on page 194; and notational samples begin on page 198.

PARENTHETICAL DOCUMENTATION

The parenthetical method that follows is recommended by the Modern Language Association of America (MLA) in the *MLA Handbook for Writers of Research Papers*. This method gives an abbreviated reference to a source parenthetically in text and lists full information about the source in a separate section called "Works Cited."

When documenting sources in text, include only the author and page number in parentheses. If the author's name is mentioned in the text, include only the page number of the source. The parenthetical citation should include no more information than is necessary to enable the reader to relate it to the corresponding entry in the list of "Works Cited." When referring to an entire work rather than to a particular page in a work, mention the author's name in the text, and omit the parenthetical citation.

The following passages contain sample parenthetical citations.

Examples

Edward A. Feigenbaum of Stanford University, an artificial intelligence researcher, has produced a program called MYCIN that is able to prescribe antibiotics and another called DENDRAL that "analyzes mass spectra and produces highly probable molecular structures" (Inose and Pierce 142).

According to Inose and Pierce, the term artificial intelligence may be a misnomer; they point out that from the first, computers have done some things better than humans, and they apparently think that humans will continue to do some things better than computers (143).

When placing parenthetical citations in text, insert them between the closing quotation mark or the last word of the sentence (or clause) and the period or other punctuation. Use the spacing shown above. If the parenthetical citation follows an extended quotation or paraphrase, however, place it outside the last sentence of the quotation or paraphrase, with two spaces between the period and the first parenthesis. Within the citation itself, allow one space (no punctuation) between the name of the author and the page number. Don't use the word *page* or its abbreviation.

If you are citing a page or pages of a multivolume work, give the volume number, followed by a colon, space, and page number: (Jones 2: 53). If the entire volume is being cited, follow the punctuation in the two samples below:

Nonstellar celestial objects include clusters, nebulae, galaxies, quasi-stellar objects, radio sources, and X-ray sources (Hirshfield and Sinnott, vol. 2).

Although dwarf elliptical systems are the most common type of galaxy in the universe, their low luminosity causes them to be outnumbered by spiral galaxies in any catalogue of galaxies above even a relatively low apparent brightness (Hirshfield and Sinnott 2: xxxi).

If your list of "Works Cited" includes more than one work by the same author, include the title of the work (or a shortened version if the title is long) in the parenthetical citation, unless you mention it in the text. If, for example, your list of "Works Cited" included more than one work by David Landes, a proper parenthetical citation for his book *Revolution in Time: Clocks and the Making of the Modern World* would appear as in the following sample:

```
The tremendous success of Timex watches in the 1950s is attributable
to drastic simplification of design; standardization of parts, making
them interchangeable even between plants; and thorough-going automa-
tion of the manufacturing process (Landes, Revolution in Time 340).
```

Use only one space between the title and the page number.

The works cited in the sample passages given would be listed alphabetically, as follows, in a "Works Cited" section:

Hirshfield, Alan, and Roger W. Sinnott, eds. Sky Catalogue 2000.0. 2 Vols. Cambridge, MA: Sky Publishing, 1982.

Inose, Hiroshi, and John R. Pierce. Information Technology and Civilization. New York: Freeman, 1984.

Landes, David S. Revolution in Time: Clocks and the Making of the Modern World. Cambridge, MA: Belknap-Harvard UP, 1983.

The following section explains the content and format of entries in a list of "Works Cited." The formal style described is that recommended in the *MLA Handbook for Writers of Research Papers*.

CITATION FORMAT FOR WORKS CITED

The list of "Works Cited" should begin on the first new page following the end of the text. Each new entry should begin at the left margin, with the second and subsequent lines within an entry indented five spaces. Single-space within entries, and double-space between them.

AUTHOR

Entries appear in alphabetical order by the author's last name (by the last name of the first author if the work has more than one author). Works by the same author should be alphabetized by the first major word of the title (following *a*, *an*, or *the*). If the author is a corporation, the entry should be alphabetized by the name of the corporation; if the author is a government agency, entries are alphabetized by the government, followed by the agency (for example, "United States. Dept. of Health and Human Services."). Some sources require more than one agency name (for example, "United

States. Dept. of Labor. Bureau of Labor Statistics.''). If no author is given, the entry should begin with the title and be alphabetized by the first significant word in the title. (Articles in reference works, like encyclopedias, are sometimes signed with initials; you will find a list of the contributors' initials and full names elsewhere in the work, probably near the introductory material or the index.)

After the first listing for an author, put three hyphens in place of the name for subsequent entries with the same author. An editor's name is followed by the abbreviation, *ed.*

McNeill, William H. The Pursuit of Power. Chicago: U of Chicago P, 1982.

---, The Rise of the West. Chicago: U of Chicago P, 1963.

TITLE

The second element is the title of the work. Capitalize the first word and each significant word thereafter. Underline the title of a book or pamphlet. Place quotation marks around the title of an article in a periodical, an essay in a collection, and a paper in a proceedings. Each title should be followed by a period.

PERIODICALS

For an article in a periodical (journal, magazine, or newspaper), the volume number, date (for a magazine or newspaper, simply the date), and the page numbers should immediately follow the title of the periodical. See the sample entries below for the proper punctuation.

SERIES OR MULTIVOLUME WORKS

For works in a series and multivolume works, the name of the series and the series number of the work in question, or the number of volumes, should follow the title. If the edition used is not the first, the edition should be specified.

PUBLISHING INFORMATION

The final elements of the entry for a book, pamphlet, or conference proceedings are the place of publication, publisher, and date of publication. If any of these cannot be found in the work, use the abbreviations n.p. (no publication place), n.p. (no publisher), and n.d.

(no date), respectively. For familiar reference works, list only the edition and year of publication.

SAMPLE ENTRIES (MLA STYLE)

Book, One Author

Landes, David S. Revolution in Time: Clocks and the Making of the Modern World. Cambridge, MA: Belknap-Harvard UP, 1983.

Book, Two or More Authors

Inose, Hiroshi, and John R. Pierce. Information Technology and Civilization. New York: Freeman, 1984.

Book, Corporate Author

CompuServe Incorporated. CompuServe Information Service: User's Guide. Columbus, OH: CompuServe Incorporated, 1985.

Work in an Edited Collection

Gibson, Ralph. "High Contrast Printing." Darkroom. Ed. Eleanor Lewis. N.p.: Lustrum Press, 1977. 63-76.

Book Edition, If Not the First

Gibson, H. Lou. Photography by Infrared: Its Principles and Applications. 3rd ed. New York: Wiley, 1978.

Translated Work

Texereau, Jean. How to Make a Telescope. Trans. Allen Strickler. 2nd English ed. Richmond, VA: Willmann-Bell, 1984.

Multivolume Work

Hirshfield, Alan, and Roger W. Sinnott, eds. Sky Catalogue 2000.0. 2 vols. Cambridge, MA: Sky Publishing, 1982.

Work in a Series

Armstrong, Joe E., and Willis W. Harman. Strategies for Conducting Technology Assessments. Westview Special Studies in Science, Technology, and Public Policy. Boulder, CO: 1980.

Report

Gould, John D. Composing Letters with Computer-Based Text Editors. IBM Computer Science Research Report RC 8446 (#36750). Yorktown Heights, NY: 1980.

Thesis or Dissertation

Ross, Brant Arnold. "Flexible Engineering Software: An Integrated Workstation Approach to Finite Element Analysis." Diss. Brigham Young U. 1985.

Encyclopedia Article

Heginbotham, Wilfred Brooks. "Robot Devices." Encyclopaedia Britannica: Macropaedia. 15th ed. 1982.

Proceedings

Crawford, David L. Instrumentation in Astronomy IV. Proc. of the Society of Photo-Optical Instrumentation Engineers. 8-10 March, 1982, Tucson, AZ. Bellingham, WA: SPIE, 1982.

Paper in a Proceedings

Carr, Marilyn. "Appropriate Technology: Theory, Policy and Practice." Fundamental Aspects of Appropriate Technology. Proc. of the International Workshop on Appropriate Technology. 4-7 Sept. 1980. Ed. J. de Schutter and G. Bemer. Delft, Neth.: Delft UP, 1980. 145-53.

Paper Presented at a Conference

Madson, Elizabeth Ann. "Use of Writer's Workbench Software." Midwest MLA Annual Meeting, Bloomington, IN, 3 November 1984.

Journal Article

MacDonald, Nina H., Lawrence T. Frase, Patricia S. Gingrich, and Stacey A. Keenan. "The Writer's Workbench: Computer Aids for Text Analysis." IEEE Transactions on Communications 30 (1982): 105-110.

Magazine Article

Sinfelt, John H. "Bimetallic Catalysts." Scientific American Sept. 1985: 90-98.

Anonymous Article (in Weekly Periodical)

"T&W Systems Adds Relational Database." Infoworld 20 Jan. 1986: 49.

Newspaper Article

Schmeck, Harold M., Jr. "Gene-Spliced Hormone for Growth is Cleared." New York Times 19 October 1985, national ed.: 8.

Letter from One Official to Another

Brown, Charles L. Letter to retired members of Bell System Presidents' Conference. 8 January 1982.

Letter Personally Received

Harris, Robert S. Letter to the author. 3 December 1985.

Personal Interview

Denlinger, Virgil, Assistant Chief of Police, Alexandria, VA. Personal interview. 15 December 1984.

Computer Software

Holdstein, Deborah. Composition Software Series. Computer software. Holt, Rinehart and Winston, 1985. IBM-PC, 4 disks.

REFERENCE DOCUMENTATION

The reference system of documenting sources lists the numbered documentation references at the end of the work. The references on the list are identified by numbers in the text that correspond to the numbered items in the list. The reference list cites only those works referred to in text.

The styles for references vary. The style described here is recommended by the American National Standards Institute (ANSI) in *American National Standard for Bibliographic References*, Z39.29-1977.

The citations in a reference list are arranged in numerical sequence (1, 2, 3), according to the order in which they are first mentioned in the text. The most common method for directing readers from your text to specific references is to place the reference number in parentheses after the work cited. Thus in the text, the number one in parentheses (1) after a reference to a book, article, or other work refers the reader to the first citation in the reference list. The number five in parentheses (5) refers the reader to the fifth citation in the reference list. A second number in the parentheses, separated from the first by a colon (3:27), refers to the page number of the source from which the information was taken. There are several other common methods for directing readers from your text to specific references. You can place the word *Reference*, or the abbreviation *Ref.*, within parentheses with the numbers: (Reference 1) or (Ref. 1). Or you can write the reference number as a superscript: text[1]. Sometimes you may cite your sources within a sentence in the text; then you should spell out the word *Reference*: "The data in Reference 3 include . . ." For subsequent references in the text, simply repeat the reference number of the first citation.

CITATION FORMAT FOR REFERENCES

In the numbered reference list, the number begins at the left margin, followed by a period. Then leave two spaces and begin the entry, indenting the second and subsequent lines to be even with the first. Single-space within entries, and double-space between them.

The recommended order for information groupings within an entry and for punctuation within and between these groups are shown in the samples beginning on page 194. In general, put a period between groups and at the end of the entry, a semicolon between elements within a group, and a comma between subelements or closely related elements.

AUTHOR

Authors' names should be given with the last name first. If a work has two authors, both names should be given, with the last name first, separated by a semicolon. If a work has more than two authors, the first author should be cited, last name first, followed by "and others" in square brackets. For an edited work (if you are listing the whole work, not just an article in it) begin the entry with the editor(s), giving the names as described for the authors, followed by a comma and the abbreviation, *ed.*

TITLE

Only the first word and any proper nouns should be capitalized in a book or article title. The title of an article, followed by a period, should precede the title of the journal or book in which it appears. In journal or magazine titles, abbreviations may be used, and all significant words (or abbreviations for them) should be capitalized.

In an entry for an article or paper in an edited collection, the authorship group and title group for the article are followed by the editors and title group for the collection.

PUBLISHING INFORMATION

For a book, the publishing information consists of the place of publication followed by a colon, the publisher followed by a semicolon, and the year of publication followed by a period. For a journal or magazine, the date (in year–month–day order) may either precede or follow the volume number, issue number, and pages, separated from them by a semicolon. The volume number is followed by the issue number in parentheses, a colon after the closing parenthesis, and then the pages of the article. (The ANSI style for magazine articles is the same as for journal articles.)

SAMPLE REFERENCES (ANSI STYLE)

The following examples will acquaint you with the ANSI style. If you need to list a source for which no example is given, follow the pattern from the examples that are most like it. If you are uncertain whether to include a piece of information about the source, include it. (Note that the ANSI standard does not cover personal communications, interviews, speeches, and the like for which no print, manuscript, or electronic record exists.)

Book, One Author

1. Landes, David S. Revolution in time: clocks and the making of the modern world. Cambridge, MA: Belknap/Harvard Univ. Press; 1983.

Book, Two Authors

2. Inose, Hiroshi; Pierce, John R. Information technology and civilization. New York: W. H. Freeman; 1984.

Book, Corporate Author

3. CompuServe Incorporated. CompuServe Information Service: user's guide. Columbus, OH: CompuServe Inc.; 1985.

Work in an Edited Collection

4. Broadhead, Glenn J. Style in technical and scientific writing. In: Moran, Michael G.; Journet, Debra, eds. Research in technical communication: a bibliographic sourcebook. Westport, CT: Greenwood Press; 1985: 217-252.

Book Edition, If Not the First

5. Gibson, H. Lou. Photography by infrared: its principles and applications. 3rd ed. New York: Wiley; 1978.

Multivolume Work

6. Hirshfield, Alan; Sinnott, Roger W., ed. Sky catalogue 2000.0. Cambridge, MA: Sky Publishing; 1982. 2v.

Book in a Series

7. Osborne, D. J.; Gruneberg, M.M. The physical environment at work. Chichester, Eng.: John Wiley and Sons; 1983. (Wiley series in psychology and productivity at work.)

Report

8. Zombeck, Martin V. High energy astrophysics handbook. Cambridge, MA: Smithsonian Astrophysical Observatory; 1980; Research in Space Science SAO Special Report 386.

Thesis or Dissertation

9. Weitz, Rob Roy. Nostradamus: an expert system for guiding the selection and use of appropriate forecasting techniques. Boston: Univ. of Massachusetts; 1985. 146 p. Dissertation.

Paper in a Conference Proceedings

10. Cochran, William D.; Smith, Harlan J.; Smith, William Hayden. Ultrahigh precision radial velocity spectrometer. Crawford, David L., ed. Instrumentation in astronomy IV. Proceedings of the Society of Photo-Optical Instrumentation Engineers; 1982 March 8-10; Tucson, AZ. Bellingham, WA: SPIE 315-320.

Unpublished Paper

11. Madson, Elizabeth Ann. Use of Writer's Workbench software. 1984. Unpublished draft supplied to author by Elizabeth Ann Madson.

Journal Article

12. MacDonald, Nina H. (and others). The Writer's Workbench: computer aids for text analysis. IEEE Transactions on Communications. 1982 January; 30(1): 105-110.

Computer Software

13. Homer: a computerized revision program (Computer program). Cohen, Michael; Lanham, Richard. New York: Scribner's; 1983. Disk; for 48K Apple II and II+, 64 K Apple IIe.

NOTATIONAL DOCUMENTATION

Notes in publications have two uses: (1) to provide background information in publications or explanations that would interrupt the flow of thought in the text and (2) to provide documentation references. (For guidance on footnotes used with tables, see **tables**.)

EXPLANATORY FOOTNOTES

Explanatory or content notes are useful when the basis for an assumption should be made explicit but spelling it out in the text might make readers lose the flow of an argument. Because explanatory or content notes can be distracting, however, they should be kept to a minimum. If you cannot work the explanatory or background material into your text, it may not belong there. Lengthy explanations should be placed in an appendix. (See **appendix/appendixes/appendices**.)

DOCUMENTATION NOTES

Notes that document sources can appear as either endnotes or footnotes. Endnotes are placed in a separate section at the end of a report, article, chapter, or book; footnotes (including explanatory notes) are placed at the bottom of a text page. Endnotes are easy to type, but readers may find them inconvenient to find and difficult to correlate with the text. Footnotes are tricky to type unless you have a **word-processing** program that automatically allocates the right amount of space at the bottom of each page. They are, however, convenient for readers.

Use superscript numbers in the text to refer readers to notes, and number them consecutively from the beginning of the report, article, or chapter to the end. Place each superscript number at the end of a sentence, **clause**, or **phrase**, at a natural pause point like this,[1] right after the **period**, **comma**, or other punctuation mark (except a **dash**, which the number should precede[2]—as here).

FOOTNOTE FORMAT

To type footnotes, leave two line spaces beneath the text on a page, and indent the first line of each footnote five spaces. Begin each note with the appropriate superscript number, and skip one space between the number and the text of the footnote. If the footnote runs longer than one line, the second and subsequent lines should begin at the left margin. Single-space within each footnote, and double-space between footnotes.

Example

```
Assume that this is the last line of text on a page.

     1 Begin the first footnote at this position. When it runs longer
than one line, begin the second and all following lines at the left
margin.

     2 The second footnote follows the same spacing as the first.
Single-space within footnotes, and double-space between them.
```

ENDNOTE FORMAT

Endnotes should begin on a separate page after the end of the text entitled "Works Cited" or "References." The individual notes should be typed as for footnotes. Single-space within notes, and double-space between them.

DOCUMENTATION NOTE STYLE

The following guidelines for documentation note style are based on those provided in the Modern Language Association's *MLA Handbook for Writers of Research Papers*.

Documentation notes give the author's full name, title, and publication information for the source, including the page or pages from which material was taken. This information is given in full in the first note for the source, and the second and subsequent notes give abbreviated author and title information.

In a footnote or endnote, the information is given in a somewhat different order and uses somewhat different punctuation than in a bibliography or a list of "References" or "Works Cited." The author's name is given in normal order (Jane F. Smith) rather than last name first; the elements of the entry are separated by commas rather than periods; and the publication information for a book is enclosed in parentheses. The first reference to a book appears as follows:

[1] David S. Landes, Revolution in Time: Clocks and the Making of the Modern World (Cambridge, MA: Belknap-Harvard UP, 1983) 340.

There is no punctuation mark between the title and the opening parenthesis or between the closing parenthesis and the page number, and the page number is not preceded by the word *page* or its abbreviation. Compare this sample note with how the same book would be listed in a bibliography or list of "Works Cited":

Landes, David S. Revolution in Time: Clocks and the Making of the Modern World. Cambridge, MA: Belknap-Harvard UP, 1983.

SAMPLE ENTRIES (MLA STYLE)

The following sample notes for first references are based on the format style of the *MLA Handbook*:

Book with a Corporate Author

[2] Borland International, Turbo Pascal Version 2.0 Reference Manual (Scotts Valley, CA: Borland, 1984) 157.

Work in an Edited Collection

[3] Ralph Gibson, "High Contrast Printing," Darkroom, ed. Eleanor Lewis (N.p.: Lustrum, 1977) 74.

Report

[4] Martin V. Zombeck, High Energy Astrophysics Handbook, Research in Space Science SAO Special Report No. 386 (Cambridge, MA: Smithsonian Astrophysical Observatory, 1980) 123.

Paper Presented at a Conference

[5] Elizabeth Ann Madson, "Use of Writer's Workbench Software," Midwest MLA Annual Meeting, 3 November 1984.

Journal Article

[6] Nina H. MacDonald, Lawrence T. Frase, Patricia S. Gingrich, and Stacey A. Keenan, "The Writer's Workbench: Computer Aids for Text Analysis," IEEE Transactions on Communications 30 (1982): 107.

Magazine Article

[7] John H. Sinfelt, "Bimetallic Catalysts," Scientific American Sept. 1985: 95.

Thesis or Dissertation

[8] Arnold Brant Ross, "Flexible Engineering Software: An Integrated Workstation Approach to Finite Element Analysis," diss., Brigham Young U, 1985, 96.

For periodicals, the date (enclosed in parentheses if the periodical is a journal) is followed by a colon.

SECOND AND SUBSEQUENT REFERENCES

Notes for second and subsequent references to works should contain only enough information to allow the reader to relate it to the note for the first reference. Generally, the author's name and page number will suffice:

9 Landes 343.

Even if the second reference immediately follows the first, put the author's last name and the page number as shown. (The abbreviations ibid. and op. cit., which you may encounter in older books and articles, are no longer recommended as ways to refer to previous citations.)

If two or more works by the same author are being cited, include a shortened version of the title in the second and subsequent notes. Insert a comma between the author's name and the title, with no punctuation between the title and the page number.

BIBLIOGRAPHY FORMAT (MLA STYLE)

The following list shows how each of the works cited in the sample notes would appear in a bibliography:

Book with a Corporate Author

Borland International. <u>Turbo Pascal Version 2.0 Reference Manual</u>. Scotts Valley, CA: Borland, 1984.

Work in an Edited Collection

Gibson, Ralph. "High Contrast Printing." <u>Darkroom</u>. Ed. Eleanor Lewis. N.p.: Lustrum, 1977.

Report

Zombeck, Martin V. <u>High Energy Astrophysics Handbook</u>. Research in Space Science SAO Special Report No. 386. Cambridge, MA: Smithsonian Astrophysical Observatory, 1980.

Paper Presented at a Conference

Madson, Elizabeth Ann. "Use of Writer's Workbench Software." Midwest MLA Annual Meeting, 3 November 1984.

Journal Article

MacDonald, Nina H., Lawrence T. Frase, Patricia S. Gingrich, and Stacey A. Keenan. "The Writer's Workbench: Computer Aids for

Text Analysis." IEEE Transactions on Communications 30 (1982): 105-110.

Magazine Article

Sinfelt, John H. "Bimetallic Catalysts." Scientific American Sept. 1985: 90-98.

Thesis or Dissertation

Ross, Arnold Brant. "Flexible Engineering Software: An Integrated Workstation Approach to Finite Element Analysis." Diss. Brigham Young U, 1985.

STYLE MANUALS

Many professional societies, publishing companies, and other organizations publish manuals that prescribe bibliographic reference formats for their publications or publications in their fields. For an annotated bibliography of style manuals issued by commercial publishers, government agencies, and university presses in fields ranging from agriculture to zoology, see Howell, John Bruce. Style manuals of the English-speaking world. Phoenix: Oryx Press; 1983.

The following is a list of style manuals for various fields:

Biology

Council of Biology Editors, Style Manuals Committee. CBE Style manual: a guide for authors, editors, and publishers in the biological sciences. 5th ed. Bethesda: Council of Biology Editors; 1983.

Chemistry

American Chemical Society. Handbook for authors of papers in American Chemical Society publications. Washington, DC: American Chemical Society; 1978.

Geology

United States Geological Survey. Suggestions to authors of the reports of the United States Geological Survey. 6th ed. Washington, DC: Government Printing Office; 1978.

Mathematics

American Mathematical Society. A manual for authors of mathematical papers. 7th ed. Providence, RI: American Mathematical Society; 1980.

Medicine

International Steering Committee of Medical Editors. Uniform requirements for manuscripts submitted to biomedical journals. Annals of Internal Medicine. 1978 January; 90:95-99.

Physics

American Institute of Physics, Publications Board. Style manual for guidance in the preparation of papers. 3rd ed. New York: American Institute of Physics; 1978.

Psychology

American Psychological Association. Publication manual of the American Psychological Association. 3rd ed. Washington, DC: American Psychological Association; 1983.

General

American national standard for bibliographic references. New York: American National Standards Institute; 1977. ANSI Z39.29-1977.

Chicago manual of style. 13th ed. Chicago: University of Chicago Press; 1982.

Turabian, K. L. A manual for writers of term papers, theses, and dissertations. 4th ed. Chicago: University of Chicago Press; 1973.

double negatives

A double negative is the use of an additional negative word to reinforce an expression that is already negative. It is an attempt to emphasize the negative, but the result is only awkward. Double negatives should never be used in writing.

> CHANGE I *haven't* got *none.*
> TO I have *none.*

Barely, *hardly*, and *scarcely* cause problems because writers sometimes do not recognize that these words are already negative.

> CHANGE I *don't hardly* ever have time to read these days.
> TO I *hardly* ever have time to read these days.

Not unfriendly, *not without*, and similar constructions are not double negatives because in such constructions two negatives are meant to suggest the gray area of meaning between negative and positive. Be careful, however, how you use these constructions; they are often confusing to the reader and should be used only if they serve a purpose.

EXAMPLES He is *not unfriendly*. (meaning that he is neither hostile nor friendly)
It is *not without* regret that I offer my resignation. (implying mixed feelings rather than only regret)

The **correlative conjunctions** *neither* and *nor* may appear together in a clause without creating a double negative, so long as the writer does not attempt to use the word *not* in the same clause.

CHANGE It was *not*, as a matter of fact, *neither* his duty *nor* his desire to fire the man.
TO It was *neither*, as a matter of fact, his duty *nor* his desire to fire the man.
OR It was *not*, as a matter of fact, *either* his duty *or* his desire to fire the man.

CHANGE He did *not neither* care about *nor* notice the error.
TO He *neither* cared about *nor* noticed the error.

Negative forms are full of traps that often entice inexperienced writers into errors of **logic**, as illustrated in the following example:

EXAMPLE There is *nothing* in the book that has *not* already been published in some form, but some of it is, I believe, very little known.

In this sentence, "some of it," logically, can refer only to "*nothing* in the book that has *not* already been published." The sentence can be corrected in one of two ways. The pronoun *it* can be replaced by a specific noun.

EXAMPLE There is nothing in the book that has not already been published in some form, but some of the *information* is, I believe, very little known.

Or the idea can be stated positively. (See also **positive writing**.)

EXAMPLE Everything in the book has been published in some form, but some of it is, I believe, very little known.

drawings

A drawing is useful when you wish to focus on details or relationships that a **photograph** cannot capture. A drawing can emphasize

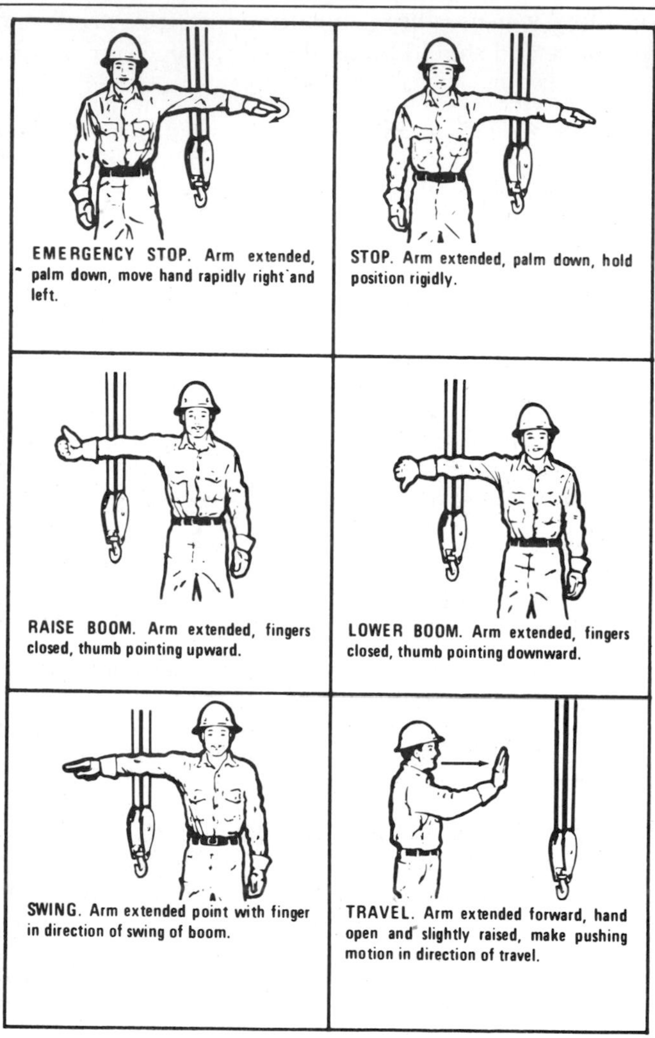
EMERGENCY STOP. Arm extended, palm down, move hand rapidly right and left.
STOP. Arm extended, palm down, hold position rigidly.
RAISE BOOM. Arm extended, fingers closed, thumb pointing upward.
LOWER BOOM. Arm extended, fingers closed, thumb pointing downward.
SWING. Arm extended point with finger in direction of swing of boom.
TRAVEL. Arm extended forward, hand open and slightly raised, make pushing motion in direction of travel.

Figure 3. Hand Signals for Crane Operation

Source: Harnischfeger Corporation

Figure 1   Drawing That Illustrates Instructions

the significant part of a mechanism, or its function, and omit what is not significant. However, if the precise details of the actual appearance of an object are necessary to your **report** or document, a photograph is essential.

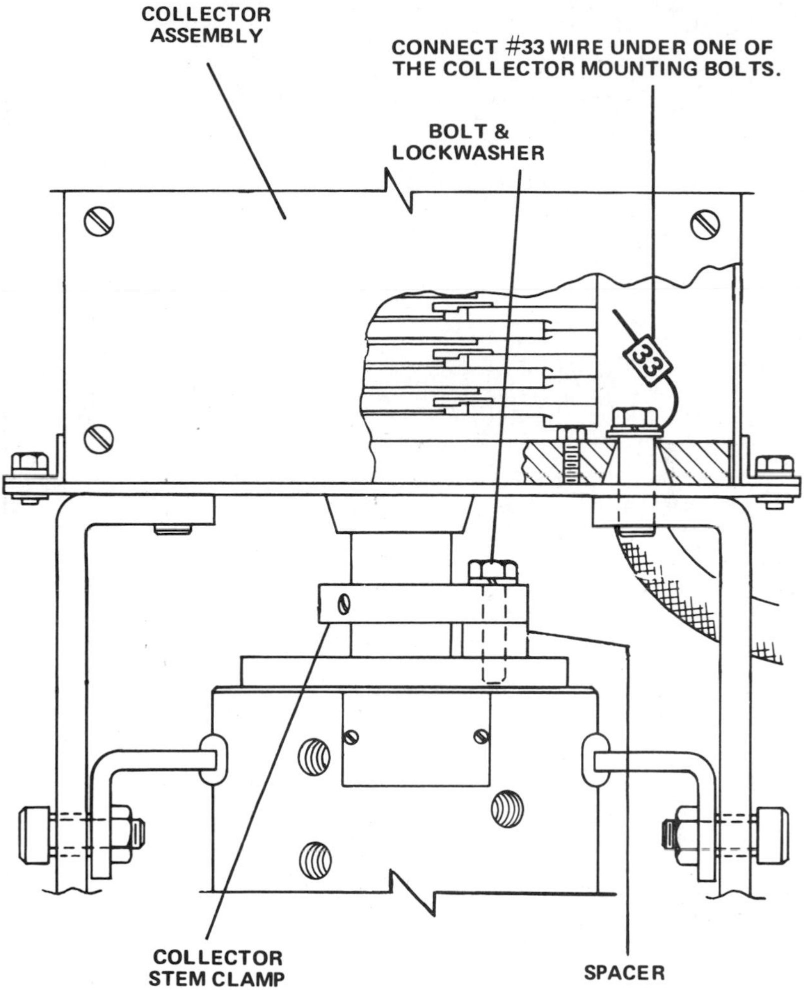

Figure 4. Detail View—Collector
Source: Harnischfeger Corporation

Figure 2 Cutaway Drawing

There are various types of drawings, each with unique advantages. The type of drawing used for an **illustration** should be determined by the specific purpose it is intended to serve. If your **reader** needs an impression of an object's general appearance or an overview of a series of steps or directions, a conventional drawing of the

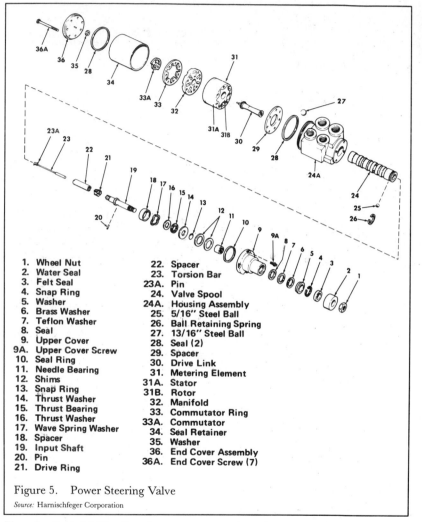

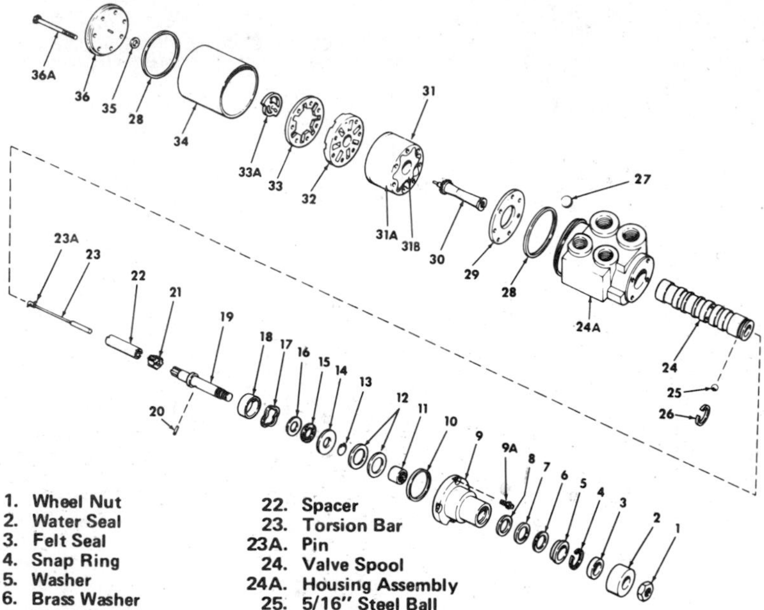

1. Wheel Nut
2. Water Seal
3. Felt Seal
4. Snap Ring
5. Washer
6. Brass Washer
7. Teflon Washer
8. Seal
9. Upper Cover
9A. Upper Cover Screw
10. Seal Ring
11. Needle Bearing
12. Shims
13. Snap Ring
14. Thrust Washer
15. Thrust Bearing
16. Thrust Washer
17. Wave Spring Washer
18. Spacer
19. Input Shaft
20. Pin
21. Drive Ring
22. Spacer
23. Torsion Bar
23A. Pin
24. Valve Spool
24A. Housing Assembly
25. 5/16" Steel Ball
26. Ball Retaining Spring
27. 13/16" Steel Ball
28. Seal (2)
29. Spacer
30. Drive Link
31. Metering Element
31A. Stator
31B. Rotor
32. Manifold
33. Commutator Ring
33A. Commutator
34. Seal Retainer
35. Washer
36. End Cover Assembly
36A. End Cover Screw (7)

Figure 5. Power Steering Valve

Source: Harnischfeger Corporation

Figure 3 Exploded-View Drawing

type illustrated by Figure 1 will suffice. Where it is necessary to show the internal parts of a piece of equipment in such a way that their relationship to the overall equipment is clear, a cutaway drawing is necessary, as in Figure 2. To show the proper sequence in which parts fit together or when it is essential to show the details of each individual part, use an exploded-view drawing such as Figure 3.

TIPS FOR CREATING AND USING DRAWINGS

Many organizations have their own **format** specifications. In the absence of such specifications, the following tips will be helpful:

1. Give the drawing a clear title and a figure number (if figure numbers are used), both of which should be centered below the drawing.
2. Place the source or courtesy line, if necessary, in the lower left corner.
3. Show the equipment from the point of view of the person who will use it.
4. When illustrating a subsystem, show its relationship to the larger system of which it is a part.
5. Draw the different parts of an object in proportion to one another, unless you indicate that certain parts are enlarged.
6. When a sequence of drawings is used to illustrate a process, arrange them from left to right or from top to bottom.
7. Label parts in the drawing so that text references to them are clear.
8. Depending on the complexity of what is shown, labels may be placed on the parts themselves, or the parts may be given letter or number symbols, with an accompanying key (see Figure 3).

due to/because of

Due to (meaning "caused by") is acceptable following a **linking verb**.

EXAMPLES His short temper was *due to* work strain.
His short temper, which was *due to* work strain, made him inefficient.

Due to is not acceptable, however, when it is used with a nonlinking verb to replace *because of*.

CHANGE He went home *due to* illness.
TO He went home *because of* illness.

E

each

When *each* is used as a **subject**, it takes a singular **verb** or **pronoun**.

EXAMPLES *Each* of the reports *is* to be submitted ten weeks after *it* is assigned.

Each worked as fast as *his* ability would permit.

When *each* occurs after a plural subject with which it is in apposition, it takes a plural verb or pronoun. (See also **agreement**.)

EXAMPLE The reports *each have* white embossed titles on *their* covers.

each and every

Although the phrase *each and every* is commonly used in speech in an attempt to emphasize a point, the phrase is redundant and should be eliminated from your writing. Replace it with *each* or *every*.

CHANGE *Each and every* part was accounted for.
TO *Each* part was accounted for.
OR *Every* part was accounted for.

economic/economical

Economic refers to the production, development, and management of material wealth. *Economical* simply means "not wasteful or extravagant."

EXAMPLES The strike at General Motors had an *economic* impact on the country.

Since gasoline will be in short supply, drivers should be as *economical* as possible with its use.

e.g./i.e.

The **abbreviation** *e.g.* stands for the Latin *exempli gratia*, meaning "for example." Since a perfectly good English expression exists for the same use (*for example*), there is no need to use a Latin expression or abbreviation except in notes and illustrations where you need to save space. In addition, many people confuse *e.g.* with *i.e.*, which means "that is." The abbreviation *e.g.* does not save enough space to justify possible misunderstanding. Avoid *e.g.* in your writing.

CHANGE Some terms of the contract (*e.g.*, duration and job classification) were settled in the first two bargaining sessions.
TO Some terms of the contract (*for example*, duration and job classification) were settled in the first two bargaining sessions.
OR Some terms of the contract, *such as* duration and job classification, were settled in the first two bargaining sessions.

The abbreviation *i.e.* stands for the Latin *id est*, meaning "that is." Since *that is* is a perfectly good English expression, there is no need to use a Latin expression or its abbreviation. In addition, many people confuse *i.e.* with *e.g.*, never being certain which means what. This abbreviation does not save enough space to justify having one person misunderstand its meaning; avoid *i.e.* in your writing.

CHANGE We were a fairly heterogeneous group; *i.e.*, there were managers, foremen, and vice-presidents at the meeting.
TO We were a fairly heterogeneous group; *that is*, there were managers, foremen, and vice-presidents at the meeting.

If you must use *i.e.* or *e.g.*, punctuate them as follows. If *i.e.* or *e.g.* connects two **independent clauses**, precede it with a **semicolon** and follow it with a **comma**, as in the above example. If *i.e.* or *e.g.* connects a **noun** and **appositives**, a comma should precede and follow it.

EXAMPLES We were a fairly heterogeneous group, *i.e.*, managers, foremen, and vice-presidents.
We were a fairly heterogeneous group; *i.e.*, we were managers, foremen, and vice-presidents.

elegant variation

The attempt to avoid repeating a **noun** in a **paragraph** by substituting other words (often pretentious **synonyms**) is called *elegant variation*. Avoid this practice—repeat a word if it says what you mean, or use a suitable **pronoun**. (See also **affectation** and **long variants**.)

CHANGE The *use* of modules in the assembly process has increased production. *Modular utilization* has also cut costs.

TO The *use* of modules in the assembly process has increased production. The *use* of modules has also cut costs.

OR The use of modules in the assembly process has increased production and also cut costs.

(See also **thesaurus**.)

ellipses

When you omit words in quoted material, use a series of three spaced periods—called *ellipsis dots*—to indicate the omission. However, such an omission should not detract from or alter the essential meaning of the sentence.

EXAMPLES Technical material distributed for promotional use is sometimes charged for, particularly in high-volume distribution to schools, although prices for these publications are not uniformly based on the cost of developing them. (without omission)

Technical material distributed for promotional use is sometimes charged for . . . although prices for these publications are not uniformly based on the cost of developing them. (with omission)

When the omitted portion comes at the beginning of the sentence, begin the **quotation** with a lowercase letter.

EXAMPLES "When the programmer has determined a system of runs, he must create a system flowchart to provide a picture of the data flow through the system." (without omission)

The letter states that the programmer ". . . must create a system flowchart to provide a picture of the data flow through the system." (with omission)

When the omission comes at the end of a sentence and you continue the quotation following the omission, use four periods to indicate both the final period to end the sentence and the omission.

EXAMPLES In all publications departments except ours, publications funds—once they are initially allocated by higher management—are controlled by publications personnel. Our company is the only one to have nonpublications people control funding within a budgeted period. In addition, all publications departments control printing funds as well. (without omission)

In all publications departments except ours, publications funds—once they are initially allocated by higher management—are controlled by publications personnel. . . . In addition, all publications departments control printing funds as well. (with omission)

Use a full line of periods across the page to indicate the omission of one or more **paragraphs**.

EXAMPLES A computer system operates with two types of programs: the software programs supplied by the manufacturer and the programs created by the user. The manufacturer's software consists of a number of programs that enable the system to perform complicated manipulations of data from the relatively simple instructions specified in the user's program.

Programmers must take a systematic approach to solving any data-processing problem. They must first clearly define the problem, and then they must define a system of runs that will solve the problem. For example, a given problem may require a validation and sort run to handle account numbers or employee numbers, a computation and update run to manipulate the data for the output reports, and a print run to produce the output reports.

When a system of runs has been determined, the programmer must create a system flowchart to provide a picture of the data flow through the system. (without omission)

A computer system operates with two types of programs: the software programs supplied by the manufacturer and the programs created by the user. The manufacturer's software consists of a number of programs that enable the system to perform complicated manipulations of data from the relatively simple instructions specified in the user's program.

. . . .

When a system of runs has been determined, the programmer must create a system flowchart to provide a picture of the data flow through the system. (with omission)

Do not use ellipsis dots for any purpose other than to indicate omission. Advertising copywriters, particularly those not skilled at achieving the effect they strive for with words, often use ellipsis dots to substitute for all marks of **punctuation**—and sometimes even use them where no punctuation is needed. This practice achieves no positive effect, and it is a poor one to emulate.

eminent/imminent/immanent

Someone or something that is *eminent* is outstanding or distinguished.

EXAMPLE She is an *eminent* scientist.

If something is *imminent*, it is about to happen.

EXAMPLE He paced nervously, knowing that the committee's decision was *imminent*.

Immanent, meaning "inherent" or "indwelling," is used chiefly by theologians and philosophers to refer to the theory that a deity pervades and sustains all material existence.

EXAMPLE The theologian suggested that God is *immanent* in all life forms.

emphasis

Emphasis is the principle by which ideas in writing are stressed according to their importance. The first step in achieving emphasis is to write simply and concisely. If a sentence expresses the writer's meaning simply and directly, it has the proper emphasis. But there are also mechanical means of achieving emphasis: by position within a sentence, **paragraph**, or **report**; by the use of **repetition**; by selection of sentence type; by varying the length of sentences; by the use of climactic order within a sentence; by **punctuation** (use of the **dash**); by the use of **intensifiers**; by the use of mechanical devices, such as **italics** and **capital letters**; and by direct statement (using such terms as *most important* and *foremost*).

Emphasis can be achieved by position because the first and last words of a sentence stand out in the **reader**'s mind.

CHANGE Because they reflect geological history, moon craters are important to understanding the earth's history.

TO Moon craters are important to understanding the earth's history because they reflect geological history.

OR Because moon craters reflect geological history, they are important to understanding the earth's history.

Notice that the revised versions of the sentence emphasize moon craters simply because the term is in the front part of the sentence. Similarly, the first and last sentences in a paragraph and the first and last paragraphs in a report or paper tend to be the most emphatic to the reader.

EXAMPLE Energy does far more than simply make our daily lives more comfortable and convenient. Suppose you wanted to stop—and reverse—the economic progress of this nation. What would be the surest and quickest way to do it? Find a way to cut off the nation's oil resources! Industrial plants would shut down, public utilities would stand idle, all forms of transportation would halt. The country would be paralyzed, and our economy would plummet into the abyss of national economic ruin. *Our economy, in short, is energy-based.*

—*The Baker World* (Los Angeles: Baker Oil Tools, Inc., 1964) p. 2.

Another way to achieve emphasis is to follow a very long sentence, or a series of long sentences, with a very short one.

EXAMPLE We have already reviewed the problem the bookkeeping department has experienced during the past year. We could continue to examine the causes of our problems and point an accusing finger at all the culprits beyond our control, but in the end it all leads to one simple conclusion. *We must cut costs.*

Emphasis can be achieved by the repetition of key words and phrases.

EXAMPLE Similarly, atoms *come and go* in a molecule, but the molecule *remains*; molecules *come and go* in a cell, but the cell *remains*; cells *come and go* in a body, but the body *remains*; persons *come and go* in an organization, but the organization *remains*.

—Kenneth Boulding, *Beyond Economics* (Ann Arbor: University of Michigan Press, 1968), p. 131.

Different emphasis can be achieved by the selection of a **compound sentence, complex sentence**, or **simple sentence**.

EXAMPLES The report turned in by the police detective was carefully illustrated, and it covered five pages of single-spaced copy. (This compound sentence carries no special emphasis because it contains two coordinate independent clauses.)

The police detective's report, which was carefully illustrated, covered five pages of single-spaced copy. (This complex sentence emphasizes the size of the report.)

The carefully illustrated report turned in by the police detective covered five pages of single-spaced copy. (This simple sentence emphasizes that the report was carefully illustrated.)

Emphasis can be achieved by a climactic order of ideas or facts within a sentence.

EXAMPLE Over subsequent weeks the industrial relations department worked diligently, management showed tact and patience, and the employees finally accepted the new policy.

Emphasis can be achieved by setting an item apart with a **dash**.

EXAMPLE Here is where all the trouble begins—in the American confidence that technology is ultimately the medicine for all ills.

Emphasis can be achieved by the use of intensifiers (*most, very, really*), but this technique is so easily abused that it should be used only with caution.

EXAMPLE The final proposal is *much* more persuasive than the first.

Emphasis can be achieved by such mechanical devices as italics (underlining) and capital letters. But this technique is also easily abused and should be used with caution.

EXAMPLE When we add the commonplace situations that allow these systems to function for hundreds of researchers *simultaneously* and on all five continents, the power of this new information medium is even more remarkable.

However, do not use all capital letters to try to show emphasis. Doing so can cause confusion because capital letters are used for so many other reasons.

The word *do* (or *does*) may be used for emphasis, but this is also easily overdone, so proceed with caution. Compare the following examples:

EXAMPLES You believe weekly staff meetings are essential, don't you? (unemphatic)
You *do* believe weekly staff meetings are essential, don't you? (emphatic)

Direct address may also be used for emphasis in **correspondence**.

EXAMPLE John, I believe we should rethink our plans.

(See also **subordination**.)

English, varieties of

There are two broad varieties of written English: standard and nonstandard. These varieties are determined through **usage** by those who write in the English language. Standard English (also called American edited English) is used to carry on the daily business of the nation. It is the language of business, industry, government, education, and the professions. Standard English is characterized by exacting standards of **punctuation** and capitalization, by accurate **spelling**, by exact **diction**, by an expressive vocabulary, and by knowledgeable usage choices. Nonstandard English, on the other hand, is the language of those not familiar with the standards of written English. This form of English rarely appears in printed material except when it is used for special effect by fiction writers. Nonstandard English is characterized by inexact or inconsistent punctuation, capitalization, spelling, diction, and usage choices. Both standard and nonstandard English find their way into everyday speech and writing in the forms of the following subcategories.

COLLOQUIAL

Colloquial English is spoken standard English or writing that attempts to re-create the flavor of this kind of speech by using words and expressions common to casual conversation. Colloquial English is appropriate to some kinds of writing (personal letters, notes, and the like) but not to most technical writing.

VERNACULAR

Vernacular English is the spoken form of the language, as opposed to its written form. It is the form used by the majority of those who speak the language. To write "in the vernacular" is to imitate this kind of language. Ordinarily, such writing is confined to fiction. Vernacular can also refer to the manner of expression common to a trade or profession (one might speak of "the legal vernacular"), although **jargon** more clearly expresses this meaning.

DIALECT

Dialectal English is a social or regional variety of the language that is comprehensible to people of that social group or region but that may be incomprehensible to outsiders. Dialect, which is usually nonstandard English, involves distinct **word choices**, grammatical forms, and pronunciations. Technical writing, because it aims at a broad audience, should be free of dialect.

LOCALISM

A localism is a word or phrase that is unique to a geographical region. For example, the words *poke, sack*, and *bag* all denote "a paper container," each term being peculiar to a different region of the United States. Such words should normally be avoided in writing because knowledge of their meanings is too narrowly restricted.

SLANG

Slang refers to a manner of expressing common ideas in new, often humorous or exaggerated, ways. Slang often finds new use for familiar words ("He *crashed* early last night"—meaning that he went to bed early) or coins words ("He's a *kook*"—meaning that he is offbeat, unconventional). Slang expressions usually come and go very quickly. The fact that a fair number of words in the present standard English vocabulary were once considered slang indicates that when a slang word fills a legitimate need it is accepted into the language. *Skyscraper, bus,* and *date* (as in "to go on a date"), for example, were once considered slang expressions. Although slang may have a valid place in some writing, particularly in fiction, be careful not to let it creep into technical writing.

BARBARISM

A barbarism is any obvious misuse of standard words, grammatical forms, or expressions. Examples include *ain't, irregardless, done got, drownded*.

equal/unique/perfect

Logically, *equal* (meaning "having the same quantity or value as another"), *unique* (meaning "one of a kind"), and *perfect* (meaning "a state of highest excellence") are **absolute words** and therefore should not be compared. However, colloquial usage of *more* and *most* as **modifiers** of *equal, unique*, and *perfect* is so common that an absolute prohibition against such use is impossible.

EXAMPLE Yours is a *more unique* (or *perfect*) coin than mine.

Some writers try to overcome the problem by using *more nearly equal (unique, perfect)*. When **clarity** and preciseness are critical, the use of **comparative degrees** with *equal, unique*, and *perfect* can be misleading. The best rule of thumb is to avoid using the comparative degrees with absolute terms.

CHANGE Ours is a more *equal* percentage split than theirs.
TO Our percentage split is 51–49; theirs is 54–46.

-ese

The **suffix** *-ese* is used to designate types of **jargon** or certain languages or literary **styles** (official*ese*, journal*ese*, Chin*ese*, Pentagon*ese*).

etc.

Etc. is an **abbreviation** for the Latin *et cetera*, meaning "and others"; therefore, *etc.* should not be used with *and*.

CHANGE He brought pencils, pads, erasers, a calculator, *and etc.*
TO He brought pencils, pads, erasers, a calculator, *etc.*

Do not use *etc.* at the end of a list or series introduced by the **phrases** *such as* or *for example* because these phrases already indicate that there are other things of the same category that are not named.

CHANGE He brought camping items, *such as* backpacks, sleeping bags, tents, *etc.*
TO He brought backpacks, sleeping bags, tents, *etc.*
OR He brought camping items, *such as* backpacks, sleeping bags, and tents.

In careful writing, *etc.* should be used only when there is logical progression (1, 2, 3, etc.) and when at least two items are named. It is often better to avoid *etc.* altogether, however, because there may be confusion as to the identity of the "class" of items listed.

EXAMPLE He brought backpacks, sleeping bags, tents, and other camping items.

euphemism

A euphemism is a word that is an inoffensive substitute for one that could be distasteful, offensive, or too blunt.

EXAMPLES *remains* for *corpse*
passed away for *died*
marketing representative for *salesman*
previously owned or *preowned* for *used*

Used judiciously, a euphemism might help you avoid embarrassing or offending someone. Overused, however, euphemisms can hide the facts of a situation (such as *incident* for *accident*) or be a form of **affectation**. (See also **word choice**.)

everybody/everyone

Both *everybody* and *everyone* are usually considered singular and so take singular **verbs** and **pronouns**.

EXAMPLES *Everybody is* happy with the new contract.
Everyone here *eats* at 11:30 a.m.
Everybody at the meeting made *his* proposals separately.
Everyone went *his* separate way after the meeting.

One exception is when the meaning is obviously plural.

EXAMPLE *Everyone* laughed at my sales slogan, and I really couldn't blame *them*.

Another exception is when the use of singular verbs and pronouns would be offensive by implying sexual bias; in such a situation, it is better to use plural verbs and pronouns or to use the expression "his or her" than to offend.

CHANGE *Everyone* went *his* separate way after the meeting.
TO *They* all went *their* separate ways after the meeting.
OR Everyone went *his or her* separate way after the meeting.

Although normally written as one word, *everyone* is written as two words when each individual in a group should be emphasized.

EXAMPLES *Everyone* here comes and goes as he pleases.
Every one of the team members contributed to this discovery.

(See also **indefinite pronouns**.)

ex post facto

The expression *ex post facto*, which is Latin for "after the fact," refers to something that operates retroactively. The term is normally used to refer to laws and is best avoided in other contexts.

CHANGE The contract will be in effect *ex post facto* of the negotiations.
TO The contract will be in effect after the negotiations.

exclamation marks

The purpose of the exclamation mark (!) is to indicate the expression of strong feeling. It can signal surprise, fear, indignation, or excitement. In technical writing, the exclamation mark is normally used only in cautions and warnings. It cannot make an argument more convincing, lend force to a weak statement, or call attention to an intended irony—no matter how many are stacked like fence posts at the end of a sentence.

USES

The most common use of an exclamation mark is after a word (**interjection**), **phrase**, **clause**, or sentence to indicate surprise. Interjections are words that express strong emotion.

EXAMPLES Ouch! Wow! Oh! Stop! Hurry!

An exclamation mark can also be used after a whole sentence, or even an element of a sentence.

EXAMPLES The subject of this meeting—please note it well!—is our budget deficit.
How exciting is Stravinsky's *The Rite of Spring!*

An exclamation mark is sometimes used after a title that is an exclamatory word, phrase or sentence.

EXAMPLES "Our information Retrieval System Must Change!" is an article by Richard Moody.
The Cancer with No Cure! is a book by Wilbur Moody.

When used with **quotation marks**, the exclamation mark goes outside, unless what is quoted is an exclamation.

EXAMPLE The boss yelled, "Get in here!" Then Ben, according to Ray, "jumped like a kangaroo"!

executive summaries

The purpose of an executive summary is to consolidate the principal points of a **report** in one place. It must cover the information in the report in enough detail to reflect accurately its contents but concisely enough to permit an executive to digest the significance of the report without having to read it in full. They are called *executive summaries* because the intended audience is the busy executive who may have to make funding, personnel, or policy decisions based on findings or recommendations reported for a project.

The executive summary is a comprehensive restatement of the document's purpose, **scope**, methods, results, **conclusions**, findings, and recommendations. The executive summary condenses the entire work or explains how the results were obtained or why the recommendations were made. It simply states the results and recommendations, providing only enough information for a reader to decide whether to read the entire work.

Because they are comprehensive, executive summaries tend to be proportional in length to the larger work they summarize. The typical summary is 10 percent of the length of the report.

The following sample executive summary is adapted from a thirty-page report that describes how electric utilities in a select group of foreign countries provide engineering expertise to control room operators for round-the-clock shift work at nuclear power plants. (See facing page).

The executive summary is usually organized according to the sequence of chapters or sections of the report it summarizes. Note that the sample summary mirrors the structure of the report, as can be seen from its **table of contents**. (A sample table of contents is shown on page 226, following the sample executive summary).

The executive summary should be written so that it can be read independently of the report. It must not refer by number to figures, **tables**, or references contained elsewhere in the report. Executive summaries do occasionally contain a figure, table, or footnote, a practice appropriate as long as that information is integral to the summary. Because executive summaries are frequently read in place of the full report, all uncommon **symbols**, **abbreviations**, and **acronyms** must be spelled out. (For guidance on the location of executive summaries in reports, see **formal reports** and **abstracts**.)

ADDITIONAL GUIDELINES ON WRITING AN EXECUTIVE SUMMARY

1. Write the executive summary after completing the report.
2. Avoid using technical terminology if your readers will include people not familiar with the topic.

EXECUTIVE SUMMARY

Introduction

Purpose and scope

This report describes the experiences and practices of utilities in a selected group of foreign countries with providing engineering expertise on shift in nuclear power plants.[1] The report also discusses the extent to which engineering expertise is made available and the alternative models of providing such expertise. The implications of the foreign experience for U.S. plants is described particularly with reference to the shift technical adviser position and to proposed shift engineer position.

Method

The relevant information for this study came from the open literature, interviews with utility staff, and utility reports.

Major findings

The major conclusions that emerge from this study include:

- The basis for the initial decision of whether to include graduate engineers in the operations shift complement is unclear for most countries. However, reasons for changes in a given system have been identified. Generally, changes have been introduced in response to specific problems (such as high turnover rates among crew members, or an accident situation seen as related to shift expertise).

----

[1]For the purposes of this discussion, "engineering expertise on shift" refers to the use of a university-degreed engineer on shift.

- Two primary models have been used to provide engineering expertise on shift:
    1. graduate engineer as line manager of shift operations (usually in a shift supervisor position)
    2. nonsupervisor graduate engineer position on shift
- The comparison of these two models did not indicate that one system inherently functions more effectively than the other. However, each alternative appears to affect the following specific areas differently:
    1. crew relationships and performance
    2. labor supply, recruitment, and retention
    3. system implementation problems
- Data were not available to analyze whether alternative approaches to providing engineering expertise on shift could be directly linked to plant safety in terms of accident frequency and severity. Indirect relationships to operational safety appear to be primarily through impacts on operations fuctioning (for example, turnover, crew relationships) that ultimately are likely to affect safe operations.
- The determination of which alternative model would be most appropriate for a given country needs to be made from a systems perspective.

Findings

- Decisions about engineering expertise on shift should not be made independently of staffing patterns and organizational design regarding the following issues:

1. effects on imcumbent reactor operations staff of changed career opportunities
2. the availability of degreed engineers and reactor operators
3. career paths for graduate engineers at nuclear power plants
4. organizational relationships between the engineering and operations sections of nuclear power plants

Engineering Expertise Alternatives

In the countries surveyed, essentially two approaches are employed to make engineering expertise available on shift: (1) a graduate engineer occupies a line management position, and (2) a specific engineering position was created to provide expertise to the operations staff. Both approaches are relevant to issues under consideration in the United States. In the United States, proposals have been made to require that all licensed reactor operators have engineering degrees. This proposed requirement would apply to reactor operators, senior reactor operators, and shift supervisors. Proposals have also been made to create a position for a degreed shift engineer in addition to the current requirement for a shift technical adviser.

Currently, Spain, Italy, and Japan do not, as a rule, use graduate engineers on shift. On the other hand, Canada, the Federal Republic of Germany, and the United Kingdom typically do have an engineering graduate on shift in operations line management position or have made the decision to establish this policy. The use of a separate engineer position on shift is found in Switzerland, France, and Sweden.

Implications of Alternative Models

Two basic models of providing engineering expertise on shift are used: (1) graduate engineers in line management operations positions and (2) a specific engineering position of shift. Given the limited data for evaluating the various approaches used by the countries, no clear evidence emerges that one system inherently functions more effectively than the other. However, the specific features of the organizational approaches used by these countries appear to affect (1) crew relationships and performance, (2) labor supply, recruitment, and retention, and (3) implementation and transition issues.

Conclusions and Recommendation

The foreign experiences described in this report suggest some key issues relevant to assessing the present U.S. requirement of shift technical adviser and alternative proposals, including creation of a shift engineer position and requirement of baccalaureate engineering degrees for shift supervisors. Either model, operations line management or a separate engineering position, involves some trade-offs in regard to system advantages.

One important issue concerns whether the separation of engineers from routine operational functions enhances or detracts from their abilities to problem-diagnose and make appropriate decisions in situations of operational failure. No sufficient evidence exists to answer this question; both options have adherents based on positive experience with a given system.

Using a degreed engineer as shift supervisor appears to have advantages for crew integration and the combination of technical expertise and operations experience in the major position of authority on shift. However, significant problems have been experienced with the retention of graduate engineers as shift supervisors. Additionaly, if this requirement were imposed in the United States, it would limit the career prospects for incumbent reactor operators, which could have adverse consequences on crew performance.

Recommendation Creating a separate nonsupervisory engineering position could have advantages in providing a job function and career path that more fully utilize engineering expertise than the shift supervisor position does. In addition, such a position does not require a change in the established recruitment and career path open to existing reactor operators.

TABLE OF CONTENTS

3. Make the summary concise but not brusque. Be especially careful not to omit transitional words and **phrases** (such as *however, moreover, therefore, for example,* and *in summary*).
4. Finally, introduce no information not discussed in the report.

explaining a process (see **process explanation**)

expletives

An expletive is a word that fills the position of another word, phrase, or clause. *It* and *there* are the usual expletives.

EXAMPLE *It* is certain that he will go.

In this example, the expletive *it* occupies the position of subject in place of the real subject, *that he will go*. Although expletives are sometimes necessary to avoid **awkwardness**, they are commonly overused and most sentences can be better stated without them.

CHANGE *There are* several reasons that I did it.
TO I did it for several reasons.

CHANGE *There were* many orders lost for unexplained reasons.
TO Many orders were lost for unexplained reasons.
OR We lost many orders for unexplained reasons.

In addition to its usage as a grammatical term, the word *expletive* means an exclamation or oath, especially one that is profane.

explicit/implicit

An explicit statement is one expressed directly, with precision and clarity. An implicit meaning may be found within a statement, even though it is not directly expressed.

EXAMPLES His directions to the new plant were *explicit*, and we found it with no trouble.
Although he did not mention the nation's financial condition, the danger of an economic recession was *implicit* in the President's speech.

exposition

Exposition, or expository writing, informs the **reader** by presenting facts and ideas in direct and concise language that is usually not adorned with colorful or figurative words and **phrases**. Expository writing attempts to explain to the reader what its subject is, how it works, and how it relates to something else. Exposition is aimed at the reader's understanding, rather than at his or her imagination or emotions; it is a sharing of the writer's knowledge with the reader. The most important function of exposition is to give accurate and complete information to the reader and analyze it.

Figure 1 explains the use of software in a computer system by defining it, explaining how it works, and showing how it functions with the hardware equipment:

> A major component of an electronic data processing system is the software—the programs supplied by the manufacturer to direct the computer's internal operation. The software supplied with the NCR Century 50 System is an extensive package that includes the operating system, utility routines, applied programs, and the NEAT/3 Compiler. These programs have been designed to perform specific functions but, at the same time, to interact with one another to increase processing efficiency and to avoid duplication.
>
> To conserve valuable memory space, a large portion of the software package remains on disc; only the most frequently used portion resides in internal memory all of the time. The disc-resident software is organized into small modules that are called into memory as needed to perform specific functions.
>
> The memory-resident portion of the operating system maintains strict control of processing. It consists of routines, subroutines, lists, and tables that are used to perform common program functions, such as processing input and output operations, calling other software routines from disc as needed, and processing errors.
>
> The disc-resident portion of the operating system contains routines that are used less frequently in system operation, such as the peripheral-related software routines that are used for correcting errors encountered on the various units, and the log and display routines that record unusual operating conditions in the system log. The disc-resident portion of the operating system also contains Monitor, the software program that supervises the loading of utility routines and the user's programs.
>
> —*NCR Century Elementary Systems Manual* (Dayton: NCR Corporation, 1974), p. 10.

Figure 1 Expository Writing

Because it is the most effective **form of discourse** for explaining difficult subjects, exposition is widely used in technical writing. To write exposition, you must have a thorough knowledge of your subject. As with all writing, how much of that knowledge you pass on to your reader should depend on the reader's needs and your **objective**.

F

fact

Expressions containing the word *fact* ("due to the *fact* that," "except for the *fact* that," "as a matter of *fact*," or "because of the *fact* that") are often wordy substitutes for more accurate terms.

CHANGE *Due to the fact that* the sales force has a high turnover rate, sales have declined.
TO *Because* the sales force has a high turnover rate, sales have declined.

The word *fact* is, of course, valid when facts are what is meant.

EXAMPLE Our research has brought out numerous *facts* to support your proposal.

Do not use the word *fact*, however, to refer to matters of judgment or opinion.

CHANGE *It is a fact that* sales are poor in the Midwest because of insufficient market research.
TO In my opinion, sales are poor in the Midwest because of insufficient market research.
OR Our analysis of the statistics indicates that sales are poor in the Midwest because of insufficient market research.
OR We infer from our statistics that sales are poor in the Midwest because of insufficient market research.

(See also **logic**.)

feasibility reports

When the managers of an organization plan to undertake a new project—a move, the development of a new product, an expansion, or the purchase of new equipment—they try to determine the project's chances for success. A feasibility report is the study conducted to help them make this determination. This **report** presents evidence about the practicality of the proposed project: How much will it cost? Is sufficient manpower available? Are any legal or other special requirements necessary? Based on the evidence, the writer of the feasibility report recommends whether or not the project should be carried out. Management officials then consider the recommendation.

The most efficient way to begin work on a feasibility report is to state clearly and concisely the purpose of the study.

EXAMPLES The purpose of this study is to determine what type and how many new vans should be purchased to expand our present delivery fleet.

This study will determine which of three possible sites should be selected for our new warehouse.

A clearly worded statement of purpose acts as an effective guide for gathering and organizing information for the study. It both states the objective and defines the **scope** of the study. In a feasibility study, the scope includes the alternatives for accomplishing the purpose and the criteria by which each alternative will be examined.

A firm needing a word-processing system, for example, might conduct a feasibility study to determine which system would best suit its requirements. The firm's requirements would establish the criteria by which each alternative system is evaluated. The following example shows how the preliminary topic outline for such a study might be organized:

I. Purpose: To determine which word-processing system would best serve our office needs.
II. Alternatives: Word-processing systems of various vendors.
III. Criteria:
   A. Tasks the equipment must perform.
      1. Ability to work with repetitive elements.
      2. Ability to accommodate extensive revision.
   B. Costs.
      1. Base vs. rental.
      2. Installation.
      3. Maintenance.
      4. Training time.

In writing a feasibility report, you must first identify the alternatives and then evaluate each against your established criteria. After completing these analyses, summarize them in a **conclusion**. This summary of relative strengths and weaknesses usually points to one alternative as the best, or most feasible. Make your recommendation on the basis of this conclusion.

The structure of a feasibility report depends on its scope. Some studies are brief and informal; others are detailed and formal.

Although the order of elements may vary, every feasibility report should contain the following sections: (1) an introduction, (2) a body, (3) a conclusion, and (4) a recommendation.

INTRODUCTION

The introduction should state the purpose of the report, describe the problems that led to it, and include any pertinent background information. You may discuss the scope or extent of the report. You might also discuss procedures or methods used in the analyses of alternatives: Were any special techniques used to gather or analyze information? Was information obtained by interviews, an examination of financial records, computer analyses, or other means? Any limitations on the study should be noted here: Does a deadline have to be met? Were there any restrictions on how data could be obtained?

BODY

The body of the report should present a detailed evaluation of all alternatives under consideration. Evaluate each alternative according to your established criteria. Ordinarily, each evaluation would comprise a separate section of the report.

CONCLUSION

The conclusion should summarize the evaluation of each alternative, usually in the order in which they are discussed in the body of the report. In your conclusion you may also want to interpret the detailed evaluations.

RECOMMENDATION

This section must state the alternative that best meets the criteria.

SAMPLE FEASIBILITY REPORT

The sample feasibility report shown in Figure 1 opens with an **introduction** that states the purpose of the report, the problem that prompted the study, the alternatives that were examined, and the criteria used. The body presents a detailed discussion of each alternative, particularly in terms of the criteria stated in the introduction. The conclusion draws together and summarizes the details in the body of the study. The recommendation section suggests the course of action that the company should take.

INTRODUCTION

The purpose of this report is to determine which of two proposed computer processors would best enable the Jonesville Engineering and Manufacturing Branch to increase its data-processing capacity and thus to meet its expanding production requirements.

Problem

In October 19-- the Information Systems and Support Group at Jonesville put the MISSION System into operation. Since then the volume of processing transactions has increased fivefold (from 1,000 to 5,000 updates per day). This increase has severely impaired system response time from less than 10 seconds in 1978 to 120 seconds on average at present. Degraded performance is also apparent in the backlog of batch-processing transactions. During a recent check 70 real-time and approximately 2,000 secondary transactions were backlogged. In addition, the ARC 98 Processor that runs MISSION is nine years old and frequently breaks down. Downtime caused by these repairs must be made up in overtime. In a recent 10-day period in January, processor downtime averaged 25 percent during working hours (7:30 a.m. to 6:00 p.m.). In February the system was down often enough that the entire plant production schedule was endangered.

Finally, because the ARC 98 cannot keep up with the current workload, the following new systems, all essential to increased plant efficiency and productivity, cannot be implemented: shipping and billing, labor collection, master scheduling, and capacity planning.

Scope

Two alternative solutions to provide increased processing capacity have been investigated: (1) purchase of a new ARC 98 Processor to supplement the first, and (2) purchase of a Landmark I Processor to replace the current ARC 98. The two alternatives will be evaluated primarily according to cost

Figure 1 Feasibility Report

and, to a lesser extent, according to expanded capacity for future operations.

PURCHASING A SECOND ARC 98 PROCESSOR

This alternative would require additional annual maintenance costs, salary for an additional computer operator, increased energy costs, and a one-time construction cost for a new facility to house the processor.

Annual maintenance costs	$ 45,000
Annual salary for computer operator	16,000
Annual increased energy costs	7,500
Annual operating costs	$ 68,500
Facility cost (one-time)	$ 50,000
Total first-year cost	$118,500

These costs for the installation and operation of another ARC 98 Processor are expected to produce the following anticipated savings in hardware reliability and system readiness.

Hardware Reliability

A second ARC 98 would reduce current downtime periods from four to two per week. Downtime recovery averages 30 minutes and affects 40 users. Assuming that 50 percent of users require the system at a given time, we determined that the following reliability savings would result:

2 downtimes × 30 minutes × 40 users × 50% × $9.00/hour overtime × 52 weeks = $9,360 (annual savings)

System Readiness

Currently, an average of one day of batch processing per week cannot be completed. This gap prevents online system readiness when users report to work and affects all users at least one hour per week. Improved productivity would yield these savings:

40 users × 1 hour/week × $6.00/hour average wage rate × 52 weeks = $12,480 (annual savings)

Summary of Savings

Hardware reliability	$ 9,360
System readiness	12,480
Total annual savings	$21,840

Costs and Savings for ARC 98 Processor

Costs	
Annual	$ 68,500
One-time	50,000
First-year total	$118,500
	– 50,000
Annual total	$ 68,500
Savings	
Hardware reliability	$ 9,360
System readiness	12,480
Total annual savings	$21,840
Annual Costs Less Savings	
Annual costs	$ 68,500
Annual savings	– 21,840
Net additional annual operating cost	$ 46,660

ARC 98 Capacity

By adding a second ARC 98 processor, current capacity will be doubled. Each processor could process 2,500 transactions per day while cutting response time from 120 seconds to 60 seconds. However, if new systems essential to increased plant productivity are added to the MISSION System, efficiency could be degraded to its present level in the next three to five years. This estimate is based on the assumption that the new systems will add between 250 and 500 transactions per day immediately. These figures could increase tenfold in the next several years if current rates of expansion continue.

PURCHASING A LANDMARK I PROCESSOR

This alternative will require additional annual maintenance costs, increased energy costs, and a one-time facility adaptation cost.

Annual maintenance costs	$45,000
Annual energy costs	6,500
Annual operating costs	$51,500
Cost of adapting existing facility	$15,300
Total first-year cost	$66,800

These costs for installation of the Landmark I Processor are expected to produce the following anticipated savings in hardware reliability, system readiness, and staffing for the Information Systems and Services Department.

Hardware Reliability

Annual savings will be the same as those for the ARC 98 Processor: $9,360.

System Readiness

Annual savings will be the same as those for the ARC 98 Processor: $12,480.

Wages for the Information Systems and Services Department

New system efficiencies would permit the following wage reductions in the department:

One computer operator (wages and fringe benefits)	$16,000
One-shift overtime premium (at $100/week × 52 weeks)	5,200
Total annual wage savings	$21,200

Summary of Savings

Hardware reliability	$ 9,360
System readiness	12,480
Wages	21,200
Total annual savings	$43,040

Cost and Savings for Landmark I Processor

Costs	
Annual	$ 51,500
One-time	15,300
First-year total	$ 66,800
	– 15,300
Annual total	$ 51,500
Savings	
Hardware reliability	$ 9,360
System readiness	12,480
Wages	21,200
Total annual savings	$43,040
Annual Costs Less Savings	
Annual costs	$ 51,500
Annual savings	– 43,040
Net additional annual operating cost	$ 8,460

Landmark I Capacity

The Landmark I processor can process 5,000 transactions per day with an average response time of 10 seconds per transaction. Should the volume of future transactions double, the Landmark I could process 10,000 transactions per day without exceeding 20 seconds per transaction on average. This increase in capacity over the present system would permit implementation of plans to add four new systems to MISSION.

CONCLUSION

A comparison of costs for both systems indicates that the Landmark I would cost $36,380 less in first-year costs.

ARC 98 Costs	
Net additional operating	$46,660
One-time facility	50,000
First-year total	$96,660

Landmark I Costs	
Net additional operating	$ 8,460
One-time facility	15,300
First-year total	$23,760

Installation of a second ARC 98 Processor will permit the present information-processing systems to operate relatively smoothly and efficiently. It will not, however, provide the expanded processing capacity that the Landmark I Processor would for implementing new subsystems essential to improved production and record keeping.

RECOMMENDATION

The Landmark I Processor should be purchased because of the initial and long-term savings and because its expanded capacity will allow the addition of essential systems.

female

Female is usually restricted to scientific, legal, or medical contexts (a *female* patient or suspect). Keep in mind that this term sounds cold and impersonal. The terms *girl, woman*, and *lady* are acceptable substitutes in other contexts; however, be aware that these substitute words have connotations involving age, dignity, and social position. (See also **male**.)

few/a few

In certain contexts, *few* carries more negative overtones than the phrase *a few* does.

EXAMPLES They have *a few* scruples. (positive)
They have *few* scruples. (negative)
There are *a few* good things about your report. (positive)
There are *few* good things about your report. (negative)

fewer/less

Fewer refers to items that can be counted (count nouns).

EXAMPLES A good diet can mean *fewer* colds.
Fewer members took the offer than we expected.

Less refers to mass quantities or amounts (mass nouns).

EXAMPLES *Less* vitamin C in your diet may mean more, not *fewer*, colds.
The crop yield decreased this year because we had *less* rain than necessary for an optimum yield.

figuratively/literally

These two words are often confused. *Literally* means "really" and should not be used in place of *figuratively*, which means "metaphorically." Do not say that someone "literally turned green with envy" unless that person actually changed color.

EXAMPLES In the winner's circle the jockey was, *figuratively* speaking, ten feet tall.
When he said, "Let's run it up the flagpole," he did not mean it *literally*.

Avoid the use of *literally* to reinforce the importance of something.

CHANGE She was *literally* the best of the group.
TO She was the best of the group.

figures of speech

A figure of speech is an imaginative **comparison**, either stated or implied, between two things that are basically unlike but have at least one thing in common. If a device is cone shaped with an opening at the top, for example, you might say that it looks like a volcano.

Technical people may find themselves using figures of speech to clarify the unfamiliar by relating a new and difficult concept to one with which the **reader** is familiar. In this respect, figures of speech help establish a common ground of understanding between the specialist and the nonspecialist. Technical people may also use figures of speech to help translate the abstract into the concrete; in the process of doing so, figures of speech also make writing more colorful and graphic.

Although figures of speech are not used extensively in technical writing, a particularly apt figure of speech may be just the right tool when you must explain or describe a complex concept. A figure of speech must be appropriate, however, to achieve the desired effect.

CHANGE Without the fuel of tax incentives, our economic engine would operate less efficiently. (It would not operate at all without fuel.)

TO Without the fuel of tax incentives, our economic engine would sputter and die. (This is not only apt, but it also states a rather dry fact in a colorful manner—always a desirable objective.)

A figure of speech must also be consistent to be effective.

CHANGE We must get our research program back *on the track*, and we are counting on you to *carry the ball*. (inconsistent)

TO We must get our research program back on the track, and we are counting on you to do it. (inconsistency removed)

A figure of speech should not, however, attract more attention to itself than to the point the writer is making.

EXAMPLE The whine of the engine sounded like ten thousand cats having their tails pulled by ten thousand mischievous children.

Trite figures of speech, which are called **clichés**, defeat the purpose of a figure of speech—to be fresh, original, and vivid. A surprise that comes "like a bolt out of the blue" is not much of a surprise. It is better to use no figure of speech than to use a trite one.

TYPES OF FIGURES OF SPEECH

Analogy is a comparison between two objects or concepts that shows ways in which they are similar. It is very useful in technical writing, especially when you are writing to an educated but nontechnical audience. In effect, analogies say "A is to B as C is to D." The resemblance between these concepts is partial but close enough to provide a striking way of illuminating the relationship the writer wishes to establish.

EXAMPLE Pollution (A) is to the environment (B) as cancer (C) is to the body (D).

Antithesis is a statement in which two contrasting ideas are set off against each other in a balanced syntactical structure.

EXAMPLES Man proposes, but God disposes.
Art is long, but life is short.

Hyperbole is gross exaggeration used to achieve an effect or **emphasis**.

EXAMPLE He *murdered* me on the tennis court.

Litotes are understatements, for emphasis or effect, achieved by denying the opposite of the point you are making.

EXAMPLES Einstein was no dummy.
Fifty dollars is not a small price for a book.

Metaphor is a figure of speech that points out similarities between two things by treating them as though they were the same thing. Metaphor states that the thing being described *is* the thing to which it is being compared.

EXAMPLE He is the sales department's *utility infielder*.

Metonymy is a figure of speech that uses one aspect of a thing to represent it, such as *the red, white, and blue* for the American flag, *the blue* for the sky, and *wheels* for an automobile. This device is common in everyday speech because it gives our expressions a colorful twist.

EXAMPLE *The hard hat* area of the labor force was especially hurt by unemployment.

Simile is a direct comparison of two essentially unlike things, linking them with the word *like* or *as*.

EXAMPLE His feelings about his business rival are so bitter that in recent conversations with his staff he has returned to the subject compulsively, *like a man scratching an itch*.

Personification is a figure of speech that attributes human characteristics to nonhuman things or abstract ideas. One characteristically speaks of the *birth* of a planet and the *stubbornness* of an engine that will not start.

EXAMPLE Early tribes of human beings attributed scientific discoveries to gods. To them, fire was not a *child of man's brain* but a gift from Prometheus.

fine

When used in expressions such as "I feel *fine*" or "a *fine* surf," *fine* is colloquial. The colloquial use of *fine*, like that of *nice*, is too vague for technical writing. In writing, the word *fine* should retain the sense of "refined," "delicate," or "pure."

EXAMPLES A *fine* film of oil covered the surface of the water.
There was a *fine* distinction between the two possible meanings of the disputed passage.
Fine crystal is currently made in Austria.

finite verbs (see **verbs**)

first/firstly

Firstly—like *secondly, thirdly . . . lastly*—is an unnecessary attempt to add the *-ly* form to an **adverb**. *First* is an adverb in its own right, and sounds much less stiff than *firstly*.

CHANGE *Firstly*, we should ask for an estimate.
TO *First*, we should ask for an estimate.

flammable/inflammable/nonflammable

Both *flammable* and *inflammable* mean "capable of being set on fire." Since the *-in* **prefix** usually causes the word following to take its opposite meaning (*incapable, incompetent*), *flammable* is preferable to *inflammable* because it avoids possible misunderstanding.

EXAMPLE The cargo of gasoline is *flammable*.

Nonflammable is the opposite, meaning "not capable of being set on fire."

EXAMPLE The asbestos suit was *nonflammable*.

flowcharts

A flowchart is a diagram of a process that involves stages, with the sequence of stages shown from beginning to end. The flowchart presents an overview of the process that allows the **reader** to grasp the essential steps of the process quickly and easily. The process being illustrated could range from the steps involved in assembling a bicycle to the stages by which bauxite ore is refined into aluminum ingots for fabrication.

Flowcharts can take several forms to represent the steps in a process. They can consist of labeled blocks (Figure 1), pictorial representations (Figure 2), or standardized symbols (Figure 3); the items in any flowchart are always connected according to the sequence in which the steps occur. The normal direction of flow in a chart is left to right or top to bottom. When the flow is otherwise, be sure to indicate it with arrows.

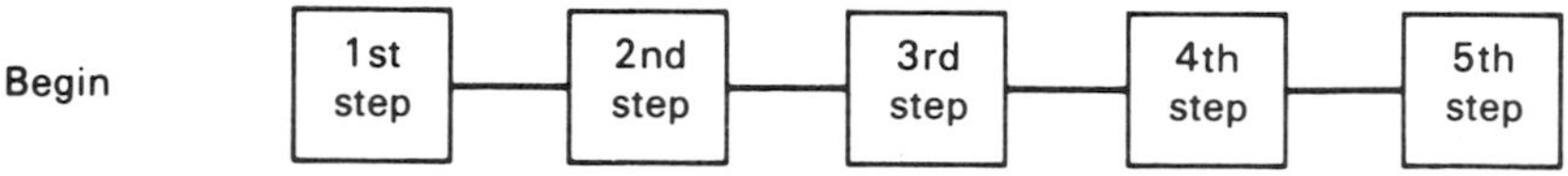

Figure 1 Example of a Block Flowchart

Flowcharts that document computer programs and other information-processing procedures use standardized **symbols**. The standards are set forth in *U.S.A. Standard Flowchart Symbols and Their Usage in Information Processing*, published by the American National Standards Institute, publication X3.5. When creating a flowchart, follow the guidelines:

1. Label the flowchart clearly and concisely.
2. Assign the chart a figure number if it is being used in a document that contains five or more **illustrations**.
3. With labeled blocks and standardized symbols, use arrows to show the direction of flow only if the flow is opposite to the normal direction. With pictorial representations, use arrows to show the direction of all flow.
4. Label each step in the process, or identify it with a conventional symbol. Steps can also be represented pictorially or by captioned blocks.
5. Include a key if the flowchart contains symbols your reader may not understand.
6. Leave adequate white space on the page. Do not crowd your steps and directional arrows too close together.
7. As with other illustrations, place the flowchart as near as possible to that portion of the text that refers to it.

(See the guidelines in **illustrations** for integrating flowcharts into the text.)

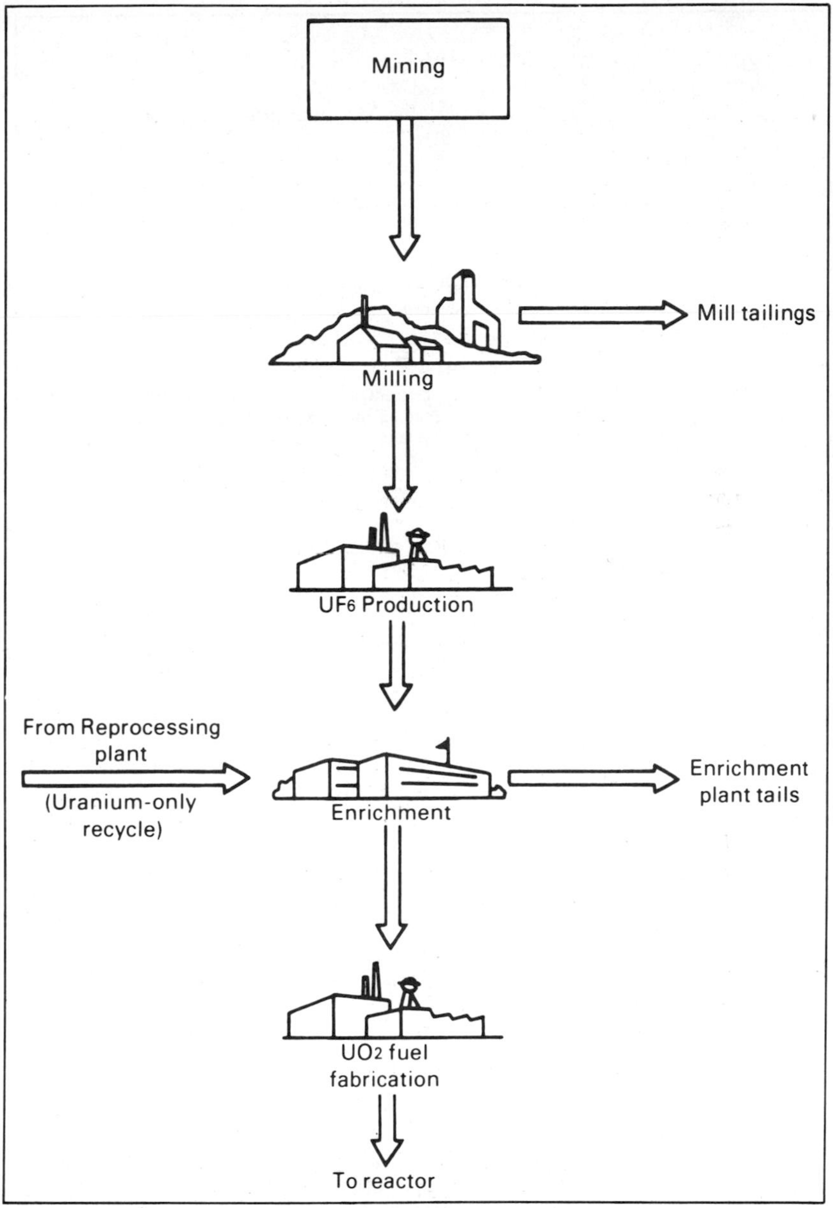

Figure 2 Pictorial Flowchart of Light-Water Reactor Uranium Fuel Cycle

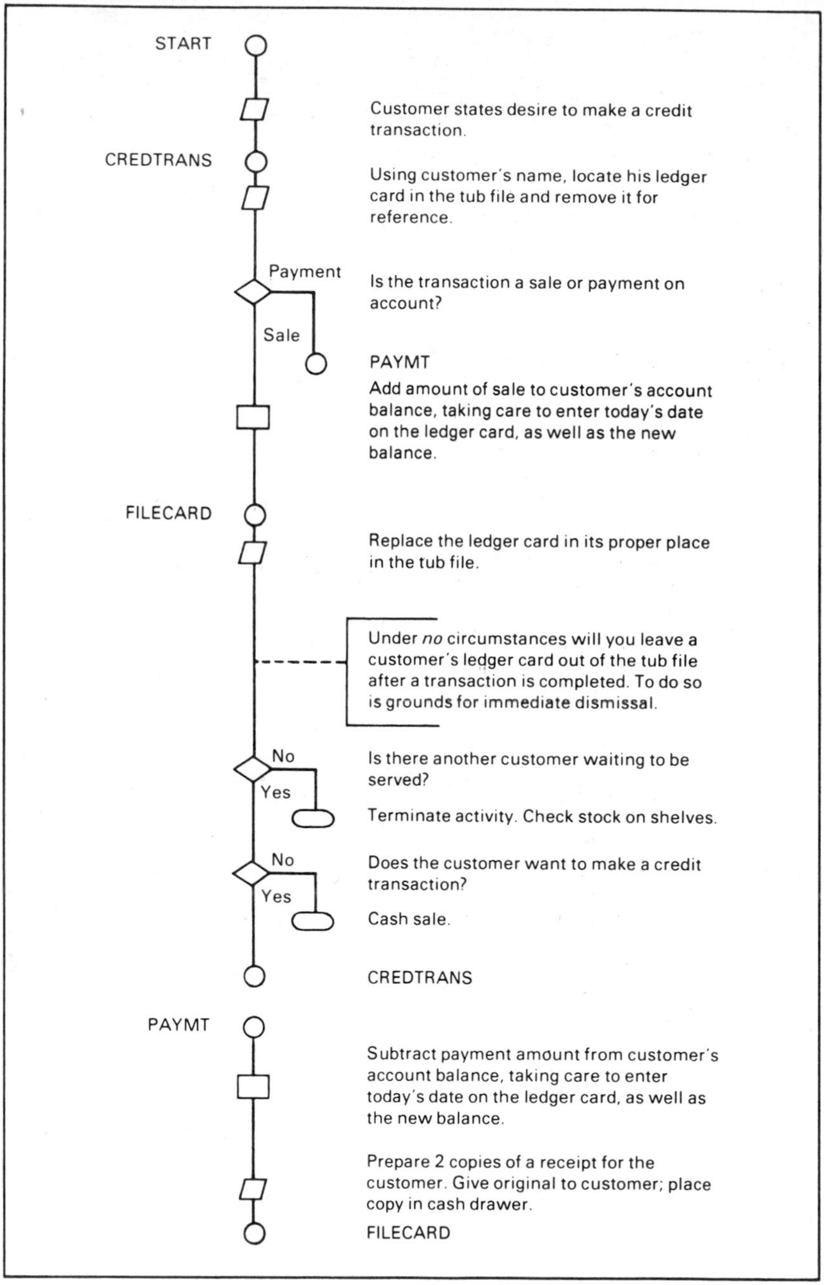

START
Customer states desire to make a credit transaction.
CREDTRANS
Using customer's name, locate his ledger card in the tub file and remove it for reference.
Payment
Is the transaction a sale or payment on account?
Sale
PAYMT
Add amount of sale to customer's account balance, taking care to enter today's date on the ledger card, as well as the new balance.
FILECARD
Replace the ledger card in its proper place in the tub file.
Under *no* circumstances will you leave a customer's ledger card out of the tub file after a transaction is completed. To do so is grounds for immediate dismissal.
No
Is there another customer waiting to be served?
Yes
Terminate activity. Check stock on shelves.
No
Does the customer want to make a credit transaction?
Yes
Cash sale.
CREDTRANS
PAYMT
Subtract payment amount from customer's account balance, taking care to enter today's date on the ledger card, as well as the new balance.
Prepare 2 copies of a receipt for the customer. Give original to customer; place copy in cash drawer.
FILECARD

Figure 3 Flowchart of a Credit Transaction

footnotes/end notes (see **documenting sources**)

forceful/forcible

Although *forceful* and *forcible* both are **adjectives** meaning "characterized by or full of force," *forceful* is usually limited to persuasive ability and *forcible* to physical force.

EXAMPLES The thief made a *forcible* entry into my apartment.
John made a *forceful* presentation at the committee meeting.

foreign words in English

The English language has a long history of borrowing words from other languages. Most of these borrowings occurred so long ago that we seldom recognize the borrowed terms (also called *loan words*) as being of foreign origin. To check the origin of a word, consult the etymology portion of its **dictionary** entry.

EXAMPLES whiskey (Gaelic), animal (Latin), church (Greek)

GUIDELINES FOR USE OF FOREIGN WORDS IN ENGLISH

Foreign expressions should be used only if they serve a real need. The overuse of foreign words in an attempt to impress your **reader** or to be elegant is **affectation**. Your goal of effective communication can be accomplished only if your reader understands what you write; choose foreign expressions, therefore, only when they make an idea clearer.

Words not fully assimilated into the English language are set in **italics** if printed (underlined in typed manuscript).

EXAMPLES *sine qua non, coup de grâce, in res, in camera*

Words and abbreviations that have been fully assimilated need not be italicized. When in doubt, consult a current dictionary.

EXAMPLES cliché, etiquette, vis-à-vis, de facto, résumé, etc., i.e., e.g.

As foreign words become current in English, their plural forms give way to English plurals.

EXAMPLES *formulae* becomes *formulas*
agenda becomes *agendas*

In addition, accent marks tend to be dropped from words (especially from French words) the longer they are used in English. But because not all foreign words shed their accent marks with the passage of time, consult a dictionary whenever you are in doubt. (See also **e.g./i.e.** and **etc.**)

foreword/forward

Although the pronunciation is the same, the spellings and meanings of these two words are quite different. The word *foreword* is a **noun** meaning "introductory statement at the beginning of a book or other work."

EXAMPLE The department chairman was asked to write a *foreword* for the professor's book.

The word *forward* is an **adjective** or **adverb** meaning "at or toward the front."

EXAMPLES Move the lever to the *forward* position on the panel. (adjective)
Turn the dial until the needle begins to move *forward*. (adverb)

foreword/preface

The terms *foreword* and *preface* are occasionally used interchangeably but are usually differentiated. A foreword is an optional introductory statement about a book or **formal report** written by someone other than the author. The writer of the foreword is usually an authority in the field, whose name and affiliation and the date the statement was written appear at the end of the foreword. The foreword may discuss the purpose of the study, but it generally provides background information about the study's significance or places the study in the context of other works written in the field. If the work was done under contract to a sponsoring organization, the manager of the project at the sponsoring organization may indicate how the work pertains to specific programs or goals of that organization. The foreword always precedes the preface when a work has both.

The preface is an optional introductory statement to a book or formal report, is usually written by the author, but is not signed. The preface may announce the purpose, background, and scope of the work. Typically, it highlights the relationship of the work to a given project or program and discusses any special circumstances leading to the study. A preface may also specify the audience for whom the work is intended. It may contain acknowledgments of help received during the course of the project or in the preparation of the publication. Works that acknowledge help received but that do not have a preface typically include a section entitled "Acknowledgments." Such a section appears on a separate page following the table of contents and preceding the *introduction*. Finally, a preface may cite permission obtained for the use of copyrighted works.

If you do not have a foreword or preface, place the type of information they typically contain, if it is essential, in the introduction. (For guidance about the location of a foreword and preface in a report, see **formal report**.) For an example of a preface, see the preface of this text at the beginning of the book.

formal reports

Formal reports are the written accounts of major projects. Projects that are likely to produce formal reports include research into new developments in a field, explorations of the advisability of launching a new product or an expanded service, or an end-of-year review of developments within an organization. The scope and complexity of the project will determine how long and how complex the report should be. Most formal reports—certainly those that are long and complex—require a carefully planned structure that offers the **readers** an easy-to-recognize guide to the material in the reports. Such aids as a **table of contents**, a list of **illustrations**, and an **abstract** make the information in the report more accessible. Making a formal topic outline that lists the major facts and ideas in the report and indicates their relationship to one another should help you write a well-organized report.

Most formal reports are divided into three major parts—front matter, body, and back matter—each of which contains a number of elements. Just how many elements are needed for a particular report depends on the subject, the length of the report, and the kinds of material covered.

ORDER OF ELEMENTS IN A FORMAL REPORT

The number and arrangement of the elements in a formal report may vary. Many companies and governmental and other institutions have a preferred style for formal reports and furnish guidelines that staff members must follow. If your employer has prepared a set of style guidelines, follow it; if not, use the format recommended in this chapter. The following list includes most of the elements a formal report might contain:

Front Matter

Title page
Abstract
Table of contents
List of figures
List of tables
Foreword
Preface
List of abbreviations and symbols

Body

Executive summary
Introduction
Text (including headings)
Conclusions
Recommendations
References

Back Matter

Bibliography
Appendixes
Glossary
Index

FRONT MATTER

The front matter, which includes all the elements that precede the body of the report, serves several purposes: it gives the reader a general idea of the author's purpose in writing the report; it indicates whether the report contains the kind of information the reader is looking for; and it lists where in the report the reader can find specific chapters, **headings**, illustrations, and **tables**. Not all formal reports require every element of front matter. A title page and table of contents are usually mandatory, but whether an abstract, a preface, and lists of figures, tables, **abbreviations**, and **symbols** are included will depend on the **scope** of the report and its intended audience. The front matter pages are numbered with lower case roman numerals. Throughout the report, page numbers should be centered near the bottom of each page.

Title Page. The formats of title pages for formal reports vary, but the page should include the following information:

1. The full **title** of the report. The title should indicate the topic and announce the scope and **objective** of the report. Titles often provide the only basis on which readers can decide whether to read a report. Titles that are too vague or too long not only hinder readers but also can prevent efficient filing and information retrieval by librarians and other information specialists. Follow these guidelines when creating the title:

   - Do not use "Report on . . .," "Technical Report on . . .," or "XYZ Corporation Report on . . .," in the title, since the fact that the information appears in a report will be self-evident to your reader.
   - Do not use abbreviations in the title. Use **acronyms** only when the report is intended for an audience familiar enough with the topic that the acronym will not confuse them.
   - Do not include the period covered by a report in the title; include that information in a subtitle:

     EFFECTS OF PROPOSED HIGHWAY
CONSTRUCTION ON PROPERTY VALUES
Annual Report, 19--

2. The name of the writer, principal investigator, or compiler. Frequently, contributors simply list their names and almost never list their academic degrees. Sometimes they identify themselves by their job title in the organization (Jane R. Doe, Cost Analyst; Jack T. Doe, Head, Research and Development). Sometimes contributors identify themselves by their tasks in contributing to the report (Jane R. Doe, Compiler; Jack T. Doe, Principal Investigator).
3. The date or dates of the report. For one-time reports, list the date when the report is to be distributed. For periodic reports, which may be issued monthly or quarterly, list in a subtitle the time period that the present report covers. Elsewhere on the title page, list the date when the report is to be distributed.

4. The name of the organization for which the writer works.
5. The name of the organization to which the report is being submitted, if the work is being done for a customer or client.

These categories are standard on most title pages, but some organizations may require additional information.

The title page, although unnumbered, is considered page i. The subsequent pages of the front matter begin with page ii.

Abstract. An abstract, which normally follows the title page and is page ii, is a condensed version of a longer piece of writing that summarizes and highlights the major points, enabling the prospective reader to decide whether to read the entire report. Usually 200 to 250 words or less, abstracts must make sense independently of the works they summarize. (For a complete discussion of abstract writing supported with samples, see **abstracts**.)

Table of Contents. A table of contents lists all the major **headings** or sections of the report in their order of appearance, along with their page numbers. It lists all front matter and back matter except the title page and the table of contents itself. The table of contents begins on a new page, numbered page iii, because it follows the title page (page i) and the abstract (page ii). By convention, the table of contents and its page number are not listed in the table of contents page.

Along with the abstract, a table of contents permits the reader to assess the value of a report. It also aids a reader who may want to look at only certain sections of the report. For this reason, the wording of chapter and section titles in the table of contents should be identical to those in the text. (For a sample contents page, see **table of contents**.) The table of contents need not list all headings in the report. Generally no more than three levels are listed. However, when any heading of a given level is listed, all headings of that level must also be listed in the contents.

Sometimes the table of contents is followed by lists of the figures and tables contained in the report. These lists should always be presented separately, and a page number should be given for each item.

List of Figures. Figures include all **illustrations—drawings, photographs, maps**, charts, and **graphs**—contained in the report. When a report contains more than five figures, they should be listed, along with their page numbers, in a separate section immediately following the table of contents. The section should be entitled "List of Figures" and should begin on a new page. Figure numbers, titles, and page numbers should be identical to those in the text. In most reports figures are numbered consecutively with arabic numbers.

List of Tables. When a report contains more than five tables, they should be listed, along with their titles and page numbers, in a separate section entitled "List of Tables," immediately following the list of figures (if there is one). Tables are numbered consecutively with arabic numbers throughout most reports.

Foreword. A foreword is an optional introductory statement written by someone other than the author. The foreword may discuss the purpose of the report but generally provides background information about its significance or places it in the context of other works written in the field.

The foreword always precedes the preface when the report contains both. (See also **foreword/preface**.)

Preface. The preface is an optional introductory statement, usually written by the author, that announces the purpose, background, and scope of the report. Sometimes the preface specifies the audience for whom the report is intended; it may also highlight the relationship of the report to a given project or program. A preface may contain acknowledgements of help received during the course of the project or in the preparation of the report. Finally, it may cite permission obtained for the use of copyrighted works. If the report does not require a preface, place this type of information, if it is essential, in the introduction.

The preface follows the table of contents (and the lists of figures or tables if they are included) and it begins on a separate page entitled "Preface." (See also **foreword/preface**.)

List of Abbreviations and Symbols. When the abbreviations and symbols used in the report are numerous and there is a chance that the reader will not be able to interpret them, the front matter should include a list of all symbols and abbreviations (including **acronyms**) and what they stand for in the report. Such a list, which follows the preface, is particularly helpful for reports on specialized subjects whose audience will include nonspecialists. Such a list, appropriately titled, begins on a new page.

BODY (TEXT)

The body is that portion of the report in which the author introduces the subject, describes in detail the methods and procedures used in the study, demonstrates how results were obtained, and draws conclusions on which any recommendations are based.

The first page of the body is numbered 1. This and all subsequent pages in the report are usually numbered in arabic numbers, unlike the front matter, which is numbered in lowercase roman numerals.

Executive Summary. The body of the report begins with an executive summary that provides a more complete overview of a report than either a descriptive or an informative abstract does. The summary states the purpose and nature of the investigation and gives major findings, conclusions, and—if any are to be made—recommendations. It also provides an account of the procedures used to conduct the study. Although more complete than an abstract, the summary should not contain a detailed description of the work on which the findings, conclusions, and recommendations were based. The size of an executive summary is proportional to the size of the report; a rule of thumb is that the length of the executive summary should be approximately 10 percent of the length of the report.

Some summaries are written to follow the organization of the report. Others highlight the findings, conclusions, and recommendations by summarizing them first before going on to discuss procedures or methodology.

A summary enables people who may not have time to read a lengthy report to scan its primary points and then decide whether they need to read the entire report. Because they are often intended for busy executives, summaries are sometimes called *executive summaries*. Addressed primarily to this audience, summaries often appear as the first section of the body of a report. Like abstracts, summaries should not contain tables, illustrations, or bibliographic citations. The summary is considered part of the body of the report. (For a more detailed discussion of summaries and a sample summary, see **executive summaries**.)

Introduction. The purpose of an introduction is to give the readers any general information they must have in order to understand the detailed information in the rest of the report. An introduction sets the stage for the report, providing a frame of reference for the details contained in the body of the report. In writing the introduction, you need to state the subject, the purpose, the scope, and the way you plan to develop the topic. You may also describe how the report will be organized. (For a detailed discussion of introductions, with samples, see **introductions**.)

Text. Generally the longest section of the report, the text (or body) presents the details of how the topic was investigated, how the problem was solved, how the best choice from among alternatives was selected, or whatever else the report covers. This information is often clarified and further developed by the use of illustrations and tables and may be supported by references to other studies.

Most formal reports have no single best organization. How the text is organized will depend on the topic and on how you have investigated it (survey, analysis, feasibility study, laboratory experiment, or whatever). The text is ordinarily divided into several major sections, comparable to the chapters in a book. These sections are then subdivided to reflect the logical divisions in your main sections. (For a discussion of the writing process, see "The Five Steps of Successful Writing" at the beginning of this book.)

Conclusions. The conclusion section pulls together the results of your study in one place. Here you tell the reader what you have learned and its implications. You show how the results follow from the study objectives and method. You should also point out any unexpected results. (For a fuller discussion of this section of a report, see **conclusions**.)

Recommendations. Recommendations, which are sometimes combined with the conclusions section, state what course of action should be taken based on the results of the study. Of the three possible locations for a new warehouse, which is the best? Which make of delivery van should be purchased to replace the existing fleet? Which contractor has submitted the best proposal for remodeling the company cafeteria? Which type of word-processing equipment should the office buy? Should the equipment be leased rather than purchased? The recommendations section says, in effect, "I think we should purchase this, or do that, or hire them."

The emphasis is on the verb *should.* Recommendations advise the reader on the best course of action based on the researcher's findings. Generally someone up the organizational ladder, or a customer or client, makes the final decision about whether to accept the recommendations.

References. If in your report, you refer to material in or quote directly from a published work or other research source, you must provide a list of references in a separate section called *references.* If your employer has a preferred reference style, follow it; otherwise, use the guidelines provided in **documenting sources**.

For a relatively short report, the references should go at the end of the report. For a report with a number of sections or chapters, the reference section should fall at the end of each major section or chapter. In either case, every reference section should be labeled as such and should start on a new page. If a particular reference appears in more than one section or chapter, it should be repeated in full in each appropriate reference section.

BACK MATTER

The back matter of a formal report contains supplemental information, such as where to find additional information about the topic (**bibliography**) and how the information in the report can be easily located (**index**), clarified (**glossary**), and explained in more detail (appendix).

Bibliography. A **bibliography** is a list, usually in alphabetical order, of all sources that were consulted in researching the report but that are not cited in the text. A bibliography may not be necessary, however, when the reference listing contains a complete list of sources. When a report has both, the bibliography follows the reference section. This section is often entitled "Works Cited," especially when the sources include nonprint information like computer codes, microfiche, or video material.

Like the other elements in the front and back matter, the bibliography starts on a new page and is labeled by name. (For detailed information about creating a bibliography, see **bibliography**.)

Appendixes. An appendix contains information that clarifies or supplements the text. This type of information is placed at the back of the report because it is too detailed or voluminous to appear in the text without impeding the orderly presentation of ideas. Material typically placed in an appendix includes long charts and supplementary graphs or tables, copies of **questionnaires** and other material used in gathering information, texts of interviews, pertinent correspondence, and explanations too long for explanatory footnotes but helpful to the reader who is seeking further assistance or clarification.

A report may have one or more appendixes; generally, each appendix contains one type of material. For example, a report with two appendixes may have a questionnaire in one and a detailed computer printout tabulating the questionnaire's results in the other.

Place the first appendix on a new page directly after the bibliography. Additional appendixes also begin on a new page. Identify each with a title and a heading. Appendixes are ordinarily labeled Appendix A, Appendix B, and so on. If your report has only one appendix, simply label it Appendix, followed by its title. To call it Appendix A would imply that an Appendix B will follow. (For a detailed discussion of appendixes, see **appendix/appendixes/appendices**.)

Glossary. A glossary is a list of selected terms that are defined and explained. Include a glossary only if your report contains many words and expressions that will be unfamiliar to your intended readers. Arrange the terms alphabetically, with each entry beginning on a new line. The definitions then follow each term, dictionary style. The glossary, labeled as such, appears directly after the appendix and begins on a new page.

Even though the report may contain a glossary, you should also define all unfamiliar terms or terms having a special meaning in your study when they are first mentioned in the text. (See **defining terms**.)

Index. An index is an alphabetical list of all the major topics discussed in the report. It cites the pages where each topic can be found and thus allows readers to find information on topics quickly and easily. The index is always the final section of the report. The index to this volume can be found at the back of the book.

An index is an optional finding device in a report because a detailed table of contents usually gives readers adequate information about topics covered. Indexes are most useful for reference works like handbooks and manuals. (For guidance on creating an index, see **indexing**.)

formal writing style (see **style**)

format

Format has at least two distinct, but related, meanings: it can refer (1) to the sequence in which information is presented in a publication, and (2) to the physical arrangement of information on the page and the general physical appearance of the finished publication. The first meaning applies to the standard arrangement or layout of information in many of the following types of job-related writing:

- **formal reports**
- **memorandums**
- **questionnaires**

- **proposals**
- **correspondence**
- **trip reports**
- **résumés**
- **test reports**
- **laboratory reports**

These types of writing are characterized by format conventions that govern where each section will be placed. In **formal reports**, for example, the **table of contents** precedes the preface but follows the title page and **abstract.** Likewise, although variations exist, letters typically are written according to the following standard pattern:

- letterhead
- date
- inside address
- salutation
- body
- complimentary close
- signature
- typed name and title
- additional information (initials, enclosure, and "carbon copy" notation)

These conventions also pertain to the use of **heads**, **illustrations**, and **indentation**. (For detailed information about numbered and unnumbered heading systems in reports and manuals, see **heads**.)

In recent years, a body of information has been published that expands the concept of format to encompass all typographic and graphic principles used to depict information in a document. The expanded scope of this notion now falls under the general term *document design*.

Document design principles help writers, editors, and graphics designers create publications that are visually distinct and attractive so that they are easier for readers to understand. The publications can range from application forms and consumer contracts, on the one hand, to **technical manuals** and books, on the other. Specific document-design goals include finding the best methods to accomplish the following:

- to distinguish primary from secondary ideas
- to set off examples from the text
- to call out headings

- to call attention to certain parts of a document
- to make information easy to find both the first and subsequent times it is looked for
- to arrange every element to lead the eye progressively from one part to the next
- to promote the reader's acceptance of material
- to provide visual relief
- to make information accessible to diverse audiences

Finally, they also consider the final "packaging" of the document: cover design, binding options, index tabs, and the size and grade of paper.

For detailed information about document-design principles and applications, begin by consulting the following publications, each of which contains a **bibliography** for further information:

1. Felker, Daniel B., ed. *Document Design: A Review of the Relevant Research*. Washington, D.C.: American Institutes for Research, 1980.
2. Felker, Daniel B., et al. *Guidelines for Document Designers*. Washington, D.C.: American Institutes for Research, 1981.

former/latter

Former and *latter* should be used to refer to only two items in a sentence or paragraph.

EXAMPLE The president and his trusted aide emerged from the conference, the *former* looking nervous and the *latter* looking downright glum.

Because these terms make the **reader** look back to previous material to identify the reference, they impede reading and are best avoided.

CHANGE Model 190432 had the necessary technical modifications, whereas model 129733 lacked them, the *former* being this year's model and the *latter* last year's.

TO Model 190432, this year's model, had the necessary technical modifications; model 129733, last year's model, did not.

forms of discourse

There are four forms of discourse: exposition, description, persuasion, and narration. **Exposition** is the straightforward presentation of facts and ideas with the **objective** of informing the **reader**; **description** is an attempt to re-create an object or situation with words so that the reader can visualize it mentally; **persuasion** attempts to convince the reader that the writer's point of view is the correct or desirable one; and **narration** is the presentation of a series of events in chronological order. These types of writing rarely exist in pure form; rather, they usually appear in combination.

formula/formulae/formulas

The plural form of *formula* is either *formulae* (a Latin derivative) or *formulas*. *Formulas* is more common, however, since it is the more natural English plural. (See also **foreign words in English**.)

EXAMPLE You may present the underlying theory in your introduction, but save proofs or *formulas* for the body of your report.

fortuitous/fortunate

When an event is *fortuitous*, it happens by chance or accident and without plan. Such an event may be lucky, unlucky, or neutral.

EXAMPLE My encounter with the general manager in Denver was entirely *fortuitous*; I had no idea he was there.

When an event is *fortunate*, it happens by good fortune or happens favorably.

EXAMPLE Our chance meeting had a *fortunate* outcome.

functional shift

Many words shift easily from one **part of speech** to another, depending on how they are used. When they do, the process is called a functional shift, or a shift in function.

EXAMPLES It takes ten minutes to *walk* from the sales office to the accounting department. (verb)
The long *walk* from the sales office to the accounting department reduces efficiency. (noun)

Let us *run* the new data through the computer. (verb)
May I see the last computer *run*? (noun)

I talk to the Chicago office on the *telephone* every day. (noun)
Sometimes I have to call from a *telephone* booth. (adjective)
He will *telephone* the home office from London. (verb)

After we discuss the project, we will begin work. (conjunction)
After lengthy discussions, we began work. (preposition)
The partners worked well together forever *after*. (adverb)

Do not arbitrarily shift the function of a word in an attempt to shorten a **phrase** or expression. Such attempts at achieving **conciseness** can result in the worst kind of **jargon**, as in the following examples:

EXAMPLES In medical jargon, an *attending physician* becomes an "attending." (a shift from an adjective to a noun)
In computer jargon, *to pass a signal through a computer logic gate* becomes "to gate." (a shift from a noun to a verb)
In nuclear energy jargon, a *reactor containment building* becomes a "containment." (a shift from an adjective to a noun)

Avoid these shortcut versions; they are unnecessary. Linguistically acceptable ways for saying the same things already exist. If you are unsure about how to use similar words and expressions, consult a **dictionary** for both the word's meaning and its part of speech. Identifying the word's part of speech will allow you to determine quickly whether or not you are using it properly. (See also **new words**.)

G

garbled sentences

A garbled sentence is one that is so tangled with structural and grammatical problems that it cannot be repaired by simply replacing words or rewriting **phrases**. The following nearly unreadable sentence appeared in an actual **memorandum**:

EXAMPLE My job objectives are accomplished by my having a diversified background which enables me to operate effectively and efficiently, consisting of a degree in mechanical engineering, along with twelve years of experience, including three years in Staff Engineering-Packaging sets a foundation for a strong background in areas of analyzing problems and assessing economical and reasonable solutions.

A garbled sentence often results from an attempt to squeeze too many ideas into one sentence. Do not try to patch such a sentence; rather, analyze the ideas it contains, list them in a logical sequence, and then construct one or more entirely new sentences. An analysis of the previous example yields the following five ideas:

EXAMPLES My job requires that I analyze problems and find economical and workable solutions to them.
My diversified background helps me accomplish my job.
I have a mechanical engineering degree.
I have twelve years of job experience.
Three of these years have been in Staff Engineering-Packaging.

Using these ideas, the writer might have described his job as follows:

EXAMPLE My job requires that I analyze problems and find economical and workable solutions to them. Both my training and my experience help me achieve this goal. Specifically, I have a mechanical engineering degree and twelve years of job experience, three of which have been in the Staff Engineering-Packaging Department.

Notice that the revised job description contains three sentences with logical development and clear **transitions**. Such analysis and rewriting is useful for any sentence or **paragraph** that you cannot repair with minor editing. (See also **clarity**.)

gender

In grammar, gender is a term for the way words are formed to designate sex. The English language provides for recognition of three genders: masculine, feminine, and neuter (to designate those objects that have no definable sex characteristics). The gender of most words can be identified only by the choice of the appropriate **pronoun** (*he, she, it*). Only these pronouns and a select few **nouns** (*waiter/waitress*) reflect gender.

Gender is important to writers because they must be sure that nouns and pronouns within a grammatical construction agree in gender. A pronoun, for example, must agree with its noun antecedent in gender. We must refer to a woman as *she* or *her*, not as *it*; to a man as *he* or *him*, not as *it*; to a barn as *it*, not *he* or *she*.

An antecedent that includes both sexes, such as *everyone* and *student*, has traditionally taken a masculine pronoun. Since this **usage** might be offensive by implying sexual bias, however, it is better to use plural nouns and pronouns.

> CHANGE Every *employee* should be aware of *his* insurance benefits.
> TO All *employees* should be aware of *their* insurance benefits.

It is best (though not always possible) to avoid the awkward "his or her," "he or she" type of construction. (See also **he/she**.)

general-to-specific method of development

In a general-to-specific **method of development**, you begin with a general statement and then provide facts or examples to develop and support that statement. For example, if you begin a **report** with the general statement "Companies that diversify are more successful than those that do not," the remainder of the report would offer examples and statistics that prove to your **reader** that companies that diversify are, in fact, more successful than companies that do not.

A **memorandum** or report organized in a general-to-specific sequence discusses only one point—the point made in the opening general statement. All other information in the memo or report supports the general statement, as shown in Figure 1.

	Locating Additional Circuit Suppliers
General statement	Based on information presented at the Supply Committee meeting on April 14, we recommend that the company locate additional suppliers of integrated circuits. Several related events make such an action necessary. Our current supplier, ABC Electronics, is reducing its output. Specifically, we can expect a reduction of between 800 and 1,000 units per month for the remainder of this fiscal year. The number of units should stabilize at 15,000 units per month thereafter.
Supporting information	Domestic demand for our calculators continues to grow. Demand during the current fiscal year is up 25,000 units over the last fiscal year. Sales Department projections for the next five years show that demand should peak next year at 50,000 units and then remain at that figure for at least the following four years. Finally, our overseas expansion into England and West Germany will require additional shipments of 5,000 units per quarter to each country for the remainder of this fiscal year. Sales Department projections put calculator sales for each country at double this rate, or 20,000 units in a fiscal year, for the next five years.

Figure 1 General-to-Specific Method of Development

gerunds

A gerund is a **verbal** ending in *-ing* that is used as a **noun**.

EXAMPLE *Smelting* is a technique used to extract metal from ore.

A gerund may be used as a **subject**, a direct **object**, an object of a **preposition, a subjective complement**, or an **appositive**.

EXAMPLES *Estimating* is an important managerial skill. (subject)
I find *estimating* difficult. (direct object)
We were unprepared for their *coming*. (object of a preposition)
Seeing is *believing*. (subjective complement)
My primary departmental function, *programming*, occupies about two-thirds of my time on the job. (appositive)

Only the possessive form of a noun or **pronoun** should precede a gerund.

EXAMPLES *John's* working has not affected his grades.
His working has not affected his grades.

Do not confuse a gerund with a present **participle**, which has the same form but functions as an **adjective.**

EXAMPLES *Operating* the machine is difficult. (gerund, used as a noun)
Please provide clearer *operating* instructions. (participle, used as an adjective)

glossary

A glossary is a selected list of terms defined and explained for a particular field of knowledge. An alphabetical glossary, such as is often found at the end of a textbook or a **formal report**, can be helpful for quick reference.

If you are writing a report that will go to people who are not familiar with many of your technical terms, you may want to include a glossary. Inclusion of a glossary does not relieve you of the responsibility of defining in the text any terms you are certain your reader will not know, however. If you do, keep the entries concise and be sure they are so clear that any reader can understand the definitions. (See also **defining terms**.)

EXAMPLES *Amplitude Modulations:* Varying the amplitude of a carrier current with an audio-frequency signal.
Carbon Microphone: A microphone that uses carbon granules as a means of varying resistance with sound waves.
Overmodulation: Distortion created when the amplitude of the modulator current is greater than the amplitude of the carrier current.

gobbledygook

Gobbledygook is writing that suffers from an overdose of traits guaranteed to make it stuffy, pretentious, and wordy. These traits include the overuse of big and mostly **abstract words**, inappropriate **jargon**, stale expressions, **euphemisms, jammed modifiers**, and deadwood. Gobbledygook is writing that attempts to sound official (officialese), legal (legalese), or scientific; it tries to make a "natural elevation of the geosphere's outer crust'' out of a molehill. Consider the following statement from an auto repair release form:

CHANGE I hereby authorize the above repair work to be done along with the necessary material and hereby grant you and/or your employees permission to operate the car or truck herein described on streets, highways, or elsewhere for the purpose of testing and/or inspection. An express mechanic's lien is hereby acknowledged on above car or truck to secure the amount of repairs thereto.

TO You have my permission to do the repair work listed on this work order and to use the necessary material. You may drive my vehicle to test its performance. I understand that you will keep my vehicle until I have paid for all repairs.

Translated into straightforward English, the statement gains in clarity what it loses in pomposity without losing its legal meaning. Today, many organizations are working to eliminate gobbledygook. In fact, the Document Design Center in Washington, D.C., publishes *Simply Stated*, a newsletter that chronicles these efforts. Subscriptions are free, and the address is Document Design Center, American Institutes for Research, 1055 Thomas Jefferson Street N.W., Washington, D.C. 20007.

Gobbledygook (also called ''bombast'' and ''pedantic writing'') is packed with unusual words when more common ones would be clearer; it is heavy with foreign words when their English equivalents would be more appropriate; and it generally stresses trivial matters in the hope that the **reader** will be impressed with the writer's mental agility. George Orwell parodied such writing in the essay ''Politics and the English Language,'' from which the following excerpt is taken:

EXAMPLE Objective consideration of contemporary phenomena compels the conclusion that success or failure in competitive activities exhibits no tendency to be commensurate with innate capacity, but that a considerable element of the unpredictable must invariably be taken into account.

The following revision of Orwell's passage demonstrates the clarity of direct writing:

EXAMPLE Success or failure today depends as much on chance as on your capabilities.

(See also **affectation, conciseness/wordiness**, and **word choice**.)

good/well

The confusion about the use of *good* and *well* can be cleared up by remembering that *good* is an **adjective** and *well* is an **adverb**.

EXAMPLES John presented a *good* plan.
The plan was presented *well*.

However, *well* can also be used as an adjective to describe someone's health.

EXAMPLES She is not a *well* woman.
Jane is looking *well*.

(See also **bad/badly**.)

government proposals

A **proposal** is a company's offer to provide goods or services to a potential buyer within a certain amount of time and at a specified cost. Government proposals are usually prepared as a result of either an Invitation for Bids or a Request for Proposals that has been issued by a government agency.

INVITATION FOR BIDS

An Invitation for Bids is inflexible; the rules are rigid and the terms are not open to negotiation. Any proposal prepared in response to an Invitation for Bids must adhere strictly to its terms.

The Invitation for Bids clearly defines the quantity and type of an item that a government agency intends to purchase. It is prepared at the request of the agency that will use the item, such as the Department of Defense. An advertisement indicating that the government intends to purchase the item is published in an official government publication, such as the *Commerce Business Daily* (available from the Superintendent of Documents, Washington, D.C. 20402). This publication is required reading for government sales-oriented organizations. The goods or services to be procured are defined in the Invitation for Bids by references to performance standards called **specifications**. If, for instance, the item to be purchased is a truck, the important characteristics of that truck (height, weight, speed, carrying capacity, and so on) will be listed in the machine specification. More than one specification may apply to a single purchase; that is, one specification may apply to the item, another to the manuals, another to welding procedures, and so forth. Bidders must be prepared to prove that their product will meet all requirements of all specifications.

A specification is restrictive, binding the bidder to produce an item that meets the exact requirements of the specification. For example, if a truck is the specified item, a company may not bid to furnish a cargo helicopter, even though the helicopter might do the job more efficiently than the truck. An Invitation for Bids requires the bidder to furnish a specific product that meets the specification parameters, and nothing else will be accepted or considered. An alternative suggestion will render the bid "nonresponsive," and it will be rejected.

Bearing in mind that the product will be tested, measured, and evaluated to see that it does, in fact, meet the requirements of the specification, price is the main criterion that enters into the selection of the vendor. However, an organization with superior engineering skills can often supply a product that meets the specification at a price well below that of a less sophisticated competitor. When a high degree of technological competence, unusual facilities, or other valid requirements make it necessary, the procuring agency may require the bidder to possess certain minimum qualifications in order to be considered; a paper clip manufacturer employing ten people, for instance, could not qualify to bid on a procurement of military aircraft.

REQUEST FOR PROPOSALS

In contrast to the Invitation for Bids, the Request for Proposals is flexible. It is open to negotiation, and it does not necessarily specify exactly what goods or services are required. Often, a Request for Proposals will define a problem and allow those who respond to it to suggest possible solutions. As an example, if a procuring agency wanted to develop a way to make foot soldiers more mobile (capable of covering difficult terrain at high speed), a Request for Proposals would normally be the means used to find the best method and select the most qualified vendor. In some instances, Requests for Proposals are presented in two or more stages, the first being development of a concept, the second being a "prototype" machine or device, and the third being the manufacture of the device selected. The procedure for preparing a proposal in response to a Request for Proposals is as follows.

The procuring agency defines the problem and publishes it as a Request for Proposals in one or more business journals. Companies interested in government business scan the appropriate publications daily. Upon finding a project of interest, the sales department of such a company obtains all available information from the procuring agency. It then presents the data to management for a decision on whether the company is interested in the project. If the decision is positive, the corporate technical staff is assigned the task of developing a "concept" to accomplish the task posed by the Request for Proposals. The technical staff normally considers several alternatives, selecting one that combines feasibility and price. The staff's concept is presented to management for a decision on whether the company wishes to present a proposal to the requesting government agency.

Assuming that the decision is to proceed, preparation of a proposal is the next step. At a minimum, the proposal should provide a clear-cut statement of the problem and the proposed solution. It should include data to show that the company is financially and technologically sound (this may require a résumé of the qualifications of the people in charge and an **organizational chart** to show the chain of command). A discussion of the company's manfacturing capabilities and any other advantages it may have is also advisable. However, in dealing with government agencies, bear in mind that unnecessarily elaborate and costly proposals may be construed as a lack of cost consciousness.

In many instances, the final cost to the government is determined by negotiation after the best overall concept has been selected. However, if your concept is one that provides a simple and inexpensive solution to the problem, cost is an advantage that you should point out. If your concept is expensive, explain the benefits of advanced engineering, speed, increased capacity, and so forth.

Further information about the government procurement process can be obtained from the *Armed Service Procurement Regulations* (available from the U.S. Government Printing Office, Washington, D.C. 20402).

The excerpt presented in Figure 1 comes from a proposal that was prepared in response to a Request for Proposals. The rest of the proposal continues to explain the design of the engine's cylinder heads, bearings, crankshaft, pistons, cylinder liners, connecting rod, and reversing mechanism.

Advantages of the Proposed 5,000 H.P. Design

Statement of problem

The Navy Department has indicated to Burdlorn Manufacturing Company a need for main diesel propulsion engines to be used in conjunction with gas turbines on a program with the code name of CODAG. We have been advised that the power requirement for this engine is 5,000 horsepower, plus 10 percent overload. This engine must be contained within a space 24 feet long, 11 feet high, and 7 feet wide and must not exceed 60,000 pounds in weight. In addition, at a 5,000 horsepower operating level, it is not expected that the engine will require replacement or renewal of wearing parts or other major components more frequently than every 5,000 hours. The proposed design changes will result in a very light-weight engine (12 pounds/BHP) but one with durability characteristics that will not be equaled by any engine throughout the world in this compact high-horsepower output.

Brief statement of solution

5,000 HORSEPOWER ENGINE

Company's technical capability

Burdlorn Manufacturing Company has completed shop tests on NOBS 72393, addenda 12 and 13. During these tests, the Navy engine serial GR-1047-0836 was operated at loads up to 4,950 brake horsepower (just 1 percent short of the 5,000 horsepower goal). As a result of these tests, we are satisfied that the 16-cylinder, 11″ bore, 12″ stroke V-engine, with a rating of 5,000 horsepower at 1,000 rpm, is well within practical limits. On the basis of an intensive review of the service history of the four engines installed on the LST-1176 and all of the research and development work at the Burdlorn factory, coupled with a substantial amount of analytical design work, Burdlorn Manufacturing Company recommends that certain redesigns be effected, based on the above information.

As a result of the recent tests at loads up to 4,950 horsepower that have been run in our factory, Burdlorn Manufacturing Company has successfully dealt with the thermodynamic problems of air/fuel requirements, air flow, heat transfer, and other related problems. In proposing design criteria, which are conservative to insure the desired reliabil-

Figure 1 Government Proposal

ity, we have applied sufficient analysis to the individual components to make specific recommendations.

Advantages to be outlined

In order to summarize findings that we feel the Navy needs in order to evaluate our proposal, we will outline in numerical sequence the improvements and advantages based on our total experience with the 11″ bore, 12″ stroke, 16-cylinder V-engine.

1. Frame. Burdlorn Manufacturing Company recommends that the present 17 inches center distance from one cylinder to the adjacent cylinder be increased by 3 inches to provide 20 inches center distance from cylinder to cylinder in a given bank. The additional 3 inches per cylinder will mean an overall increase in engine length of 24 inches. This will mean an engine length of approximately 22 feet with the engine-driven auxiliary equipment. This will be well within the 24-foot maximum covered by the Navy specifications. The additional 3 inches will provide the following advantages.

   a. The principal load carrying member in the frame, which is the athwartship plate, will be increased in thickness from ½ inch to ¾ inch.
   b. Additional clearance is provided for welding, and machining, particularly on the intermediate deck.
   c. Additional length is provided for the main and crankpin journals. This will increase the bearing areas for the main and crankpin bearings and thus gain the advantage of reduced bearing loads.

Minor problem and proposed solution

The results of our experimental stress analysis have indicated a substantial bending movement in the top deck of the present design. Accordingly, Burdlorn Manufacturing Company proposes to increase the thickness of the top deck. The present design uses forgings. We are proposing to use controlled quality steel castings of substantially heavier construction to reduce the stress levels.

Burdlorn Manufacturing Company has done a substantial amount of stress analysis on a similar frame on our Model FS-13½″ × 16½″ V-engine. On the basis of this experience, we recommend a minor redesign of the main bearing saddles to create a more even stress distribution in transmitting the load through the athwartship plates to the main bearing saddles. These main bearing saddles would be made from steel castings such as our present 13½" × 16½" engines currently use.

To provide for reduction in stress levels in the sidesheets of the engine, we propose increasing the plate thickness from ½ inch to ⅝ inch. In addition, we have found through extensive testing of the Navy engine and our 13½" × 16½" V-engine and also by exhaustive photoelastic stress analysis techniques that the present "hourglass" design results in a stress concentration factor at the "waist" of the frame. Accordingly, we propose a modification in this shape that will reduce the stress concentration factor due to this shape effect (see Figure 9).

Introduction to illustration

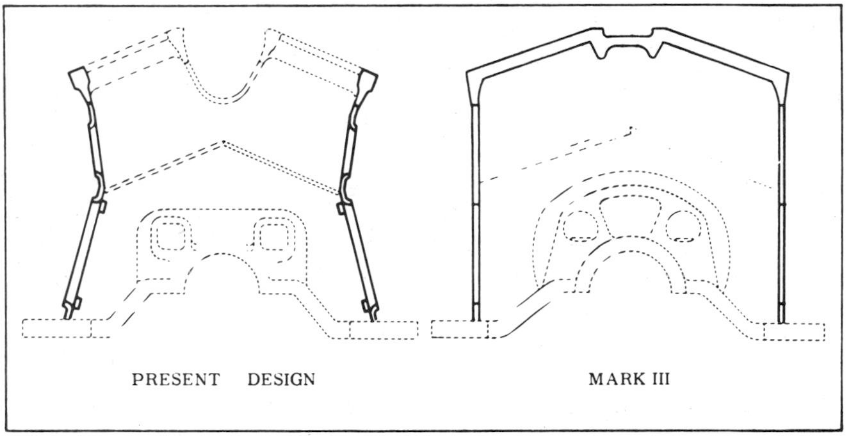

Figure 9. Frame Profiles

Another problem and proposed solution with illustration

One problem experienced repeatedly on the LST-1176 was associated with the bosses for the handhole covers. Stress analysis work has shown a high stress concentration factor around these bosses because of unfused weld roots and also the increased localized stiffness caused by these bosses. Accordingly, we have designed handhole covers held in place by clamping action, which do not impose additional stress in the sidesheet due to the torquing down of the bolts against the bosses. We therefore gain the advantage by this direct means of eliminating a superimposed stress of 5,000 psi (see Figure 10).

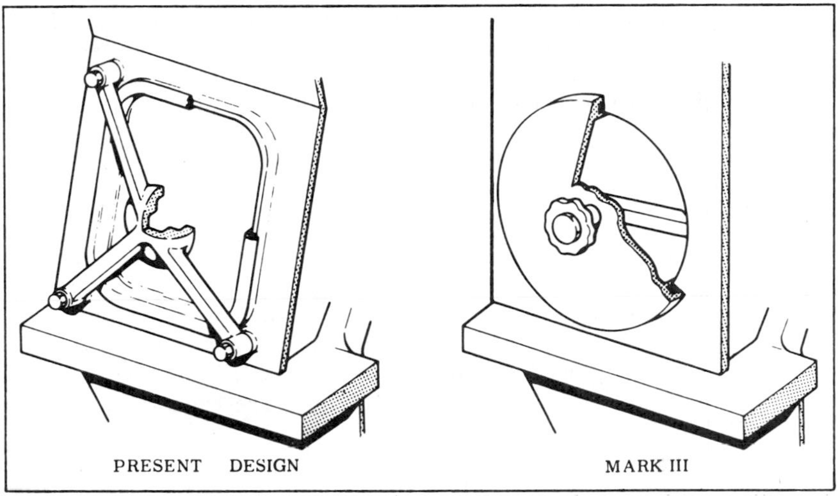

Figure 10. Side Cover Details

An additional problem experienced on earlier frames was cracking in the air manifold. The present design provides for the air header to be welded in as an integral stress member of the frame. A number of problems are associated with this type of design, some of which are stress concentration factors in the welding together of the top deck, the air header, and the athwartship plate. A second problem associated with this type of construction is the thermogradient experienced because the air manifold temperature is regulated to approximately 100°F., whereas other heat sinks at higher temperature levels create thermal stresses in the frame that are difficult to manage. Accordingly, Burdlorn proposes a separate and independent air header that will be bolted to the frame in such a manner as not to contribute to the overall stresses or stress concentration factors of the frame weldment.

2. Valves. During the 1,000 rpm test of the engine under Navy contract NOBS 72393, a wear rate of approximately .050 inch per 1,000 hours was experienced on the inlet valves. The wear rate on the exhaust valves was approximately .010 inch per 1,000 hours. A total wear of .100 inch would be considered acceptable for this size valve and

insert. Accordingly, it will be appreciated that a wear life of more than 5,000 hours could be expected on the exhaust valve, but a much shorter life was experienced on the inlet valve. A careful analysis of all available information has revealed only two differences between the intake valve and the exhaust valve that would contribute to this difference in wear rates. These two differences are as follows:

a. The exhaust cam is driven through a hollow shaft that is 4 inches O.D. by 2½ inches I.D. The inlet cam is driven by a solid shaft that runs within the hollow exhaust camshaft and is 2½ inches in diameter. It will be recognized that the relative stiffness of these two shafts in torsion is in the ratio of 5½:1 as revealed by our calculations (see Figure 11).

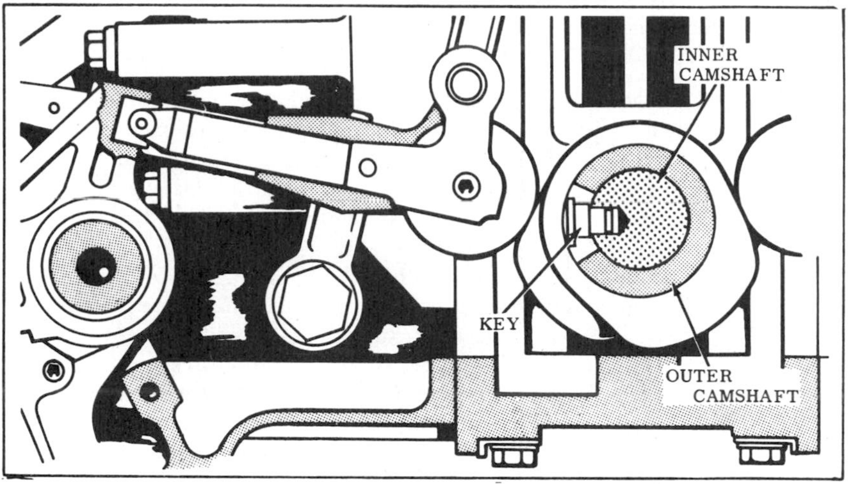

Figure 11. Present Inner and Outer Drive Shafts for Inlet and Exhaust Cams

b. The lift profiles of the intake and exhaust cams are identical; however, there is a greater dwell on the exhaust cam than there is on the inlet cam. The inlet cam has 15 degrees dwell, whereas the exhaust cam has 50 degrees dwell. The longer dwell on the exhaust cam, which cannot be incorporated in the inlet cam due to timing requirements, provides additional time in which the vibratory amplitudes in the valve train dampen or

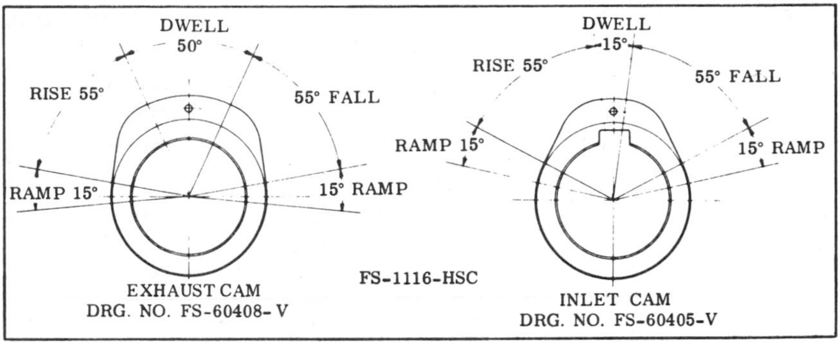

Figure 12. Comparison of Inlet and Exhaust Cam Profiles

> attenuate to a greater extent than the 15 degrees dwell provides on the inlet cam (see Figure 12). . . .

grammar

Grammar is the systematic description of the way words work together to form a coherent language; in this sense, it is an explanation of the structure of a language. However, grammar is popularly taken to mean the set of "rules" that governs how a language ought to be spoken and written; in this sense, it refers to the **usage** conventions of a language.

These two meanings of grammar—how the language functions and how it ought to function—are easily confused. To clarify the distinction, consider the expression *ain't*. Unless used purposely for its colloquial flavor, *ain't* is unacceptable to careful speakers and writers because a convention of usage prohibits its use. Yet taken strictly as a **part of speech**, the term functions perfectly well as a **verb**; whether it appears in a declarative sentence ("I ain't going.") or an interrogative sentence ("Ain't I going?"), it conforms to the normal pattern for all verbs in the English language. Although we may not approve of its use in a sentence, we cannot argue that it is ungrammatical.

To achieve **clarity**, writers need to know both grammar (as a description of the way words work together) and the conventions of usage. Knowing the conventions of usage helps writers select the appropriate over the inappropriate word or expression. A knowledge of grammar helps them diagnose and correct problems arising from how words and **phrases** function in relation to one another. For example, knowing that certain words and phrases function to modify other words and phrases gives the writer a basis for correcting those **modifiers** that are not doing their job. Understanding **dangling modifiers** helps the writer avoid or correct a construction that obscures the intended meaning. In short, an understanding of grammar and its special terminology is valuable for writers chiefly because it enables them to recognize problems and thus communicate clearly and precisely.

graphs

A graph presents numerical data in visual form. This method has several advantages over presenting data in **tables** or within the text. Trends, movements, distributions, and cycles are more readily apparent in graphs than they are in tables. By providing a means for ready comparisons, a graph often shows a significance in the data not otherwise immediately apparent. Be aware, however, that although graphs present statistics in a more interesting and comprehensible form than tables do, they are less accurate. For this reason, they are often accompanied by tables giving the exact figures. The main types of graphs are line graphs, bar graphs, pie graphs, and picture graphs. This entry also includes a compact discussion of computer graphics and provides guidelines for integrating graphs into the text.

LINE GRAPHS

The line graph shows the relationship between two sets of numbers by means of points plotted in relation to two axes drawn at right angles. The points, once plotted, are connected to one another to form a continuous line. In this way, what was merely a set of dots having abstract mathematical significance becomes graphic, and the relationship between the two sets of figures can easily be seen.

The line graph's vertical axis usually represents amounts, and its horizontal axis usually represents increments of time, as shown in Figure 1.

Line graphs with more than one line are common because they allow for comparisons between two sets of statistics for the same period of time. In creating such graphs, be certain to identify each line

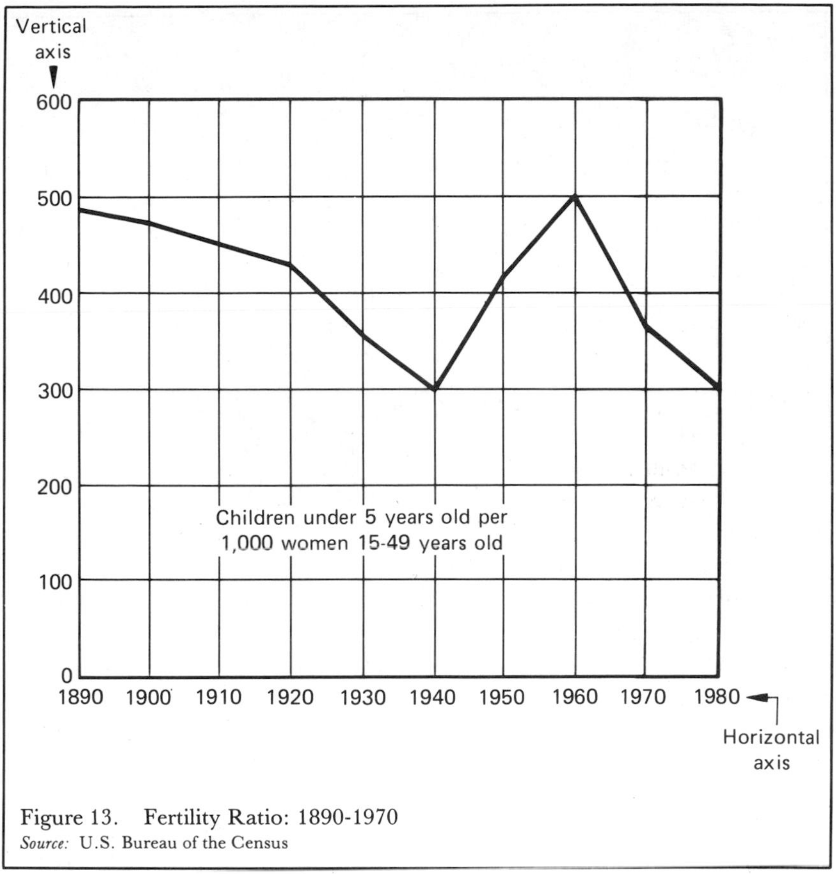

Figure 1 Single-Line Graph

with a label or a legend, as shown in Figure 2. The difference between the two lines can be emphasized by shading the space between them.

Tips on Preparing Line Graphs

1. Give the graph a title that describes the data clearly and concisely. Either center the title below the graph or align it with the left margin. If the title is too long to fit within the margins of the graph on one line, break the title into two or more lines of roughly equal length. In this case, the figure number should

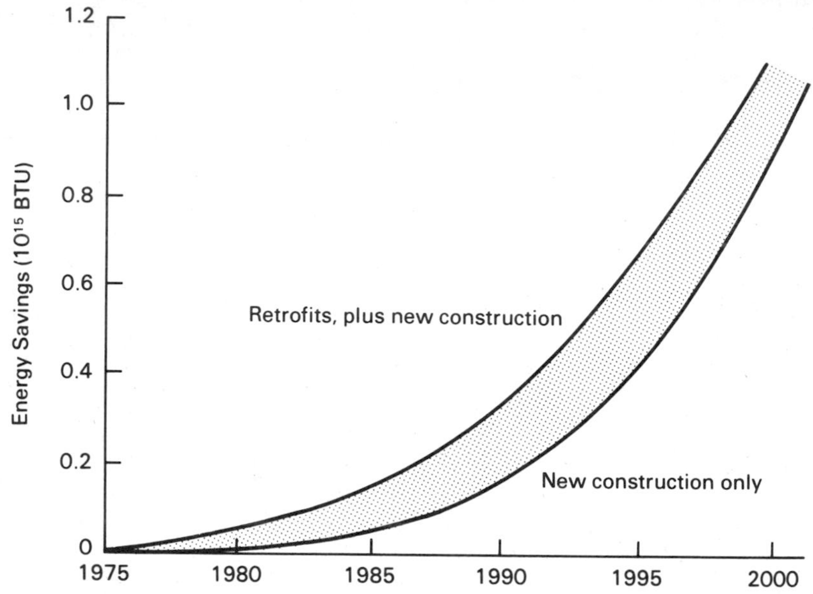

Figure 14. Projected energy savings for residences heating by solar power
Source: Electric Power Research Institute

Figure 2 Double-Line Graph with Difference Shaded

precede the first line of the title, and the other lines should begin directly beneath the first word of the title. See Figure 9.

2. Assign a figure number if your **report** includes five or more **illustrations**. The figure number precedes the title, as in the examples throughout this entry.
3. Indicate the zero point of the graph (the point where the two axes meet). If the range of data shown makes it inconvenient to begin at zero, insert a break in the scale, as in Figure 3.
4. Graduate the vertical axis in equal portions from the least amount at the bottom to the greatest amount at the top. Ordinarily, the caption for this scale is placed at the upper left.
5. Graduate the horizontal axis in equal units from left to right. If a caption is necessary, center it directly beneath the scale.
6. Graduate the vertical and horizontal scales so that they give an accurate visual impression of the data, since the angle at which

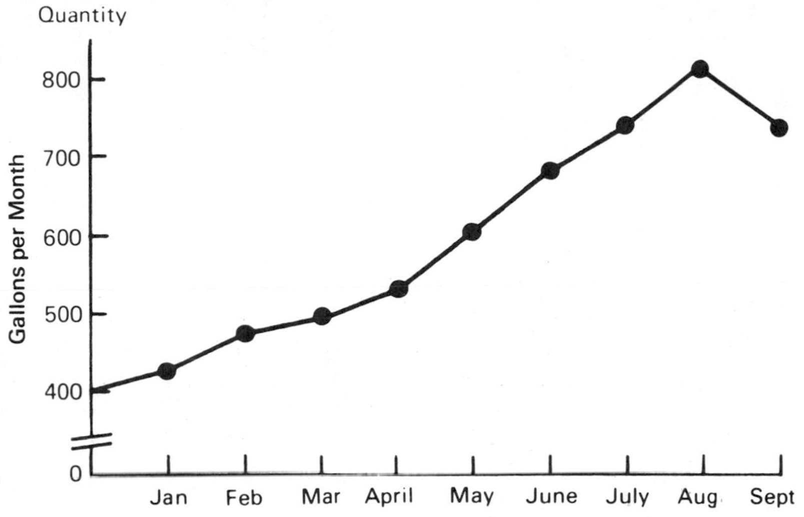

Figure 15. Surge Tank Leakage Rate, January–September

Figure 3 Line Graph with Vertical Axis Broken

the curved line rises and falls is determined by the scales of the two axes. The curve can be kept free of distortion if the scales maintain a constant ratio with each other. See Figures 4 and 5.

7. Hold grid lines to a minimum so that curved lines stand out. Since precise values are usually shown in a table of data accompanying a graph, detailed grid lines are unnecessary. Note the increasing clarity of the three graphs in Figures 6, 7, and 8.
8. Include a key (which lists and explains **symbols**) when necessary, as in Figure 7. At times a label will do just as well, as in Figure 8.
9. If the information comes from another source, include a source line below and on the same left margin as the caption, as in Figure 1 in this entry.
10. Place explanatory footnotes below the graph, to the lower left.
11. Place all lettering horizontally if possible.

BAR GRAPHS

Bar graphs consist of horizontal or vertical bars of equal width but scaled in length to represent some quantity. They are commonly

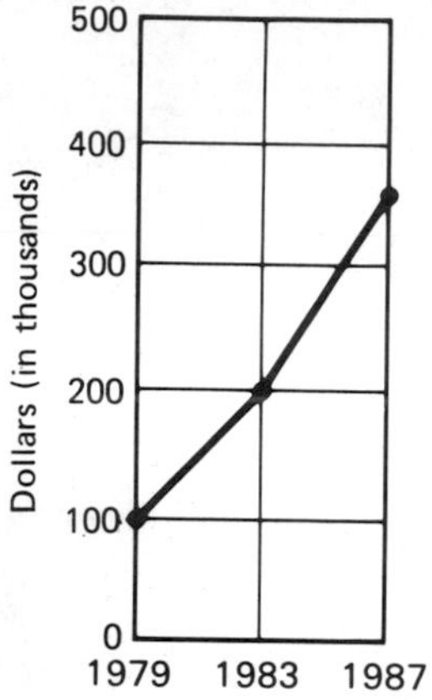

Figure 16. Growth in Engineering Research Funds

Figure 4   Distorted Curve

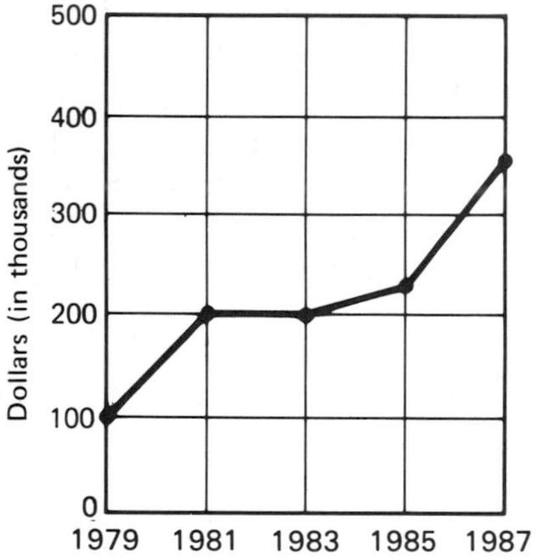

Figure 17. Growth in Engineering Research Funds

Figure 5   Distortion-Free Curve

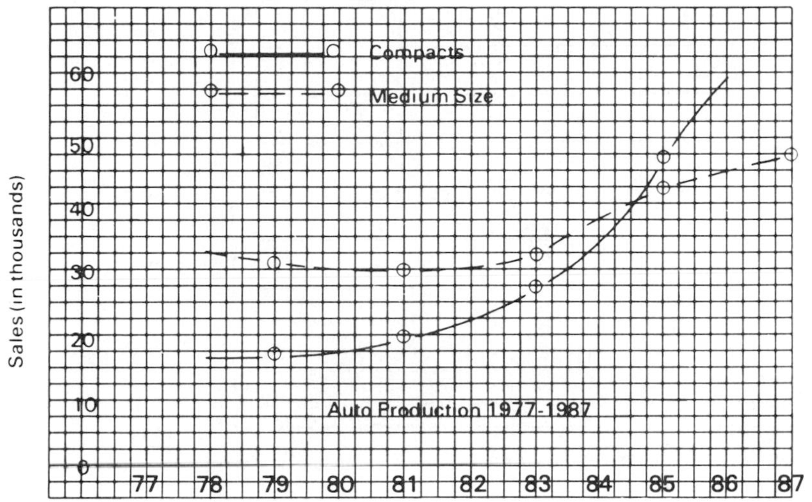

Figure 18. Production

Figure 6 Line Graph That Is Difficult to Read

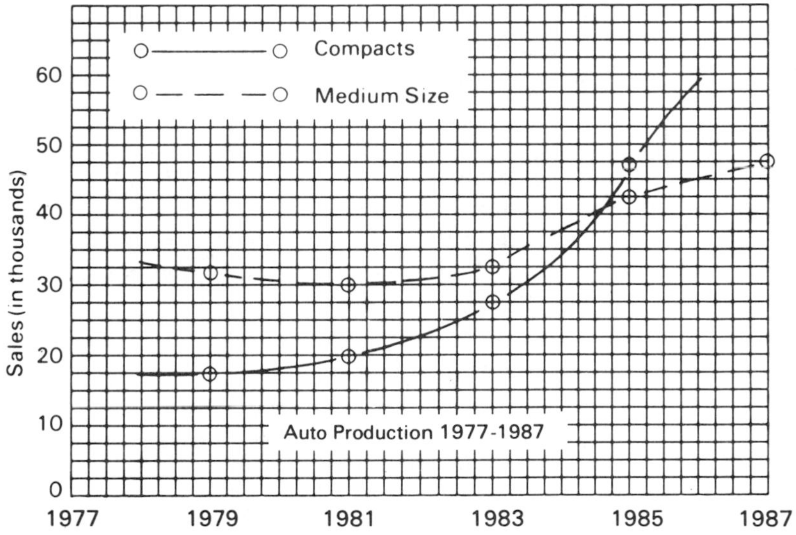

Figure 19. Auto Production

Figure 7 A More Legible Version of Figure 6

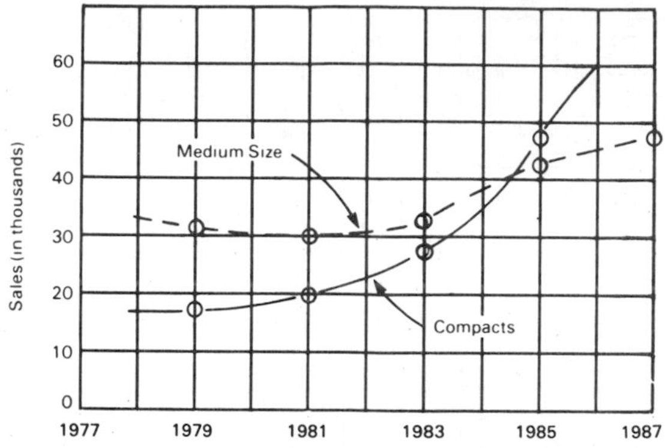

Figure 20. Auto Production, 1977–1987.

Figure 8 A Clear Version of Figure 6

used to show (1) quantities of the same item at different times, (2) quantities of different items for the same time period, or (3) quantities of the different parts of an item that make up the whole.

Figure 9 is an example of a bar graph showing varying quantities of the same item at the same time. Here each bar, which represents a different quantity of the same item, begins at the left scale. The left scale also provides additional information in the form of the percentage of the whole population each bar (and therefore each state) represents.

Some bar graphs show the quantities of different items for the same period of time. See Figure 10. (A bar graph with vertical bars is also called a column graph.)

Bar graphs can also show the different portions of an item that make up the whole. Here the bar is equivalent to 100 percent. It is then divided according to the appropriate proportions of the item sampled. This type of graph can be constructed vertically or horizontally and can indicate more than one whole when comparisons are necessary. See Figures 11 and 12.

If the bar is not labeled, each portion must be marked clearly by shading or crosshatching. Include a key that identifies the subdivisions.

PIE GRAPHS

A pie graph presents data as wedge-shaped sections of a circle. The circle must equal 100 percent, or the whole, of some quantity (a tax

Supercity*	1960 Population	Land Area (Square Miles)	Population Density (People per Square Mile)
New York City/ Northeastern New Jersey/ Southern Connecticut	17,606,680	3,656	4,616
Los Angeles/Long Beach/ San Bernardino/Riverside	10,184,611	2,187	4,657
Chicago/Northwest Indiana/ Aurora/Elgin/Joliet	7,212,778	1,675	4,306
Philadelphia/ Wilmington/Trenton	4,779,796	1,270	3,764
San Francisco/San Jose	4,434,650	1,122	3,952
Detroit	3,809,327	1,044	3,649
Boston/Brockton/Lowell/ Lawrence/Haverhill	3,225,386	1,064	2,975
Miami/Fort Lauderdale/ Hollywood/West Palm Beach	3,103,729	816	3,804
Washington, D.C.	2,763,105	807	3,424
Houston/ Texas City/Lamarque	2,521,856	1,169	2,157
Cleveland/Akron/ Lorain/Elyria	2,493,475	989	2,521
Dallas/Fort Worth	2,451,390	1,280	1,915
St. Louis	1,848,590	597	3,096
Pittsburgh	1,810,038	713	2,539
Seattle/ Everett/Tacoma	1,793,612	672	2,669
Minneapolis/St. Paul	1,787,564	980	1,824
Baltimore	1,755,477	523	3,357
San Diego	1,704,352	611	2,789
Atlanta	1,613,357	905	1,783
Phoenix	1,409,279	641	2,199
Tampa/St. Petersburg	1,354,249	527	2,570
Denver	1,352,070	439	3,080
Cincinnati/Hamilton	1,228,438	465	2,642
Milwaukee	1,207,008	496	2,433
Newport News/Hampton/ Norfolk/Portsmouth	1,099,360	614	1,790
Kansas City	1,097,793	589	1,864
New Orleans	1,078,299	230	4,688
Portland	1,026,144	349	2,940
Buffalo	1,002,285	266	3,768

*Cities in bold type lost population in the 1970s.

Figure 14. Urban areas of 1 million or more population, with land area and population density. Source: *Scientific American.*

Figure 9 Bar Graph of Varying Quantities of the Same Item at the Same Time

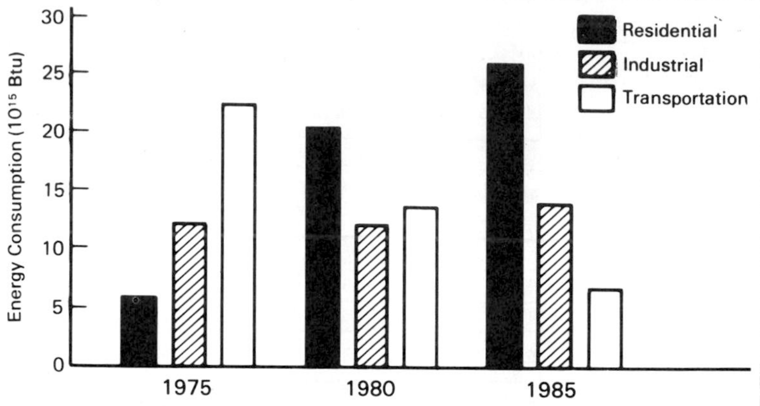

Figure 22. U.S. Energy Demand.

Figure 10 Bar Graph of Quantities of Different Items for the Same Period (shown for three separate years)

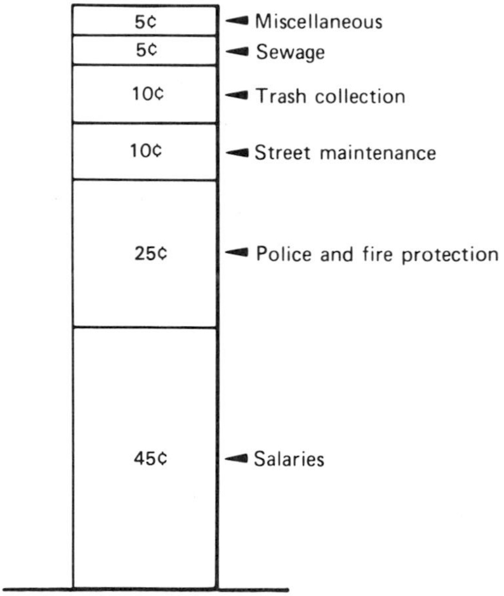

Figure 23. Your Municipal Tax Dollar

Figure 11 Bar Graph of Quantities of Different Parts Making Up a Whole

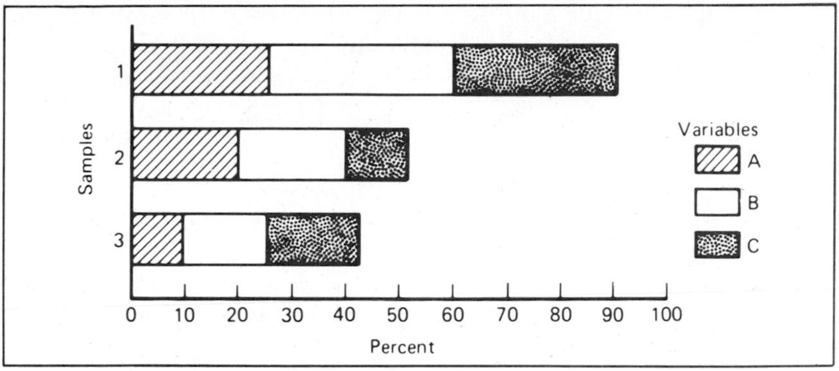

Figure 12 Bar Graph Showing Variables in Three Samples

dollar, a bus fare, the hours of a working day), with the wedges representing the various ways in which the whole is divided. In Figure 13, for example, the circle stands for a city tax dollar, and it is divided into units equivalent to the percentage of the tax dollar spent on various city services.

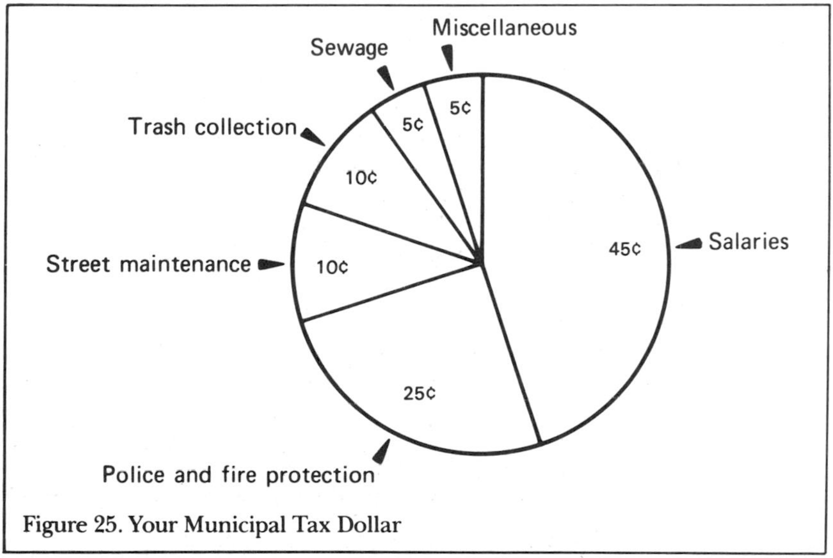

Figure 13 Pie Graph

Pie graphs provide a quicker way of presenting the same information that can be presented in a table; in fact, a table often accompanies a pie graph with a more detailed breakdown of the same information.

Tips on Preparing Pie Graphs

1. The complete 360° circle is equivalent to 100 percent; therefore, each percentage point is equivalent to 3.6°
2. To make the relative percentages clear, begin at the 12 o'clock position and sequence the wedges clockwise, from largest to smallest.
3. If you shade the wedges, do so clockwise and from light to dark.
4. Keep all labels horizontal, and most important, give the percentage value of each wedge.
5. Finally, check to see that all wedges, as well as the percentage values given for them, add up to 100 percent.

Although pie graphs have strong visual impact, they also have drawbacks. If more than five or six items are presented, the graph looks cluttered. Also, since they usually present percentages of something, they must often be accompanied by a table listing precise statistics. Further, unless percentages are shown on the sections, the reader cannot compare the values of the sections as accurately as with a bar graph. (The terms that identify each segment of the graph are referred to as *callouts*.)

PICTURE GRAPHS

Picture graphs are modified bar graphs that use picture symbols of the item presented. Each symbol corresponds to a specified quantity of the item. See Figure 14. Note that precise figures are included, since the graph can present only approximate figures.

Tips on Preparing Picture Graphs

1. Make the symbol self-explanatory.
2. Have each symbol represent a single unit.
3. Show larger quantities by increasing the number of symbols rather than by creating a larger symbol. (It is difficult to judge relative sizes accurately.)
4. Consult a good graphics manual for details about preparing graphs. See **reference books** for a list of useful manuals.

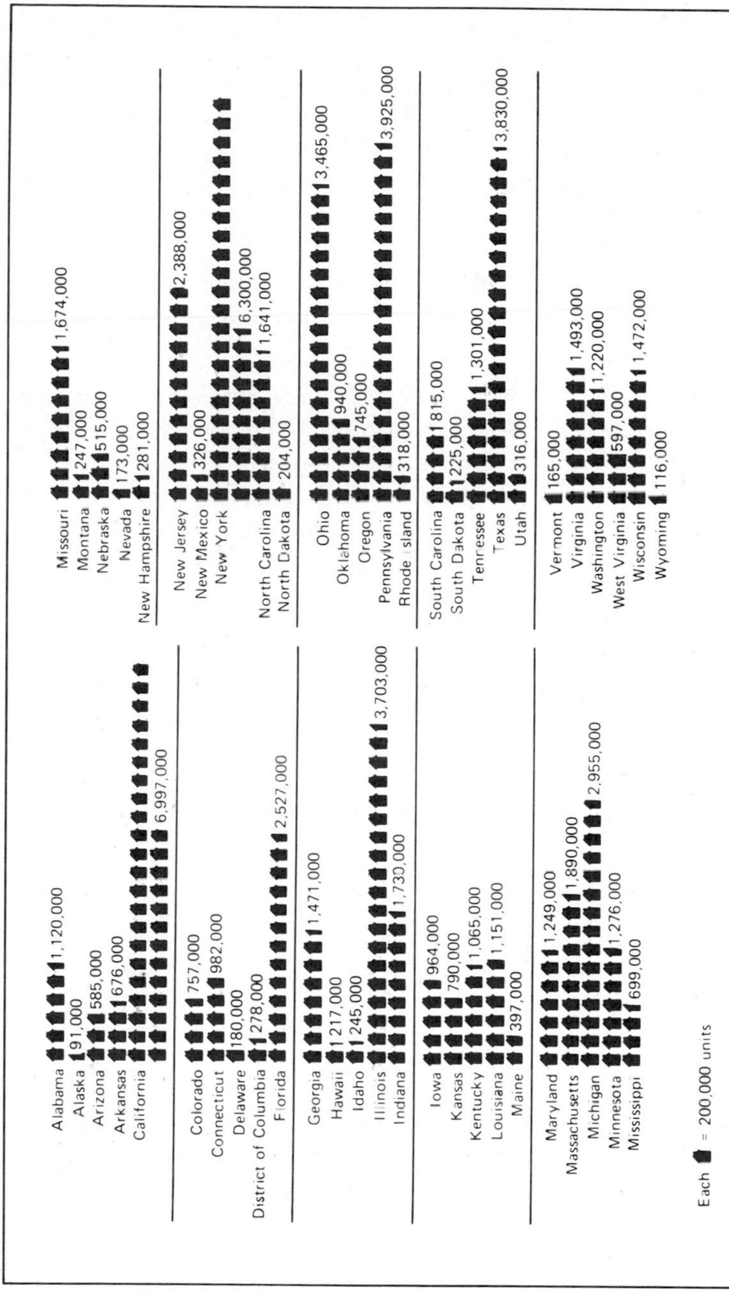

Figure 26. Number of Housing Units, by States: 1970
Source: Bureau of the Census

Figure 14 Picture Graph

COMPUTER GRAPHICS

Many computer systems, including microcomputers, have sophisticated graphics programs that allow you to create, in seconds, complex six-color graphs, such as multiple bar graphs and pie graphs. Computer graphics help you display and interpret data at a glance, resulting in faster decision making. By integrating computer graphics and word-processing programs, you can rapidly insert your visual aids anywhere in the text of a report, delete them, or store them for later use. With the right software, you can create all of the illustrations shown in this entry. Some other features of computer graphics are the capability of:

- using an electronic stylus (a pen with an electronic signal) for drawing pictures displayed on a monitor.
- mapping densities and concentrations in different colors.
- using shading and shadows to create three-dimensional effects.
- creating and revising mathematical models, animations, or simulated events.
- using computer-assisted design (CAD) to direct and automate computer-assisted manufacturing (CAM).

IDENTIFYING GRAPHS AND INTEGRATING THEM INTO THE TEXT

Observe the following guidelines when you plan to integrate graphs into your text:

1. Give each graph a title that describes the data clearly and concisely. Either center the title below the graph or align it with the left margin. If the title is too long to fit within the margins of the graph on one line, break the title into two or more lines of roughly equal length. The figure number should precede the first line of the title; the other lines should begin directly below the first word of the title.
2. Assign a figure number if your report contains five or more graphs. The figure number precedes the title, as in the examples throughout this entry.
3. Relate the graph to the text around it by referring to it by figure number and title.
4. Locate the graph as close as possible to its first mention in the text. However, the graph should never precede its first text mention; its appearance without an introduction will confuse readers.

5. If the graph is central to the discussion, illuminating or strongly reinforcing it, include the graph in the text. But if the graph is lengthy, detailed, and peripheral to the discussion, place it in an appendix. However, even material in an appendix should be mentioned or referred to in the text; otherwise, the information in the appendix is superfluous. (See **appendix/appendixes/appendices**.)
6. The amount of text discussion each graph requires will vary with its importance to the discussion. A graph showing an important trend over time may be the central focus of an entire discussion. The level of complexity of the graph will also affect the discussion, as will how knowledgeable your readers are. Nonexperts require lengthier explanations than experts do, as a rule.

H

half

Half a, half an, and *a half* all are correct idiomatic uses of the word *half*.

EXAMPLES It takes *a half* second for the signal to reach your receiver.
Wait *half a* minute before stirring the solution.
Allow *half an* hour for the compound to solidify.

A half a (or *a half an*) is colloquial, however, and should be avoided in writing.

CHANGE The subroutine is executed in *a half a* minute.
TO The subroutine is executed in *half a* minute.

he/she

Since there is no singular **personal pronoun** in English that refers to both sexes, the word *he* has traditionally been used when the sex of the antecedent is unknown.

EXAMPLE Whoever is appointed [either a man or woman] will find *his* task difficult.

Because the use of a masculine **pronoun** when both sexes are being referred to is now recognized as offensive, it is better to rewrite the sentence in the plural (or avoid use of a pronoun altogether) than to offend.

EXAMPLES *Employees* should take advantage of *their* insurance benefits.
Whoever is appointed will find *the* task difficult.

You could also use the **phrases** *he or she* and *his or her*.

EXAMPLE *Whoever* is appointed will find *his or her* task difficult.

Unfortunately, *he or she* and *his or her* are clumsy when used repeatedly; the best advice is to reword the sentence to use a plural pronoun. Be sure also to change to its plural form the **noun** to which the pronoun refers. (See also **plurals, gender**, and **Ms./Miss/Mrs.**)

CHANGE An engineer cannot do *his or her* job until *he or she* understands the concept.

TO Engineers cannot do *their* jobs until *they* understand the concept.

headers and footers

A header in a **report, technical manual**, or **specification** is identifying information carried at the top of each page. The header normally contains the **topic** (or topic and subtopic) dealt with in that section of the report, manual, or specification (specifications also include the identification numbers of sections and **paragraphs**).

A footer is identifying information carried at the bottom of each page. The footer generally contains the date of the document, the page number, and sometimes the manual name and section title.

Although the types of information included in headers and footers vary greatly from one organization to the next, the example in Figure 1 is fairly typical (see page 290).

heads

Heads (also called headings) are **titles** or subtitles within the body of a long piece of writing that serve as guideposts for the **reader**. They divide the material into manageable segments, call attention to the main topics, and signal changes of topic. If a **formal report, proposal**, or other document you are writing is long or complicated, you may need several levels of heads to indicate major divisions, subdivisions, and even smaller units of those. In extremely technical material as many as five levels of heads may be appropriate, but as a general rule it is rarely necessary (and usually confusing) to use more than three.

The best advice on the effective use of heads is to be sure you prepare a good topic outline before writing your draft and then use the major topic listings of the topic outline (without roman numerals and capital letters). In a short document, you probably will use only the major divisions of your topic outline as heads, but in a longer document, you may use both the major and minor divisions. Be careful, however, not to use too many heads; four or five on a page would certainly be too many under most circumstances.

If your document requires only one level of head, you may use any form (all **capital letters**, initial capital letters only, and the like).

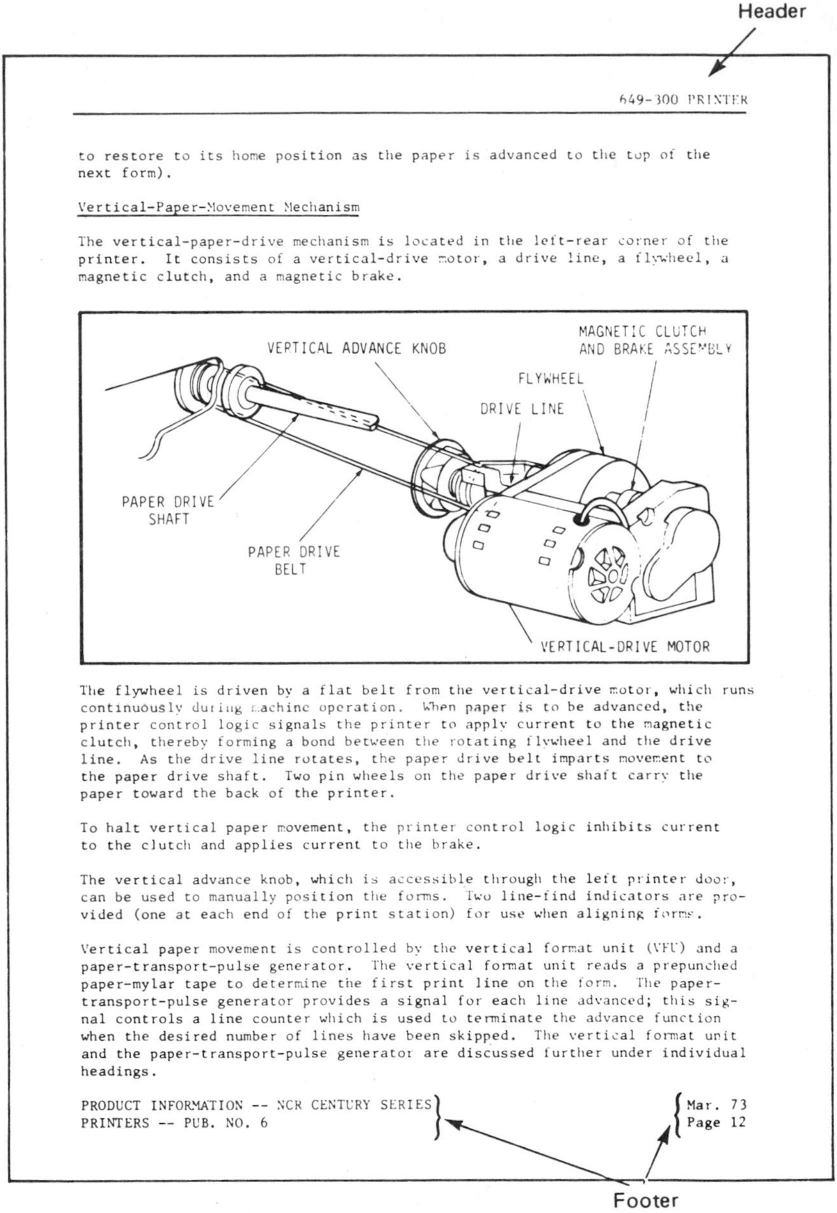

649-300 PRINTER

to restore to its home position as the paper is advanced to the top of the next form).

Vertical-Paper-Movement Mechanism

The vertical-paper-drive mechanism is located in the left-rear corner of the printer. It consists of a vertical-drive motor, a drive line, a flywheel, a magnetic clutch, and a magnetic brake.

The flywheel is driven by a flat belt from the vertical-drive motor, which runs continuously during machine operation. When paper is to be advanced, the printer control logic signals the printer to apply current to the magnetic clutch, thereby forming a bond between the rotating flywheel and the drive line. As the drive line rotates, the paper drive belt imparts movement to the paper drive shaft. Two pin wheels on the paper drive shaft carry the paper toward the back of the printer.

To halt vertical paper movement, the printer control logic inhibits current to the clutch and applies current to the brake.

The vertical advance knob, which is accessible through the left printer door, can be used to manually position the forms. Two line-find indicators are provided (one at each end of the print station) for use when aligning forms.

Vertical paper movement is controlled by the vertical format unit (VFU) and a paper-transport-pulse generator. The vertical format unit reads a prepunched paper-mylar tape to determine the first print line on the form. The paper-transport-pulse generator provides a signal for each line advanced; this signal controls a line counter which is used to terminate the advance function when the desired number of lines have been skipped. The vertical format unit and the paper-transport-pulse generator are discussed further under individual headings.

PRODUCT INFORMATION -- NCR CENTURY SERIES PRINTERS -- PUB. NO. 6

Mar. 73
Page 12

Figure 1 Header and Footer

For example, capitalizing the first letter of all the main words (as in a title) and then underlining them is very common.

On page 292, the example (Figure 1), greatly shortened, is intended only to illustrate the use of the heads—in this case, three levels (beyond the title of the entire report).

As you can infer from this example, the heads you use grow naturally from the divisions and subdivisions prepared in **outlining**. If more of the preceding example were given, the heads would look like this:

CHICAGO
Location
Transportation
Rail Transportation
Sea Transportation
Air Transportation
Truck Transportation
Labor Supply
Engineering Personnel
Manufacturing Personnel
Outside and Consulting Services
Tax Structure
Living Conditions
Housing
Education
Recreation
Climate
MINNEAPOLIS
Location
Transportation

The decimal numbering system uses a combination of numbers and decimal points to subordinate levels of heads in a report. The following outline shows the correspondence between different levels of headings and the decimal numbers used:

1. FIRST-LEVEL HEAD
1.1 Second-Level Head
1.2 Second-Level Head
1.2.1 Third-Level Head
1.2.2 Third-Level Head

Interim Report of the Committee to Investigate New Factory Locations

The committee initially considered thirty possible locations for the proposed new factory. Of these, twenty were eliminated almost immediately for one reason or another (unfavorable tax structure, remoteness from rail service, inadequate labor supply, etc.). Of the remaining ten locations, the committee selected for intensive study the three that seemed most promising: Chicago, Minneapolis, and Salt Lake City. These three cities we have now visited, and our observations on each of them follow.

CHICAGO

Of the three cities, Chicago presently seems to the committee to offer the greatest advantages, although we wish to examine these more carefully before making a final recommendation.

Location

Though not at the geographical center of the United States, Chicago is centrally located in an area that contains more than three-quarters of the U.S. population. It is within easy reach of our corporate headquarters in New York. And it is close to several of our most important suppliers of components and raw materials—those, for example, in Columbus, Detroit, and St. Louis.

Transportation

Rail Transportation. Chicago is served by the following major railroads. . .

Sea Transportation. Except during the winter months when the Great Lakes are frozen, Chicago is an international seaport. . . .

Air Transportation. Chicago has two major airports (O'Hare and Midway) and is contemplating building a third. Both domestic and international air cargo service are available. . . .

Transportation by Truck. Virtually all of the major U.S. carriers have terminals in Chicago. . . .

Figure 1 The Use of Heads

1.2.2.1 Fourth-Level Head
1.2.2.2 Fourth-Level Head
1.3 Second-Level Head
1.3.1 Third-Level Head
1.3.2 Third-Level Head
2. FIRST-LEVEL HEAD

Although the second-, third-, and fourth-level heads are indented in an outline or table of contents, as headings they are flush with the left margin in the body of the report.

Note that the first-level heads are typed in all-capital letters; the second-level heads and all subsequent levels of heads are typed in capital and lowercase letters. That is, the first letter of the first word and of every important word is capitalized; prepositions and conjunctions of five letters or more are also capitalized. Every head is typed on a separate line, with an extra line of space above and below the head.

There is no one correct format for heads. Sometimes a company settles on a standard format, which everyone within the company is then expected to follow. Or a customer for whom a report or proposal is being prepared may specify a particular format. In the absence of such guidelines, the system used in the example should serve you well. Note the following formal characteristics: (1) The first-level head is in all capital letters, typed flush to the left margin on a line by itself and separated by a line space from the material it introduces. (2) The second-level head is in capital and lowercase letters, also typed flush to the left margin on a line by itself and also separated by a line space from the material it introduces. (3) The third-level head is in capital and lowercase letters, but it is "run in" right on the same line with the first sentence of the material it introduces. Therefore, it is followed by a period to set it apart from what follows, and it is underlined or italicized so that it will stand out clearly on the page.

The following are some important points to keep in mind about heads:

1. They should signal a shift to a new topic (or, if they are lower-level heads, a new "subtopic" within the larger topic).
2. Within the larger unit they subdivide, all heads at any one level should be consistent in their relationship to the topic of the larger unit. For instance, notice in the extended list of heads in our ex-

ample that the first-level heads ("CHICAGO" and "MINNEAPOLIS") both bear the same relationship to the larger topic of the report (possible factory locations). When the discussion narrows to Chicago, all of the second-level heads within that unit bear the same relationship to the first-level head ("CHICAGO"): they refer to Chicago's location, its transportation, its labor supply, its tax structure, and its living conditions. And the third-level heads conform to the same principle within their units.

3. The fact that one unit at a particular level is subdivided by lower-level heads does not mean that *every* unit at that particular level must also include lower-level heads. In the example, notice that under the first-level head "CHICAGO" some of the second-level units (for instance, "Transportation") are further subdivided by third-level heads and some (for instance, "Location") are not.
4. All heads at any one level within the same next-larger unit should be parallel with one another in structure (see **parallel structure**). For instance, note in the example that all of the heads are **nouns** or noun **phrases.**
5. Too many heads, or too many levels of heads, can be as bad as too few. A highway map that showed only New York and San Francisco would not be very helpful, even to the traveler driving from New York to San Francisco. But a map that tried to show every street and every building in every town along the way would not be much better. Keep in mind the needs of your reader.
6. If your document needs a **table of contents**, create it from your heads and subheads. Be sure that the wording in your table of contents is the same as the wording in your text.
7. The head does not substitute for the discussion; the text should read as if the head were not there. In particular, don't use a **pronoun** in the text to refer to a noun in the head.
8. Heads should be specific and concise without being too cryptic.

healthful/healthy

If something is *healthful*, it promotes good health. If something is *healthy*, it has good health. Carrots are *healthful*; people can be *healthy*.

EXAMPLES Yogurt is *healthful.*
We provide *healthful* lunches in the cafeteria because we want our employees to be *healthy*.

helping verbs

A helping verb (also called an auxiliary verb) is a **verb** that is added to a main verb to help it indicate **mood, tense**, and **voice**. In doing so, it also forms a verb **phrase**.

EXAMPLES I *am* going.
I *will* go.
I *should have* gone.
I *should have been* gone.
The order *was* given.

All helping verbs (*be, have, do, shall, will, may, can, must, ought*) can function alone as main verbs, but they usually function as helping verbs in verb phrases. Notice that the principal helping verbs are the various forms of the main verbs *be, have*, and *do*.

herein/herewith

The words *herein* and *herewith* are a type of business **jargon** that should be avoided.

CHANGE I have *herewith* enclosed a copy of the contract.
TO I have enclosed a copy of the contract.

house organ articles (see **newsletter articles**)

hyperbole

Hyperbole is exaggeration to achieve an effect or **emphasis**. It is a device used to emphasize or intensify a point by going beyond the literal facts. It distorts to heighten effect.

EXAMPLES They *murdered* us at the negotiating session.
The hail was like *boulders*.

Used cautiously, hyperbole can magnify an idea without distorting it; therefore, it plays a large role in advertising.

EXAMPLE A box of Clenso detergent can clean a *mountain of clothes*.

Everyday speech abounds in hyperbole.

EXAMPLES I'm *starved*. (meaning "hungry")
I'm *dead*. (meaning "tired")

Because technical writing needs to be as accurate and precise as possible, always avoid hyperbole. (See also **figures of speech**.)

hyphens

Although the hyphen functions primarily as a **spelling** device, it also functions to link and to separate words; in addition, it occasionally replaces the **preposition** *to* (0–100 for 0 *to* 100). The most common use of the hyphen, however, is to join **compound words.**

EXAMPLES able-bodied, self-contained, carry-all, brother-in-law

A hyphen is used to form compound **numbers** from twenty-one to ninety-nine and fractions when they are written out.

EXAMPLES twenty-one, one forty-second

When in doubt about whether or where to hyphenate a word, check your **dictionary.**

HYPHENS USED WITH MODIFIERS

Two-word and three-word unit **modifiers** that express a single thought are hyphenated when they precede a **noun** (an *out-of-date* car, a *clear-cut* decision). If each of the words can modify the noun without the aid of the other modifying word or words, however, do not use a hyphen (a *new digital* computer—no hyphen). Also, if the first word is an **adverb** ending in *-ly*, do not use a hyphen (a *hardly* used computer, a *badly* needed micrometer).

The presence or absence of a hyphen can alter the meaning of a sentence.

EXAMPLE We need a biological waste management system.
COULD MEAN We need a biological-waste management system.
OR We need a biological waste-management system.

A modifying **phrase** is not hyphenated when it follows the noun it modifies.

EXAMPLE Our office equipment is *out of date*.

A hyphen is always used as part of a letter or number modifier.

EXAMPLES 5-cent, 9-inch, A-frame, H-bomb

In a series of unit modifiers that all have the same term following the hyphen, the term following the hyphen need not be repeated throughout the series; for greater smoothness and brevity, use the term only at the end of the series.

CHANGE The third-floor, fourth-floor, and fifth-floor rooms have recently been painted.
TO The third-, fourth-, and fifth-floor rooms have recently been painted.

HYPHENS USED WITH PREFIXES AND SUFFIXES

A hyphen is used with a **prefix** when the root word is a **proper noun**.

EXAMPLES pre-Sputnik, anti-Stalinist, post-Newtonian

A hyphen may optionally be used when the prefix ends and the root word begins with the same vowel. When the repeated vowel is *i*, a hyphen is almost always used.

EXAMPLES re-elect, re-enter, anti-inflationary

A hyphen is used when *ex-* means "former."

EXAMPLES ex-partners, ex-wife

A hyphen may be used to emphasize a prefix.

EXAMPLE He was *anti-everything*.

The **suffix** *-elect* is hyphenated.

EXAMPLES president-elect, commissioner-elect

OTHER USES OF THE HYPHEN

To avoid confusion, some words and modifiers should always be hyphenated. *Re-cover* does not mean the same thing as *recover*, for example; the same is true of *re-sent* and *resent*, *re-form* and *reform*, *re-sign* and *resign*.

Hyphens should be used between letters showing how a word is spelled.

EXAMPLE In his letter he misspelled *believed* b-e-l-e-i-v-e-d.

Hyphens identify prefixes, suffixes, or written syllables.

EXAMPLE *Re-*, *-ism,* and *ex-* are word parts that cause spelling problems.

A hyphen can stand for *to* or *through* between letters and numbers.

EXAMPLES pp. 44-46
the Detroit-Toledo Expressway
A-L and M-Z

Finally, hyphens are used to divide words at the end of a line. Avoid dividing words whenever possible, but, if you must, use the following guidelines for hyphenation:

Divide

- Between syllables (but leave at least three letters on each line)

  EXAMPLE let-ter

- Between the compounding parts of compound words

  EXAMPLE time-table

- After a single-letter syllable in the middle of the root word

  EXAMPLE sepa-rate

- After a prefix

  EXAMPLE pre-view

- Before a suffix

  EXAMPLE cap-tion

- Between two consecutive vowels with separate sounds

  EXAMPLE gladi-ator

Do Not Divide

- A word that is pronounced as one syllable

  EXAMPLE shipped

- A contraction

  EXAMPLE you're

- An **abbreviation** or **acronym**

  EXAMPLE AOPA

If a word is spelled with a hyphen, divide it only at the hyphen break, unless dividing a word there would confuse the reader because the hyphen is essential to the word's meaning.

EXAMPLE The accidents had all occurred during the post-
blast period.

(See also **syllabication** and **jammed modifiers**.)

I

idea

The word *idea* means the "mental representation of an object of thought."

EXAMPLE He didn't accept my idea about how to reorganize the branch.

Use *impression* or *intention* when they convey more precisely what you mean to say.

CHANGE We started out with the *idea* of returning early.
TO We started out with the *intention* of returning early.

CHANGE She left me with the *idea* that he was in favor of the proposal.
TO She left me with the *impression* that he was in favor of the proposal.

idioms

Idioms are groups of words that have a special meaning apart from their literal meaning. Someone who "runs for office" in the United States, for example, need not be a track star. The same person would "stand for office" in England. In both nations, the individual is seeking public office; only the idioms of the two countries differ. This difference indicates why idioms give foreign writers and speakers trouble.

EXAMPLES The judge *threw the book* at the convicted arsonist.
The company's legal advisers debated whether to *stand fast* or to *press forward* with the lawsuit.

The foreigner must memorize such expressions; they cannot logically be understood. The native writer has little trouble understanding idioms and need not attempt to avoid them in writing, provided the **reader** is equally at home with them. Idioms are often helpful shortcuts, in fact, that can make writing more vigorous and natural.

If there is any chance that your writing might be translated into another language or read in other English-speaking countries, eliminate obvious idiomatic expressions that might puzzle readers. For example, the owner's manual for a Volvo automobile was obviously translated into English with British drivers in mind. Where the American expression would be "filling up," there are references to "topping up" various fluids; where Americans would expect "windshield wipers," they find "windscreen wipers"; and the rear-view mirror has an "antidazzle knob" instead of an "antiglare adjustment" for night driving. Such idiomatic expressions can be an annoying hindrance to clear understanding.

Idioms are the reason that certain **prepositions** follow certain **verbs**, **nouns**, and **adjectives**. Since there is no sure system to explain such **usages**, the best advice is to check in a **dictionary**. The following are some common pairings that give writers trouble:

absolve from (responsibility)
absolve of (crimes)
accede to
accordancc with
according to
accountable for (actions)
accountable to (a person)
accused by (a person)
accused of (a deed)
acquaint with

adapt for (a purpose)
adapt to (a situation)
adapt from (change)
adept in
adhere to
adverse to
affinity between, with
agree on (terms)
agree to (a plan)
agree with (a person)

angry with (a person)
angry at, about (a thing)
approve of
apply for (a position)
apply to (contact)
argue for, against (a policy)
argue with (a person)
arrive in (a city, country)
arrive at (a specific location, conclusion)
based on, upon

blame for (an action)
blame on (a person)
compare to (things that are similar but not the same kind)
compare with (things of the same kind to determine similarities and differences)
comply with
concur in (consensus)
concur with (a person)
conform to
consist of
convenient for (a purpose)
convenient to (a place)
correspond to, with (a thing)
correspond with (a person)

deal with
depend on
deprive of
devoid of
differ about, over (an issue)

differ from (a thing)
differ with (a person)
differ on (amounts, terms)
different from

disagree on (an issue, plan)
disagree with (a person)
disappointed in
disapprove of
disdain for
divide between, among
divide into (parts)
engage in
exclude from
expect from (things)
expect of (people)
expert in

identical with, to
impatient for (something)
impatient with (someone)
imply that
impose on
improve on
inconsistent with
independent of

infer from
inferior to
inseparable from
mediate between, among

necessary for (an action)
necessary to (a state of being)
occupied by (things, people)
occupied with (actions)
opposite of (qualities)
opposite to (positions)
part from (a person)
part with (a thing)
proceed with (a project)
proceed to (begin)
proficient in
profit by (things)
profit from (actions)
prohibit from
qualify as (a person)
qualify by (experience, actions)
qualify for (a position, award)

rely on
responsibility for, of
reward for (an action)
reward with (a gift)
similar to
surrounded by (people)
surrounded with (things)
talk to (a group)
talk with (a person)
wait at (a place)
wait for (a person, event)
wait on (a customer)
yield to

i.e. (see **e.g./i.e.**)

illegal/illicit

If something is *illegal*, it is prohibited by law. If something is *illicit*, it is prohibited by law or *custom*. *Illicit* behavior may or may not be *illegal*, but it does violate custom or moral codes and therefore usually has a clandestine or immoral **connotation**.

EXAMPLES Their *illicit* behavior caused a scandal, and the district attorney charged that *illegal* acts were committed.

Explicit sexual material may be *illicit* without being *illegal*.

illustrations

The objective of using an illustration is to help your **reader** absorb the facts and ideas you are presenting. When used well, an illustration can convey an idea that words alone could never really make clear. Notice how the description in Figure 1 depends on its illustration:

EXAMPLE The 682-100 Card Reader, a serial Reader that uses photoelectric sensing for data detection, can read a maximum of 300 cards per minute, with the actual rate dependent upon the processor's demand for input. Each time the processor requests that a card be input, the Reader completes one cycle of the following concurrent steps: (1) the card in the read station moves past the photoelectric cells to the output hopper (the photoelectric cells sense the data in the card for transmission to the processor's main memory), (2) the card in the ready station moves to the read station,and (3) a card feeds from the input hopper to the ready station.

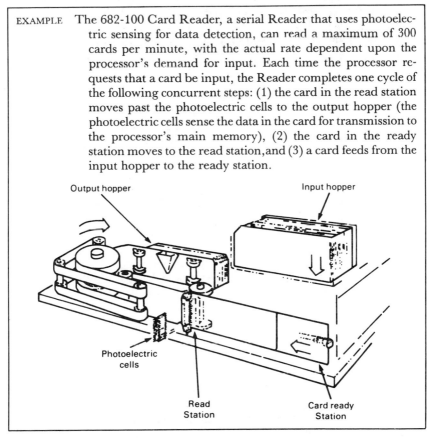

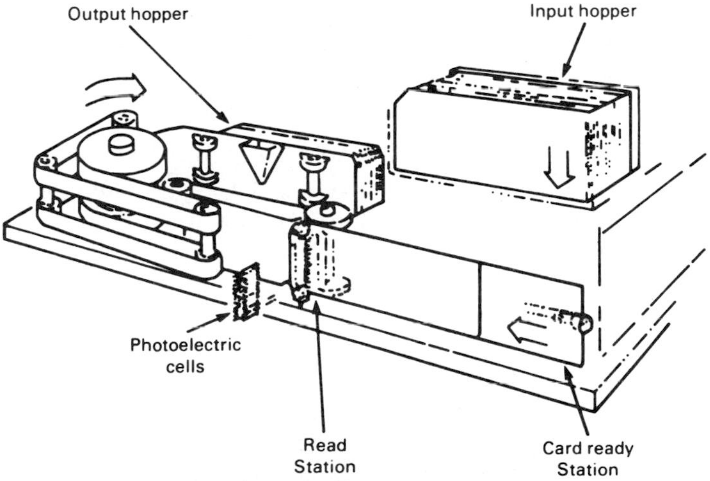

Figure 1 Drawing Illustrating a Process

Illustrations should never be used as ornaments, however; they should always be functional working parts of your writing. Be careful not to overillustrate; use an illustration only when you are sure that it makes a direct contribution to your reader's understanding of your subject. When creating illustrations, you must always consider your objective and your reader. You would use a different illustration of an x-ray machine, for example, for a high school science class than you would for a group of medical doctors. Many of the attributes of good writing—simplicity, **clarity, conciseness**, directness—are equally important in creating and using illustrations.

The most common types of illustrations are **photographs, graphs, tables, drawings, flowcharts, organizational charts, schematic diagrams**, and **maps**. Your material will normally suggest one of these types when an illustration is needed.

TIPS FOR CREATING AND USING ILLUSTRATIONS

Each type of illustration has its unique strengths and weaknesses, and these are discussed in the specific text entry for each type. The guidelines presented here apply to most visual material you use to supplement the information in your text. Following these tips should help you create and present your visual material to good effect.

1. Keep the information as brief and simple as possible.
2. Try to present only one type of information in each illustration.
3. Label or caption each illustration carefully.
4. Include a key that identifies all **symbols**, when necessary.
5. Specify the proportions used or include a scale of relative distances, when appropriate.
6. If possible, make the lettering horizontal for easy reading.
7. Keep terminology consistent. Do not refer to something as a "proportion" in the text and as a "percentage" in the illustration.
8. Allow enough white space around and within the illustration.
9. Position the illustration as close as possible to the text that refers to it; however, an illustration should normally not appear ahead of the first text reference to it.
10. Be sure the significance of an illustration is clear in the text.
11. If figure numbers or table numbers are used, as they should be if five or more illustrations or tables are used, number the illustrations or tables consecutively.

12. If more than five illustrations or tables appear in a **report**, list them by title, together with figure and page numbers, under a separate heading ("List of Figures" or "List of Tables") following the **table of contents.**

IDENTIFYING ILLUSTRATIONS AND INTEGRATING THEM INTO THE TEXT

Presented with clarity and consistency, illustrations can help your reader focus on key portions of your report. Be aware, though, that even the best illustration can only supplement the text. Your writing must carry the major burden of providing the context for the illustration and pointing out its significance. To do so effectively, adhere to the following guidelines:

1. Give each illustration a concise title that clearly describes the information. Either center the title below the illustration or align it with the left margin. If the title is too long to fit within the margins of the illustration on one line, break the title into two or more lines of roughly equal length. The figure number should precede the first line of the title; the other lines should begin directly below the first word of the title.
2. Assign a figure number if your report contains five or more illustrations. The figure number precedes the title. If figure numbers are used, number the illustrations consecutively as Figure 1, Figure 2, and so on.
3. If more than five illustrations appear in a **formal report**, list them by title, together with figure and page numbers, under a separate heading ("List of Figures") following the report's **table of contents**.
4. Relate the illustration to the text around it by referring to it by figure number and title.
5. Locate the illustration as close as possible to its first mention in the text. However, the illustration should never precede its first text mention; its appearance without an introduction will confuse readers.
6. If the illustration is central to the discussion, illuminating or strongly reinforcing it, place the illustration in the text. But, if the illustration is lengthy, detailed, and peripheral to the discussion, place it in an appendix. However, even material in an appendix should be mentioned or referred to in text; otherwise, the

information in the appendix is superfluous. (See **appendix/appendixes/appendices**.)

7. The amount of discussion that each illustration requires will vary with its importance to the text. An illustration showing an important feature or system may be the central focus of an entire discussion. The complexity of the illustration will also affect the discussion, as will how knowledgeable your readers are. Nonexperts require lengthier explanations than experts do, as a rule.

imply/infer

If you *imply* something, you hint or suggest it. If you *infer* something, you reach a conclusion on the basis of evidence.

EXAMPLES His memo *implied* that the project would be delayed.
The general manager *inferred* from the memo that the project would be delayed.

in/into

In means "inside of"; *into* implies movement from the outside to the inside.

EXAMPLE The equipment was *in* the test chamber, so she sent her assistant *into* the chamber to get it.

in order to

The **phrase** *in order to* is sometimes essential to the meaning of a sentence.

EXAMPLE If the vertical scale of a graph line would not normally show the zero point, use a horizontal break in the graph *in order to* include the zero point.

The phrase *in order to* also helps control the **pace** of a sentence, even though it is not essential to the meaning of the sentence.

EXAMPLE The committee must know the estimated costs *in order to* evaluate the feasibility of the project.

Most often, however, the phrase *in order to* is just a meaningless filler phrase that is dropped into a sentence without thought.

CHANGE *In order to* start the engine, open the choke and throttle and then press the starter.
TO To start the engine, open the choke and throttle and then press the starter.

Search for these thoughtless uses of the phrase *in order to* in your writing and eliminate them.

in terms of

When used to indicate a shift from one kind of language or terminology to another, the **phrase** *in terms of* can be useful.

EXAMPLE *In terms of* gross sales, the year has been relatively successful; however, *in terms of* net income, it has been discouraging.

When simply dropped into a sentence because it comes easily to mind, however, the phrase *in terms of* is meaningless **affectation**.

CHANGE She was thinking *in terms of* subcontracting much of the work.
TO She was thinking *about* subcontracting much of the work.
OR She was thinking *of* subcontracting much of the work.

inasmuch as/insofar as

Inasmuch as, meaning "because," is a weak connective for stating causal relationships.

CHANGE *Inasmuch as* the heavy spring rains delayed construction, the office building will not be completed on schedule.
TO *Because* the heavy spring rains delayed construction, the office building will not be completed on schedule.

Insofar as, meaning "to the extent that," should be reserved for that explicit **usage** rather than being used to mean *since*.

EXAMPLE *Insofar* as the report deals with causes of highway accidents, it will be useful in preparing future plans for highway construction.

CHANGE *Insofar as* you are here, we will review the material.
TO *Since* you are here, we will review the material.

increasing-order-of-importance method of development

When you want the most important of several ideas to be freshest in your **reader's** mind at the end of your writing, organize your information by increasing order of importance. This sequence begins with the least important point or fact, then moves to the next least important, and builds finally to the most important point at the end. Increasing order of importance is often used when the writer is leading up to an unpleasant on unexpected conclusion.

Writing organized by increasing order of importance has the disadvantage of beginning weakly, with the least important information, which can cause your reader to become impatient before reaching your main point. But when you build a case in which the ideas lead, point by point, to an important **conclusion**, increasing order of importance is an effective method of **organization. Reports** on production or personnel goals are often arranged by this method, as are **oral presentations**.

In the example given in Figure 1, the writer begins with the least productive source of applicants and builds up to the most productive source.

incredible/incredulous

Something that is unbelievable or nearly unbelievable is *incredible*.

EXAMPLE The precision of the new instrument is truly *incredible*.

Incredulous means that a person is skeptical or unbelieving about some thing or some event.

EXAMPLE He was *incredulous* at the reports of the instrument's precision.

indefinite adjectives

Indefinite adjectives are so called because they do not designate anything specific about the **nouns** they modify.

EXAMPLES *some* circuit boards, *any* branches, *all* aircraft engines

Notice that the **articles** *a* and *an* are included among the indefinite adjectives. (See also **a, an**.)

MEMORANDUM

To: William D. Vane, Vice President for Operations
From: Harry Matthews, Personnel Department HM
Date: December 3, 19--

Subject: Recruiting Qualified Electronics Technicians

To keep our company staffed with qualified electronics technicians, we will have to redirect our recruiting program.

In the past dozen years, we have relied heavily on the recruitment of skilled veterans. An end to the military draft, as well as attractive reenlistment bonuses for skilled technicians now in uniform, has all but eliminated veterans as a source of trained employees. Our attempts to reach this group through ads in service newspapers and daily newspaper ads has not been successful recently. I think that the want ads should continue, although each passing month brings fewer and fewer veterans as job applicants.

Our in-house apprentice program has not provided the needed personnel either. High-school enrollments in the area are continually dropping. Each year, fewer high-school graduates, our one source of trainees for the apprentice program, enter the shop. Even vigorous Career Day recruiting has yielded disappointing results. The number of students interested in the apprentice program has declined proportionately as school enrollment goes down.

The local and regional technical schools produce the greatest number of qualified electronics technicians. These graduates tend to be highly motivated, in part because many have obtained their education at their own expense. The training and experience they have received in the tech program, moreover, are first-rate. Competition for these graduates is keen, but I recommend that we increase our recruiting efforts and hire a larger share of this group than we have been doing.

I would like to meet with you soon to discuss the details of a more dynamic recruiting program in the technical schools. I am certain that with the right recruitment campaign we can find the skilled personnel essential to our expanding role in electronics products and service.

mo

Figure 1 Increasing Order of Importance

EXAMPLE It was *an* hour before the needle of the gauge finally approached the danger area.

For recommendations on the placement and **usage** of adjectives, see **adjectives**.

indefinite pronouns

Indefinite pronouns do not specify a particular person or thing; they specify any one of a class or group of persons or things. *Any, another, anyone, anything, both, each, either, everybody, few, many, most, much, neither, none, several, some*, and *such* are indefinite pronouns. Most indefinite **pronouns** require singular **verbs**.

EXAMPLES If *either* of the vice-presidents *is* late, we will delay the meeting.
Neither of them *was* available for comment.
Each of the writers *has* a unique style.
Everyone in our department *has* completed the form.

Some indefinite pronouns require plural verbs.

EXAMPLES *Many are* called, but *few are* chosen.
Several of them *are* aware of the problem.

A few indefinite pronouns may take either plural or singular verbs, depending on whether the **nouns** they stand for are plural or singular.

EXAMPLES *Most* of the employees *are* pleased, but *some are* not.
Most of the oil *is* imported, but *some is* domestic.

indentation

Type that is indented is set in from the margin. The most common use of indentation is at the beginning of a **paragraph**, where the first line is usually indented five spaces in a typed manuscript and one inch in a longhand manuscript.

EXAMPLE A hydraulic crane is not like a lattice boom friction crane in one very important way. In most cases, the safe lifting capacity of a lattice boom crane is based on the weight needed to tip the machine. Therefore, operators of friction machines sometimes depend on signs that the machine might tip to warn them of impending danger.

This is a very dangerous thing to do with a hydraulic crane. Hydraulic crane ratings are based on the strength of the material of the boom (and other components). Therefore, the hydraulic crane operator who waits for signs of tipping to warn him of an overloaded condition will often bend the boom or cause severe damage to his machine before any signs of tipping occur.

Another use of indentation is in **outlining**, in which each subordinate entry is indented under its major entry.

EXAMPLE II. Types of Outlines
A. The Sentence Outline
1. A sentence outline provides order and establishes the relationship of topics to one another to a greater degree than a topic outline.

Finally, a block **quotation** may be indented within a manuscript instead of being enclosed in **quotation marks**; it must, in this case, be single-spaced. Set off the indented material from the text by indenting five spaces from the left margin and five spaces from the right margin and by double-spacing above and below the passage.

EXAMPLE *The Operator's Manual* states:

> This is a very dangerous thing to do with a hydraulic crane. Hydraulic crane ratings are based on the strength of the material of the boom (and other components). . . .
>
> A hydraulic crane is not like a lattice boom friction crane in one very important way. In most cases, the safe lifting capacity of the lattice boom crane is based on the weight needed to tip the machine. Therefore, operators of friction machines sometimes depend on signs that the machine might tip to warn them of impending danger.

This warning was given because many operators were accustomed to . . .

independent clauses

An independent clause is a group of words containing a **subject** and a **predicate** that expresses a complete statement and that could stand alone as a separate sentence.

EXAMPLE *Production was not as great as we had expected*, although we met our quota.

In the example the independent **clause** is italicized and could clearly stand alone as a complete sentence. The second, unitalicized clause is a **dependent clause**; although it contains a subject and predicate, it does not express a complete thought and could not stand alone as a sentence.

indexing

An index is an alphabetical list of all the major topics discussed in a written work. It cites the pages where each topic can be found and thus allows **readers** to find information on particular topics quickly and easily. The index falls at the very end of the work.

The key to compiling a useful index is selectivity. Instead of listing every possible reference to a topic, select references to passages where the topic is discussed fully or where a significant point is made about it. For actual index entries choose those words or **phrases** that best represent a topic. For example, the key terms in a reference to the development of legislation about environmental impact statements would probably be *legislation* and *environmental impact statement*, not *development*. Key terms are those that a reader would most likely look for in an index. In selecting terms for index entries, use chapter or section titles only if they include such key words. For index entries on **tables** and **illustrations**, use the key words in their titles.

COMPILING AN INDEX

Do not attempt to compile an index until the final manuscript is completed, because terminology and page numbers will not be accurate before then. The best way to compile your list of topics is with 3 × 5-inch index cards. Read through your written work from the beginning; each time a key term appears in a significant context, enter the term and its page number on a separate index card. An index entry can consist solely of a main entry and its page number.

EXAMPLE Aquatic monitoring programs, 42

An index entry can also include a main entry, subentries, and even sub-subentries, as in Figure 1. A subentry indicates pages where a specific subcategory or subdivision of a main topic can be found. When compiling index entries on cards, enter a subentry on its own

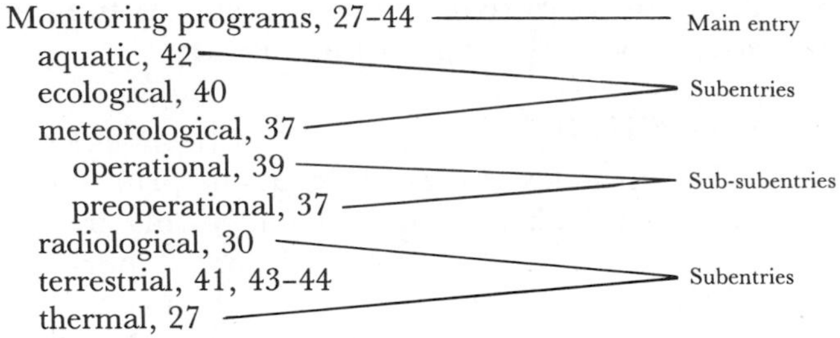
Monitoring programs, 27–44
aquatic, 42
ecological, 40
meteorological, 37
operational, 39
preoperational, 37
radiological, 30
terrestrial, 41, 43–44
thermal, 27
Main entry
Subentries
Sub-subentries
Subentries

Figure 1 Index Entry with Main Entry, Subentries, and Sub-subentries

separate card, but include the main entry as well. Also put a sub-subentry on a separate index card, but include the entry and subentry to which it is related. You must include all this information so that you will be able to tell where each card belongs when you arrange the cards in alphabetical order. (See Figure 2.)

When you have completed this process for the entire work, arrange the cards alphabetically and type the index, two columns to a

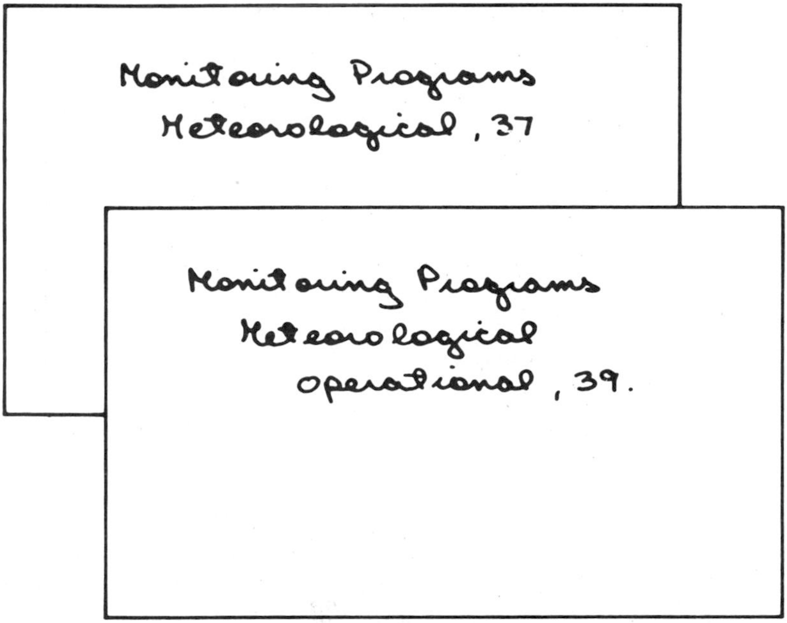

Figure 2 Sample Index Cards

page, from the cards. Be sure to arrange subentries and sub-subentries alphabetically beneath the proper main entry.

WORDING INDEX ENTRIES

The first word of an index entry should be the principal word because the reader will look for topics alphabetically by their main words. Selecting the right word to list first is of course easier for some topics than for others. For instance, *Tips on repairing electrical wire* would not be a suitable index entry because a reader looking for information on electrical wire would not look under the word *tips*. Whether you select *electrical wire; wire, electrical;* or *repairing electrical wire* depends on the purpose and **scope** of your report. Ordinarily, an entry with two key words, like *electrical wire*, should be indexed under each word. (When entering words on index cards you make out one card for *electrical wire* and another for *wire, electrical.*) An index entry should be written as a **noun** or a noun **phrase** rather than as an **adjective** alone.

CHANGE Electrical
    grounding, 21
    insulation, 20
    repairing, 22
    size, 21
    wire, 20–22
TO Electrical wire, 20–22
    grounding, 21
    insulation, 20
    repairing, 22
    size, 21

CROSS-REFERENCING

Cross-references in an index guide the reader to other pertinent topics in the text. A reader looking up *technical writing*, for example, might find a cross-reference to *memorandum*. Cross-references do not include page numbers; they merely direct the reader to another index entry, which of course gives pages. There are two kinds of cross-references: *see* references and *see also* references.

See references are most commonly used with topics that can be identified by several different terms. Listing the topic page numbers by only one of the terms, the indexer then lists the other terms throughout the index as *see* references.

EXAMPLE Economic costs. *See* Benefit-cost analyses

See references also direct readers to index entries where a topic is listed as a subentry.

EXAMPLE L-shaped fittings. *See* Elbows, L-shaped fittings

See also references indicate other entries that include additional information on a topic.

EXAMPLE Ecological programs, 40–49
See also Monitoring programs

STYLING THE INDEX

Capitalize the first word of a main entry and all other terms normally capitalized in the text. Do not capitalize the first words in subentries and sub-subentries unless they appear that way in the text. Italicize (underline) the cross-reference terms *see* and *see also*.

Place each subentry in the index on a separate line, indented from its main entry. Indent sub-subentries from the preceding subentry. These indentations allow readers to scan a column quickly for pertinent subentries or sub-subentries.

Separate entries from page numbers with **commas**. Type the index in a double-column format, as is done in the index to this book.

indirect objects (see **objects**)

indiscreet/indiscrete

Indiscreet means "lacking in prudence or sound judgment."

EXAMPLE His public discussion of the proposed merger was *indiscreet*.

Indiscrete means "not divided or divisible into parts."

EXAMPLE The separate units, once combined, become *indiscrete*.

(See also **discreet/discrete**.)

individual

Avoid using *individual* as a **noun** if *person* is more appropriate.

CHANGE Several *individuals* on the panel did not vote.
TO Several *persons* on the panel did not vote.
OR Several *people* on the panel did not vote.

CHANGE She is an *individual* who says what she thinks.
TO She is a *person* who says what she thinks.
OR She says what she thinks.

Individual is most appropriate when used as an **adjective** to distinguish a single person from a group.

EXAMPLE The *individual* employee's obligations to the firm are detailed in the booklet that describes company policies.

(See also **persons/people**.)

infinitive phrases

An infinitive phrase consists of an **infinitive** (usually with *to*) and any **modifiers** or **complements**.

EXAMPLE The product must be able *to resist repeated hammer blows*.

Do not confuse a **prepositional phrase** beginning with *to* with an infinitive phrase. In the infinitive phrase, the *to* is followed by a **verb**; in the prepositional phrase, *to* is followed by a **noun** or **pronoun**.

EXAMPLES We went *to the building site*. (prepositional phrase)
Our firm tries *to provide a comprehensive training program*. (infinitive phrase)

An infinitive phrase may function as a noun, an **adjective**, or an **adverb**.

EXAMPLES His goal is *to become sales manager*. (noun)
The need *to increase sales* should be obvious. (adjective)
We must work *to increase sales*. (adverb)

An infinitive phrase that functions as a noun may serve within the sentence as **subject, object,** complement or **appositive**.

EXAMPLES *To form her own company* was her lifelong ambition. (subject)
They want *to know when we can begin the project.* (direct object)
His objective is *to live well*. (subjective complement)
His ambition, *to form his own company*, is soon to be realized. (appositive)

The implied subject of an introductory infinitive phrase should be the same as the subject of the sentence. If it is not, the **phrase** is a **dangling modifier**. In the following example, the implied subject of the infinitive is *you*, or *one*, not *practice*.

CHANGE *To learn shorthand*, practice is needed.
TO *To learn shorthand*, you must practice.
OR *To learn shorthand*, one must practice.

infinitives

An infinitive is the bare, or uninflected, form of a **verb** (*go, run, fall, talk, dress, shout*) without the restrictions imposed by **person** and **number**. Along with the **gerund** and **participle**, it is one of the non-finite verb forms. The infinitive is generally preceded by the word *to*, which, although not an inherent part of the infinitive, is considered to be the sign of an infinitive.

EXAMPLES It is time *to go* to work.
We met in the conference room *to talk* about the new project.

An infinitive is a **verbal** and may function as a **noun**, and **adjective**, or an **adverb**.

To expand is not the only objective. (noun)
These are the instructions *to follow*. (adjective)
The company struggled *to survive*. (adverb)

The infinitive may reflect two **tenses**: the present and (with a **helping verb**) the present perfect.

EXAMPLES to go (present tense)
to have gone (present perfect tense)

The most common mistake made with infinitives is to use the present perfect tense when the simple present tense is sufficient.

CHANGE I should not have tried *to have gone* so early.
TO I should not have tried *to go* so early.

Infinitives formed with the root form of **transitive verbs** can express both active and (with a helping verb) passive **voice**.

EXAMPLES to hit (present tense, active voice)
to have hit (present perfect tense, active voice)
to be hit (present tense, passive voice)
to have been hit (present perfect tense, passive voice)

SPLITTING INFINITIVES

A split infinitive is one in which an adverb is placed between the sign of the infinitive, *to*, and the infinitive itself. Because they make up a grammatical unit, the infinitive and its sign are better left intact than separated by an intervening adverb.

CHANGE *To* initially *build* the table in the file, you could input transaction records containing the data necessary to construct the record and table.

TO *To build* the table in the file initially, you could input transaction records containing the data necessary to construct the record and table.

However, it may occasionally be better to split an infinitive than to allow a sentence to become awkward, ambiguous, or incoherent.

CHANGE She agreed immediately to deliver the toxic materials. (Could be interpreted to mean that she agreed immediately.)

TO She agreed to immediately deliver the toxic materials. (No longer ambiguous.)

informal writing style (see **style**)

ingenious/ingenuous

Ingenious means "marked by cleverness and originality," and *ingenuous* means "straightforward" or "characterized by innocence and simplicity."

EXAMPLES Wilson's *ingenious* plan, which streamlined production in Department L, was the beginning of his rise within the company.

I believe that the *ingenuous* co-op students bring some freshness into the company.

inquiry letters and responses

An *inquiry letter* may be as simple as requesting a free brochure or as complex as asking a financial consultant to define the specific requirements for floating a multimillion-dollar bond issue.

There are two broad categories of inquiry letters. One kind provides a benefit (or potential benefit) to the reader: you may ask, for instance, for information about a product that a company has recently advertised. Be clear, concise, and accurate in stating your inquiry. The second kind of inquiry letter primarily benefits the writer; an example is a request to a public utility for information on the energy-related project you are developing. This kind of letter requires the use of persuasion and special consideration of your **reader's** needs.

WRITING AN INQUIRY LETTER

Your **objective** in writing the letter will probably be to obtain, within a reasonable period of time, answers to specific questions. You will be more likely to receive a prompt, helpful reply if you make it easy for the reader to respond by following these guidelines:

1. Keep your questions concise but specific and clear.
2. Phrase your questions so that the reader will know immediately what type of information you are seeking, why you are seeking it, and how you will use it.
3. If possible, present your questions in a numbered list.
4. Keep the number of questions to a minimum, to save the reader's time.
5. Offer some inducement for the reader to respond, such as promising to share the results of your research.
6. Promise to keep responses confidential, especially if you are asking the recipient to complete a questionnaire.

At the end of the letter, thank the reader for taking the time and trouble to respond, and do not forget to include the address to which the material is to be sent. Your chances of getting a reply will improve if you enclose a stamped, self-addressed return envelope, especially if you are writing to someone who is self-employed. Letter 1 is a typical inquiry letter.

RESPONSE TO AN INQUIRY LETTER

When you receive an inquiry letter, first read it quickly to determine whether you are the right person in your organization to answer it—that is, whether you are the one who has both the information and authority to respond. If you are in a position to answer, do so as promptly as you can, and be sure to answer every question the writer has asked. How long your responses should be, and how much technical language you should use depend, of course, on the nature of the question and on what information the writer has provided about himself or herself. Even if the writer has asked a question for which the answer is obvious or a question that seems foolish,

P.O. Box 113
University of Dayton
Dayton, OH 45409
March 11, 19--

Ms. Jane Metcalf
Engineering Services
Miami Valley Power Company
P.O. Box 1444
Miamitown, OH 45733

Dear Ms. Metcalf:

Could you please send me some information on heating systems for an all-electric, energy-efficient, middle-priced house that our systems design class at the University of Dayton is designing. The house, which contains 2,000 square feet of living space (17,600 cubic feet), meets all the requirements stipulated in your brochure "Insulating for Efficiency."

We need the following information:

1. The proper size heat pump to use in this climate for such a home;
2. The wattage of the supplemental electrical furnace that would be required for this climate; and
3. The estimated power consumption, and current rates, of these units for one year.

We will be happy to send you a copy of our preliminary design report. Thank you very much.

Sincerely yours,

Kathryn J. Parsons

Kathryn J. Parsons

Letter 1 Inquiry Letter

answer it as completely as you can, and do so courteously. You may point out that the reader has omitted or misunderstood a particular piece of information or has in some other way introduced an error, but be tactful in your correction so that he or she will not feel foolish or ignorant. And at the end of the letter, offer to provide further assistance, if necessary.

Sometimes a letter of inquiry sent to a large company arrives at the desk of a staff member who realizes that he or she is not the employee best able to answer the letter. If you have received a letter that you feel you cannot answer, find out who can. Then forward the letter to that person. This person should state in the first paragraph of his or her letter that although the letter was addressed to you, it is being answered by someone else in the firm because he or she has the information needed to respond to the inquiry. Letter 2 indicates to the inquirer that her letter has been forwarded and Letter 3 shows a typical response to an inquiry. (See also **correspondence**.)

inside/inside of

In the **phrase** *inside of*, the word *of* is redundant and should be omitted.

CHANGE The switch is just *inside of* the door.
TO The switch is just *inside* the door.

Using *inside of* to mean "in less time than" is colloquial but should be avoided in writing.

CHANGE They were finished *inside of* an hour.
TO They were finished *in less than* an hour.

insoluble/unsolvable

Although *insoluble* and *unsolvable* are sometimes used interchangeably to mean "incapable of being solved," careful writers distinguish between them. *Insoluble* means "incapable of being dissolved." *Unsolvable* means "incapable of being solved."

EXAMPLES The plastic is *insoluble* in most household solvents.
Until yesterday, the production problem seemed *unsolvable*.

MIAMI VALLEY POWER COMPANY
P.O. BOX 1444
MIAMITOWN, OH 45733

(513) 264-4800

March 15, 19--

Ms. Kathryn J. Parsons
P.O. Box 113
University of Dayton
Dayton, OH 45409

Dear Ms. Parsons:

Thank you for inquiring about the heating system we would recommend for use in homes designed according to the specifications outlined in our brochure "Insulating for Efficiency."

Since I cannot answer your specific questions, I have forwarded your letter to Mr. Michael Stott, Engineering Assistant in our development group. He should be able to answer the questions you have raised.

Sincerely,

Jane E. Metcalf

Jane E. Metcalf
Director of Public Information

JEM/mk
cc: Michael Stott

Letter 2 Letter Indicating That the Inquiry Has Been Forwarded

MIAMI VALLEY POWER COMPANY
P.O. BOX 1444
MIAMITOWN, OH 45733

(513) 264-4800

March 24, 19--

Ms. Kathryn Parsons
P.O. Box 113
University of Dayton
Dayton, OH 45409

Dear Ms. Parsons:

Jane Metcalf has forwarded your letter of March 11 about the house that your systems design class is designing. I can estimate the insulation requirements of a typical home of 17,600 cubic feet, as follows:

1. We would generally recommend, for such a home, a heat pump capable of delivering 40,000 BTUs. Our model AL-42 (17 kilowatts) meets this requirement.
2. With the efficiency of the AL-42, you would not need a supplemental electrical furnace.
3. Depending on usage, the AL-42 unit averages between 1,000 and 1,500 kilowatt hours from December through March. To determine the current rate for such usage, check with the Dayton Power and Light Company.

I can give you an answer that would apply specifically to your house only with information about its particular design (such as number of stories, windows, and entrances). If you would send me more details, I would be happy to provide more precise figures--your project sounds interesting.

Sincerely,

Michael Stott

Michael Stott
Engineering Assistant

MS/mo

Letter 3 Response to an Inquiry

instructions

When you explain how to perform a specific task—however simple or complex it may be—you are giving instructions. Written instructions that are based on clear thinking and careful planning should enable a **reader** to carry out the task successfully.

To write instructions that are accurate and easily understandable, you first must thoroughly understand the task you are describing. Otherwise, your instructions could prove confusing or even dangerous. If you are unfamiliar with the task you are describing, you should watch someone who is familiar with the task go through each step. As you watch, ask questions about any step that is not clear to you. Such direct observation should help you to write instructions that are exact, complete, and easy to follow.

Keep in mind your reader's level of knowledge and experience. Is he or she at all skilled in the kind of task for which you are writing instructions? If you know that your reader has a good background in the topic, you might feel free to use fairly specialized words. If, on the other hand, he or she has little or no knowledge of the subject, you would more appropriately use simple, everyday language—and avoid specialized terms as much as possible.

If the task requires any special tools or materials, say so at the beginning of the instructions. List all essential equipment at the beginning, in a section labeled "Tools Required" or "Materials Required."

The clearest, simplest instructions are written as commands, in the imperative **mood**.

CHANGE The operator should raise the access lid. (indicative)
TO Raise the access lid. (imperative)

Phrase instructions concisely but not as if they were telegrams. You can make sentences shorter by leaving out **articles** (*a, an, the*), some **pronouns** (*you, this, these*), and some **verbs**, but sometimes such sentences sound very **telegraphic** and are difficult to understand. The first version of the following instruction for cleaning a computer punch card assembly, for example, is confusing and difficult to understand.

CHANGE Pass card through punch area for debris.
TO Pass *a* card through *the* punch area *to clear away* any debris.

The meaning of the **phrase** *for debris* needed to be made clearer. The revised instruction is easily and quickly understood.

One good way to make instructions easy to follow is to divide them into short, simple steps. Be sure to order the steps in the proper sequence. Steps can be organized in one of two ways: with numbers or with words. You can label each step with a sequential number.

EXAMPLE 1. Connect each black cable wire to a brass terminal. . . .
2. Attach one 4-inch green jumper wire to the back. . . .
3. Connect both jumper wires to the bare cable wires. . . .

Or you can use words that indicate time or sequence.

EXAMPLE *First*, determine what the customer's problem is with the computer. *Next*, observe the system in operation. *At that time*, question the operator until you are sure that the problem has been explained completely.

Do not sacrifice clarity in an effort to be concise. Plan ahead for your reader. If the instructions in step 2 will affect a process in step 9, say so in step 2. Otherwise, your reader may reach step 9 before discovering that an important piece of information—that should have been given in advance—is missing.

Sometimes your instructions have to make clear that two operations must be performed at the same time. Either state this fact in an **introduction** to the specific instructions or include both operations in the appropriate step.

CHANGE 4. Hold the CONTROL key down.
5. Press the BELL key before releasing the CONTROL key.

TO 4. While holding the CONTROL key down, press the BELL key.

In any operation certain steps must be performed with more preciseness than others. Alert your reader to those steps that require especially precise timing or measurement. Also warn of potentially hazardous steps or materials. If the instructions call for materials that are flammable or that give off noxious fumes, let the reader know before he or she reaches the step for which the material is needed.

If your instructions involve a great many steps, break them into stages, each with a separate head so that each stage begins again with step 1. Using heads as dividers is especially important if your reader is likely to be performing the operation as he or she reads the instructions.

Clear and well-planned **illustrations** can make even complex instructions easily understandable. Illustrations often simplify instructions by reducing the number of words necessary to explain a process or procedure. Appropriate **drawings** and diagrams will enable your reader to identify parts and the relationships between parts more easily than will long explanations. They will also free you, the writer, to focus on the steps making up the instructions rather than on the descriptions of parts. Not all instructions require illustrations, of course. Whether or not illustrations will be useful depends on your reader's needs as well as on the nature of the project. Instructions for inexperienced readers need to be more heavily illustrated than do those for experienced readers.

To test the accuracy and **clarity** of your instructions, ask someone who is not familiar with the operation to follow your written directions. A novice will quickly spot missing steps or point out passages that should be worded more clearly.

The instructions in Figure 1 guide the reader through the steps of "streaking" a saucer-sized disk of material (called *agar*) used to grow bacteria colonies. The object is to thin out the original specimen (the inoculum) so that the bacteria will grow in small, isolated colonies. The streaking process makes certain that part of the saucer is inoculated heavily, whereas its remaining portions are inoculated progressively more lightly. The streaking is done by hand with a thin wire, looped at one end for holding a small sample of the inoculum. (See also **process explanation**.)

insure/ensure/assure

Insure, ensure, and *assure* all mean "make secure or certain." *Assure* refers to persons, and it alone has the **connotation** of setting a person's mind at rest. *Ensure* and *insure* also mean "make secure from harm." Only *insure* is widely used in the sense of guaranteeing the value of life or property.

EXAMPLES I *assure* you that the equipment will be available.
We need all the data to *ensure* the success of the project.
We should *insure* the contents of the building.

INSTRUCTIONS

Distribute the inoculum over the surface of the agar in the following manner:

(1) Beginning at one edge of the saucer, thin the inoculum by streaking back and forth over the same area several times, sweeping across the agar surface until approximately one quarter of the surface has been covered.

(2) Sterilize the loop in an open flame.

(3) Streak at right angles to the originally inoculated area, carrying the inoculum out from the streaked areas onto the sterile surface with only the first stroke of the wire. Cover half of the remaining sterile agar surface.

(4) Sterilize the loop.

(5) Repeat as described in step (3), covering the remaining sterile agar surface.

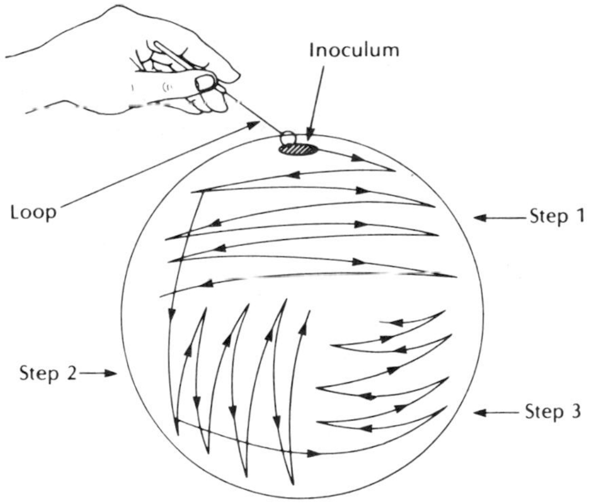

Figure 1 Illustrated Instructions

intensifiers

Intensifiers are **adverbs** that emphasize degree, such as *very, quite, rather, such,* and *too*. Although they serve a legitimate and necessary function, they can also seduce the unwary writer who is not on guard against overusing them. Too many intensifiers weaken your writing. When revising your draft, either eliminate intensifiers that do not make a definite contribution, or replace them with specific details.

CHANGE The team was *quite* happy to receive the *very* good news that it had been awarded a *rather* substantial monetary prize for its design.

TO The team was happy to receive the good news that it had been awarded a $5,000 prize for its design.

The difference is not that the first example is wrong and the second one right but that the intensifiers in the first example add nothing to the sentence and are therefore superfluous.

Some words (such as *unique, perfect, impossible, final, permanent, infinite,* and *complete*) do not logically permit intensification because they do not permit degrees of comparison. Although **usage** often ignores this logical restriction, the writer should be aware that to ignore it is, strictly speaking, to defy the basic meanings of these words. (For more detailed information, see **comparative degree**.)

CHANGE It was *quite* impossible for the part to fit into its designated position.

TO It was impossible for the part to fit into its designated position.

interface

An *interface* is the "surface providing a common boundary between two bodies or areas." The bodies or areas may be physical (the *interface* of piston and cylinder) or conceptual (the *interface* of mathematics and statistics). Do not use *interface* as a substitute for the **verbs** *cooperate, interact,* or even *work*.

CHANGE The Water Resources Department will *interface* with the Department of Marine Biology on the proposed project.

TO The Water Resources Department will *work* with the Department of Marine Biology on the proposed project.

interjections

An interjection is a word or **phrase** of exclamation that is used independently to express emotion or surprise or to summon attention. *Hey! Ouch!* and *Wow!* are strong interjections. *Oh, well*, and *indeed* are mild ones. An interjection functions much as do *yes* or *no*, in that it has no grammatical connection with the rest of the sentence in which it appears. When an interjection expresses a sudden or strong emotion, punctuate it with an **exclamation mark**.

EXAMPLE His only reaction was a resounding *Wow*!

Because they get their main expressive force from sound, interjections are more common in speech than in writing. They are rarely appropriate to technical writing.

internal proposals

The purpose of an internal proposal is to suggest a change or an improvement within an organization. For example, one might be written to propose initiating a new management reporting system, expanding the cafeteria service, or changing the policy on parking privileges. An internal proposal, usually in a memo format, is prepared by a person or a department and is sent to a higher-ranking person who has the authority to accept or reject the proposal.

In the **opening** of a **proposal**, you must establish that a problem exists that needs a solution. If the person to judge the proposal is not convinced that there is a problem, your proposal will have no chance of success. Notice how the following opening to a proposal states its problem directly:

> To: Harriet Sullivan, Office Supervisor
> From: Christina Lopez, Administrative Assistant
>
> Date: December 3, 19—
>
> Subject: Purchase of a Bently XL-100 System
>
> The number of direct mailings requiring an original copy for each recipient is increasing. The cost of calling in temporary help to get out such mailings is very high, and the normal office work suffers in our effort to get out the direct mailings. In addition, the pressure causes much confusion and many frayed nerves.

The body of a proposal should offer a practical solution to the problem. In building a case for a solution, be as specific as possible. When it is appropriate, include (1) a breakdown of costs; (2) information about equipment, material, and personnel requirements; and (3) a schedule for completing the task. Such information can help your **readers** to think about the proposal and thus may stimulate them to act. Consider the following example, which continues the proposal introduced above:

> The following details show the savings and efficiency of the Bently XL-100 System compared with our present method of handling the typing for our direct mailings. Based on these details, I suggest we consider purchasing this system.
>
> COST SAVINGS
>
> Purchase of the Bently XL-100 System would result in savings of over $2,500 in the first year, as shown below:

Purchase price of Bently XL-100	$1,495.00	
Cost of one-year supply of XL-100 magnetic cards	238.00	
	$1,733.00	TOTAL
Cost of temporary typists from Wilson Secretarial Agency last year	$3,700.00	
Temporary typewriter rental from World Office Machines last year	560.00	
	$4,260.00	TOTAL
Savings for the first year	$2,527.00	

> Since purchase of the Bently XL-100 would be a one-time investment, the savings would increase to over $3,000.00 a year after the first year—even considering the cost of magnetic cards and repairs.
>
> EFFICIENCY
>
> The Bently XL-100 System types without error and four times faster than a human typist, once programmed. Because it requires only one operator, the XL-100 would eliminate the

> need for temporary typists. It would also eliminate the confusion caused by our training these typists while attempting to meet our own deadlines. In short, we could carry on with our normal office routine. A normal routine would not only improve our employees' morale but also increase their efficiency.

The **conclusion** should be brief but must tie everything together. It is a good idea to restate recommendations here. Be careful, of course, to conclude in a spirit of cooperation, offering to set up a meeting, supplying further facts and figures, or providing any other assistance that might solve the problem, as is done in the following example:

> On the basis of these details, I recommend that we consider purchasing a Bently XL-100 System to handle our direct mail program.
>
> Enclosed is a brochure that describes the Bently XL-100 System in detail. I will be happy to provide any additional details about the system as I understand it, at your request.

A proposal to commit large sums of money is a common internal proposal. Although such internal proposals have various names, *capital appropriations proposal* is a common and descriptive name for them.

The introduction to such a proposal should provide any background information your reader might need in order to make the decision in question. The introduction should also briefly describe any feasibility study you may have conducted, the conclusions the study led you to reach, and the decision you made on the basis of those conclusions.

The body of the proposal should identify all the equipment or property you are asking to acquire and should state the economic justification for the expenditure involved. If you need to purchase equipment, be sure to list *all* of it so that later you don't have to prepare another proposal to purchase the rest of it. Include an economic justification of either the calculated return on investment or the internal rate of return. Your proposal should include all of the following items that are applicable to your particular request:

1. A brief description of the item to be purchased.
2. The alternative choices you considered before deciding on the one you are recommending.

3. The specific products or projects to be supported by the item.
4. The volume of usage that the item would receive over a specified period of time, such as the next two years.
5. If the company already has some of the item, an explanation of why you require an additional purchase.
6. The life expectancy of the item and of the product or project it is to support.
7. Any future products or projects that the item could be used to support.
8. Whether a leasing arrangement is available and more economical than purchasing the item.
9. The impact on the company if the item is not acquired.
10. The latest feasible date of acquisition, including the impact that a ninety-day delay would have.
11. The calculated return on investment, or the internal rate of return.

With your capital appropriations request, you should also submit a cover **memorandum**. The memo, which should be no more than one page long, should give a brief description of the item and the reason it is needed. Figure 1 is such a capital appropriations proposal.

To: Ronald Wepner, General Manager
From: Nevil Broadmoor, Data Center Manager

Date: June 14, 19—

Subject: Expansion of the Data Center

Problem

Because the data center is now operating at its maximum capacity, we have found ourselves having to decline new business. By expanding our equipment, staff, and facility we could attract new customers while offering our present customers additional and faster services. I have outlined a plan to expand the data center that would allow us to increase our business—and thus our net income—about $300,000 a year.

REQUIREMENTS

Equipment. To provide the computer time required to process new applications, we need to reduce our present time requirements. By upgrading our processor to a V8565M, upgrading our Century software to VRX software, and adding four 658 disc units, we would reduce the time required for our present applications by as much as 50 percent. The addition of six online terminals and a multiplexer would reduce our program maintenance time as much as 60 percent, which would allow our staff to convert applications to take advantage of the new software.

Solution

Staff. To convert present applications to VRX software and to develop and support new applications, we need to increase our programming staff by three people (one senior programmer and two junior programmers).

Facility. To house the additional staff, we need to expand our office area and to purchase additional office furniture for the six online terminals we would place in the programmers' offices. The V8565M requires less space than the present processor, however, so we could rearrange the computer room to provide space for the multiplexer and additional disc units.

BENEFITS

On the basis of the business available in this area, we could expect to increase our net income by $300,000 a year. We would be able to attract new customers—as well as new business from our present customers' customers—by offering additional and faster services. The increase in our operating budget would actually be small because of the reduction in time requirements provided by the new equipment.

COSTS

Equipment. The purchase of new equipment—reduced by the trade-in allowance on our old processor—would be $383,000.

Cost Breakdown

Processor	$250,000
Integrated Disc Controller	20,000
Disc Units (four @ $20,000 each)	80,000
Integrated Multiplexer	10,000
Online Terminals (six @ $3,000 each)	18,000
VRX Software	5,000

Budget. The annual increase in our operating budget would be $86,800.

Salaries	$75,000
Benefits	6,800

Office Supplies	1,000
Utilities	1,000
Equipment Maintenance	3,000

Facility. The cost of expanding our facility would be $33,000.

Office Area	$20,000
Computer Room	10,000
Furniture	3,000

SCHEDULE

To achieve the expected net income increase, we would need to meet the following schedule in the expansion.

July 1–Order the required equipment.
September 1–Begin recruiting efforts.
September 15–Start expanding the office area.
October 1–Order the required office furniture.
November 1–Start soliciting new business.
November 5–Finish expanding office area.
November 5–Fill programming positions.
November 10–Start new applications design.
November 10–Start conversion effort.
December 1–Install new equipment.
December 15–Implement the first new customer application.

Conclusion

I feel that this expansion plan would enable us to meet the requirements of the business in this area. I have looked at the alternative equipment and believe that the V8565M offers better performance than the other hardware available in this price range. The design of the V8565M would allow us to grow with a minimum of effort as the business in this area grows. I would like to meet with you Friday to discuss this plan and will be happy to explain any details or to gather any additional information that you may desire. Please let me know if we can get together then to discuss this proposed expansion of the data center.

Figure 1 Internal Proposal

interrogative adverbs (see **adverbs**)

interrogative pronouns (see **pronouns**)

interviewing for information

The **scope** of coverage that you have established for your writing project may require more information (or more current information) than is available in written form (in the library, in corporate specifications, in trade journals, and so on). When you reach this point, consider a personal interview with an expert in your subject.

A discussion of interviewing can be divided into three parts: (1) determining the proper person to interview, (2) preparing for the interview, and (3) conducting the interview.

DETERMINING THE PROPER PERSON TO INTERVIEW

Many times your subject, or your **objective** in writing about the subject, logically points to the proper person to interview. If you were writing about the use of a freeway onramp metering device on Highway 103, for example, you would want to interview the Director of the Traffic Engineering Department; if you were writing about the design of the device, you would want to interview the manufacturer's design engineer. You may interview a professor who specializes in your subject, in addition to getting leads from him or her as to others you might interview. Other sources available to help you determine the appropriate person to interview are (1) the city directory in the library, (2) professional societies, (3) the yellow pages in the local telephone directory, or (4) a local firm that is involved with all or some aspects of your subject.

PREPARING FOR THE INTERVIEW

After determining the name of the person you want to interview, you must request the interview. You can do this either by telephone or by letter, although a letter may be too slow to allow you to meet your deadline.

Learn as much as possible about the person you are going to interview and about the company or agency for which he works. When you make contact with your interviewee, whether by letter or by telephone, explain (1) who you are, (2) why you are contacting him, (3) why you chose him for the interview, (4) the subject of the interview,

(5) that you would like to arrange an interview at his convenience, and (6) that you will allow him to review your draft.

Prepare a list of specific questions to ask your interviewee. The natural temptation for the untrained writer is to ask general questions rather than specific ones. "What are you doing about air pollution?" for example, may be too general to elicit specific and useful information. "Residents in the east end of town complain about the black smoke that pours from your east-end plant; are you doing anything to relieve the problem?" is a more specific question. Analyze your questions to be certain that they are specific and to the point.

CONDUCTING THE INTERVIEW

When you arrive for the interview, be prepared to guide the discussion. The following list of points should help you to do so:

1. Be pleasant, but purposeful. You are there to get information, so don't be timid about asking leading questions on the subject.
2. Use the list of questions you have prepared, starting with the less controversial aspects of the topic to get the conversation started and then going on to the more controversial aspects.
3. Let your interviewee do the talking. Don't try to impress him with your own knowledge of the subject on which he is presumably an expert.
4. Be objective. Don't offer your opinions on the subject. You are there to get information, not to debate.
5. Some answers prompt additional questions; ask them. If you do not ask these questions as they arise, you may find later that you have forgotten to ask them at all.
6. When the interviewee gets off the subject, be ready with a specific and direct question to guide him or her back onto the track.
7. Take only memory-jogging notes that will help you recall the conversation later. Do not ask your interviewee to slow down so you can take detailed notes. To do so would not only be an undue imposition on the interviewee's time but might also disturb or destroy his or her train of thought.
8. The use of a tape recorder is often unwise, since it often puts people on edge and requires tedious transcription after the interview. On the other hand, a tape recorder does allow you to listen more intently instead of taking notes. If you do use one, do not let it lure you into relaxing so that you neglect to ask crucial questions.

Immediately after leaving the interview, use your memory-jogging notes to help you mentally review the interview and record your detailed notes. Do not postpone this step. No matter how good your memory, you will forget some important points if you do not do this at once.

interviewing for a job

A job interview may last for thirty minutes, or it may take several hours; it may be conducted by one person or by several, either at one time or in a series of interviews. Since it is impossible to know exactly what to expect, it is important to be as well prepared as possible.

BEFORE THE INTERVIEW

Learn as much as you can about your potential employer before the interview. What kind of business is it? Is the company locally owned? Is it a nonprofit organization? If the job is public employment, at what level of government is it? Does the business provide a service and, if so, what kind? How large is the firm? Is the owner self-employed? Is it a subsidiary of a larger operation? Is it expanding? Where will you fit in? This kind of information can be obtained from current employees, from company publications, or from back issues of the local newspaper's business section (available at the public library). You should try to learn about the company's size, sales volume, products and new products, credit rating, and subsidiary companies from its **annual reports**, and from other business reference sources such as *Moody's Industrials, Dunn and Bradstreet, Standard and Poor's*, and *Thomas' Register*. Such preparation will help you to speak knowledgeably about the firm or industry as well as to ask intelligent questions of the interviewer.

It is a good idea to try to anticipate the questions an interviewer might ask and to prepare your answers in advance. The following is a list of some typical questions posed in job interviews:

What are your short-term and long-term occupational goals?
What are your major strengths and weaknesses?
Do you work better alone or with others?
Why do you want to work for this company?
How do you spend your free time?

DURING THE INTERVIEW

Promptness is very important. Be sure to arrive for an interview at the appointed time. It is usually a good idea, in fact, to arrive ahead of time, since you may be asked to fill out an application before meeting the interviewer. Take along your **résumé**, even if the company already has a copy. For one thing, the interviewer may want another copy; for another, the résumé contains most of the information required on an application.

Remember that the interview will actually begin before you are seated. What you wear and how you act will be closely observed. The way you dress matters: It is usually best to dress conservatively and to be well groomed.

Remain standing until you are offered a seat. Then sit up straight—good posture suggests self-assurance—and look at the interviewer, trying to appear relaxed and confident. It is natural to be nervous during an interview, but be careful to remain alert. Listen carefully and make an effort to remember especially important information. (See **listening**.) Do not attempt to take extensive notes during an interview, although it is acceptable to jot down a few facts and figures.

When answering questions, don't ramble or stray from the subject. Take a minute to think before you answer difficult questions; not only will the time help you collect your thoughts, but it will also make you appear careful in your answer. Say only what you must in order to answer each question and then stop; however, avoid giving just yes and no answers, which usually don't permit the interviewer to learn enough about you. Some interviewers allow a silence to fall just to see how you will react. The burden of conducting the interview is the interviewer's, not yours—and he or she may interpret it as a sign of insecurity if you rush in to fill a void in the conversation. But if such a silence would make you uncomfortable, be ready to ask an intelligent question about the company.

Highlight your qualifications for the job, but admit obvious limitations as well. Remember also that the job, the company, and the location must be right for you. Ask about such factors as opportunity for advancement, fringe benefits (but don't create the impression that your primary interest is security), educational opportunity and assistance, and community recreational and cultural activities (if the job would require you to relocate).

If the interviewer overlooks important points, bring them up. But if possible, let the interviewer mention salary first. If you are forced to bring up the subject, ask it as a straightforward question. Knowing the prevailing salaries in your field will make you better prepared to discuss salary. It is usually unwise to bargain, especially if you are a recent graduate. Many companies have inflexible starting salaries for beginners.

Interviewers look for self-confidence and an understanding, on the part of the candidate, of the field in which he or she is applying. Less is expected of a beginner, but even a newcomer must show some knowledge of the field. One way to impress your interviewer is to ask questions about the company that are related to your line of work. Interviewers respond favorably to people who can communicate easily and present themselves well. Jobs today require interactions of all kinds: person-to-person, department-to-department, division-to-division.

At the conclusion of the interview, thank your interviewer for his or her time. Indicate that you are interested in the job (if true), and try tactfully to get an idea of when you can expect to hear from the company.

AFTER THE INTERVIEW

When you leave, jot down pertinent information you learned during the interview. (This information will be especially helpful in comparing job offers.) A day or two later, send the interviewer a brief note of thanks, saying that you find the job attractive (if true) and feel you can fill it well. Letter 1 is typical. (See also **job search, résumés, listening, application letters**, and **acceptance letters**.)

intransitive verbs (see **verbs**)

introductions

The purpose of an introduction is to give your readers enough general information about your topic to enable them to understand the detailed information in the body of the paper or report. The introduction should explain the context out of which the work grew and establish the reasons that it was written. If you don't need to set the stage in this way, turn to the entry on **openings** for a variety of inter-

2647 Sitwell Road
Charlotte, NC 28210
March 17, 19--

Mr. F. E. Vallone
Personnel Manager
Calcutex Industries, Inc.
3275 Commercial Park Drive
Raleigh, NC 27609

Dear Mr. Vallone:

Thank you for the informative and pleasant interview we had last Wednesday. Please extend my thanks to Mr. Wilson of the Servocontrol Group as well. I came away from our meeting most favorably impressed with Calcutex Industries.

I find the position to be an attractive one and feel confident that my qualifications would enable me to perform the duties to everyone's advantage.

I look forward to hearing from you soon.

Sincerely yours,

Philip Ming

Letter 1 Job Interview

esting ways to begin without writing an introduction. If you are writing a **newsletter article**, an **annual report**, a **proposal**, a **memorandum**, or something similar, you will probably find one of the various types of openings more appropriate than an introduction. If you are preparing a **formal report**, a **laboratory report**, a **technical manual**, a **specification**, a **journal article**, a conference paper, or some similar writing project, however, you will need to present your readers with some general information before getting into the body of your writing.

You may find that you need to write one kind of introduction for a **report** or **journal article** and a different kind for a **technical manual** or **specification.**

WRITING INTRODUCTIONS FOR REPORTS AND JOURNAL ARTICLES

In writing an introduction for a report or journal article, you need to state the subject, the purpose of your report or article, its scope, and the way you plan to develop the topic.

Stating the Subject. In stating the subject, it may be helpful to the reader if you include a small amount of information on the history or theory of the subject. In addition to stating the subject, you may need to define it if it is one with which some of your readers may be unfamiliar. (See **definition method of development**.)

Stating the Purpose. The statement of purpose in your introduction should function for your article or paper much as a **topic sentence** functions in a **paragraph**. It should make your readers aware of your goal as they read your supporting statements and examples. It should also tell them why you are writing about the subject: whether your material provides a new perspective or clarifies an existing perspective.

Stating the Scope. Stating the **scope** of your article or report tells your reader how much or how little detail to expect. Does your article or report present a broad survey of your **topic**, or does it concentrate on one facet of the topic? Keep the statement of scope brief and simple, however; lack of restraint could lead you to write an **abstract** of your article or paper instead of a simple statement of scope in your introduction.

Stating the Development of the Subject. In a larger work, it may be helpful to your reader if you state how you plan to develop the subject. Is the work an analysis of the component parts of some whole?

Is the material presented in chronological sequence? Or does it involve an inductive or deductive approach? Providing such information makes it easier for your readers to anticipate how the subject will be presented (see **methods of development**), and it gives them a basis for evaluating how you arrived at your **conclusions** or recommendations.

The following introduction is from a journal article:

> The recent sharp increase in crude oil prices, coupled with the current drive toward energy independence in the United States, has suddenly made the development of processes for the large-scale production of synthetic fuels from domestic coal of crucial importance. Methanol, which is one such fuel that can be derived from coal, has projected manufacturing costs that, on an energy-equivalent basis, are roughly comparable to the other two likely conversion productions: synthetic gasoline and substitute natural gas. In addition to the possibility of producing methanol from coal, it has been suggested that methanol might be produced from the currently flared Middle East natural gas. These two potential sources could result in large quantities of methanol becoming available for fuel usage within the next few years.
>
> Subject, its background and its value
>
> One potential use for methanol is as a motor fuel, either in the "pure" form or as a gasoline blending component. The possible use of pure methanol is complicated by the lack of interchangeability between methanol and gasoline; the two fuels require very different carburetor settings. For vehicles used in captive fleet operations methanol has definite possibilities. However, this would only have a small impact on the overall consumption of motor fuel. Wide-spread use of pure methanol as a motor fuel is clearly a long-range proposition.
>
> Scope of article
>
> A detailed discussion of the use of pure methanol as a motor fuel is beyond the scope of this article; an evaluation of the use of methanol as an extender of gasoline is of greater interest because this use of methanol could have a greater impact near term. The advocates of this application for methanol point to better fuel economy, lower exhaust emissions, and better performance as compelling reasons for including methanol in motor gasoline at the earliest possible date. Fuel economy and lower emissions stand out as being particularly important factors for consideration since they are receiving a great deal of attention in today's climate of energy shortages and environmental awareness. While these claimed advantages for metha-

nol use appear to be valid in certain cases, recent experiments conducted at this laboratory indicate that they are only significant for the older cars with richer carburetion, which are rapidly disappearing from the roads. The main purpose of this article is to put fuel economy and emission issues, as they relate to methanol–gasoline blends, into proper perspective. A secondary purpose is to draw attention to some product quality considerations that are associated with the use of methanol and have sometimes been overlooked.

Primary and secondary purposes of article

—E. E. Wigg, "Methanol as a Gasoline Extender: A Critique," *Science* 186 (November 1974), 785–790. Copyright © 1974 by the American Association for the Advancement of Science.

The following sample introduction comes from a report that presents a method for analyzing the efficiency with which cooling ponds can dissipate heat from nuclear power plants:

Subject and definition

The ultimate heat sink refers to the complex of water sources necessary to safely operate, shut down, and cool a nuclear power plant. Cooling ponds, spray ponds, and mechanical draft cooling towers are examples of some types of ultimate heat sinks currently in use.

The U.S. Nuclear Regulatory Commission has set forth the following positions on the design of ultimate heat sinks. They must do the following:

1. be able to dissipate the heat of a design-basis accident (for example, a loss-of-coolant accident) of one unit plus the heat of a safe shutdown and cooldown of all other units it serves,
2. provide a 30-day supply of cooling water at or below the design-basis temperature for all safety-related equipment, and
3. be capable of performing under the meteorologic conditions leading to the worst cooling performance and under the conditions leading to the highest water loss.

Purpose and scope

Method of development

This report identifies a procedure that may be used to select the most severe combinations of meteorologic parameters that control surface cooling pond heat transfer and evaporative water loss. In this procedure, a long weather record, usually available from the National Weather Service, is reviewed and used to predict the period for which either pond temperature or water loss would be maximized for a hydraulically simple cooling pond. The principle of linear superposition is assumed, which allows the peak ambient pond temperature to be superimposed on the peak "excess" temperature due to plant heat rejection. This procedure determines the timing within the weather record of the peak ambient pond temperature. The true peak can then be determined in a subsequent, more rigorous calculation.

Maximum evaporative water loss is determined by picking the thirty-day continuous period of the record that has the highest evaporation losses and by assuming that all heat rejected by the plant results in the evaporation of pond water.

To be effective, the data-scanning procedure requires a data record on the order of tens of years in length. Since these data will usually come from somewhere other than the site (such as a nearby airport), methods to compare these data with the limited onsite data are developed so that the adequacy, or at least the conservatism, of the offsite data can be established. Conservative correction factors to be added to the final results are suggested.

WRITING INTRODUCTIONS FOR TECHNICAL MANUALS AND SPECIFICATIONS

In writing an introduction for a technical manual or set of specifications, identify the topic and its primary purpose or function in the first sentence or two. Be specific, but do not get into details. Your introduction sets the stage for the entire document, and it is important that your reader get from the introduction a broad frame of reference. Only later, when your reader has some understanding of the overall topic and can appreciate its details in proper perspective, should you introduce technical details. The following is an example:

> The System Constructor is a program that can be used to create operating systems for a specific range of microcomputer systems. The constructor selects requested operating software modules from an existing file of software modules and combines those modules with a previously compiled application program to create a functional operating system designed for a specific hardware configuration. . . .
>
> The constructor selects the requested software modules, generates the necessary linkage between the modules, and generates the necessary control tables for the system according to parameters specified at run time. These parameters can either be input as online responses to sequentially presented display messages or prepared on punched cards and input to the constructor in a batch mode.

How technical your introduction should be depends on your readers: What is their technical background? What kind of information are they seeking in the manual or specification? A computer programmer, for example, has different interests in an application program or utility routine than does an operator—and the topic should be introduced accordingly. The assumptions you can make without providing explanations and the terminology you can use without providing definitions differ greatly from one reader to another.

Remember that this is an introduction to a topic with which your readers may be totally unfamiliar. Concepts and techniques that may be new to them need to be explained at a general level. Do not lose your readers by using terms and **acronyms and initialisms** that they cannot be expected to recognize. Some writers make a point of writing their introduction last, so that they are sure exactly which terms need defining.

You may encounter a dilemma that is common in technical writing: Though you can't explain topic A until you have explained topic B, you can't explain topic B before explaining topic A. The solution is to explain both topics in broad, general terms in the introduction. Then, when you need to write a detailed explanation of topic A, you will be able to do so because your reader will know just enough about both topics to be able to understand your detailed explanation.

EXAMPLE The NEAT/3 programming language, which treats all peripheral units as file storage units, allows your program to perform data input or output operations depending on the specific unit. Peripheral units from which your program can only input data are referred to as *source units*. Units to which your program can only output data are referred to as *destination units* or as combination *source-destination units*.

Thus, when the writer needs to explain any one of the three units in detail, readers will have at least a general knowledge of the topic.

TIPS ON WRITING INTRODUCTIONS

Not every document needs an introduction. When your readers are already familiar with the major elements of your topic, you can save them time and hold their interest better by getting on to the details. If you do not need a full-scale introduction, turn to the entry on openings for a variety of interesting ways to begin without an introduction.

Make your **point of view** toward the subject obvious to the readers. Do not assume that readers knowledgeable about your topic will instinctively know your point of view toward it. Be brief, if that will suffice, but be explicit.

Finally, consider writing the introduction last. Many writers find it helpful to write the introduction last because they feel that only then do they have a full enough perspective on the writing to introduce it adequately.

inverted sentence order (see **sentence contruction** and **sentence variety**)

investigative reports

When a person investigates, or checks into, a particular topic, he or she might write up the results in an investigative report. For example, the report could summarize the findings of an opinion survey, of a product evaluation, or of a marketing study. It could review the published work on a particular topic or compare the different procedures used to perform the same operation. An investigative report gives a precise analysis of its topic and then offers the writer's conclusions and recommendations.

Open such a **report** with a statement of its purpose. In the body of the report, first define the **scope** of your investigation. If the report is on a survey of opinions, for example, you might need to indicate the number of people and the geographical areas surveyed as well as the income categories, the occupations, the age groups, and possibly even the racial groups of those surveyed. Include any information that is pertinent in defining the extent of the investigation. Then report your findings and, if necessary, discuss their significance. End your report with your conclusions and, if appropriate, any recommendations.

Notice that the sample investigative report shown in Figure 1 opens by stating its purpose and ends with its recommendations.

irregardless/regardless

Irregardless is nonstandard English because it expresses a double negative. The word *regardless* is already negative, meaning, "unmindful." Always use *regardless* or *irrespective*.

MEMORANDUM

To: Noreen Rinaldo, Training Manager
From: Charles Lapinski, Senior Instructor *CL*
Date: February 14, 19--

Subject: Addison Corporation's Basic English Program

As you requested, I have investigated Addison's program to determine whether we might also adopt such a program. The purpose of the Addison Basic English course is to teach foreign mechanics who do not speak or read English to understand repair manuals written in a special 800-word vocabulary called "Basic English," and thus eliminate the need for Addison to translate its manuals into a number of different languages. The Basic English Program does not attempt to teach the mechanics to be fluent in English but, rather, to recognize the 800 basic words that appear in the repair manuals.

The course does not train mechanics. Students must know, in their own language, what a word like <u>torque</u> means; the course simply teaches them the English term for it. As prerequisites for the course, students must have a basic knowledge of their trade, must be able to identify a part in an illustrated parts book, must have served as a mechanic on Addison products for at least one year, and must be able to read and write in their own language.

Students are given the specially prepared instruction manual, an illustrated book of parts and their English names, and a pocket reference containing all 800 words of the Basic English vocabulary plus the English names of parts (students can write the corresponding word in their language beside the English words and then use the pocket reference as a bilingual dictionary). The course consists of thirty two-hour lessons, each lesson introducing approximately 27 words. No effort is made to teach pronunciation; the course teaches only recognition of the 800 words, which include 450 nouns, 70 verbs, 180 adjectives and adverbs, and 100 articles, prepositions, conjunctions, and pronouns.

The 800-word vocabulary enables the writers of the manuals to provide mechanics with any information that might be required, because the area of communication is strictly limited to maintenance, inspection, troubleshooting, safety, and the operation of Addison equipment. All nonessential words (such as <u>apple</u>, <u>father</u>, <u>mountain</u>, and so on) have been eliminated, as have most synonyms (for example, <u>under</u> appears, but <u>beneath</u> does not).

<u>Conclusions</u>

I see three possible ways in which we might be able to use some or all of the elements of the Basic English Program: (1) in the preparation of all our student manuals, (2) in the preparation of student manuals for the international students in our service school, or (3) as Addison uses the program.

I think it would be unnecessary to use the Basic English methods in the preparation of student manuals for <u>all</u> our students. Most of our students are English-speaking people to whom an unrestricted vocabulary presents no problem.

In conjunction with the preparation of student manuals for international students, the program might have more appeal. Students would take the Basic English course either before coming to this country to attend school or after arriving but before beginning their technical training.

As for our initiating a Basic English Program similar to Addison's, we could create our own version of the Basic English vocabulary and write our service manuals in it. Since our product lines are much broader than Addison's, however, we would need to create illustrated parts books for each of the different product lines.

ja

Figure 1 Investigative Report

CHANGE *Irregardless* of the difficulties, we must increase the strength of the outer casing.

TO *Regardless* of the difficulties, we must increase the strength of the outer casing.

it

The **pronoun** *it* has a number of uses. First, it can refer to a preceding **noun** that names an object or idea or to a baby or animal whose sex is unknown or unimportant to the point. *It* should have a clear antecedent.

EXAMPLES Darwinism made an impact on nineteenth-century American thought. *It* even influenced economics.

The Moodys' baby is generally healthy, but *it* has a cold at the moment.

It can also serve as an **expletive**.

EXAMPLES *It* is necessary to sand the hull before you paint it.

It is a truth universally acknowledged that there is no such thing as a free lunch.

Be on guard against overusing expletives, however.

CHANGE *It* is seldom that we go.

TO We seldom go.

(See also **its/it's**.)

italics

Italics (indicated on the typewriter by underlining) is a style of type used to denote **emphasis** and to distinguish foreign expressions, book titles, and certain other elements. *This sentence is printed in italics.* You may need to italicize words that require special emphasis in a sentence.

EXAMPLE Contrary to projections, sales have *not* improved since we started the new procedure.

Do not overuse italics for emphasis, however.

CHANGE This will hurt *you* more than *me*.

TO This will hurt you more than me!

TITLES

Italicize the titles of books, periodicals, newspapers, movies, and paintings.

EXAMPLES The book *Statistical Methods* was published in 1981.
The *Cincinnati Enquirer* is one of our oldest newspapers.
Chemical Engineering is a monthly journal.

Abbreviations of such titles are italicized if their spelled-out forms would be italicized.

EXAMPLE The *Journal of QA Technology* is an informative publication.

Titles of chapters or articles within publications and titles of reports are placed in **quotation marks**, not italicized.

EXAMPLE "Clarity, the Technical Writer's Tightrope" was an article in *Technical Communications*.

Titles of holy books and legislative documents are not italicized.

EXAMPLE The Bible and the Magna Carta changed the history of Western civilization.

Titles of long poems and musical works are italicized, but titles of short poems and musical works and songs are enclosed in quotation marks.

EXAMPLES Milton's *Paradise Lost* (long poem)
Handel's *Messiah* (long musical work)
T.S. Eliot's "The Love Song of J. Alfred Prufrock" (short poem)
Leonard Cohen's "Suzanne" (song)

PROPER NAMES

The names of ships, trains, and aircraft (but not the companies that own them) are italicized.

EXAMPLE They sailed to Africa on the Onassis *Clipper* but flew back on the TWA *New Yorker*.

Craft that are known by model or serial designations are exceptions. These are not italicized.

EXAMPLES DC-7, Boeing 747

WORDS, LETTERS, AND FIGURES

Words, letters, and figures discussed as such are italicized.

EXAMPLES The word *inflammable* is often misinterpreted.
I should replace the *s* and the *6* keys on my old typewriter.

FOREIGN WORDS

Foreign words that have not been assimilated into the English language, are italicized.

EXAMPLES *sine qua non, coup de grâce, in res, in camera*

Foreign words that have been fully assimilated into the language, however, need not be italicized.

EXAMPLES cliché, etiquette, vis-à-vis, de facto, siesta

When in doubt about whether or not to italicize a foreign word, consult a current **dictionary**. (See also **foreign words in English**.)

SUBHEADS

Subheads in a report are sometimes italicized—on a typewriter, they are of course underlined. (See also **heads**.)

EXAMPLE There was no publications department as such, and the writing groups were duplicated at each plant or location. Wellington, for example, had such a large number of publications groups that their publication efforts can only be described as disorganized. Their duplication of effort must have been enormous.

Training Writers

We are certainly leading the way in developing first-line managers (or writing supervisors) who are not only technically competent but can also train the writers under their direction and be responsible for writing quality as well.

its/it's

Be careful never to confuse these two words: *Its* is a possessive **pronoun**, whereas *it's* is a **contraction** of *it is*.

EXAMPLE *It's* important that the factory meet *its* quota.

Although nouns normally form the possessive by the addition of an **apostrophe** and an *s*, the contraction of *it is (it's)* has already used that device; therefore, the possessive form of the pronoun *it* is formed by adding only the *s*.

J

jammed modifiers

Some writing is unclear or difficult to read because it contains jammed modifiers, or strings of **modifiers** preceding **nouns.**

> CHANGE Your *staffing level authorization reassessment* plan should result in a major improvement.

In this sentence the noun *plan* is preceded by four modifiers; this string of modifiers slows the **reader** down and makes the sentence awkward and clumsy. Jammed modifiers often result from an overuse of **jargon** or **vogue words.** Occasionally, they occur when writers mistakenly attempt to be concise by eliminating short **prepositions** or connectives—exactly the words that help to make sentences clear and readable. See how breaking up the jammed modifiers makes the previous example easier to read.

> TO Your plan for the reassessment of staffing-level authorizations should result in a major improvement.
>
> OR Your plan to reassess authorizations for staffing levels should result in a major improvement.

(See also **conciseness, gobbledygook,** and **telegraphic style.**)

jargon

Jargon is a highly specialized technical slang that is unique to an occupational group. If all your **readers** are members of a particular occupational group, jargon may provide a time-saving and efficient means of communicating with them. For example, finding and correcting the errors in a computer program is referred to by programmers as "debugging." If you have any doubt that your entire reading audience is a part of this group, however, avoid using jargon.

When jargon becomes so specialized that it applies only to one company or a subgroup of an occupation, it is referred to as *shop talk.* Obviously, shop talk is appropriate only for those familiar with its special vocabulary and should be reserved for speech or informal **memorandums.**

When jargon becomes enmeshed in a tangle of **absract words,** many of them pseudoscientific or pseudolegal, it becomes an **affec-**

tation known as **gobbledygook.** Jargon, used to this extreme, is "language more technical than the ideas it serves to express."*

Jargon and shop talk are similar in many ways to "lingo" and "cant," which are terms used to describe language that is intended to exclude outsiders from a select circle of the initiated.

job descriptions

Most large companies and many small ones specify, in a formal job description, the duties of and requirements for many of the jobs in the firm. Job descriptions fill several important functions: They provide information on which equitable salary scales can be based; they help management determine whether all responsibilities within a company are adequately covered; and they let both prospective and current employees know exactly what is expected of them. Together, all of the job descriptions in a firm present a picture of the organization's structure.

Sometimes plant or office supervisors are given the task of writing the job descriptions of the employees assigned to them. In many organizations, though, an employee may draft his or her own job description, which the immediate superior then checks and approves.

A FORMAT FOR WRITING JOB DESCRIPTIONS

Although job description formats vary from organization to organization, they commonly contain the following sections:

Accountability. This section identifies, by title, the person or persons to whom the employee reports.

Scope of responsibilities. This section provides an overview of the primary and secondary functions of the job and states, if applicable, who reports to the employee.

Specific duties. This section gives a detailed account of the specific duties of the job, as concisely as possible.

Personal requirements. This section lists the education, training, experience, and licensing required or desired for the job.

TIPS FOR WRITING JOB DESCRIPTIONS

If you have been asked to prepare a job description for your position, the following guidelines should help:

*Susanne K. Langer, *Mind: An Essay on Human Feeling* (Baltimore: Johns Hopkins University Press, 1967), p. 36.

1. Before attempting to write your job description, make a list of all the different tasks you do for a week or a month. Otherwise, you will almost certainly leave out some of your duties.
2. Focus on content. Remember that you are describing your job, not yourself.
3. List your duties in decreasing order of importance. Knowing how your various duties rank in importance makes it easier to set valid job qualifications.
4. Begin each statement of a duty with a **verb,** and be specific. Write "Answer and route incoming telephone calls" rather than "Handle telephone calls."
5. Review existing job descriptions that have been successful.

The job description shown in Figure 1, which is a typical one, never mentions the person holding the job described; it focuses, instead, on the job and on the qualifications *any* person must possess to fill the position. (See pages 354–355.)

For additional and detailed information on the use and creation of job descriptions, see: Joseph J. Famularo, *Organization Planning Manual,* New York: American Management Association, 1979.

job interviews (see **interviewing for a job**)

job search

To begin an effective search for a job, you must know essentially three things: (1) yourself (your specific skills, your goals, and so forth), (2) the job you are searching for (geographical area, likely industries, likely companies, and so on), and (3) where to get all the information you need to conduct an effective job search.

YOURSELF

Do some homework on yourself—some real soul searching. Decide excactly what your goals are. What would you like to do in the immediate future? When you have answered that, think about the kind of work you would like to be doing two years from now. Then think seriously about what you would like to be doing five years from now. When you have pinned down these answers, you have established your goals.

You must also know exactly what your strongest skills are, for they are your most salable products. You must know your appropriate

Manager, Technical Publications
Acme Electrical Corporation

ACCOUNTABILITY

Reports directly to the Vice President, Customer Service.

SCOPE OF RESPONSIBILITIES

The Manager of Technical Publications is expected to plan, coordinate, and supervise the design and development of technical publications and documentation required in the support of the sale, installation, and maintenance of Acme products. The manager is responsible for the administration and morale of the staff. The supervisor for instruction manuals and the supervisor for parts manuals report directly to the manager.

SPECIFIC DUTIES

Direct an organization presently composed of twenty people (including two supervisors), over 75 percent of whom are writing professionals and graphics specialists.
Screen, select, and hire qualified applicants for the department.
Prepare a formal program designed to orient writing trainees to the production of reproducible copy and graphic arts.
Evaluate the performance of departmental members and determine salary adjustments for all personnel in the department.
Plan documentation to support new and existing products.
Determine the need for subcontracted publications and act as a purchasing agent when they are needed.
Offer editorial advice to supervisors.
Develop and manage an annual budget for the Technical Publications Department.
Cooperate with the Engineering, Parts, and Service Departments to provide the necessary repair and spare parts manuals upon the introduction of new equipment.
Serve as a liaison between technical specialists, the publications staff, and field engineers.
Recommend new and appropriate uses for the department within the company.
Keep up with new techniques in printing processes, typesetting, computerized text editing, art, and graphics and use them to the advantage of Acme Electrical Corporation where applicable.

PERSONAL REQUIREMENTS

B.S.E.E. (or equivalent electrical background) desired.
Minimum of three years professional writing experience with a general knowledge of graphics and production.
Minimum of two years management experience with a knowledge of the general principles of management.
Must be conversant with the needs of support people, technical people, and customers.

Figure 1 Job Description

level of entry into a job requiring those skills (entry level for the inexperienced, intermediate level for the moderately experienced, and senior level for the very experienced). Then you must know where you would be most likely to find a need for those skills (geographical location, professional vocation, which industries, which companies).

Determine how you are different, or better qualified, for a job than the other people who are likely to apply for it. Are you expert in your field? Are you especially painstaking with details? Are you particularly good with people? Are you persistent and determined in solving problems? Do you have any particular skills or talents, such as writing, public speaking, or analytical ability? If you know what makes you stand out and can communicate it during an interview, you greatly increase your chances.

THE JOB

Start with the geographical area that would most favor your job search or in which you prefer to live. Then narrow the geographical area by determining the field in which you prefer to work. Then narrow the field by identifying those specializations within that field that interest you most. Finally, pinpoint those companies within your selected specializations that need your skills and for which you feel you could work comfortably. Be sure to include small and even very small companies, since collectively, they create far more jobs than do large companies.

Thoroughly **research** each company. Visit the library to find information about the company in sources such as **annual reports,** the *Standard and Poor's Directory,* and *Thomas' Register of American Manufacturers* (see also **library research**). Ask family, friends, and even casual acquaintances about the companies. One of the key things to learn is who in the company has the authority to hire you. For a recent graduate, this may be the personnel manager; for a highly experienced professional, it is more likely to be the manager of the department doing the hiring. Your goal should be to get a face-to-face meeting with that person.

SOURCES OF INFORMATION

The information you need to help you land a job can come from a number of sources: classified ads, **inquiry letters**, trade and professional journals, school placement services, employment agencies, friends already employed within your area of interest, and others.

Classified Ads. Many employers advertise in the classified sections of newspapers. For the widest selection, look in the Sunday editions of local and big city newspapers. Although reading want ads can be tedious, an item-by-item check is necessary if you are to make a thorough search. The job you are looking for might be listed in the classified ads under various titles. For example, an accountant seeking a job might find the specialty listed under "General Accountant," "Cost Analyst," or "Budget Analyst." So play it safe—read *all* the ads.

Occasionally, newspapers print special employment supplements that provide valuable information on many facets of the job market. Watch for these. As you read the ads, note such factors as salary ranges, job locations, and job duties and responsibilities.

Watch out for fake ads. These are ads for jobs that don't exist, run by placement firms to fatten their "résumé bank" for future clout with employers. They usually tell you nothing more than a job title and a salary (usually inflated).

Blind ads (no company name, just a box number) rarely prove worthwhile. Legitimate advertisers have nothing to hide.

Do not place a "job wanted" ad yourself, because potential employers don't read them—but placement services do (generally the less reputable ones).

Inquiry Letters. If you are interested in a particular company, write and ask whether it has an opening for someone with your skills and qualifications. This is essentially an **application letter,** except that you are not responding to an advertisement for a job. Normally you should send the letter either to the director of personnel or to the specific department head; for a small company, however, you can write to the president.

Trade and Professional Journals. Many occupations have associations that publish periodicals containing listings of current job opportunities. If you are seeking a job in forestry, for example, you could check the job listing in the *Journal of Forestry,* published by the Society of American Foresters. To learn about the trade or professional associations for your occupation, consult the following reference books:

Encyclopedia of Associations
Encyclopedia of Business Information Sources
National Directory of Employment Services

For a listing of journal ads, see *900,000 Plus Jobs Annually: Published Sources of Employment Listings,* by Norman S. Fiengold and Glenda Ann Winkler, Garrett Park Press, Garrett Park, MD 20896. This document lists more than 900 journals that carry want ads.

School Placement Services. Check with the career counselors in your school's job-placement office (you can sometimes use the job-placement services of other schools as well). Government, business, and industry recruiters often visit job-placement offices to interview prospective employees; the recruiters also keep college placement offices aware of their company's current employment needs.

While you are in the placement office, ask to see a current issue of the *College Placement Annual.* This publication lists the occupational requirements and addresses of over a thousand industry, business, and government employers.

Another document you can examine in most college placement offices is the *Directory of Career Planning and Placement Offices,* published by the College Placement Council, Inc., Box 2263, Bethlehem, PA 18001.

State Employment Agencies. Most states have free employment agencies that function specifically to match applicants and jobs. If your state has one, register with the local employment office. It may have just the job you want; if not, it will keep your résumé on file and call you if such a job comes along.

Private Employment Agencies. Private employment agencies are profit-making organizations that are in business to help people find jobs—for a fee. Reputable private employment agencies give you job leads and help you organize your campaign for the job you want. They may also provide useful information on the companies doing the hiring. Choose a private employment agency carefully. Some are well established and quite reputable, but others have questionable reputations. Check with your local Better Business Bureau as well as with friends and acquaintances before signing an agreement with a private employment agency.

Who will pay the fee if you are offered a job through a private agency? Sometimes the employer will pay the agency's fee; if not, you must pay either a set fee or a percentage of your first month's salary. In general, if the employer pays the fee, the job is a prized position that is hard to fill, and the fact that the employer is paying the fee is a good indication that the employment agency is reputable. Before signing a contract, be sure you understand who is paying the fee and, if *you* are, how much you are agreeing to pay. As with any written agreement, read the fine print carefully.

Do not give a private employment agency "exclusive handling." If you do, and you find *yourself* a job independent of the agency, you may still have to pay it a fee.

Other Sources. When searching for a job, consult with people whose judgment you respect. Use all available resources. Recruit family members and friends to help, and alert as many people as possible to your search. Talk especially with people who are already working in your chosen field.

Local, state, and federal government agencies offer many employment opportunities. Local government agencies are listed in the white pages of your telephone directory under the name of your city, county, or state. For information about jobs with the federal government, contact the U.S. Office of Personnel Management or the Federal Job Information Center; both have branches in most major cities. The offices are listed in the blue pages of the telephone directory under "U.S. Government."

If you are a veteran, local and campus Veterans' Administration offices can provide material on special placement programs for veterans. Such agencies will supply you with the necessary information about the particular requirements or entrance tests for your occupation.

The following registers might also be helpful:

Federal Career Opportunities, for government jobs, published biweekly by Federal Research Service, Inc., 370 Maple Avenue West, Box 1059, Vienna, VA 22180. This document lists over 3,200 available federal jobs in the United States and overseas.

International Employment Hotline, for overseas jobs, published monthly by International Employment Hotline, P.O. Box 6170, McLean, VA 22106.

The NELS Monthly Bulletin, for jobs in criminal justice, published by the National Employment Listing Service, Criminal Justice Center, Sam Houston State University, Huntsville, TX 77341.

Executive search firms are also available for those who consider themselves to be executives—such firms are not, of course, for recent college graduates. For the executive job searcher, the following documents might be useful:

Executive Employment Guide, published by the American Management Association, 135 W. 50th St., New York, NY 10020.

Recruiting and Research Report, published by Kenneth Cole, Box 895, Naperville, IL 60566.

Directory of Executive Recruiters, published by Consultant News, Templeton Rd., Fitzwilliam, NH 03447.

The Association of Executive Recruiting Consultants, 151 Railroad Avenue, Greenwich, CT 06830, also publishes a list of 60 top executive recruiters.

Keep a loose-leaf notebook with *dated* job ads, copies of letters of applications and résumés, notes sent in regard to interviews, and the names of important contacts. This notebook can act as a future resource and as a tickler file. (See also **résumé, interviewing for a job,** and **acceptance letter.**)

journal articles

A journal article is an article written on a specific subject for a professional periodical. These periodicals are often the official publications of professional societies. *Technical Communication,* for example, is an official voice of the Society for Technical Communication. Other professional publications include *Electrical Engineering Review, Chemical Engineering, Journal of Geophysical Research,* and *Nucleonics Week.* There are hundreds of periodicals, of course, ranging from scholarly journals that publish articles on advanced research in highly specialized areas to industrial and business periodicals that publish material of more general interest.

From time to time in your career, you may wish to write a journal article about something you know that would be of interest to others in your field. Such an article makes your work more widely known, provides favorable publicity for your employer, gives you a sense of satisfaction, and may even improve your chances for professional advancement.

PLANNING THE ARTICLE

When thinking about writing such an article, ask yourself the following questions:

- Is your work or your knowledge of the subject original? If not, what is there about your approach that justifies publication?
- Will the significance of the article justify the time and effort needed to write it?
- What parts of your work, project, or study are most appropriate to include in the article?

To help you answer these questions, learn about the periodical or periodicals to which you will send the article, and consult your colleagues for advice. Once you have decided on several journals, consider the following factors about each:

- the professional interests of its readership.
- the size of its readership.
- the professional reputation of the journal.
- the appropriateness of your article to the journal's goals, as stated on the "masthead" page.
- the frequency with which its articles are cited in other journals.

After you have settled on the right journal, read back issues to find out such information as the amount and kind of details that the articles include, their length, and the typical writing **style.**

If your subject involves a particular project, you should begin work on it, ideally, when the project is in progress. Doing so will permit you to devote short periods during the project to writing various sections of the article, thereby writing the draft in manageable increments. This practice makes the writing integral to the project and may even reveal any weaknesses in the design or details of the project, such as the need for more data.

As you plan your article, decide whether to invite one or more coauthors to join you. Coauthors can add strength and substance to the paper but can add complications as well. To minimize potential problems, one of you should take the responsibility of being the primary author. That person creates the **outline,** makes assignments, sets up a schedule, and keeps the various parts in perspective. The primary author must also take responsibility for ensuring that the finished paper reads as smoothly as if it had been written by one person. Review the entry **revision** before revising for such consistency.

GATHERING THE DATA

As you gather information, take notes from all the sources available to you, such as the following:

- published material on the subject.
- your own experience.
- notebooks recorded during your research.
- survey results.
- progress reports.
- test data.
- performance records.
- patent disclosures.

You should begin your **research** with a careful review of the literature to establish what has been published about your topic. A review of the relevant information in your field can be insurance against writing an article that in essence has already been published. (Some articles, in fact, begin with a **literature review.**) As you compile this information, record your references in full; that is, include all the information you would need to document the source (see **documenting sources**).

Be aware that published material may contain a number of flaws, such as incomplete and inaccurate citations; irrelevant citations; references to hard-to-find books, articles, and other material; and two slightly different versions of the same article published in separate journals. (See **library research** for advice about finding published sources.)

ORGANIZING THE DRAFT

Some journals (particularly in the sciences) use a prescribed organization for the major sections, such as the following:

- Introduction.
- Materials and methods.

- Results.
- Discussion.
- Literature cited.

If the major organization is not prescribed, choose and arrange the various sections of the draft in a way that shows your results to best advantage.

In any case, the best guarantee of a logically organized article is a good outline. In addition to shaping your information in a logical order, outlining helps you organize your thinking. Moving topics about in an outline to produce the best organizational arrangement is easier and faster than moving whole paragraphs and pages of text in a manuscript. If you have coauthors, it is *essential* that the writing team work from a common, well-developed outline; otherwise, coordinating the various writers' work will be impossible, and the parts produced by each will not fit together logically as a whole.

As you prepare the outline, use the following questions to guide your thinking:

- What is the primary thesis?
- What are the key ideas?
- What details need to be included, considering your readers and your purpose?
- What data should be presented in illustrations or tables?

Once you have written your outline, however, consider it to be flexible; you may well change and improve your organization as you write the rough draft.

PREPARING SECTIONS OF THE ARTICLE

Introduction. The purpose of an introduction is to give your **readers** enough general information about your topic for them to be able to understand the detailed information in the body of the article. The introduction should include the following:

- the purpose of the article.
- a definition of the problem examined.
- the scope of the article.
- the rationale for your approach to the problem or project and the reasons you rejected alternative approaches.
- previous work in the field, including other approaches described in previously published articles.

Emphasize what is new and different about your approach, especially if you are not dealing with a new concept. Show the overall significance of your project or approach by explaining how it fills a need, solves a current problem, or offers a useful application. For detailed guidance on writing introductions, see **introductions.**

Conclusion. The conclusion section pulls together your results and findings and interprets them based on the purpose of the study and the methods used to conduct it. Thus, the conclusion is the focal point of the article, the goal toward which the study was aimed. Your conclusions must grow out of the evidence for the findings in the body of the article. They must also be consistent with the scope of information presented in the introduction. For detailed guidance on writing conclusions, see **conclusions.**

Abstract. After the article is written, you will need to write an abstract. Be sure to follow any instructions provided by the journal on writing abstracts, and review previously published abstracts in that journal. Prepare your abstract carefully, since it will be the basis on which many other researchers will decide whether to read the work in full. Abstracts are often published independently in abstract journals (see **reference books**) and are used as a source of terms (called *keywords*) used to index, by subject, the original article for computerized information-retrieval systems. Abstract journals and computer searches allow researchers to review a great deal of literature in a short time. (For details on writing abstracts, see **abstracts.**)

Illustrations and Tables. Use illustrations and tables wherever they are appropriate, but design each for a specific purpose:

- to describe a function.
- to show an external appearance.
- to show internal construction (with cutaway or exploded-view drawings, for example).
- to display statistical data.
- to indicate trends.

Consider working your rough illustrations into your outline so that you decide on your illustrations as you write your manuscript. Used effectively and appropriately, illustrative material can clarify information in ways that text alone cannot. (For details, see **illustrations** and **tables.**)

Heads. The use of **heads** throughout your manuscript is important because they break the manuscript into manageable portions.

They allow journal readers to understand the development of your topic and pinpoint sections of particular interest to them. The main headings in your outline will often become the headings in your final manuscript.

References. This section of the article lists the sources used or quoted in the article. The specific format for listing these sources varies from field to field. Usually the journal to which the article is submitted will specify the form the editors require for citing sources. (For a detailed discussion of using and citing sources, see **documenting sources** and **quotations.**)

PREPARING THE MANUSCRIPT

Some journals make available, or at least recommend, a "style sheet" with detailed guidelines on the style and **format** of the article. Such style sheets often include specific instructions about how the manuscript should be typed, the style for **abbreviations, symbols,** and units (like the International System of Units), how to handle **mathematical equations,** how many copies to submit, and the like. The following guidelines are typical:

- Double space the manuscript, using at least one-inch margins all around, and number each page.
- Make sure that the captions for figures and tables are specific, accurate, and self-explanatory. Use labels (known as *callouts*) with those illustrations that need them.
- Provide clean, even, line work and accurately worded axes and labels for your **drawings** if the journal will not be redrawing them.
- Place mathematical equations, either typed or legibly printed in black ink, on separate lines in the text, and number them consecutively.
- If you include **photographs,** use glossy-finish black and white prints. Both the photograph and the contrast should be of good quality. Identify each photograph on the back; don't write on the face of it, and don't write on the back of it with a heavy hand. If necessary, indicate which edge of the photograph is the top.

OBTAINING PUBLICATION CLEARANCE

After the manuscript has been prepared, submit a copy to your employer for review before sending it to the journal. This review will make sure that no proprietary information or other kinds of trade secrets have been inadvertently revealed and that classified or other restricted information has not been disclosed. Likewise, secure ahead of time permission to print information for which someone else holds the **copyright.** At times, your employer may require that the published article contain a statement indicating approval or clearance for publication.

THE REFEREE PROCESS

Many journals use reviewers called referees—experts in the field—to evaluate manuscripts before they are accepted for publication. Their appraisals not only help maintain the quality of the journal but also help prospective authors by finding weaknesses and errors that can be corrected before publication.

The editor of the journal acts as an arbiter between the author and referees. Because they are anonymous (as the author is to them), they can be as critical as necessary without concern that the author knows their identities. The editor may send the referees' comments back verbatim or may only summarize them. These comments must be evaluated carefully and objectively. If you find misunderstanding or bias, bring it to the editor's attention. The editor then will make the final decision on the adequacy of the author's response or revision.

After such a review, you may be asked to revise your article before the journal will accept it for publication. Do not let this request discourage you, as the revision will probably be beneficial.

judicial/judicious

Judicial is a term that pertains only to law.

EXAMPLE The *judicial* branch is one of the three branches of the United States government.

Judicious refers to careful or wise judgment.

EXAMPLE We intend to convert to the new refining process by a series of *judicious* steps.

K

kind of/sort of

Kind of and *sort of* should be used in writing only to refer to a class or type of things.

EXAMPLE They used a special *kind of* metal in the process.

Do not use *kind of* or *sort of* to mean "rather," "somewhat," or "somehow."

CHANGE It was *kind of* a bad year for the firm.
TO It was a bad year for the firm.

know-how

An informal term for "special competence or knowledge," *know-how* should be avoided in **formal writing.**

CHANGE He has great technical *know-how*.
TO He has great technical *skill.*

(See also **English, varieties of.**)

L

laboratory reports

A laboratory report communicates information acquired from a laboratory test or investigation. (Simpler tests use the less formal **test report** form.) Laboratory reports should state the reason the laboratory investigation was conducted, the equipment and procedures used, any problems encountered, any conclusions reached, and any recommendations based on the conclusions.

A laboratory report often places special **emphasis** on the equipment and the procedures used in the investigation because these two factors can be critical in determining the accuracy of the **data** obtained.

Present the results of the laboratory investigation clearly and concisely. Write complete sentences, however; **sentence fragments** are not usually sufficient for clear communication. (Use the Checklist of the Writing Process to write and revise your **report.**) Read the entries on **graphs** and **tables** if your report requires graphic or tabular presentation of data.

Each laboratory usually establishes the **organization** of its laboratory reports. Figure 1 on pages 368–370 is a typical example of a laboratory report.

lay/lie

Lay is a transitive verb (a **verb** that has a direct **object** to complete its meaning) and means "place" or "put." Its present **tense** form is *lay.*

EXAMPLE We *lay* the foundation of the building one section at a time.

The past tense form of *lay* is *laid.*

EXAMPLE We *laid* the first section of the foundation on the 27th of June.

The perfect tense form of *lay* is also *laid.*

EXAMPLE Since June we *have laid* all but two sections of the foundation.

PCB Exposure from Oil Combustion Wayne County Professional Fire Fighters

Submitted to:
Mr. Philip Landowe
President, Wayne County Professional Fire-Fighters Association
Wandell, IN 45602

Submitted by:
Analytical Laboratories, Incorporated
Mr. Arnold Thomas
Certified Industrial Hygienist
Mr. Gary Seabolm
Laboratory Manager
Environmental Analytical Services
1220 Pfeiffer Parkway
Indianapolis, IN 46223

February 28, 19—

INTRODUCTION

Waste oil used to train fire fighters was suspected of containing polychlorinated biphenyls (PCB). According to information provided by Mr. Philip Landowe, President of the Wayne County Professional Fire-Fighters Association, it has been standard practice in training fire fighters to burn 20-100 gallons of oil in a diked area of approximately 25-50M^3. Fire fighters would then extinguish the fire at close range. Exposure would last several minutes, and the exercise would be repeated two or three times each day for one week.

Oil samples were collected from three holding tanks near the training area in Englewood Park on November 11, 19—. To determine potential fire-fighter exposure to PCB, bulk oil analyses were conducted on each of the samples. In addition, the oil was heated and burned to determine the degree to which PCB is volatized from the oil, thus increasing the potential for fire-fighter exposure via inhalation.

TESTING PROCEDURES

Bulk oil samples were diluted with hexane, put through a clean-up step, and analyzed in electron-capture gas chromatography. The oil from the underground tank that contained PCB was then exposed to temperatures of 100° C without ignition and 200° C with ignition. Air was passed over the enclosed sample during heating, and volatized PCB was trapped in an absorbing me-

Figure 1 Laboratory Report

dium. The absorbing medium was then extracted and analyzed for PCB released from the sample.

RESULTS

Bulk oil analyses are presented in Table 1. Only the sample from the underground tank contained detectable amounts of PCB. Aroclor 1260, containing 60 percent chlorine, was found to be present in this sample at 18 μg/g. Concentrations of 50 μg/g PCB in oil are considered hazardous. Stringent storage and disposal techniques are required for oil with PCB concentrations at these levels.

Results for the PCB volatization study are presented in Table 2. At 100° C, 1μg PCB from a total of 18 μg PCB (5.6 percent) was released to the air. Lower levels were released at 200° C and during ignition, probably as a result of decomposition. PCB is a mixture of chlorinated compounds varying in molecular weight; lightweight PCBs were released at all temperatures to greater degree than the high molecular weight fractions.

TABLE 1. Bulk Oil Analysis

Source	*Sample #*	*PCB Content (μg/g)*
Underground Tank (11′ deep)	#6062	18*
Circle Tank (3′ deep)	6063a 6063b	<1 <1
Square Pool (3′ deep)	6064a 6064b	<1 <1

*Aroclor 1260 is the PCB type. This sample was taken for volatilization study.

TABLE 2. PCB Volatilization Study for the 11-Foot Deep Underground Tank*

Outgassing Temp. (°C)	*Outgasing Time* (Min)	*Sample Outgassed* (g)	*PCB Total* (μg)	*PCB Outgassed* (μg)
100	30	1	18	1
200	30	1	18	0.6
200 with ignition	30	1	18	0.2

*Bulk analysis of 18 μg/g PCB

DISCUSSION AND CONCLUSIONS

At a concentration of 18 μg/g, 100 gallons of oil would contain approximately 5.5 grams of PCB. Of the 5.5 grams of PCB, about 0.3 grams would be released to the atmosphere under the worst conditions.

The American Conference of Governmental Industrial Hygienists has established a TLV* of 0.5 μg/M^3 air for a PCB containing 54 percent Cl as a time-weighted average over an 8-hour workshift and has stipulated that exposure over a 15-minute period should not exceed 1 mg/M^3. The 0.3 gram of released PCB would have to be diluted to 600 M^3 air to result in a concentration of 0.5 mg/M^3 or less. Since the combustion of oil lasted several minutes, a dilution to more than 600 M^3 is likely; thus exposure would be less than 0.5 mg/M^3. Since an important factor in determining exposure is time and the fire fighters were exposed only for several minutes at intermittent intervals, adverse effects from long-term exposure to low-level concentrations of PCB should not be expected.

It should be stressed, however, that these conclusions are based solely on oil containing 18 μg/g PCB. If, on previous occasions, the PCB content of the oil was much higher, greater exposure could have occurred. PCB is a known liver toxin and has also been classified as a suspected carcinogen. Although the primary route of entry into the body is by inhalation, PCB can be absorbed through the skin. PCB can cause a skin condition known as chloracne which results from a clogging of the pores. This condition, which is often associated with a secondary infection, should not occur at the PCB concentration found in this oil. A clinical test exists for determining if PCB has been absorbed by the liver.

In summary, because exposure to this oil was limited and because PCB concentrations in the oil were low, it is unlikely that exposure from inhalation would be sufficient to cause adverse health effects. However, we cannot rule out the possibility that excessive exposure may have occurred under certain circumstances, based on factors such as excessive skin contact and the possibility that higher-level PCB concentrations in the oil could have been used earlier. The practice of using this oil should be terminated.

*The Threshold Limit Value (TLV) is the safe average concentration that most individuals can be exposed to in an 8-hour day.

Lay is frequently confused with *lie,* which is an intransitive verb (a verb that does not require an object to complete its meaning) meaning "recline" or "remain." Its present tense form is *lie.*

EXAMPLE Injured employees should *lie* down and remain still until the doctor arrives.

The past tense form of *lie* is *lay.* (This form causes the confusion between *lie* and *lay.*)

EXAMPLE The injured employee *lay* still for approximately five minutes.

The perfect tense form of *lie* is *lain.*

EXAMPLE The injured employee *had lain* still for approximately five minutes when the doctor arrived.

leave/let

As a **verb,** *leave* should never be used in the sense of "allow" or "permit."

CHANGE *Leave* me do it my way.
TO *Let* me do it my way.

As a **noun,** however, *leave* can mean "permission granted."

EXAMPLE Employees are granted a *leave* of absence if they have a chronic illness.

lend/loan

Lend or *loan* each may be used as a **verb,** but *lend* is more common.

EXAMPLES You can *loan* them the money if you wish.
You can *lend* them money if you wish.

Unlike *lend, loan* can be a **noun.**

EXAMPLE We made arrangements at the bank for a *loan.*

letters (see **correspondence**)

libel/liable/likely

The term *libel* refers to "anything circulated in writing or pictures that injures someone's good reputation." When someone's reputation is injured in speech, however, the term is *slander.*

EXAMPLES When an editorial charged our board of directors with bribing a representative, the board sued the newspaper for *libel*.
If the mayor supports the bribery charge in his speech tonight, we will accuse him of *slander*.

The term *liable* means "legally subject to" or "responsible for."

EXAMPLE Employers are held *liable* for their employees' decisions.

In technical writing, *liable* should retain its legal meaning. Where a condition of probability is intended, use *likely*.

CHANGE Rita is *liable* to be promoted.
TO Rita is *likely* to be promoted.

library classification systems

Libraries classify and arrange books and journals by call numbers. Call numbers are normally based on the Dewey Decimal System or the Library of Congress system. These systems can help you in any library research by leading you to the place in the library where your topic can be found. If you were looking for books about technology, for example, you could go directly to the 600–699 books or the T shelves, depending on which system the library used.

The Dewey Decimal System divides all books and journals by sets of numbers into the following ten general subject categories:

000–099 General Works
100–199 Philosophy
200–299 Religion
300–399 Social Sciences
400–499 Language
500–599 Pure Sciences
600–699 Technology (Applied Sciences)
700–799 Fine Arts
800–899 Literature
900–999 History

Each of these general categories is in turn divided into ten smaller categories. For example, the Technology classification (600–699) is broken down in the following way:

600–609 Technology
610–619 Medical Sciences

620–629 Engineering
630–639 Agriculture
640–649 Home Economics
650–659 Business and Business Methods
660–669 Chemical Technology and Industrial Chemistry
670–679 Manufacturing (metal, textile, paper, and so on)
680–689 Miscellaneous Manufacturers (hardware, furniture, and so on)
690–699 Building Construction

The Library of Congress System is divided by letters into the following major subject categories:

A General Works
B Philosophy and Religion
C History and Auxiliary Sciences
D Universal History and Topography
E/F American History
G Geography, Anthropology, Folklore, and so on
H Social Sciences
J Political Science
K Law
L Education
M Music
N Fine Arts
P Language and Literature
Q Science
R Medicine
S Agriculture
T Technology
U Military Service
V Naval Service
Z Library Science and Bibliography

Technology and related topics come under the Technology category (T), as follows:

T Technology (General)
TA Engineering (General). Civil engineering (General)
TC Hydraulic engineering
TD Environmental technology. Sanitary engineering
TE Highway engineering. Roads and pavements

TF Railroad engineering and operation
TG Bridge engineering
TH Building construction
TJ Mechanical engineering and machinery
TK Electrical engineering. Electronics. Nuclear engineering
TL Motor vehicles. Aeronautics. Astronautics
TN Mining engineering. Metallurgy
TP Chemical technology
TR Photography
TS Manufactures
TS Packaging
TT Handicrafts. Arts and crafts
TX Home economics

library research

The key tools for doing library research include the card catalog, the periodical indexes, the bibliographies, the computer-search services, and the reference books. For help in using any of these, as well as in using the library in general, ask a librarian.

THE CARD CATALOG

The card catalog contains a listing of each book a library owns. It is usually found in cabinets near the reference or circulation section, though some larger libraries keep their card catalogs in bound books or on computer terminals. Go to the card catalog to determine whether the library has the book you need and where to find it.

Most books have three types of cards in the catalog—an *author card* and a *title card* (found in the Author/Title Index) and a *subject card* (found in the Subject Index). Cards are arranged alphabetically: Author cards are filed by the author's last name; title cards are filed by the first important word in the title (excluding *a, an,* and *the*); and subject cards are filed by the first word of the general subject. All three cards include the author, the title, the publisher, the date and place of publication, the number of pages, and the major **topics** covered. They indicate if a book has an **index,** a **bibliography,** or any **illustrations.** At the bottom are subject headings that lead you to more materials on the same topic.

Each card also includes, in the upper left-hand corner, the *call number,* which indicates where the book is kept in the library. The call number is also written on the book's spine, which enables you to

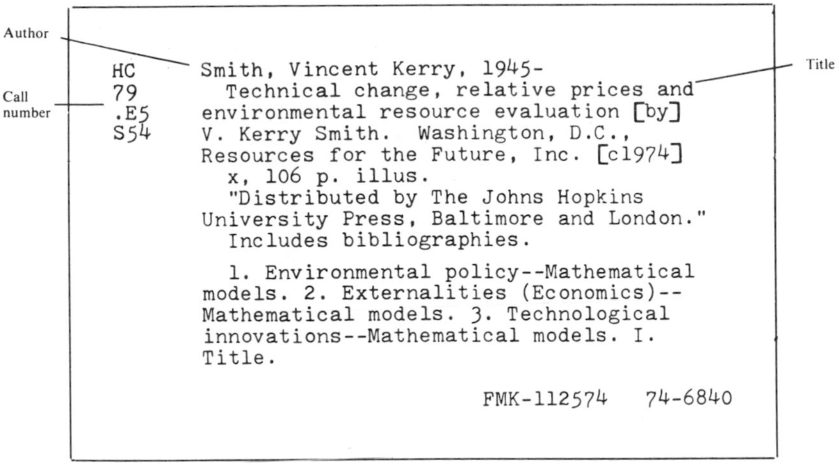

HC 79 .E5 S54

Smith, Vincent Kerry, 1945-
Technical change, relative prices and environmental resource evaluation [by] V. Kerry Smith. Washington, D.C., Resources for the Future, Inc. [c1974]
x, 106 p. illus.
"Distributed by The Johns Hopkins University Press, Baltimore and London."
Includes bibliographies.

1. Environmental policy--Mathematical models. 2. Externalities (Economics)--Mathematical models. 3. Technological innovations--Mathematical models. I. Title.

FMK-112574 74-6840

Author Card

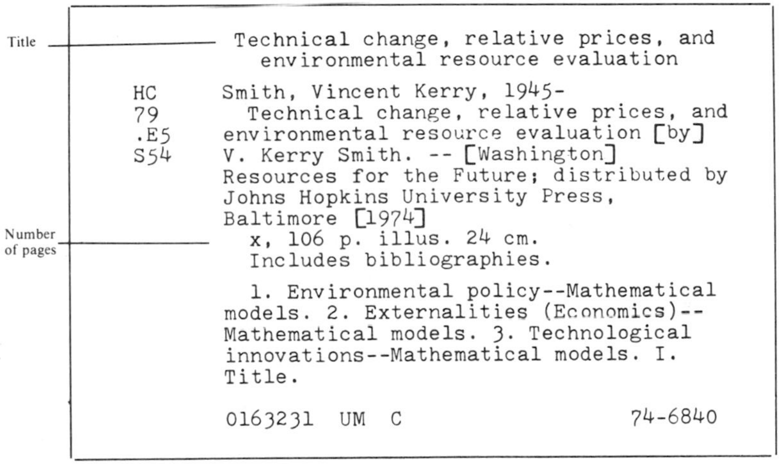

Technical change, relative prices, and environmental resource evaluation

HC 79 .E5 S54

Smith, Vincent Kerry, 1945-
Technical change, relative prices, and environmental resource evaluation [by] V. Kerry Smith. -- [Washington] Resources for the Future; distributed by Johns Hopkins University Press, Baltimore [1974]
x, 106 p. illus. 24 cm.
Includes bibliographies.

1. Environmental policy--Mathematical models. 2. Externalities (Economics)--Mathematical models. 3. Technological innovations--Mathematical models. I. Title.

0163231 UM C 74-6840

Title Card

find the book on the shelf. Call numbers are actually the book's classification according to the Dewey Decimal System, the Library of Congress System, or the library's own system. (For details of the first two systems, see **library classification systems.**)

A card catalog also contains *see* cards and *see also* cards. *See* cards indicate that the heading you are looking under is not the one used for your topic and refer you instead to another heading. For exam-

Subject categories

ENVIRONMENTAL POLICY--MATHEMATICAL
MODELS.

HC
79
.E5
S54

Smith, Vincent Kerry, 1945-
Technical change, relative prices, and environmental resource evaluation [by] V. Kerry Smith. -- [Washington] Resources for the Future; distributed by Johns Hopkins University Press, Baltimore [1974]
x, 106 p. illus. 24 cm.
Includes bibliographies.

Subjects covered and other subjects in catalog with related information

1. Environmental policy--Mathematical models. 2. Externalities (Economics)--Mathematical models. 3. Technological innovations--Mathematical models. I. Title.

0163231 UM C 74-6840

Library of Congress Catalog Card Number

Subject Card

ple, if you looked up *instrumentation,* a *see* card might refer you to *control systems. See also* cards list additional headings related to the topic. Under *mining,* for example, a *see also* card might refer you to *metallurgy.*

BIBLIOGRAPHIES AND PERIODICAL INDEXES

Bibliographies and periodical indexes contain lists of books and articles. Bibliographies list books, periodicals, and other research materials published in a particular subject area—for example, in business or engineering. Periodical indexes list journal, magazine, and newspaper articles. (*Periodicals* are publications that are issued at regular intervals—daily, weekly, monthly, and so on.) For a list of some useful bibliographies and periodical indexes, see **reference books**.

Instructions for using bibliographies and periodical indexes are found in the front of each volume (or the first volume of a series). There you will find a key to the **abbreviations** and **symbols** used as well as an explanation of the way that information is arranged in the book. You will also find a list of the specific topics covered and, for periodical indexes, a list of which newspapers, magazines, and journals are included. The following are some indexes that are useful for technical writing:

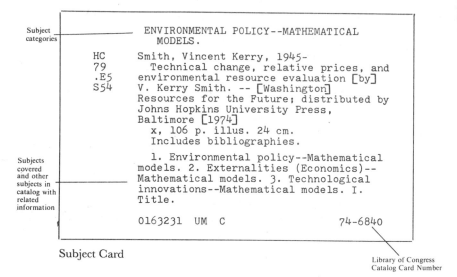

Applied Science and Technology Index, 1913– (monthly)
Monthly Catalog of U.S. Government Publications, 1895–
New York Times Index, 1851– (bimonthly)

Libraries generally keep periodicals in a separate room or area. Current issues are often in one area, back issues (which are bound in volumes) in another. Back issues of some journals and most newspapers are generally in microform. If your library does not have the periodicals you need, you can get them through interlibrary loan, a service that permits you to borrow books and buy photocopied periodical articles from other libraries. Consult your librarian for specific details of the system.

COMPUTER-SEARCH SERVICES AND INDEXES

Many academic, industrial, and medical libraries now offer computer-search services. Indexes are electronically stored as data bases in a computer. Librarians can then quickly find, or retrieve, information from the indexes. This computer-search service is an automated method of preparing a bibliography; it provides you with a computer printout of a list of citations, some with **abstracts**, appropriate to your subject. Data bases are available in medicine, business, psychology, biology, management, engineering, environmental studies, and many other subjects. Ordinarily there is a charge for this service, but the cost is relatively low when the speed and convenience of the search are taken into account.

Computer-search services are usually located in the reference section of the library. (Searches are almost always conducted by reference librarians skilled in searching computer data bases.) At some large universities, these services are located in satellite libraries. The reference staff can provide information about the range of subjects in the data base, the types and cost of search services, and procedures for using the service.

It is generally necessary to set up an appointment with a reference librarian to arrange for the search. Some libraries provide a form to fill out in advance that asks for a brief statement of the research topic and for any pertinent information necessary or helpful for conducting the search.

After completing the search, the librarian generally meets with the researcher to go over the printout. He or she can help you evaluate the citations for their relevance to your project and can also assist you in finding them. Some libraries provide with the printout a brief guide that indicates where books, periodicals, and microform sources on the citation list are located.

REFERENCE WORKS

Additional sources of helpful information include encyclopedias, specialized dictionaries, handbooks and manuals, statistical sources, and atlases.

Encyclopedias. Alphabetically arranged collections of articles, encyclopedias are often illustrated and usually published in multivolume sets. Some encyclopedias cover a wide range of general subjects, and others specialize in a particular subject. General encyclopedias provide the kind of overview that can be helpful to someone new to a particular subject. The articles in general encyclopedias are useful sources of background information and can be especially helpful in defining the terminology essential for understanding the subject. Some articles contain bibliographies, which list additional sources. Three major general encyclopedias are the following:

Encyclopedia Americana (30 volumes)
Encyclopaedia Britannica (24 volumes)
New Columbia Encyclopedia (1 volume)

Specialized encyclopedias provide detailed information on a particular field of knowledge. Their treatment of a subject is rather thorough, and so the researcher ought to have some background information on the subject in order to use the encyclopedia's information to full advantage. Specialized encyclopedias contain comprehensive bibliographies. There are specialized encyclopedias on many subjects; the following two are especially useful to technical writers:

Encyclopedia of Chemical Technology
McGraw-Hill Encyclopedia of Science and Technology

Dictionaries. Alphabetical arrangements of words with information about their forms, pronunciations, origins, meanings, and uses, dictionaries are essential to all writers. For a list of major abridged and unabridged dictionaries, look in the **dictionary** entry of this book. Specialized dictionaries define those terms used in a particular field—for example, technology, architecture, geology. Their definitions are generally more complete and are more likely to be current.

The following are three specialized dictionaries that may be helpful to some technical writers:

Harris, C. M., ed. *Dictionary of Architecture and Construction*
McGraw-Hill Dictionary of Scientific and Technical Terms
Monkhouse, F. J., ed. *A Dictionary of Geography*

Handbooks and manuals. Compilations of frequently used information in a particular field of knowledge, handbooks and manuals are usually single volumes that generally provide such information as brief definitions of terms or concepts, explanations and details about particular organizations, numerical data in **graphs** and **tables, maps,** and the like. Handbooks and manuals are useful sources of fundamental information on a particular subject, although they are most valuable for someone with a basic knowledge of the topic, particularly in scientific or technical fields. Every field has its own handbook or manual; the following are some useful ones:

United States Government Printing Office Style Manual
CRC Handbook of Chemistry and Physics
Environmental Regulation Handbook
Merritt, F. S., ed. *Building Construction Handbook*
U.S. Government Manual

Statistical sources. Collections of numerical data, these are the best sources for such information as the height of the Washington Monument; the population of Boise, Idaho; the cost of living in Aspen, Colorado; and the annual number of motorcycle fatalities in the United States. The answers to many statistical questions can be found in almanacs and encyclopedias. The answers to difficult or comprehensive questions, however, are most likely to be found in works devoted exclusively to statistical data, a selection of which follows:

American Statistics Index (1978– with monthly, quarterly, and annual supplements. Lists and abstracts all statistical publications issued by agencies of the U.S. government, including periodicals, reports, special surveys, and pamphlets.)

U.S. Bureau of the Census. *County and City Data Book* (1952– issued every five years. Includes a variety of data from cities, counties, metropolitan areas, and the like. Arranged by geographic and political areas, it covers such topics as climate, dwellings, population characteristics, school districts, employment, and city finances.)

U.S. Bureau of the Census. *Statistical Abstract of the United States* (1979- annual. Includes statistics on the U.S. social, political, and economic condition and covers broad topics such as population, education, public land, and vital statistics. Some state and regional data.)

Atlases. Collections of maps, atlases are classified into two categories based on the type of information their maps present—general atlases show physical and political boundaries, and thematic atlases give special information, such as climate, population, natural resources, or agricultural products. The following are several general atlases:

Hammond Medallion World Atlas
National Geographic Atlas of the World
Rand McNally New Cosmopolitan World Atlas
The Times Atlas of the World
U.S. Geological Survey. *The National Atlas of the United States of America*

MICROFORMS

Some library source materials may require you to use microforms. Reduced photographic images of printed pages, microforms are used by many libraries for storing magazines, newspapers, and other materials. Because they are reduced, microforms must be magnified by machines called microreaders in order to be read. The most common kinds of microforms are microfilm and microfiche.

Microfilm. Rolls of 35-millimeter film, usually with four printed pages per frame, microfilm must be read on manually or electrically operated machines that advance the roll, frame by frame, for viewing.

Microfiche. A flat sheet of film, usually 4 × 6 inches, that can contain as many as 98 pages, microfiche is read on a microreader, where it can be moved horizontally or vertically from frame to frame for viewing.

If you are unsure about where to get microform materials or how to use microreaders, ask a librarian. Microform materials and microreaders are usually located in a special section of the library. Some microreaders are equipped with photocopying devices that for a fee permit you to make paper copies of microforms.

TIPS FOR DOING LIBRARY RESEARCH

Once you find sources with the information you need and actually get your hands on those sources, you must be conscientious about keeping track of them. Besides taking careful notes of any information they contain, you should methodically note down certain information about the sources themselves.

When you get the book or magazine or microform, there are certain shortcuts that can help you decide quickly whether it has information that will be useful to your research. In a book it is best to start by looking over the **table of contents** and then by reading the **introduction, conclusion, summary,** and **index.**

Does the book have an index or a bibliography? For a magazine article, it helps first to scan the **heads,** to get an idea of its major topics. Is the magazine an authoritative one, or is it brief and topical in its scope? With microforms, you can quickly flip through the text also looking for major topics. No matter what form your source takes, always consider its date: How current is the information? Timeliness is important in any research; you don't want to use out-of-date information.

Prepare a 3 × 5-inch note card for any source that you include in your research. The card should include the source's call number, author, and publication information (publisher, city, year). Then, when you compile your bibliography, you will have all necessary information on these cards. The accompanying figures are two examples of bibliography cards. (See also **note-taking.**)

-like

The **suffix** *-like* is sometimes added to **nouns** to make them into **adjectives.** The resulting **compound word** is hyphenated only if it is unusual or might not immediately be clear.

EXAMPLES childlike, lifelike, dictionary-like, computer-like
Her new assistant works with machine-*like* efficiency.

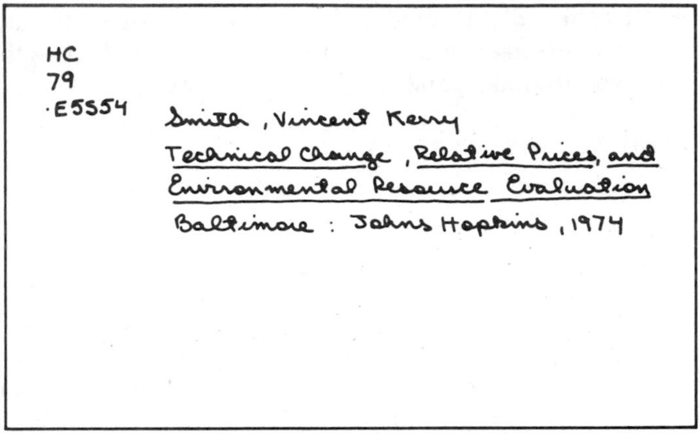

HC
79
.E5554

Smith, Vincent Kerry
Technical Change, Relative Prices, and Environmental Resource Evaluation
Baltimore: Johns Hopkins, 1974

Bibliography Card for a Book

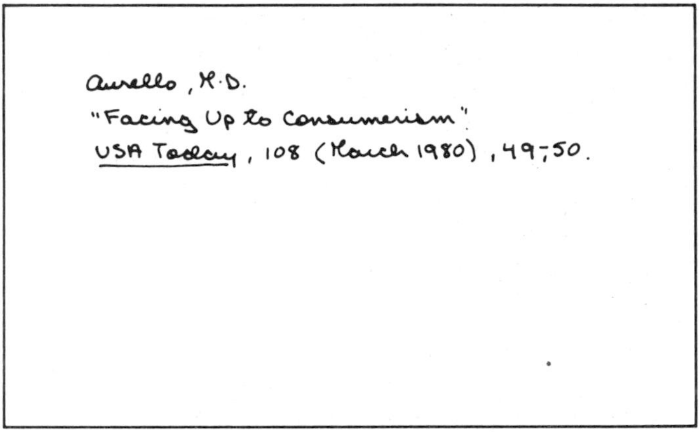

Aurello, M.D.
"Facing Up to Consumerism."
USA Today, 108 (March 1980), 49-50.

Bibliography Card for a Periodical

like/as

To avoid confusion between *like* and *as,* remember that *like* is a **preposition** and *as* is a **conjunction.** Use *like* with a **noun** or **pronoun** that is not followed by a **verb.**

EXAMPLE The new supervisor behaves *like a novice.*

Use *as* before **clauses** (which contain verbs).

EXAMPLES He acted *as though he owned the company.*
He responded *as we expected he would.*

Like may be used in elliptical constructions that omit the verb.

EXAMPLE She took to architecture *like a bird to nest building.*

If the omitted portions of the elliptical construction were restored, however, *as* would be used.

EXAMPLE She took to architecture *as a bird takes to nest building.*

linking verbs

A **verb** that functions primarily to link the **subject** to a **noun** or **modifier** is called a *linking verb.* The most common linking verb is a form of the verb *be.*

EXAMPLE Each group *is* important to the firm's overall performance.

However, many other verbs may also function as linking verbs, including *become, remain,* and *seem.*

EXAMPLE He *became* an industrial engineer.

Verbs that convey the senses (*look, feel, sound, taste, smell*) often perform only a linking function.

EXAMPLE The repair crew *looked* tired.

A few verbs may function either as linking verbs or as transitive or intransitive verbs. When *be* may be substituted, the verb is probably a linking verb.

EXAMPLES He *grew* angry. (linking verb, equivalent in meaning to *became*)
He *grew* African violets in his office. (transitive verb)
The African violets *grew.* (intransitive verb)

(See also **subjective complements.**)

listening

Listening is a learned skill; it is not automatic or instinctive. Listening requires concentration and mental effort. But it is worth the effort it takes, for good listening facilitates the exchange of ideas, opinions, and information.

The first thing to understand is that there are a number of natural human barriers to effective listening. One of them is that if the speaker says something that is contrary to your own prejudice, you may immediately start to prepare a rebuttal—and stop listening. Another is that your personal feelings about the speaker can get in your way; for example, if you dislike the speaker, your feelings may convince your subconscious mind that he or she will not have anything to say that is worth listening to—and your mind is more likely to stray. Physical barriers can also get in the way of effective listening. If you are hungry or tired, for example, you will certainly find concentrating more difficult. And finally, you may be distracted by noises and sights. The important thing is to be aware of these potential distractions and try to prevent them or, if you cannot, to overcome them.

The following guidelines may help you improve your listening skills:

1. Make a conscious effort to listen carefully, to stay involved, and to react to what is being said. It may help to try to put yourself in thc speaker's place.
2. Learn to use the spare time that results from your mind's absorbing information faster than the speaker can present it to listen more effectively and efficiently. For example, you might try to find the speaker's pattern of organization. Identify the introduction, the method being used to develop the topic, the transition (so that you will be alert for the next point), and the conclusion. Or evaluate the evidence the speaker uses to substantiate his or her points. Listen "between the lines" for meaning that may not be in the words. Or review the talk to this point. These techniques will help you not only probe the meaning of the subject but also remember key points.
3. Don't allow yourself to be distracted by the speaker's personality or mannerisms. Respond thoughtfully to the speaker's words, and avoid making judgments too quickly.
4. Be prepared. Think about the subject of a meeting or workshop ahead of time. This will help you understand and remember the material.
5. Take notes. If you are following a speaker's presentation in order to take notes, your mind will be less likely to wander or be

distracted. However, your note taking must enhance your listening and not detract from it. Also, if you have taken notes, you are much more likely to review what you have heard—and reviewing your notes can remedy weaknesses in your listening skills.

The best test of listening well is the ability to respond well. Get into the habit of summarizing and paraphrasing in your own words what has been said. Doing so will both reduce the probability of listening mistakes and increase retention.

lists

Lists can save **readers** time by allowing them to see at a glance specific items, questions, or directions. Lists also help readers by breaking up complex statements that include figures and by allowing key ideas to stand out.

EXAMPLE Before we agree to hold the convention at the Brent Hotel, we should make sure the hotel facilities meet the following criteria:

1. At least eight meeting rooms that can accommodate 25 people each.
2. Ballroom and dining facilities for 250 people.
3. Duplicating facilities that are adequate for the conference committee.
4. Overhead projectors, flip charts, and screens that are sufficient for eight simultaneous sessions.
5. Ground-floor exhibit area that can provide room for thirty 8 × 15-foot booths.

To confirm that the Brent Hotel is our best choice, perhaps we should take a look at its rooms and facilities during our stay in Kansas City.

Notice that all the items in this example have a **parallel structure.** In addition, all the items are balanced—that is, all points are relatively equal in importance and are of the same general length. Consider, however, the following example and its revision:

CHANGE I believe we should consider several important items at the meeting.

1. Our 19-- Production Schedule
2. The Five-Year Corporate Plan

3. We should also develop an agenda for the next meeting that includes a discussion of criteria for relocating the Westdale Division. Perhaps you can give me some tentative ideas at this meeting.
4. The draft of our Year-End Financial Report

TO Please bring to the meeting the following items:

1. Our 19-- Production Schedule
2. The Five-Year Corporate Plan
3. The draft of our Year-End Financial Report

In addition, you might give me some tentative ideas about criteria for relocating the Westdale Division. Since we must formally discuss this problem soon, we should include such a discussion in the agenda for the committee's next meeting.

In the original version, item 3 is neither parallel in structure nor balanced in importance with the other items. In the revision, item 3 is discussed separately, since it deserves a more developed treatment.

In an attempt to avoid writing **paragraphs**, some writers tend to overuse lists, however. A **memorandum** or **report** that consists almost entirely of lists, for example, can be difficult to understand, for the reader is forced to connect the separate items and mentally to provide **coherence.** Do not expect a reader to deal with unexplained lists of ideas.

To ensure that the reader understands how a list fits with the surrounding sentences, always provide adequate **transitions** before and after any lists. If you do not wish to indicate rank or sequence, which numbered lists suggest, you can use bullets, as shown in the list of tips that follows. In typography a bullet is a small *o* that is filled in with ink.

TIPS FOR USING LISTS

- List only comparable items.
- Use parallel structure throughout.
- Use only words, phrases, or short sentences.
- Provide adequate transitions before and after lists.
- Use bullets when rank or sequence is not important.
- Do not overuse lists.

literature

In technical contexts, the word *literature* applies to a body of writing pertaining to a specific field, such as finance, insurance, or computers. We commonly speak, for example, of medical literature or campaign literature.

EXAMPLE Please send me any available *literature* on computerized translations of foreign languages.

literature reviews

A literature review is a summary **report** on the **literature,** or printed material, that is available on a particular subject over a specific period of time. For example, a literature review might describe all material published within the past five years on a special technique for debugging computers, or it might describe all reports written in the past fifteen years on efforts to improve quality control at a particular company. A literature review tells the **reader** what is available on a particular subject and gives him or her an idea of what should be read in full. The review may be self-contained, or it may be part of a larger work.

Industrial research departments often prepare literature reviews, which then serve as starting points for detailed **research.** Managers in business and industry also use literature reviews to keep specialists informed of the latest developments and trends in their fields. Some **journal articles** or theses begin with a brief literature review to bring the reader up to date on current research in the area. The author can then use the review as background for his or her own discussion of the subject.

To prepare a literature review, you must begin with **library research** or even a computer search of published material on your **topic.** Since your reader may begin research on the basis of your literature review, be especially careful to cite accurately all bibliographic information. As you review each source, note the scope of the book or article and judge its value to the reader. Save all printouts of computer-assisted searches—you may wish to incorporate the sources in the **bibliography** of the final version of your literature review.

You can arrange your discussion chronologically, beginning with a description of the earliest relevant literature and progressing to the most recent. You can also subdivide the topic, discussing works on various subcategories of the topic.

Begin a literature review by defining the area to be covered and the types of works to be reviewed. For example, a literature review may be limited to articles and reports and not include any books. Remember also to make the review as concise as possible (much like an **abstract** or an **executive summary**). Mention material directly pertinent to your topic, but do not go into any details. (Your readers can go to the original literature if they want detail.) If you cite many works, number them, and provide a list of references at the end.

The passage presented in Figure 1 forms the **introduction** and the first two sections of a literature review concerning the use of acti-

Introduction

The performance of the charcoal beds that remove noble gases in the safety systems of a nuclear reactor is difficult to quantify. Knowledge of the basic parameters influencing adsorption behavior has often not been sufficient to permit the calculation of adsorption coefficients under the full range of necessary conditions. Further complicating this situation is the fact that samples of charcoal taken from sequential lots, or even from the same lot, can be significantly different in their adsorption effectiveness.

Despite this situation, existing uncertainties in analyzing the performance of fission gas adsorption systems can be reduced. These uncertainties can be reduced through better interpretation of data relating to the various parameters affecting the performance of these systems. The resulting information should provide not only a better understanding of the variables associated with adsorption system performance, but also a means of predicting their behavior under conditions that differ substantially from those described for specific nuclear facilities. The purpose of this report is to review and tabulate the published information on the adsorption of krypton and xenon to promote these goals.

Correlations with Surface Area

If the adsorption coefficients for krypton and xenon can be correlated successfully with commonly measured physical parameters, such as pore structure or the surface area of the adsorbent, then these parameters can serve as secondary standards in evaluating charcoals. However, current correlations between surface areas and adsorption coefficients of activated charcoals are not promising. As the data in Table 1.1 indicate, First, and others (Ref. 1), were unable to observe any apparent correlation between the adsorption coefficients for krypton and the surface areas of several samples of charcoal tested. Table 1.2, from Kitani, and others (Ref. 2), and Table 1.3, from Nakhutin, and others (Ref. 3), also show no apparent relationship between either the pore volume or the surface area and the adsorption coefficient for krypton. In experiments with radon, Strong and Levins (Ref. 4) observed a linear relationship between surface area and the adsorption coefficient. Why a similar relationship has not been observed for krypton and xenon is not known.

Correlations with Density

Nakhutin, and others (Ref. 3), showed that there appears to be a positive correlation between the apparent density of a charcoal sample and its adsorption coefficients for krypton and xenon. This result may be related to the observation of Kovach and Etheridge (Ref. 5) that the adsorption coefficient for krypton was a maximum if the charcoal was activated to give a surface area of 900 square meters per gram (Figure 1.1). Schroeter, and others (Ref. 6), obtained a patent for the use of dense charcoals for the adsorption of krypton and xenon. The data developed by these authors (Table 1.4) also show clearly that the adsorption coefficients for krypton and xenon are not proportional to the surface area of the adsorbent, and that high surface area and/or pore volume are not necessarily indicators of high adsorption capacity for krypton and xenon. These results are also shown in Kovach and Etheridge in Figure 1.1 (Ref. 5).

—D. W. Underhill and D. W. Moeller, *The Effects of Temperature, Moisture, Concentration, Pressure and Mass Transfer on the Adsorption of Krypton and Xenon on Activated Carbon* (Washington, D.C.: U.S. Nuclear Regulatory Commission, 1980), pp. 1-1–1-2.

Figure 1 Literature Review

vated charcoal at nuclear power plants to remove radioactivity from the noble gases krypton and xenon.

The references cited in this passage would appear in the references section at the end of the literature review, as follows:

REFERENCES

1. First, M. W., Underhill, D. W., Hall, R. R., Grubner, O., Reist, P. C., Baldwin, T. W. and Moeller, D. W. (1971): Semiannual Progress Report, Harvard Air Cleaning Laboratory, U.S. AEC Report NYO-841-24.
2. Kitani, S., Uno, S., Takada J., Takada, H., and Segawa, T. (1968): Recovery of Krypton by Adsorption Process, Japan Atomic Energy Research Institute Report (JAERI) 1167.
3. Nakhutin, I. E., Ochkin, D. V., and Linde, Yu. V. (1969): Dynamic Adsorption of Noble Gases, Russian J. Phys. Chem. 43: 1012–4.
4. Strong, K. P. and Levins, D. M. (1979): Dynamic Adsorption of Radon on Activated Carbon, U.S. ERDA Report CONF-780819: 627-39.
5. Kovach, J. L. and Etheridge, E. L. (1973): Heat and Mass Transfer of Krypton-Xenon Adsorption on Activated Carbon, U.S. AEC Report CONF-720823: 71–85.
6. Schroeter, H., Juentgen, M., Zuendorf, D., and Knoblauch, K. (1974): Process for Retarding Flowing Radioactive Noble Gases, U.S. Patent 3,803,802.

ANNOTATED BIBLIOGRAPHIES

Related to literature reviews are *annotated bibliographies,* which also give readers information about printed material. Rather than discussing the literature in **paragraphs,** however, an annotated **bibliography** lists each item in a standard form (see **documenting sources**) and then briefly describes it. This description (or *annotation*) may include the purpose of the book, its scope, the main topics covered, its historical importance, and anything else the writer thinks the **reader** should know.

The following annotation, among others, appeared in a book-length annotated bibliography. (The number, 213, appears in the original to help index the over 800 annotations.)

> 213 Day, Robert A. *How to Write and Publish a Scientific Paper.* Philadelphia: ISI, 1979. Index Bib. 160p.
>
> According to the author, "the purpose of this book is to help scientists and students of the sciences in all disciplines (but with emphasis on biology) to prepare manuscripts that will have a high probability of being accepted for publication and

of being completely understood when they are published.'' The 26 chapters begin with discussions of ''What Is a Scientific Paper?'' and how to write various parts of the paper, e.g., title, abstract, introduction, materials and methods, results, and discussion. Following chapters cover citing sources, using tables, preparing illustrations, and typing the manuscript. Chapters 15–22 deal with steps of the publishing process: submitting the manuscript, dealing with editors and reviewers, handling reprints, and reviewing the articles of others. Chapter 22 is concerned with ethics, rights, and permissions. Chapter 23 covers general problems of grammar and structure, and Chapter 24 discusses jargon. The last two chapters provide advice on using abbreviations and a short commentary by the author. The six appendixes provide guidelines for abbrevations and symbols, words and expressions to avoid, and common errors in style and spelling.

logic

Logic, or correct reasoning, is essential to convincing your **reader** that your conclusion is valid. Errors in logic can quickly destroy your credibility with your reader. Although a detailed discussion of logic is beyond the scope of this book (see **reference books** for books on logic), the following discussion points out some common errors in logic that you should watch out for in your writing.

LACK OF REASON

When a statement violates the reader's common sense, that statement is not reasonable. If, for example, you stated ''New York City is a small town,'' your reader might immediately question your logic. Common sense would suggest that a city of over eight million people is not a small town. If, however, you stated ''Although New York's population is over eight million, it is a city composed of small towns,'' your reader could probably accept the statement as reasonable—that is, if you then demonstrated the truth of the statement with reasonable examples. Always be sure that your statements are sensible and that they are supported by sound examples.

SWEEPING GENERALIZATIONS

Sweeping generalizations are statements that are too large to be supportable; they generally enlarge an observation about a small group to a generalization about an entire population.

EXAMPLES Management is never concerned about employees.
Computer instructions are always confusing.

These statements are too broad; they ignore any possibility that some companies show great concern for employee welfare or that some computer instructions are clearly written. Such generalizations are simply not true; using them will weaken your credibility. No matter how certain you are of the general applicability of an opinion, exercise great caution when using such all-inclusive expressions as *anyone, everyone, no one, all,* and *in all cases*. Otherwise, your statement is likely to sound illogical and will not be taken seriously. Certain stereotypical **phrases** are based on sweeping generalizations and should also be avoided: "typical business person," "bureaucratic mentality," and "absent-minded professor."

NON SEQUITUR

A statement that does not logically follow from a previous statement or has little bearing on the previous statement is called a *non sequitur.*

EXAMPLE I arrived at work early today, so the weather is calm.

In this example, common sense tells us that arriving early for work does not produce calm weather, or any other kind. Thus, the second part of the sentence does not follow logically from the first. Notice how much more logical the following statement is. It makes sense because each idea is logically linked to the other.

EXAMPLE The snow slowed traffic in the city, so although I left for work at the usual time, I arrived a bit late.

Of course, most non sequiturs are not so obvious as the first example. They often occur when a writer neglects to express adequately the logical links in a chain of thought.

EXAMPLE Last year our laboratory increased its efficiency by 50 percent by using part-time help. I suggest that we hire two full-time laboratory assistants.

Although both the statements are about the same subject—additional laboratory staff—a logical connection is missing. The reader is forced to guess the meaning. Does the writer mean that full-time assistants will be more efficient? Are part-time employees unavailable? Such non sequiturs can cause the reader to misunderstand what he or she is reading or to assume that it is illogical. In your own writing be careful that all points stand logically connected; non sequiturs cause difficulties for reader and writer alike.

POST HOC, ERGO PROPTER HOC

This term means literally "after this, therefore because of this," and it refers to the logical fallacy that because one event happened after another event, the first somehow caused the second.

EXAMPLE I didn't bring my umbrella today. No wonder it is now raining.

Many superstitions are based on this error in logic. In on-the-job writing, this error in reasoning usually results from a hasty conclusion that events are related, without the writer examining the logical connection between the two events.

EXAMPLE We issued the new police uniforms on September 2. Arrests then increased 40 percent.

In the example, it is not logical to assume that new uniforms would produce such a dramatic increase in arrests. To demonstrate such a claim logically would require an explanation of the special circumstances that produced the result. For example, if the new uniforms resulted in a union settlement with police officers, which in turn caused the officers to end a work slowdown, then the conclusion would be reasonable. Even in that case, however, the new uniforms per se did not increase arrests; rather, a complete chain of events produced the result.

BIASED OR SUPPRESSED EVIDENCE

A **conclusion** reached as a result of self-serving data, questionable sources, or purposely incomplete facts is illogical—and probably dishonest. If you were asked to prepare a **report** on the acceptance of a new policy among employees and you distributed **questionnaires** only to those who thought the policy was effective, the resulting evidence would be biased. If you purposely ignored employees who did not believe the policy was effective, you would also be suppressing evidence. Any writer must be concerned about the fair presentation of evidence—especially in reports that recommend or justify actions.

FACT VERSUS OPINION

Distinguish between fact and opinion. Facts include verifiable data or statements, whereas opinions are personal conclusions that may or may not be based on facts. For example, it is a verifiable fact that distilled water boils at 100°C; that distilled water tastes better than tap water is an opinion. In order to prove a point, some writers mingle facts with opinions, thereby obscuring the differences between them. This is of course unfair to the reader. Be sure always to distinguish your facts from your opinions in your writing so that your reader can clearly understand and judge your conclusions.

EXAMPLES The new milling machines produce parts that are within 2 percent of specification. (This sentence is stated as a fact and can be verified by measurement.)

The milling machine operators believe the new models are safer than the old ones. (The word *believe* identifies the statement as an opinion—later statistics on the accident rates may or may not verify the opinion as a fact.)

The opinion of experts, or their professional judgment, in their area of expertise is often accepted as valid evidence in court. In writing, too, such testimony can help you convince your reader of the logic of your conclusion. Be on guard against flawed testimonials, however. For example, a chemist may well make a statement about the safety of a product for use on humans, but the chemist's opinion would not be as valid as that of a medical researcher who has actually studied the effects of that product on humans. When you quote the opinion of an authority, be sure that his or her expertise is appropriate to your point. Make sure also that the opinion is a current one and that the expert is indeed highly respected.

LOADING

When you include an opinion in a statement and then reach conclusions that are based on that statement, you are *loading the argument.* Consider the following opening for a **memorandum:**

EXAMPLE I have several suggestions to improve the *poorly written* policy manual. First, we should change. . . .

By opening with the assumption that the manual is poorly written, the writer has loaded the statement to get readers to accept his arguments and conclusions. Yet, he has said nothing to establish *how* the manual is poorly written. In fact, it may be that only parts of the manual are poorly written. And if so, what are the exact problems? Be careful not to load arguments in your writing; conclusions reached with loaded statements are weak and ultimately unconvincing.

In your writing you should always take pains to express your ideas logically. Logic not only gives writing **coherence** but also keeps it fair and honest.

long variants

Guard against inflating plain words beyond their normal value by adding extra **prefixes** or **suffixes,** a practice that creates long variants. The following is a list of some normal words followed by their inflated counterparts:

analysis/analyzation
certified/certificated
commercial/commercialistic
connect/interconnect
finish/finalize
orient/orientate
priority/prioritization (or prioritize)
use/utilize (see also **utilize**)
visit/visitation

(See also **gobbledygook** and **word choice.**)

loose/lose

Loose is an **adjective** meaning "not fastened" or "unrestrained."

EXAMPLE He found a *loose* wire.

Lose is a **verb** meaning "be deprived of" or "fail to win."

EXAMPLE Did you *lose* the operating instructions?

lowercase and uppercase letters

Lowercase letters are small letters, as distinguished from **capital letters** (known as *uppercase* letters). The terms were coined in the early history of printing when printers kept the small letters in a "case," or tray, below the tray where they kept the capital letters. (See also **capitalization.**)

The use of all capital letters ("all caps") to emphasize important information should be confined to no more than three or four words at a time. Setting an entire passage in all caps makes it hard to read, because the shapes of capital letters are not as distinctive as are the shapes of lowercase letters or a combination of uppercase and lowercase letters. Uniformity of shape deprives readers of important visual clues about the identities of letters and thus slows them down. Instead, request **boldface** or *italic* for such passages, which makes the words stand out without affecting the unique shapes of the letters. (See also **capital letters.**)

M

malapropisms

A malapropism is a word that sounds similar to the one intended but that is ludicrously wrong in the context.

CHANGE The repairman cleaned the typewriter's *plankton*.
TO The repairman cleaned the typewriter's *platen*.

Intentional malapropisms are sometimes used in humorous writing; unintentional malapropisms can embarrass a writer. (See also **antonyms** and **synonyms**.)

male

The term *male* is usually restricted to scientific, legal, or medical contexts (a *male* patient or suspect). Keep in mind that this term sounds cold and impersonal. The terms *boy, man*, and *gentleman* are acceptable substitutes in other contexts; however, be aware that these substitute words also have **connotations** involving age, dignity, and social position. (See also **female.**)

maps

Maps can be used to show specific geographic features (roads, mountains, rivers, and the like) or to show information according to geographic distribution (population, housing, manufacturing centers, and so forth). Bear these points in mind in creating and using maps:

1. Label the map clearly.
2. Assign the map a figure number if you are using enough **illustrations** to justify use of figure numbers. (See Figure 1.)
3. Make sure all boundaries within the map are clearly identified. Eliminate unnecessary boundaries.
4. Eliminate unnecessary information from your map (if population is important, do not include mountains, roads, rivers, and the like).
5. Include a scale of miles or feet to give your **reader** an indication of the map's proportions.

6. Indicate which direction is north.
7. Show the features you want emphasized by using shading, dots, cross-hatching, or appropriate symbols when color reproduction cannot be used.
8. If you use only one color, remember that only three shades of a single color will show up satisfactorily.
9. Include a key telling what the different colors, shadings, or symbols represent.

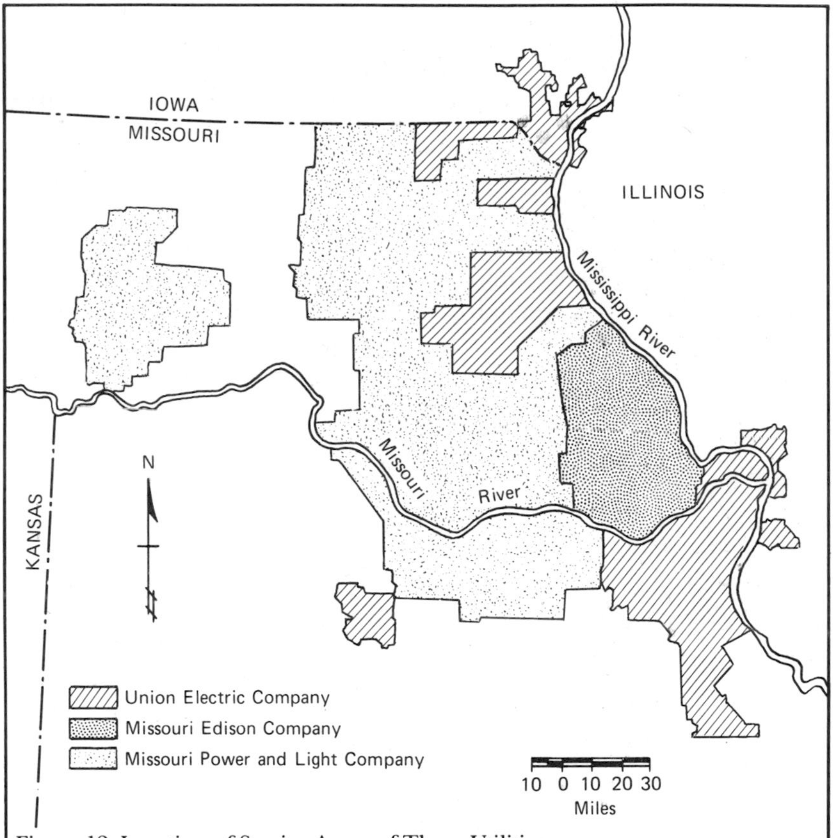

Figure 12. Location of Service Areas of Three Utilities

Source: The U.S. Nuclear Regulatory Commission

Figure 1 Sample Map

10. Place maps as close as possible to the portion of the text that refers to them.

mass nouns (see **nouns**)

mathematical equations

Material with mathematical equations can be easily prepared and easily read by following a few standard guidelines on displaying equations.

Short and simple equations, such as $x(y) = y^2 + 3y + 2$, should be set as part of the running text rather than being displayed or set on a separate line. (However, an equation or mathematical symbol should never appear at the beginning of a sentence.) If a document contains many equations that are referred to in the text, short equations should be displayed and identified with a number, as the following example shows:

$$x(y) = y^2 + 3y + 2 \qquad (1)$$

Equations are usually numbered consecutively throughout the work. Place the equation number, in **parentheses**, at the right margin, on the same line as the equation (or on the first line if the equation runs longer than one line). Leave at least four spaces between the equation number and the equation. Number displayed equations consecutively, and refer to them in the text as "Equation 1" or "Eq. 1."

POSITIONING DISPLAYED EQUATIONS

Equations that are set off from the text need to be surrounded by space. Triple-space between displayed equations and the normal text. Double-space between one equation and another and between the lines of multiline equations. Count space above the equation from the uppermost character in the equation; count space below from the lowermost character.

Type displayed equations either at the left margin or indented five spaces from the left margin, depending on their length. When a series of short equations is displayed in sequence, however, align them on their equal signs.

$$p(x,y) = \sin(x + y) \tag{2}$$
$$p(x,y) = \sin x \cos y + \cos x \sin y$$
$$p(x_0, y_0) = \sin x_0 \cos y_0 + \cos x_0 \sin y_0$$

$$q(x,y) = \cos(x + y)$$
$$= \cos x \cos y - \sin x \sin y$$
$$q(x_0, y_0) = \cos x_0 \cos y_0 - \sin x_0 \sin y_0$$

Break an equation requiring two lines at the equal sign, carrying the equal sign over to the second portion of the equation.

$$_0\!\int^1 (f_n - \tfrac{n}{r} f_n)^2 \, r \, dr \; + 2n \, _0\!\int^1 f_n \, f_n \, dr \tag{3}$$
$$= {}_0\!\int^1 (f_n - \tfrac{n}{r} f_n)^2 \, r \, dr \; + \; n f_n^2 (1)$$

If you cannot break it at the equal sign, break it at a plus sign or minus sign that is not within parentheses or **brackets**. Bring the plus sign or minus sign to the next line of the equation, which should be positioned to end near the right margin.

$$\phi(x,y,z) = (x^2 + y^2 + z^2)^{1/2} \, (x - y + z) \, (x + y - z)^2 \tag{4}$$
$$- [f(x,y,z) - 3x^2]$$

The next best place to break an equation is between parentheses or brackets that indicate multiplication of two major elements.

For equations requiring more than two lines, start the first line at the left margin, end the last line at the right margin, and center intermediate lines inside the margins. Whenever possible, break equations at operational signs, parentheses, or brackets.

EXPRESSING MATHEMATICAL EQUATIONS

Type mathematical equations rather than writing them out. Many electric typewriters have special mathematical-symbol typing elements for equations, as do the printing elements for personal computer printers. If your typewriter does not have mathematical symbols or Greek letters, use commercially produced press-on letters and symbols. If press-on material is not available, use templates to create letters and symbols in ink.

ADDITIONAL READING

Detailed guidance on the preparation of mathematical equations for publication can be found in the following handbooks and manuals:

CBE Style Manual, 5th ed. (Bethesda, Md.: Council of Biology Editors, Inc., 1983).

Chicago Manual of Style, 13th ed. (Chicago: University of Chicago Press, 1982).

Handbook for Authors of Papers in American Chemical Society Publications (Washington, D.C.: American Chemical Society, 1978).

Mathematics into Type, rev. ed. (Providence, R.I., American Mathematical Society, 1979).

maybe/may be

Maybe (one word) is an **adverb** meaning "perhaps."

EXAMPLE *Maybe* the legal staff can resolve this issue.

May be (two words) is a **verb** phrase.

EXAMPLE It *may be* necessary to ask for an outside specialist.

media/medium

Media is the plural of *medium* and should always be used with a plural verb.

EXAMPLES The *media are* a powerful influence in presidential elections.
The most influential *medium is* television.

memorandums

The memorandum is the most frequently used form of communication among members of the same organization. Called memos for short, memorandums are routinely used for internal communications of all kinds—from short notes to small **reports** and **internal proposals.** Among their many uses, memos announce policies, confirm conversations, exchange information, delegate responsibilities, request information, transmit documents, instruct employees, and report results. As this partial list illustrates, memos provide a record of decisions made and virtually all actions taken in an organization. For this reason, clear and effective memos are essential to the success of any organization. The slogan "Put it in writing" reflects the importance of memos and the care with which they should be written. The result of a carelessly prepared memo is a garbled message that could baffle **readers,** waste valuable time, produce costly errors, or irritate employees with an offensive **tone.**

Memos play a key role in the management of many organizations. Effective managers use memos to keep employees informed about company goals, motivate them to achieve these goals, and keep their morale high in the process. To achieve these ends, managers must be clear and accurate in their memos in order to gain respect and maintain credibility among their subordinates. Consider the unintended secondary messages the following notice conveys:

EXAMPLE It has been decided that the office will be open the day after Thanksgiving.

The first part of the sentence ("It has been decided") not only sounds impersonal but also communicates an authoritarian management-versus-employee tone: *somebody* decides that *you* work. The passive voice also suggests that the decision maker does not want to say, "I have decided" and thus be identified. One solution, of course, is to remove the first part of the sentence.

EXAMPLE The office will open the day after Thanksgiving.

But even this statement sounds impersonal. A better solution would be to suggest both that the decision is good for the company and that employees should be privy to (if not part of) the decision-making process.

EXAMPLE Because we must meet the December 15 deadline to be eligible for the government contract, the office will be open the day after Thanksgiving.

This version is forthright and informative and helps the employees understand the decision.

MEMO PROTOCOL

In many organizations, memos follow both written and unwritten lines of protocol. Protocol is usually reflected in the order of the recipients and those who receive copies. The following description by a public-utility employee illustrates how rank and lines of authority can be revealed in both obvious and subtle ways:

> If I put my immediate supervisor's name AND the vice-president's name on the top (of the memo's heading "To"), it says I really report to both equally. If the supervisor's name is on the top and I send a copy to the vice-president, it means that, though structurally I report to the supervisor, I have direct access to the VP.*
>
> —Carla Butenfoff, "Bad Writing Can Be Good Business," *ABCA Bulletin* 40 (June 1977): 12–13.

Although such practices vary, it is important to learn the etiquette of memos in your organization.

WRITING MEMOS

The writing process described on pages x–xvii will help you compose memos. It is particularly important to **outline** a memo, even if that means simply jotting down the points to be covered and then ranking them in a logical **method of development.** With careful preparation, memos can be both concise and adequately developed. Adequate development is crucial to ensure clarity.

CHANGE Be more careful on the loading dock.

TO To prevent accidents, follow these procedures:

1. Check . . .
2. Load only . . .
3. Replace . . .

Although the original version is concise, it is not as clear and specific as the revision. Don't assume your reader will know what you mean. State what you mean explicitly: readers aren't always careful, and some will provide their own interpretations if you are not as specific as possible.

Memos should ordinarily deal with only one subject; if you need to cover two subjects, you may wish to write two memos, since multisubject memos are difficult to file.

MEMO OPENINGS

Although methods of development may vary in memos, a memo should normally begin with a statement of its main idea. Even if your **opening** will review the background of a problem, begin with the main point.

EXAMPLE Because of our inability to serve our present and future clients as efficiently as in the past, I recommend we hire an additional claims representative and a part-time receptionist. (main idea) Last year we did not hire new staff because of the freeze on hiring. . . . (background opening)

As with other writing, if your reader is not familiar with the subject or the background of a problem, provide an introductory background paragraph. A brief background is especially important in memos that serve as records that can provide crucial information months (or even years) later. Generally, longer memos or those dealing with complex subjects benefit most from developed **introductions**. However, even when you are writing a short memo and the recipient is familiar with the situation, you need to remind your reader of the context. Readers have so much crossing their desks that they need a quick orientation. (Words that provide context are shown below in *italics*.)

EXAMPLES *As we decided after yesterday's meeting,* we need to set new guidelines for . . .

As Jane recommended, I reviewed the office reorganization plan. I like most of the features; however, the location of the receptionist and word processor . . .

The only exceptions to stating the main point first are (1) when the reader is likely to be highly skeptical and (2) when you are disagreeing with persons in positions of higher authority. In such cases, a more persuasive tactic is to state the problem first (rather than *your* solution) and then present the specific points that will support your final recommendation. (See also **persuasion.**)

LISTS AND HEADINGS

It is often a good idea to use **lists** to emphasize your points in a memo. If you are trying to convince a skeptical reader, a list of your points—from most to least persuasive—will stand out rather than being lost in a lengthy paragraph. Be careful, however, not to overuse lists. A memo that consists almost entirely of lists is difficult for readers to understand because they are forced to connect the separate items for themselves and mentally provide coherence. Lists lose their impact when they are overused.

Another useful device, particularly in long memos, is **headings.** Headings have a number of advantages:

1. They divide material into manageable segments.
2. They call attention to main topics.
3. They signal a shift in topic.

Headings also function like the parts of a **formal report** in that readers interested only in one section of a memo can easily identify that section.

WRITING STYLE

Like all **correspondence,** the level of formality in memos depends entirely on your reader and **objective.** For example, is your reader a peer, superior, or subordinate? A memo to an associate who is of equal rank and is a friend is likely to be informal and personal. However, a memo written as an internal proposal to several readers or a memo to someone two or three levels higher in your organization is likely to use the more formal **style** of a report. Consider the following original and revised versions of a subordinate's statement to a superior:

CHANGE I can't agree with your plan because I think it poses logistical problems. (informal, personal, and forceful)

TO The logistics of moving the department may pose serious problems. (formal, impersonal, and cautious)

A memo giving instructions to a subordinate will also be relatively formal and impersonal, but more direct—unless you are trying to reassure or praise. Remember, however, if you become too impersonal and formal and use many fancy words, you will create **affectation.** Managers who sound stuffy are assumed to be stuffy. They may also be regarded as rigid, lacking a nimble mind capable of moving an organization ahead. When writing to subordinates, also remember that managing does not mean ordering. An officious tone will prevent a memo from being an effective management tool.

FORMAT AND PARTS

Memo format and customs vary greatly from organization to organization. The following are some of the styles used:

preprinted half sheets (8½" × 5½")

message-and-reply forms ("speed messages")
regular 8½" × 11" letter stationery
special 8½" × 11" forms with company name or logo

Although there is no single, standard form, Figure 1 shows a typical memorandum format. Some firms also print short-purpose statements on the side or bottom of memos, with space for the writer to make a check mark.

EXAMPLES ☐ For your information
☐ For your action
☐ For your reply

You can also use this quick-response system for memos sent to numerous readers whose responses you need to tabulate.

EXAMPLE I can meet at 1 p.m. ☐
2 p.m. ☐
3 p.m. ☐

Regardless of the parts included, perhaps the one requiring the most careful preparation is the subject line or the title of a memo. Subject lines function much like the titles of reports, aiding in filing and later retrieval. Subject lines are also an important orientation when the reader first sees the memo. Therefore, it is important to make them accurate. That is, the memo should deal only with the single subject announced in the subject line, which also should be complete.

CHANGE Subject: Tuition Reimbursement
TO Subject: Tuition Reimbursement for Time Management Seminar

Remember, too, that subject lines should not substitute for an opening that provides a context for the message.

Capitalize all major words in the title of a subject line. Do not capitalize an **article, conjunction,** or **preposition** of fewer than four letters unless it is the first or last word.

The final step is signing or initialing a memo, which lets the reader know that you approve of its contents, especially if you did not type it. Where you sign or initial the memo depends on the practice of your organization: Some writers sign at the end, and others sign their initials next to their typed name. For further secretarial

PROFESSIONAL PUBLISHING SERVICES
MEMORANDUM

DATE: April 14, 19--

TO: Hazel Smith, Publications Manager

FROM: Herbert Kaufman HK

SUBJECT: Schedule for Acme Electronics Brochure

Acme Electronics has asked us to prepare a comprehensive brochure for their Milwaukee office by August 9, 19--. We have worked with electronic firms in the past, so this job should be relatively easy. My guess is that it will take nearly two months. Ted Harris has requested time and cost estimates. Fred Moore in accounting will prepare the cost estimates, and I would like you to prepare a schedule for the estimated time.

<u>Additional Personnel</u>
In preparing the schedule, check the following:

1. Production schedule for all staff writers.
2. Available free-lance writers.
3. Dependable graphic designers.

Ordinarily, we would not need to depend on outside personnel; however, since our bid for the <u>Wall Street Journal</u> special project is still under consideration, we could be pressed in June and July. We have to keep in mind staff vacations that have already been approved.

<u>Time Estimates</u>
Please give me time estimates by April 19. A successful job done on time will give us a good chance to obtain the contract to do Acme's publications for their annual stockholders' meeting this fall.

I know your staff can do the job.

lcs
Copies: Ted Harris, Senior Vice-President
Fred Moore, Accounting Manager

Figure 1 Typical Memo Format

and typing details, we recommend *The Gregg Reference Manual* by William A. Sabin (McGraw-Hill).

metaphor

Metaphor is a figure of speech that points out similarities between two things by referring to one of them as being the other. Metaphor states that the thing being described *is* the thing to which it is being compared.

> EXAMPLE *The building site was a beehive,* with iron workers still at work on the tenth floor, plumbers installing fixtures on the floors just beneath them, electricians busy on the middle floors, and carpenters putting the finishing touches on the first floor.

The use of metaphor often helps clarify complex theories or objects. For example, the life-sustaining tube connecting space-walking astronauts to their oxygen supply in the spacecraft is called an *umbilical cord.* The person who first drew this **comparison** knew that such a life support tube was new and unfamiliar, so to make its function immediately apparent he used a metaphor that compared it to something with which people were already familiar. A similar purpose was served when the term "window" was used to describe the abstract concept of the point at which a spacecraft reenters the earth's atmosphere. Here the strange and faraway was made near and familiar by the metaphorical use of a term familiar to everyone. Both of these expressions have gained widespread use because they helped clarify new and unfamiliar concepts.

A word of caution: Some writers sometimes mix metaphors, which usually results in an illogical statement.

> EXAMPLE Billingham's proposal *backfired,* and now she is *in hot water.*

This example makes an incongruous statement because a backfiring automobile does not land someone in hot water. (See also **figures of speech.**)

methods of development

One of the most important elements in any writing is its method of development. After you have completed your **research,** but before beginning your **outline,** ask yourself how you can most effectively "unfold" your **topic** for your **reader.** An appropriate method of development will make it easy for your reader to understand your topic and will move the topic smoothly and logically from an **introduction** (or **opening**) to a **conclusion.** There are several common methods of development, each best suited to particular purposes.

If you are writing a set of **instructions,** for example, you know that your readers need the instructions in the order that will enable them to perform some task. Therefore, you should use a **sequential method of development.** If you wished to emphasize the time element of a sequence, however, you could follow a **chronological method of development.**

If writing about a new topic that is in many ways similar to another, more familiar topic, it is sometimes useful to develop the new topic by comparing it to the old one—thereby enabling your readers to make certain broad assumptions about the new topic, based on their understanding of the familiar topic. By doing so, you are using a **comparison method of development.**

When describing a mechanical device, you may divide it into its component parts and explain each part's function as well as how all the parts work together. In this case, you are using a **division-and-classification method of development.** Or you may use a **spatial method of development** to describe the physical appearance of the device from top to bottom, from inside to outside, from front to back, and so on.

If you are writing a report for a government agency explaining an airplane crash, you might begin with the crash and trace backwards to its cause, or you could begin with the cause of the crash (for example, a structural defect) and show the sequence of events that led to the crash. Either way is the **cause-and-effect method of development**. You can also use this approach to develop a report dealing with the solution to a problem, beginning with the problem and moving on to the solution, or vice versa.

If you are writing about the software for a new computer system, you might begin with a general statement of the function of the total software package, then explain the functions of the larger routines within the software package, and finally deal with the functions of the various subroutines within the larger routines. You would be using a **general-to-specific method of development.** (In another situation, you might use the **specific-to-general method of development.**)

To explain the functions of the departments in a company, you could present them in a sequence that reflects their importance within the company: the executive department first and the custodial department last, with all other departments (sales, engineering, accounting, and so on) arranged in the relative order of importance they are given in that company. You would be using the **decreasing-order-of-importance method of development.** (In another situation, you might use the **increasing-order-of-importance method of development.**)

Methods of development often overlap, of course. Rarely does a writer rely on only one method of development in a written work. The important thing is to select one primary method of development and then to base your outline on it. For example, in describing the organization of a company, you would actually use elements from three methods of development: you would *divide* the larger topic (the company) into departments, present the departments *sequentially,* and arrange the departments by their *order of importance* within the company.

For example, assume that you are writing an article about a particular company. You could approach such a topic any number of ways. You could *compare* the company with its major competitor, touching upon such things as each one's historical development, market share, number of employees, revenue produced per employee, gross annual sales, net income, return on investment, research and development expenditures, debt as a percentage of assets, and relative financial condition. Or you could *divide,* or break down, the company by departments, and discuss each in turn by decreasing order of importance: the executive department, engineering, sales, production, accounting, data processing, inventory control, advertising, order processing. Or in order to discuss and illustrate the company's operations, you could follow a purchase order through its entire life cycle, from the time an advertisement is placed in a trade journal, to the actual product sale, to the receipt of the order at the factory, to the product being pulled from stock and shipped to the customer, to the invoice being received by the accounting department and the bill being sent to the customer, to the payment being received from the customer, and finally to the account being closed.

minutes of meetings

Organizations and committees keep official records of their meetings; such records are known as *minutes*. If you attend many business-related meetings, you may be asked to serve as recording secretary and write and distribute the minutes of a meeting. At each meeting, the minutes of the previous meeting are usually read aloud if printed copies of the minutes were not distributed to the members beforehand; the group then votes to accept the minutes as prepared or to revise or clarify specific items. The bylaws or **policies and procedures** of your organization may specify what must be included in your minutes of meetings. As a general guide, *Robert's Rules of Order* recommends that the minutes of meetings include the following:

1. Name of the group or committee holding the meeting.
2. Place, time, and date of the meeting.
3. Kind of meeting (a regular meeting or a special meeting called to discuss a specific subject or problem).
4. Number of members present and, for committees or boards of ten or fewer members, their names.
5. A statement that the chairperson and the secretary were present, or the names of any substitutes.
6. A statement that the minutes of the previous meeting were approved, revised, or not read.
7. A list of any reports that were read and approved.
8. All the main motions that were made, with statements as to whether they were carried, defeated, or tabled (vote postponed), and the names of those who made and seconded the motions (motions that were withdrawn are not mentioned).
9. A full description of resolutions that were adopted and a simple statement of any that were rejected.
10. A record of all ballots with the number of votes cast for and against.
11. The time that the meeting was adjourned (officially ended) and the place, time, and date of the next meeting.
12. The recording secretary's signature and typed name and, if desired, the signature of the chairperson.

Except for recording motions, which must be transcribed word for word, summarize what occurs and paraphrase discussions.

Since minutes are often used to settle disputes, they must be accurate, complete, and clear. When approved, minutes of meetings are official and can be used as evidence in legal proceedings.

Keep your minutes brief and to the point. Give complete information on each topic, but do not ramble—conclude the topic and go on to the next one. Following a set **format** will help you keep the minutes concise. You might, for example, use the heading *TOPIC*, followed by the subheadings *Discussion* and *Action Taken,* for each major point discussed.

Avoid abstractions and generalities; always be specific. If you are referring to a nursing station on the second floor of a hospital, write "the nursing station on the second floor" or "the second-floor nursing station," not just "the second floor."

Be especially specific when referring to people. Avoid using titles (the chief of the Word Processing Unit) in favor of names and titles (Ms. Florence Johnson, chief of the Word Processing Unit). And be consistent in the way you refer to people. Do not call one person *Mr.* Jarrel and another *Janet* Wilson. It may be unintentional, but a lack of consistency in titles or names may reveal a deference to one person at the expense of another. Avoid **adjectives** and **adverbs** that suggest either good or bad qualities, as in "Mr. Sturgess's *capable* assistant read the *extremely comprehensive* report to the subcommittee." Minutes should always be objective and impartial.

If a member of the committee is to follow up on something and report back to the committee at its next meeting, state clearly the person's name and the responsibility he or she has accepted.

When assigned to take the minutes at a meeting, be prepared. Bring more than one pen and plenty of paper. If convenient, you may bring a tape recorder as a backup to your notes. Have ready the minutes of the previous meeting and any other material that you may need. If you do not know shorthand, take memory-jogging notes during the meeting and then expand them with the appropriate details immediately after the meeting. Remember that minutes are primarily a record of specific actions taken. (See also **note taking.**)

Figure 1 is a sample set of minutes.

misplaced modifiers

A **modifier** is misplaced when it modifies, or appears to modify, the wrong word or phrase. It differs from a **dangling modifier** in that a

Minutes of the Regular Meeting of the Credentials Committee

DATE: April 18, 19--

PRESENT: M. Valden (Chairperson), R. Baron,
M. Frank, J. Guern, L. Kingson,
L. Kinslow (Secretary), S. Perry,
B. Roman, J. Sorder, F. Sugihana

Dr. Mary Valden called the meeting to order at 8:40 p.m. The minutes of the previous meeting were unanimously approved, with the following correction: the name of the secretary of the Department of Medicine is to be changed from Dr. Juanita Alvarez to Dr. Barbara Golden.

<u>Old Business</u>

None.

<u>New Business</u>

The request by Dr. Henry Russell for staff privileges in the Department of Medicine was discussed. Dr. James Guern made a motion that Dr. Russell be granted staff privileges. Dr. Martin Frank seconded the motion, which passed unanimously.

Similar requests by Dr. Ernest Hiram and Dr. Helen Redlands were discussed. Dr. Fred Sugihana made a motion that both physicians be granted all staff privileges except respiratory-care privileges because the two doctors had not had a sufficient number of respiratory cases. Dr. Steven Perry seconded the motion, which was passed unanimously.

Dr. John Sorder and Dr. Barry Roman asked for a clarification of general duties for active staff members with respiratory-care privileges. Dr. Richard Baron stated that he would present a clarification at the next scheduled staff meeting on May 15.

Dr. Baron asked for a volunteer to fill the existing vacancy for Emergency Room duty. Dr. Guern volunteered. He and Dr. Baron will arrange a duty schedule.

There being no further business, the meeting was adjourned at 9:15 p.m. The next regular meeting is scheduled for May 15, at 8:40 p.m.

Respectfully submitted,

Leslie Kinslow Medical Staff Secretary	Mary Valden, M.D. Chairperson

Figure 1 Sample of a Set of Minutes

dangling modifier cannot *logically* modify any word in the sentence because its intended referent is missing. The best general rule for avoiding misplaced modifiers is to place modifiers as close as possible to the words they are intended to modify. A misplaced modifier can be a word, a **phrase,** or a **clause.**

MISPLACED WORDS

Adverbs are especially likely to be misplaced because they can appear in several positions within a sentence.

EXAMPLES We *almost* lost all of the parts.
We lost *almost* all of the parts.

The first sentence means that all of the parts were *almost* lost (but they were not), and the second sentence means that a majority of the parts (*almost* all) were in fact lost. Possible confusion in sentences of this type can be avoided by placing the adverb immediately before the word it is intended to modify.

CHANGE All navigators are *not* talented in mathematics. (The implication is that *no* navigator is talented in mathematics.)
TO *Not* all navigators are talented in mathematics.

MISPLACED PHRASES

To avoid confusion, place phrases near the words they modify. Note the two meanings possible when the phrase is shifted in the following sentences:

EXAMPLES The equipment *without the accessories* sold the best. (Different types of equipment were available, some with and some without accessories.)
The equipment sold the best *without the accessories.* (One *type* of equipment was available, and the accessories were optional.)

MISPLACED CLAUSES

To avoid confusion, clauses should be placed as close as possible to the words they modify.

CHANGE We sent the brochure to four local firms *that had three-color art.*
TO We sent the brochure *that had three-color art* to four local firms.

SQUINTING MODIFIERS

A modifier "squints" when it can be interpreted as modifying either of two sentence elements simultaneously, so that the **reader** is confused about which is intended.

EXAMPLE We agreed *on the next day* to make the adjustments.

The meaning of the preceding example is unclear; it could have either of the following two senses:

EXAMPLES We agreed to *make the adjustments on the next day.*
On the next day we agreed to make the adjustments.

A squinting modifier can sometimes be corrected simply by changing its position, but often it is better to recast the sentence:

EXAMPLES We agreed that *on the next day* we would make the adjustments. (The adjustments were to be made on the next day.)
On the next day we agreed that we would make the adjustments. (The agreement was made on the next day.)

(See also **dangling modifiers.**)

mixed constructions

A mixed construction occurs when a sentence contains grammatical forms that are improperly combined. These constructions are improper because the grammatical forms are inconsistent with one another. The most common types of mixed constructions result from the following causes:

TENSE

CHANGE The pilot *lowered* the landing gear and *is approaching* the runway. (shift from past to present tense)
TO The pilot *lowered* the landing gear and *approached* the runway.
OR The pilot *has lowered* the landing gear and *is approaching* the runway.

PERSON

CHANGE The *technician* should take care in choosing *your* equipment. (shift from third to second person)
TO The *technician* should take care in choosing *his or her* equipment.

NUMBER

CHANGE My *car,* though not as fast as the others, *operate* on regular gasoline. (singular subject with plural verb form)

TO My *car*, though not as fast as the others, *operates* on regular gasoline.

VOICE

CHANGE *I will check* your report, and then *it will be returned* to you. (shift from active to passive voice)

TO *I will check* your report, and then *I will return* it to you.

(See also **parallel structure** and **agreement.**)

modifiers

Modifiers are words, **phrases,** or **clauses** that expand, limit, or make more precise the meaning of other elements in a sentence. Although we can create sentences without modifiers, we often need the detail and clarification they provide.

EXAMPLES Production decreased. (without modifiers)
Automobile production decreased *rapidly.* (with modifiers)

Most modifiers function as **adjectives** or **adverbs.** An adjective makes the meaning of a **noun** or **pronoun** more precise by pointing out one of its qualities or by imposing boundaries on it.

EXAMPLES *ten* automobiles — *this* crane
an *educated* person — *loud* machinery

An adverb modifies an adjective, another adverb, a **verb,** or an entire clause.

EXAMPLES Under test conditions, the brake pad showed *much* less wear than it did under actual conditions. (modifying the adjective *less*)

The wear was *very* much less than under actual conditions. (modifying another adverb, *much*)

The recording had hit the surface of the disc *hard.* (modifying the verb *hit*)

Surprisingly, the machine failed even after all the tests that it had passed. (modifying a clause)

Adverbs become **intensifiers** when they increase the impact of adjectives (*very* fine, *too* high) or adverbs (*rather* quickly, *very* slowly). As a rule, be cautious in using intensifiers; their overuse can lead to exaggeration and hence to inaccuracies. (See also **dangling modifiers** and **misplaced modifiers.**)

mood

The grammatical term *mood* refers to the **verb** functions—and sometimes form changes—that indicate whether the verb is intended to (1) make a statement or ask a question (indicative mood), (2) give a command (imperative mood), or (3) express a hypothetical possibility (subjunctive mood).

The indicative mood refers to an action or a state that is conceived as fact.

EXAMPLES *Is* the setting correct?
The setting *is* correct.

The imperative mood expresses a command, suggestion, request, or entreaty. In the imperative mood, the implied subject "you" is not expressed.

EXAMPLES *Install* the wiring today.
Please *let* me know if I can help.

The subjunctive mood expresses something that is contrary to fact, that is conditional, hypothetical, or purely imaginative; it can also express a wish, a doubt, or a possibility. The subjunctive mood may change the form of the verb, but the verb *be* is the only one in English that preserves many such distinctions.

EXAMPLES The senior partner insisted that he (I, you, we, they) *be* in charge of the project.
If we *were* to close the sale today, we would meet our monthly quota.
If I *were* you, I would postpone the trip.

Form change for the subjunctive is rare in most verbs other than *be.* Instead, we use **helping verbs** to show the subjunctive function. Be careful, however, not to use unnecessary verbs.

CHANGE If I *would have known* that you were here, I would have come earlier.
TO If I *had known* that you were here, I would have come earlier.
OR *Had I known* that you were here, I would have come earlier.

The advantage of the subjunctive mood is that it enables us to express clearly whether or not we consider a condition contrary to fact. If so, we use the subjunctive; if not, we use the indicative.

EXAMPLES If I *were* president of the firm, I would change several personnel policies. (subjunctive)
Although I *am* president of the firm, I don't feel that I control every aspect of its policies. (indicative)

Be careful not to shift from one to another within a sentence; to do so makes the sentence not only ungrammatical but unbalanced as well.

CHANGE *Put* the clutch in first (imperative); then you *can* put the truck in gear. (indicative)
TO *Put* the clutch in first (imperative); then *put* the truck in gear. (imperative)

Ms./Miss/Mrs.

Ms. is a convenient form of addressing a woman, regardless of her marital status, and it is now almost universally accepted. *Miss* is used to refer to an unmarried woman, and *Mrs.* is used to refer to a married woman. Some women indicate a preference for *Miss* or *Mrs.,* and such a preference should be honored. An academic or professional title (*Doctor, Professor, Captain*) should take preference over *Ms., Miss,* or *Mrs.* (See also **correspondence**.)

MS/MSS

MS (or ms) is the **abbreviation** for *manuscript.* The plural form is MSS (or mss). Do not use this abbreviation (or any abbreviation) unless you are certain that you **reader** is familiar with it.

mutual/in common

When two or more persons (or things) have something in common, they share it or possess it jointly.

EXAMPLES What we have *in common* is our desire to make the company profitable.
The fore and aft guidance assemblies have a *common* power source.

Mutual may also mean "shared" (as in *mutual* friends), but it usually implies something given and received reciprocally and is used with reference to only two persons or parties.

EXAMPLES Smith mistrusts Jones, and I am afraid the mistrust is *mutual.* (Jones also mistrusts Smith.)

Both business and government consider the problem to be of *mutual* concern.

N

narration

Narration is the presentation of a series of events in a prescribed (usually chronological) sequence. Much narrative writing explains how something happened: a laboratory or field study, a site visit, an accident, the events and decisions in an important meeting. (A scenario is a form of narrative that raises the question, "What if?" In it, the writer speculates about what is likely to happen under certain circumstances.)

Effective narration rests on two key writing techniques: the careful, accurate sequencing of events and a consistent **point of view** on the part of the narrator. Narrative sequence and essential shifts in the sequence are signaled in three ways: chronology (clock and calendar time), transitional words pertaining to time (*before, after, next, first, while, then*), and verb tenses that indicate whether something has happened (past **tense**) or is under way (present tense). The point of view indicates the writer's relation to the information being narrated as reflected in the use of **person.** Narration usually expresses either a first- or third-person point of view. First-person narration indicates that the writer is a participant ("this happened to me"), and third-person narration indicates that the writer is writing about what happened to someone or something else ("this happened to her, to them, or to it").

The narrative presented in Figure 1 reconstructs the final hours of the flight of a small aircraft that attemped to land at the airport in Hailey, Idaho. Instead, the plane crashed into the side of a mountain, killing the pilot and copilot. Because of the disastrous outcome of the flight, the investigators needed to "tell the story" in detail so that any lessons learned could be made available to other flyers. To do so, they had to recount the events as closely as possible. Thus, the chronology of the events is specified throughout the narrative in local (Hailey, Idaho) time.

To "tell the story," the narrator used the third-person point of view throughout, except for the first-person reports from witnesses. The words of each witness were chosen carefully and quoted for their bearing on what happened, from that person's vantage point.

*History of the Flight**

At 0613 m.s.t[1] on January 3, 1983, N805C, a Canadair Challenger owned and operated by the A. E. Staley Company departed Decatur, Illinois, on a flight to Friedman Memorial Airport, Hailey, Idaho. The route of flight was via Capitol, Illinois, Omaha, Nebraska, Scotts Bluff, Nebraska, Riverton, Wyoming, Idaho Falls, Idaho, direct 43°30′ north lattitude, 114°17′ west longitude.

The en route portion of the flight was uneventful, and about 35 nmi east of Idaho Falls N805C was cleared by the Salt Lake City, Utah, Air Route Traffic Control Center (ARTCCS) to descend from 39,000 feet to 22,000 feet. N805C descended to 22,000 feet and the flightcrew then requested a descent to 17,000 feet. About 35 nmi east of Sun Valley Airport, after being cleared, N805C descended to 17,000 feet. About 0901, N805C's flightcrew cancelled their flight plan and, shortly thereafter, changed the transponder from 1311, the assigned discrete code, to 1200, the visual flight rules (VFR) code. At 0901:07, the data analysis reduction tool (DART) radar data showed a 1311 beacon code at 17,000 feet about 11 nmi east of the Sun Valley Airport. At 0901:37, the DART radar data showed a 1200 VFR transponder beacon code with no altitude readout about 2 nmi west of the 1311 beacon code that was recorded at 0901:07.

At 0904:10, DART radar data recorded a 1200 code target at 13,500 feet almost directly over the Sun Valley Airport. According to an employee of the airport's fixed base operator, N805C's flightcrew called on the airport's UNICOM[2] frequency and requested a landing advisory and asked if a food order had been placed. The flightcrew then stated that

*National Transportation Safety Board, "Aircraft Accident Report: A. E. Staley Manufacturing Co., Inc., Canadair Challenger CL-600, N805C, Hailey, Idaho, January 3, 1983. National Technical Information Service, Springfield, Va., 1983.

[1]All times, unless otherwise noted, are mountain standard time based on the 24-hour clock.

[2]UNICOM. The Non-government air/ground radio communications facility which may provide airport advisory information at certain airports. The Sun-Valley UNICOM did not record, nor was it required to record or log, the time of radio communications.

Figure 1 Example of a Narration

there would "be a quick-turn," and placed a fuel request. This was the last transmission heard from N805C. The employee said that she provided the latest altimeter setting to N805C, and "since we did not have the cloud conditions in the area, I was glad when other pilots were able to give reports as they saw things from the air."

The flightcrew of Cessna Citation, N13BT, which had landed at Sun Valley about 0903, also heard N805C report "over the field." According to N13BT's pilot, N805C reported over the field "sometime during our final approach or landing." According to the pilot, the weather at the airport when he landed "was 800 (feet) overcast with 10 miles visibility. The tops of the overcast or fog bank was about 6,800 (feet) m.s.l." He said that the overcast was "solid northwest up the valley. Visibility appeared lower (to the) northwest."

About 0908, Trans Western Flight 1301, a Convair 580, landed at Sun Valley Airport. Flight 1301 had descended through a hole in the overcast about 15 nmi southwest of Bellevue, Idaho, which is about 3 nmi southeast of the Sun Valley Airport. The first officer said that he gave position reports to the Sun Valley UNICOM when the flight was 15 nmi from the airport, when it was 10 nmi from the airport over Bellevue turning on final approach for runway 31, and when it was 1 mile from the runway. The first officer said that he could see the visual approach slope indicator (VASI) lights for runway 30 during the landing approach. The captain and the first officer said that they neither saw N805C nor heard radio transmissions from N805C.

About 0900, a man who was driving his truck north on the highway between Bellevue and Hailey, Idaho, saw a twin engine, cream-colored jet, break through the clouds when he was about 2.5 miles north of Bellevue. He saw that the landing gear was down but he did not see any lights on the airplane. When the airplane appeared, "it was about 300 to 500 yards fron the west hills adjacent to the airport and about 1,000 feet from the valley floor." The airplane was in a nose-up attitude. The witness said that after the airplane descended below the clouds "and (the pilot) saw how close to the hills he was, he then started a sharp right turn." The airplane disappeared from his view into "low hanging clouds" over the northwest side of the hangar at the airport.

Between 0900 and 0930, another man, who was in the yard of his home in northeast Hailey, saw a jet airplane east of his home. The airplane was "white or silver with a blue tint."

(N805C was painted white with blue and gold stripes along the length of the fuselage and tops of the wings.) The airplane was below the clouds, and he had "a good view of the airplane for about 10 to 15 seconds." He said that the airplane had a high noseup attitude and "the wings were rocking up and down about 20°." The witness said that the clouds obscured all but the lower peaks of the mountains to the east and that after he lost sight of the airplane he thought it was "odd that the aircraft was under the cloud cover."

Shortly after 0900, a woman who was located in an apartment in southeast Hailey, heard a jet airplane fly over "in a northerly direction." She thought that this was "odd because jets don't go over us heading north from the airport. The engines sounded very loud. . . ."

A fourth witness said that, between 0900 and 0940, she heard a jet airplane overfly her house in northeast Hailey. The woman was in the living room of her house when she heard the airplane and thought that "it must be low because of the loudness of the (engine) noise," and that "the sound of the jet did not trail off as they do as they fly farther away from you. The sound stopped less than 30 seconds from the time I first heard it." At the time, the clouds were resting on and hiding the tops of the mountains to the east.

About 1030, the chief pilot of the A. E. Staley Manufacturing Company, who was to board N805C at Sun Valley, arrived at the airport. Since the airplane was overdue, he instituted inquiries to several nearby airports to determine where the airplane had landed. At 1300, he asked air traffic control to make a full communications search. At 1400, after being told that the airplane had not been found, he requested an air search. While waiting for search and rescue teams to arrive, the chief pilot rented an airplane and at about 1700, found the accident site. The impact site, elevation about 6,510 feet, was about 2.2 nmi north of Sun Valley Airport at coordinates 43°32′50″ N latitude, 114°17′35″ W longitude.

The events were sequenced as precisely as possible from the beginning of the flight until the crash site was located by referencing verified clock times and approximating those that could not be verified. The verb tenses throughout indicated a past action: *departed, canceled, cleared, called, reported*.

Important but secondary information was either mentioned in a footnote or inserted, like the color of the plane, in parentheses. Information about the color of the plane was important to the narration only because it verified a sighting by an eyewitness. Otherwise, this information would have been unnecessary.

Once a narrative is under way, it should not be interrupted by lengthy explanations or analysis. Explain only what is necessary for readers to follow the action. In the plane-crash narrative, numerous pertinent issues might have been explained or evaluated in more detail. The weather, the condition of the plane's radar equipment and engines, the pilot's professional and medical history, the airport and its personnel and equipment—all are pertinent, but each was assessed after the narrative in a separate section of the report. To have addressed each in the course of the narrative would have made the reader's task of finding out what happened unnecessarily difficult.

For further information about sequencing narratives, see **sequential** and **chronological methods of development.** Other writing strategies, like **process explanation** and **instructions,** also provide valuable guidance about how to sequence information accurately.

nature

Nature, used to mean "kind" or "sort," can—like those words—often be vague. Avoid the word in your writing. Say exactly what you mean.

CHANGE The *nature* of the engine caused the problem.
TO The compression ratio of the engine caused the problem.

needless to say

Although the **phrase** *needless to say* sometimes occurs in speech and writing, it is redundant because it always precedes a remark that is stated despite the inference that nothing further need be said.

CHANGE *Needless to say,* departmental cutbacks have meant decreased efficiency.
TO Departmental cutbacks have meant decreased efficiency.
OR Understandably, departmental cutbacks have meant decreased efficiency.

neo-

The **prefix** *neo-* is derived from a Greek word meaning "new." It is hyphenated when used with a **proper noun,** and it may be hyphenated when used with a **noun** beginning with the vowel *o*.

EXAMPLES neo-Fascism, neo-Darwinism, neo-orthodoxy
neologism, neonatal, neoplasm

new words

New words continually find their way into the language from a variety of sources. Some come from other languages.

EXAMPLES discotheque (French), skiing (Norwegian), whiskey (Gaelic)

Some come from technology.

EXAMPLES software, vinyl, nylon

Some come from scientific research.

EXAMPLES berkelium, transistor

Some come from brand or trade names.

EXAMPLES Jell-O, Teflon, Kleenex

Some, called **blend words,** are formed by combining existing words.

EXAMPLE smoke + fog = smog

Some come from **acronyms.**

EXAMPLES scuba, radar, laser

Business and technology are responsible for many new words. Some are necessary and unavoidable; however, it is best to avoid creating a new word if an existing word will do. If you do use what you believe to be a new word or expression, be sure to define it the first time you use it; otherwise, your **reader** will be confused. (See also **long variants.**)

newsletter articles

A newsletter is a company publication or an employee publication that is produced by the employer. Its primary purpose is to keep employees informed about the company and its operations and policies. Many different types of newsletters are found in business and industry—from the gossip sheet to the more sophisticated instruments of company policy that increase the employee's understanding of the company, its purpose, and the business in which it is engaged. If your company publishes a newsletter of the latter type, you may be asked to contribute an article on a subject that you are especially qualified to write about. The editor may offer some general advice but can give you little more help until you submit a draft.

Before beginning to write, consider the traditional *who, what, where, when,* and *why* of journalism (who did it? what was done? where was it done? when was it done? and why was it done?)—and then add *how,* since the *how* of your subject may be of as much interest to your fellow employees as any of the five *w*'s.

Before going any further, determine whether there is an official company policy or position on your subject. If so, adhere to it as you prepare your article. If there is no company policy on your subject, determine as nearly as you can what management's attitude is toward your subject.

Gather several fairly recent issues of your company newsletter, and study the **style** and **tone** of the writing and the company's approach to various kinds of subjects in past issues. Try to emulate these as you work on your own article. Ask yourself the following questions about your subject: What is its significance to the company? What is its significance to employees? The answers to these questions should help you establish the style, tone, and approach for your article. These answers should also heavily influence the **conclusion** you write for your article.

Research for a newsletter article frequently consists of **interviewing.** Interview everyone concerned with your subject. Get all available information and all points of view. (Be sure to give maximum credit to the maximum number of people.)

Writing a newsletter article requires a little more imagination than does writing **reports.** With a newsletter, you do not have a captive audience; therefore, you may want to provide four helpful ingredients to ensure a successful article: (1) an intriguing **title** to catch the **readers'** attention (a **rhetorical question** is often effective); (2) eye-catching **photographs** or **illustrations** that entice them to read your lead **paragraph** to see what the subject is really about; (3) a lead, or first paragraph, that is designed to encourage further reading (this paragraph generally makes the **transition** from the **title** to the down-to-earth treatment of the subject); and (4) a well-developed presentation of your subject to hold the readers' interest all the way to the conclusion.

Your **conclusion** should emphasize the significance of your subject to your company and its employees. Because of its strategic location to achieve **emphasis,** your conclusion should include the thoughts that you want your readers to retain about your subject.

In preparing your newsletter article you will find it helpful to refer to the Checklist of the Writing Process and to follow the steps listed there. Figure 1 (pp. 428–431) is an article written for a newsletter. The original included photographs and boxed quotations.

no doubt but

In the **phrase** *no doubt but,* the word *but* is redundant.

CHANGE There is *no doubt but* that he will be promoted.
TO There is *no doubt* that he will be promoted.

nominalizations

We have a natural tendency to want to make our on-the-job writing sound very formal—even "impressive." One practice that contributes to both is the use of *nominalizations,* or a weak **verb** (*make, do, conduct, perform,* and so forth) and a noun, when the verb form of the noun would communicate the same idea more effectively in fewer words.

CHANGE The Legal Department will *conduct an investigation* of the purchase.
TO The Legal Department will *investigate* the purchase.

CHANGE The quality assurance team will *perform an evaluation* of the new software.
TO The quality assurance team will *evaluate* the new software.

You may occasionally find a legitimate use for a nominalization. You might, for example, use a nominalization to slow down the **pace** of your writing. But if you use nominalizations thoughtlessly or carelessly or just to make your writing sound more formal, the result will be **affectation.**

Author! Author!

It was a number of centuries back that the Roman satirist Persius said, "Your knowing is nothing, unless others know you know."

Today at Allen-Bradley some of our engineers are subscribing to Persius's philosophy. They contend it takes more than the selling of a product for a corporation to enjoy continuing success.

"We must sell our knowledge as well," stressed Don Fitzpatrick, Commercial Chief Engineer. "We have to get our customers to think of us as knowledge experts."

Fitzpatrick's point is well made. In today's high technology business climate, it is essential to present a vanguard image as well as produce quality products. A corporation can be considered a "knowledge expert" when it "not only has a product for sale, but is the acknowledged leader in the application of that product," Fitzpatrick explained.

How does a corporation acquire an image as a "knowledge expert"?

Customer Roundtables, distributor schools, and participation by our engineers in professional societies, such as the Institute of Electrical and Electronics Engineers (IEEE), have helped Allen-Bradley to enhance its image in the industrial control market, noted Fitzpatrick. Yet, one subliminal selling tool that Fitzpatrick believes A-B engineers could sharpen to foster more influence is the writing and presentation of technical papers.

"I'm talking about technical writing. Not the kind of technical writing that explains our products or how to use them safely, but the kind of writing that is done as a method of sharing knowledge among companies who have similar concerns," Fitzpatrick explained. Such writings discuss new methods being used in the industry, or a company's difficulties working with an electrical phenomenon. They introduce state-of-the-art technology, new theories, or are tutorials on the developments in the industry to date, he added.

FITZPATRICK'S PAPER CITED

Fitzpatrick knows about technical papers. The former Purdue University professor has been preparing and presenting topics for Allen-Bradley Roundtables and distributor schools for several years. He also has been active in the presentation of papers for IEEE conferences. This past September in Houston, Fitzpatrick received a second-place award for a paper he presented at a Petrochemical Industrial Conference (PCIC) in San Diego in 1979. (PCIC is an organization of the Industry Application Society, which is part of the IEEE.) The award-winning

Figure 1 Newsletter Article

paper, entitled "Transient Phenomena in the Motor Control Distribution Systems," also was published in IEEE's renowned annual publication "IAS Transactions."

Only a few papers are accepted annually for presentation at professional conferences, noted Fitzpatrick. In turn, just a few of those presentations are published in IEEE-related "Transactions" catalogs. Occasionally a paper's topic also may generate an article in a trade publication. For these reasons preparing such papers can be plums—for both company recognition and the author's esteem among colleagues.

Cultivating peer recognition is as important as nurturing an expert image for your company in today's technological world, said Steven Bomba, Director of Advanced Technology Development. "In the area of advanced technology the only way to test if your work is current is to share it with people outside of the company," he said. "When you publish, you don't reveal proprietary information. You publish information that will be economically beneficial to your organization. Our employees should be anxious to test their sword against their competitors."

GROWING FRATERNITY HERE

Bomba and Fitzpatrick noted that there is a "new spirit for Allen-Bradley" in the area of technical writing and presentation. The number of engineers participating as guest speakers at national engineering, application, and technical conferences is a small but growing fraternity. Participation by A-B personnel in these professional conferences spurs a "psyche tune-up" and is critical to "innovation, staying current, and invention. Without it, a company lacks a reference point. Encouraging your people to write these papers provides a measure of value and stimulation," Bomba contends.

Technical papers are always noncommercial. They don't mention specific products or the company's name. The papers are bylined, however, usually with an editor's note containing information about the writer's background and employment. "If you word a paper right, you can sell your company. You can make the paper work as a subliminal public relations tool for you. One payoff for the extra time it often takes to prepare a paper is the direct influence in the public eye that is gained. It is impossible to measure the exact influence, but the exposure generally generates positive reaction," Fitzpatrick notes.

CONTACTS AT CONFERENCES

"Participation in conferences through writing of technical papers results in an employee becoming more active in professional organizations. That in turn leads to more business contacts. We're trying to broaden our market base, enter new industries. Roundtables, publishing, and conference programs are by-products, or spin-offs, that work for us. They help us to sell our ability,'' said Fitzpatrick.

They also work for the engineers and technicians who get involved in the paper's preparation. Often a paper may be co-authored or researched as a team. Engineers or technicians who take the time to research a possible topic, prepare an abstract for query, and finally write a paper upon abstract acceptance, are taking those small steps of knowledge that lead to leaps in industry technology.

''Getting published places the author in the elite 'Invisible College of Knowledge Workers,' which is a collection of technical people who transcend the boundaries of industry, cities, and nations," said Bomba. "Only 1 percent of degreed engineers write technical papers that are published," he added.

''Peer fear'' may be the greatest barrier for an engineer to pick up the pen—or word processor, as the case may be today. Walt Maslowski, Principal Engineer, Dept. 756, was an engineer for 12 years before he decided to give technical writing a whirl three years ago. He joined A-B in 1979. This past September he presented his second paper as an A-B employee at an Industry Application Society conference in Cincinnati. His latest paper, which addressed the techniques in controlling three-phase induction motor drives, received a second-place award from IAS. The article will be published in IEEE's next issue of IAS Transactions. ''There was a lot of positive response to the paper, and motor drives recognition is an area that A-B has been pursuing for some time,'' Maslowski noted.

MANY BENEFITS INTANGIBLE

''The fear of possibly writing substandard papers held me back from writing for years,'' Maslowski continued. ''What I've discovered, though, is that when you do write a qualified paper, your technical expertise becomes known. Many of your personal gains are intangible, but you meet people, open contacts, and absorb more knowledge.''

How does one decide to write a paper? ''First you should review enough of a field to get a pulse for it. Read key papers. Review key results that people are talking about. Analyze the unsolved areas—or holes—of these writings and then figure out how those holes might relate to your job,'' suggested Dave Linton, a project engineer in Dept. 756. Linton presented a tutorial on computer software design at a Digital Equipment Computer User Society symposium in early November.

Another barrier for some engineers is getting an abstract accepted, since all papers are reviewed by published professionals, or "referees." "Making that jump from the inside to the outside is hard without some middle ground to test your ideas. Some companies active in publishing have internal review systems where your co-workers give you some input and encouragement. It would be nice if Allen-Bradley had a similar system of internal publishing," Linton suggested. Bomba, Fitzpatrick, and others involved in technical writing here, concurred.

"Industry does not make the 'publish or perish' threat that is prevalent in the academic world. At A-B we really encourage our employees to contribute to our industry: to get involved," said Bomba.

Being recognized in your field builds self-confidence and generates a sense of pride and accomplishment. Writing technical papers is a good way to let "others know you know."

none

None may be considered either a singular or plural **pronoun,** depending on the context.

EXAMPLES *None* of the material *has* been ordered. (always singular with a singular noun—in this case, "material")

None of the clients *has* been called yet. (singular even with reference to a plural noun [*clients*] if the intended emphasis is on the idea of *not one*)

None of the clients *have* been called yet. (plural with a plural noun)

For **emphasis,** substitute *no one* or *not one* for *none* and use a singular **verb.**

EXAMPLE I paid the full retail price for three of your firm's machines, *no one* of which was worth the money.

(See also **agreement.**)

nonstandard English (see **English, varieties of**)

nor/or

Nor always follows *neither* in sentences with continuing negation.

EXAMPLE They will *neither* support *nor* approve the plan.

Likewise, *or* follows *either* in sentences.

EXAMPLE The firm will accept *either* a short-term *or* a long-term loan.

Two or more singular **subjects** joined by *or* or *nor* usually take a singular **verb.** But when one subject is singular and one is plural, the verb agrees with the subject nearer to it.

EXAMPLES *Neither* the manager *nor* the secretary *was* happy with the new filing system. (singular)
Neither the manager *nor* the secretaries *were* happy with the new filing system. (plural)
Neither the secretaries *nor* the manager *was* happy with the new filing system. (singular)

(See also **correlative conjunctions.**)

notable/noticeable

Notable, meaning "worthy of notice," is sometimes confused with *noticeable,* meaning "readily observed."

EXAMPLES His accomplishments are *notable.*
The construction crew is making *noticeable* progress on the new building.

note taking

The purpose of note taking is to summarize and record the information that you extract from your **research** material. The great challenge in taking notes is to condense another's thoughts in your own words without distorting the original thinking. Meeting this challenge involves careful reading on your part, because to compress someone else's ideas accurately, you must first understand them.

When taking notes on abstract ideas, as opposed to factual data, be careful not to sacrifice **clarity** for brevity. You can be brief—if you are accurate—with statistics, but notes expressing concepts can lose their meaning if they are too brief. The critical test of a note is whether after a week has passed, you will still know what the note means and from it be able to recall the significant ideas of the passage. If you are in doubt about whether or not to take a note, take it—it is much easier to discard a note you don't need than to find the source again if it is needed.

As you extract information, be guided by the **objective** of your writing and by what you know about your **readers.** (Who are they? How much do they know about your subject? What are their needs?)

Resist the temptation to copy your source word for word as you take notes. Paraphrase the author's idea or concept in your own words. But don't just change a few words in the original passage; you will be guilty of **plagiarism.** On occasion, when your source concisely sums up a great deal of information or points to a trend or development important to your subject, you are justified in quoting it verbatim and then incorporating that **quotation** into your paper. As a general rule, you will rarely need to quote anything longer than a **paragraph.** (See also **paraphrasing**.)

If a note is copied word for word from your source, be certain to enclose it in **quotation marks** so that you will know later that it is a direct quotation. In your finished writing, be certain to give the source of your quotation; otherwise, you will be guilty of plagiarism. (See also **documenting sources.**)

The mechanics of note taking are simple and, if followed conscientiously, can save you much unnecessary work. The following guidelines may help you take notes more effectively:

1. First, don't try to write down everything. Pick out the most important ideas and concepts to record as notes.
2. Create your own shorthand. Common words can be indicated by symbols. For example, you can use & for *and,* + for *plus,* w for *with,* and so on.
3. Use numbers (5, 10, 20) for numerical terms instead of writing *five, ten,* and *twenty*.
4. Leave out vowels when you can do so and still keep the word recognizable, as in vwl for *vowel*.
5. Be sure to record all vital names, dates, and definitions.
6. Mark any notes you need to question. You might put an asterisk beside any note you don't fully understand or think you might need to pursue further.
7. Check your notes for accuracy against the printed material before leaving it.

Consider the information in the following paragraph:

> Long before the existence of bacteria was suspected, techniques were in use for combatting their influence in, for instance, the decomposition of meat. Salt and heat were known to be effective and these do in fact kill bacteria or prevent them from multiplying. Salt acts by the osmotic effect of extracting water from the bacterial cell fluid. Bacteria are less easily destroyed by osmotic action than are animal cells because their cell walls are constructed in a totally different way, which makes them very much less permeable.
>
> —Roger James, *Understanding Medicine* (London: Penguin Books, 1970), p. 103

The paragraph says essentially three things:

1. Before the discovery of bacteria, salt and heat were used in combatting bacteria.
2. Salt kills bacteria by extracting water from their cells by osmosis, hence its use in curing meat.
3. Bacteria are less affected by the osmotic effect of salt than are animal cells because bacterial cell walls are less permeable.

If your readers' needs and your objective involved tracing the origin of the bacterial theory of disease, you might want to note that salt was traditionally used to kill bacteria long before people realized what caused meat to spoil, though it might not be necessary to your topic to say anything about the relative permeability of bacterial cell walls.

So that you will give proper credit when you incorporate your notes into your writing, be sure to include the following with the first note taken from a book: author, title, publisher, place and date of publication, and page number. (On subsequent notes from the same book, you will need to include only the author and page number.)

Figures 1 and 2 are examples of a first and a subsequent note taken from the paragraph on the use of salt as a preservative.

If you do not have time to take careful notes in the library, you may want to check out a book or books to review at home. If you need to take more information from periodical articles (which cannot be checked out) than can be conveniently committed to notes, you may want to make a photographic copy at the library to take home for fuller evaluation. (See also **library research.**)

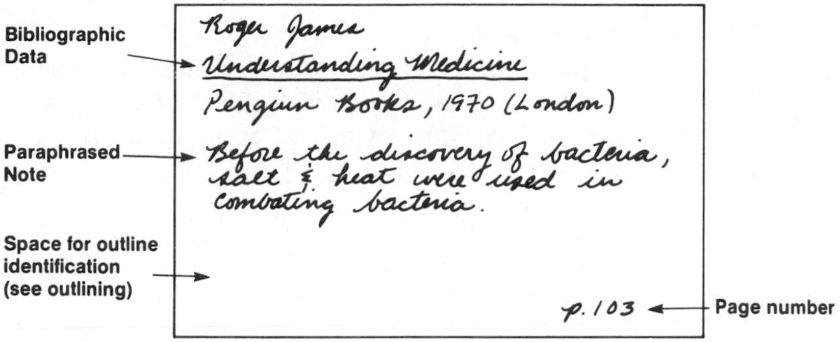

Bibliographic Data
Roger James
Understanding Medicine
Penguin Books, 1970 (London)
Paraphrased Note
Before the discovery of bacteria, salt & heat were used in combating bacteria.
Space for outline identification (see outlining)
p. 103
Page number

Figure 1 *First Note from a Source*

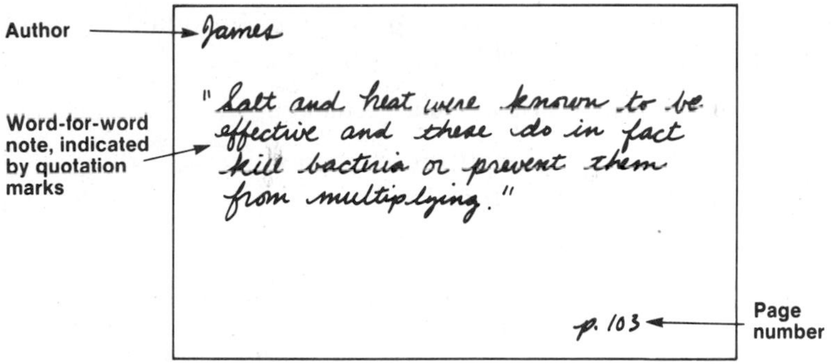

Author
James
Word-for-word note, indicated by quotation marks
"Salt and heat were known to be effective and these do in fact kill bacteria or prevent them from multiplying."
p. 103
Page number

Figure 2 *Subsequent Note from a Source*

noun clauses (see **clauses**)

nouns

A *noun* names a person, place, thing, concept, action, or quality. The two basic types of nouns are **proper nouns** and **common nouns.**

PROPER NOUNS

Proper nouns name specific persons, places, things, concepts, actions, or qualities. They are usually capitalized.

EXAMPLES Abraham Lincoln, New York, U.S. Army, Nobel Prize, Montana, Independence Day, Amazon River, Butler County, Magna Carta, June, Colby College

COMMON NOUNS

Common nouns name general classes or categories of persons, places, things, concepts, actions, or qualities. Common nouns include all types of nouns except proper nouns. Some nouns (turkey/Turkey) may be both common and proper.

EXAMPLES human, college, knife, bolt, string, faith, copper

Abstract nouns are common nouns that refer to something intangible that cannot be discerned by the five senses.

EXAMPLES love, loyalty, pride, valor, peace, devotion, harmony

Collective nouns are common nouns that indicate a group or collection of persons, places, things, concepts, actions, or qualities. They are plural in meaning but singular in form.

EXAMPLES audience, jury, brigade, staff, committee

Concrete nouns are common nouns used to identify those things that can be discerned by the five senses.

EXAMPLES house, carrot, ice, tar, straw, grease

A count noun is a type of concrete noun that identifies things that can be separated into countablc units.

EXAMPLES desks, chisels, envelopes, engines, pencils
There were four *calculators* in the office.

(See also **fewer/less** and **amount/number.**)

A mass noun is a type of concrete noun that identifies things that comprise a mass, rather than individual units, and cannot be separated into countable units.

EXAMPLES electricity, water, sand, wood, air, uranium, gold, oil, wheat, cement
The price of *silver* increased this year.

(See also **fewer/less** and **amount/number.**)

NOUN FUNCTION

Nouns may function as subjects of **verbs,** as objects of verbs and **prepositions,** as **complements,** or as **appositives.**

EXAMPLES The *metal* bent as *pressure* was applied to it. (subjects)
The bricklayer cemented the *blocks* efficiently. (direct object)
The company awarded our *department* a plaque for safety. (indirect object)
The event occurred within the *year.* (object of a preposition)
An equestrian is a *horseman.* (subjective complement)
We elected the sales manager *chairman.* (objective complement)
George Thomas, the *treasurer,* gave his report last. (appositive)

Words usually used as nouns may also be used as **adjectives** and **adverbs.**

EXAMPLES It is *company* policy. (adjective)
He went *home.* (adverb)

USING NOUNS

Forming the Possessive. Nouns form the possessive most often by adding *'s* to the names of living things and by adding an *of* **phrase** to the names of inanimate objects.

EXAMPLES The *chairman's* statement was forceful.
The keys *of the typewriter* were sticking.

However, either form may be used.

EXAMPLES The *table's* mahogany finish was scratched.
Two personal friends *of the chairman* were on the committee.

Plural nouns ending in *s* need only to add an **apostrophe** to form the **possessive case.**

EXAMPLE The *architects'* design manual contains many illustrations.

Plural nouns that do not end in *s* require both the apostrophe and the *s.*

EXAMPLE The installation of the plumbing is finished except in the *men's* room.

With group words and compound nouns, add the *'s* to the last noun.

EXAMPLES The *chairman of the board's* report was distributed.
My *son-in-law's* address was on the envelope.

To show individual possession with coordinate nouns, use the possessive with both.

EXAMPLES Both the *Senate's* and *House's* galleries were packed for the hearings.
Mary's and *John's* presentations were the most effective.

To show joint possession with coordinate nouns, use the possessive with only the last.

EXAMPLES The *Senate and House's* joint committee worked out a satisfactory compromise.
Mary and John's presentation was the most effective.

Forming the Plural. Most nouns form the plural by adding *s*.

EXAMPLE *Dolphins* are capable of communicating with people.

Those ending in *s, z, x, ch,* and *sh* form the plural by adding *es*.

EXAMPLES How many size *sixes* did we produce last month?
The letter was sent to all the *churches*.
Technology should not inhibit our individuality; it should fulfill our *wishes*.

Those ending in a consonant plus *y* form the *plural* by changing *y* to *ies*.

EXAMPLE The store advertises prompt delivery but places a limit on the number of *deliveries* in one day.

Some nouns ending in *o* add *es* to form the plural, but others add only *s*.

EXAMPLES One tomato plant produced twelve *tomatoes*.
We installed two *dynamos* in the plant.

Some nouns ending in *f* or *fe* add *s* to form the plural; others change the *f* or *fe* to *ves*.

EXAMPLES cliff/cliffs, fife/fifes, knife/knives

Some nouns require an internal change to form the plural.

EXAMPLES woman/women, man/men, mouse/mice, goose/geese

Some nouns do not change in the plural form.

EXAMPLE *Fish* swam lazily in the clear brook while a few wild *deer* mingled with the *sheep* in a nearby meadow.

Compound nouns form the plural in the main word.

EXAMPLES sons-in-law, high schools

Compound nouns written as one word add *s* to the end.

EXAMPLE Use seven *tablespoonfuls* of freshly ground coffee to make seven cups of coffee.

If in doubt about the plural form of a word, look up the word in a good **dictionary.** Most dictionaries give the plural form if it is made in any way other than by adding *s* or *es*.

nowhere near

The **phrase** *nowhere near* is colloquial and should be avoided in writing.

CHANGE His ability is *nowhere near* Jim's.
TO His ability does *not approach* Jim's.
OR His ability is *not comparable to* Jim's.
OR His ability is *far inferior* to Jim's.

number

Number is the grammatical property of **nouns, pronouns,** and **verbs** that signifies whether one thing (singular) or more than one (plural) is being referred to. Nouns normally form the plural by simply adding *s* or *es* to their singular forms.

EXAMPLES Many new business *ventures* have failed in the past ten *years*.
Partners in successful *businesses* are not always personal *friends*.

But some nouns require an internal change to form the plural.

EXAMPLES woman/women, man/men, goose/geese, mouse/mice

All pronouns except *you* change internally to form the plural.

EXAMPLES I/we
he, she, it/they

By adding an *s* or *es,* most verbs show the singular of the third **person,** present **tense,** indicative **mood.**

EXAMPLES he *stands,* she *works,* it *goes*

The verb *be* normally changes form to indicate the plural.

EXAMPLES I *am* ready to begin work. (singular)
We *are* ready to begin work. (plural)

(See also **agreement.**)

numbers

The general rule is to write numbers from zero to ten as words and numbers above ten as figures. There are, however, a number of exceptions.

One exception is that page numbers of books, as well as figure and **table** numbers, are expressed as figures.

EXAMPLES Figure 4 on page 9 and Table 3 on page 7 provide pertinent information.

Another exception is that units of measurement are expressed in figures.

EXAMPLES 3 miles, 45 cubic feet, 9 meters, 27 cubic centimeters, 4 picas

Numbers that begin a sentence should always be spelled out, even if they would otherwise be written as figures.

EXAMPLE One hundred and fifty people attended the meeting.

If spelling out such a number seems awkward, rewrite the sentence so that the number does not appear at the beginning.

CHANGE Two hundred seventy-three defective products were returned last month.
TO Last month, 273 defective products were returned.

When several numbers appear in the same sentence or **paragraph,** they should be expressed alike, regardless of other rules and guidelines.

EXAMPLE The company owned 150 trucks, employed 271 people, and rented 7 warehouses.

When numbers appear run together in the same **phrase,** write one as a figure and the other as a word.

CHANGE The order was for *12* 6-inch pipes.
TO The order was for *twelve* 6-inch pipes.

Approximate numbers are normally spelled out.

EXAMPLE More than two hundred people attended the conference.

Percentages are normally given as figures, and the word *percent* is written out, except when the number is in a table.

EXAMPLE Approximately 85 percent of the area has been seeded with ponderosa pine.

In typed manuscript, page numbers are written as figures, but chapter or volume numbers may appear either way.

EXAMPLES page 37
Chapter 2 or Chapter Two
Volume 1 or Volume One

The plural of a written number is formed by adding *s* or *es* or by dropping *y* and adding *ies,* depending on the last letter, just as the plural of any other noun is formed. (See **nouns.**)

EXAMPLES sixes, elevens, twenties

The plural of a figure may be written with *s* alone or with *'s*.

EXAMPLES 5s, 12s/5's, 12's

Do not follow a word representing a number with a figure in **parentheses** representing the same number.

CHANGE Send five (5) copies of the report.
TO Send five copies of the report.

CARDINAL AND ORDINAL NUMBERS

A cardinal number expresses a specific quantity.

EXAMPLES *one* pencil, *two* typewriters, *three* airplanes

An ordinal number expresses degree or sequence.

EXAMPLES *first* quarter, *second* edition, *third* degree

In most writing, an ordinal number should be spelled out only if it is a single word (*tenth, 312th*). Ordinal numbers can also function as **adverbs** (John arrived *first*). (See also **first/firstly.**)

DATES

The year and day of the month should be written as figures. Dates are usually written in a month–day–year sequence, in which the year may or may not be followed by a comma.

EXAMPLES The August 26, 19-- issue of *Computer World* announced the voice-recognition system.
The August 26, 19--, issue of *Computer World* announced the voice-recognition system.

In the day–month–year sequence, commas are not used.

EXAMPLE The 26 August 19-- issue of *Computer World* announced the voice-recognition system.

The **slash** form of expressing dates (8/24/87) is used in informal writing only.

TIME

Hours and minutes are expressed as figures when a.m. and p.m. follow.

EXAMPLES 11:30 a.m., 7:30 p.m.

When not followed by a.m. or p.m., however, time should be spelled out.

EXAMPLES four o'clock, eleven o'clock

FRACTIONS

Fractions are expressed as figures when written with whole numbers.

EXAMPLES 27½ inches, 4¼ miles

Fractions are spelled out when they are expressed without a whole number.

EXAMPLES one-fourth, seven-eighths

Numbers with decimals are always written as figures.

EXAMPLE 5.21 meters

ADDRESSES

Numbered streets from one to ten should be spelled out except when space is at a premium.

EXAMPLE East Tenth Street

Building numbers are written as figures. The only exception is the building number *one.*

EXAMPLES 4862 East Monument Street
One East Tenth Street

Highways are written as figures.

EXAMPLES U.S. 70, Ohio 271, I 94

numeral adjectives

Numeral adjectives identify quantity, degree, or place in a sequence. They always modify count **nouns.** Numeral adjectives are divided into two subclasses: cardinal and ordinal.

A cardinal **adjective** expresses an exact quantity.

EXAMPLES *one* pencil, *two* typewriters, *three* airplanes

An ordinal adjective expresses degree or sequence.

EXAMPLES *first* quarter, *second* edition, *third* degree, *fourth* year

In most writing, an ordinal adjective should be spelled out if it is a single word (*tenth*) and written in figures if it is more than one word (*312th*). Ordinal numbers can also function as **adverbs.**

EXAMPLE John arrived *first.*

(See also **first/firstly.**)

O

objective

What do you want your **readers** to know or be able to do when they have read your finished writing project? When you have answered this question, you have determined the *objective,* or *purpose,* of your writing project. Too often, however, beginning writers state their objectives in broad terms that are of no practical value to them. Such an objective as "to explain the Model 6000 Accounting Machine" is too general to be of any real help. But "to explain how to operate a Model 6000 Accounting Machine" is a specific objective that will help keep the writer on the right track.

The writer's objective is rarely simply to "explain" something, although on occasion it may be. You must ask yourself, "Why do I need to explain it?" In answering this question, you may find, for example, that your objective is also to persuade your reader to change his or her attitude toward the thing you are explaining.

A writer for a company magazine who has been assigned to write an article on the firm's new computer installation, in answer to the question *what,* could state the objective as "to explain the way the computer will increase efficiency and ultimately benefit employees." In answer to the question *why,* the writer might state "to ease the fear of automation that has become evident in employees."

If you answer these two questions *exactly* and put your answers in writing as your stated objective, not only will your job be made easier but also you will be considerably more confident of ultimately reaching your goal. As a test of whether you have adequately formulated your objective, try to state it in a single sentence. If you find that you cannot, continue to formulate your objective until you *can* state it in a single sentence.

Even a specific objective is of no value, however, unless you keep it in mind as you work. Guard against losing sight of your objective as you become involved with the other steps of the writing process.

objective complements

An objective complement is a **noun** or **adjective** that completes the meaning of a sentence by revealing something about the direct **ob-**

ject of a transitive **verb**; it may either describe (adjective) or rename (noun) the direct object.

EXAMPLES We painted the house *white*. (adjective)
I like my coffee *hot*. (adjective)
They call her a *genius*. (noun)

The objective complement is one of the four kinds of **complements** that complete the subject–verb relationship; the others are the direct object, the indirect **object,** and the **subjective complement.**

objects

There are three kinds of objects: direct object, indirect object, and object of a **preposition.** All objects are **nouns** or noun equivalents (**pronoun, gerund, infinitive,** noun **phrase**, noun **clause**) and can be replaced by a pronoun in the objective **case.**

DIRECT OBJECTS

The direct object answers the question "what?" or "whom?" about a **verb** and its **subject.**

EXAMPLES John built a *business*. (noun)
I like *jogging*. (gerund)
I like *to jog*. (infinitive)
I like *it*. (pronoun)
I like *what I saw*. (noun clause)
Sheila designed a new *circuit*. (*Circuit,* the direct object, answers the question, "Sheila designed *what*?")
George telephoned the *chief engineer*. (*Chief engineer*, the direct object, answers the question, "George telephone *whom*?")

A verb whose meaning is completed by a direct object is called a transitive verb.

INDIRECT OBJECTS

An indirect object is a noun or noun equivalent that occurs with a direct object after certain kinds of transitive verbs, such as *give, wish, cause, tell,* and their **synonyms** or **antonyms.** The indirect object is usually a person or persons and answers the question "to whom or what?" or "for whom or what?" The indirect object always precedes the direct object.

EXAMPLES Wish *me* success.
It caused *him* pain.
Their attorney wrote *our firm* a follow-up letter.
Give *the car* a push.
We sent the *general manager* a full report. (*Report* is the direct object. The indirect object, *general manager,* answers the question, "We sent a full report *to whom*?")
The general manager gave the *report* careful consideration. (*Consideration* is the direct object. The indirect object, *report,* answers the question, "The general manager gave careful consideration *to what*?")
The purchasing department bought *Sheila* a new oscilloscope. (*Oscilloscope* is the direct object. The indirect object, *Sheila,* answers the question, "The purchasing department bought a new oscilloscope *for whom*?")

OBJECTS OF PREPOSITIONS

For a discussion of objects of prepositions, see **prepositional phrases.**

observance/observation

An *observance* is the "performance of a duty, custom, or law"; it is sometimes confused with *observation,* which is the "act of noticing or recording something."

EXAMPLES The *observance* of Veterans Day as a paid holiday varies from one organization to another.
The laboratory technician made careful *observations* during the experiment.

OK/okay

The expression *okay* (also spelled *OK* or *O.K.*) is common in informal writing but should be avoided in more formal **correspondence** and **reports.**

CHANGE Mr. Sturgess gave his *okay* to the project.
TO Mr. Sturgess *approved* the project.
CHANGE The solution is *okay* with me.
TO The solution is *acceptable* to me.

on account of

The **phrase** *on account of* should be avoided as a substitute for *because.*

CHANGE He felt that he had lost his job *on account of* the company's switch to automated equipment.

TO He felt that he had lost his job *because* the company switched to automated equipment.

on/onto

On is normally a **preposition** meaning "supported by," "attached to," or "located at."

EXAMPLE Install the telephone *on* the wall.

Onto implies movement to a position on or movement up and on.

EXAMPLE The union members surged *onto* the platform after their leader's defiant speech.

on the grounds that/of

The **phrase** *on the grounds that*—or *on the grounds of*—is a wordy substitute for *because.*

CHANGE She left *on the grounds that* the NASA position offered a higher salary.

TO She left *because* the NASA position offered a higher salary.

one

When used as an **indefinite pronoun,** *one* may help you avoid repeating a **noun.**

EXAMPLE We need a new plan, not an old *one.*

One is often redundant in **phrases** in which it restates the noun, and it may take the proper **emphasis** away from the **adjective.**

CHANGE The computer program was not a unique *one.*

TO The computer program was not unique.

One can also be used in place of a noun or **personal pronoun** in a statement such as the following:

EXAMPLE *One* cannot ignore *one's* physical condition.

In statements of this kind, to avoid the tedious **repetition** of *one,* it is permissible to use a third-person singular personal pronoun with *one* as the antecedent.

CHANGE *One* must do *one's* best to correct the injustices *one* sees.
TO *One* must do *his* best to correct the injustices *he* sees.
OR *One* must do *her* best to correct the injustices *she* sees.

If you use *one* in this manner, be careful not to shift back and forth between *one* and *you.* Using *one* in this way is formal and impersonal, and so in any but the most formal writing you are better advised to address your **reader** directly and personally as *you.*

CHANGE *One* cannot be too careful about planning for leisure time. The cost of *one's* equipment can, for example, force *one* to work more and therefore reduce *one's* leisure time.
TO *You* cannot be too careful about planning for leisure time. The cost of *your* equipment can, for example, force *you* to work more and thus reduce *your* leisure time.

(See also **point of view.**)

one of those . . . who

A **dependent clause** beginning with *who* or *that* and preceded by *one of those* takes a plural **verb.**

EXAMPLES She is *one of those* executives *who are* concerned about their writing.
This is *one of those* policies *that make* no sense when you examine them closely.

In the preceding two examples, *who* and *that* are subjects of dependent clauses and refer to plural antecedents (*executives* and *policies*) and thus take plural verbs, *are* and *make.*

Because people so often use singular verbs in such constructions in everyday speech and very informal writing, this rule confuses many writers. The principle behind the rule becomes clearer if *among* is substituted for *one of.* Compare the following examples with the preceding ones:

EXAMPLES She is *among* those executives *who are* concerned about their writing. (You would not write, "She is among those executives who *is* concerned . . .")

This is *among* those policies *that make* no sense when you examine them closely. (You would not write "This is among those policies that *makes* no sense . . . ")

Another way to clarify the principle behind the rule is to turn the sentence around. Compare the following examples with the examples given above:

EXAMPLES Of those executives who *are* concerned about their writing, she is one.

Of those policies that *make* no sense when you examine them closely, this is one.

When the sentences are seen turned around in this rather unnatural but perfectly logical way, you would not be likely to use *is* for *are* (in the first sentence) or *makes* for *make* (in the second sentence). It may also be helpful to contrast these sentences with others in which the singular verb is the correct form.

EXAMPLES She is one executive *who is* concerned about her writing. (*Who* refers to a singular antecedent, *executive,* and is therefore singular and takes a singular verb, *is.*)

This is one policy *that makes* no sense when it is examined closely. (*That* refers to a singular antecedent *policy,* and so requires a singular verb, *makes.*)

There is one exception to the plural-verb rule stated at the beginning of this entry: If the phrase *one of those* is preceded by *the only* (*the only one* . . .), the verb in the following dependent clause should be *singular:*

EXAMPLES She is *the only one* of those executives *who is* concerned about her writing. (The verb is singular because its subject, *who,* refers to a singular antecedent, *one.* If the sentence were reversed, it would read, "Of those executives, she is the only *one who is* concerned about her writing.")

This is *the only one* of those policies *that makes* no sense when you examine it closely. (If the sentence were reversed, it would read, "Of those policies, this is the only *one that makes* no sense when you examine it closely.")

Finally, it should be repeated that the plural-verb rule given at the beginning of this entry is often ignored in everyday speech and very informal writing. Indeed, in spoken English the singular form of the verb, though technically illogical, is probably far more common than the plural. In work-related writing, in which preciseness is crucial, you should use a plural verb after *one of those who* or *one of those that.*

only

In writing, the word *only* should be placed immediately before the word or **phrase** it modifies.

CHANGE We *only* lack financial backing; we have determination.
TO We lack *only* financial backing; we have determination.

A speaker can place *only* before the **verb** and avoid **ambiguity** by stressing the word being modified; in writing, only the correct placement of the word can ensure **clarity.** Incorrect placement of *only* can change the meaning of a sentence.

EXAMPLES *Only* he said that he was tired. (He alone said he was tired.)
He *only* said that he was tired. (He actually was not tired, although he said he was.)
He said *only* that he was tired. (He said nothing except that he was tired.)
He said that he was *only* tired. (He was nothing except tired.)

openings

If your **reader** is already familiar with your subject or if what you are writing is short, you may not need to begin your writing project with a full **introduction.** You may simply want to focus the reader's attention with a brief opening.

For many types of on-the-job writing, openings that simply get to the point are quite adequate, as shown in the following examples:

Correspondence

Mr. George T. Whittier
1720 Old Line Road
Thomasbury, WV 26401

Dear Mr. Whittier:

You will be happy to know that we have corrected the error in your bank balance. The new balance shows . . .

Progress Report Letter

William Chang, M.D.
Phelps Building
9003 Shaw Avenue
Parksville, MD 21221

Dear Dr. Chang:

To date, 18 of the 20 specimens you submitted for analysis have been examined. Our preliminary analysis indicates . . .

Longer Progress Report

PROGRESS REPORT ON REWIRING THE SPORTS ARENA

The rewiring program at the Sports Arena is continuing on schedule. Although the costs of certain equipment are higher than our original bid had indicated, we expect to complete the project without exceeding our budget, because the speed with which the project is being completed will save labor costs.

Work Completed

As of August 15, we have . . .

Memorandum

To: Jane T. Meyers, Chief Budget Manager
From: Charles Benson, Assistant to the Personnel Director
Date: June 12, 19--
Subject: Budget Estimates for Fiscal Year 19--

The personnel budget estimates for fiscal year 19-- are as follows: . . .

You may also use an intriguing or interesting opening, either by itself or in conjunction with an introduction, to stimulate your reader's interest. Such types of writing as **annual reports, journal articles,** and **newsletter articles** often require an interesting opening because you do not have a captive audience for such articles and **reports.** Such openings have two purposes: to indicate the subject and to catch the interest of the reader. Several types of openings achieve both purposes. The opening should be natural, not forced; an obviously "tacked on" opening will only puzzle your reader, who will be unable to establish any meaningful connection between the opening and the body of your article or report.

STATEMENT OF THE PROBLEM

One way to give the reader the perspective of your report is to present a brief account of the problem that led to the study or project being reported.

EXAMPLE Several weeks ago a brewmaster noticed a discoloration in the grain supplied by Acme Farms, Inc. He immediately reported his discovery to his supervisor. After an intensive investigation, we found that Acme . . .

However, if the reader is familiar with the problem or if the particular problem has been solved for a long period, a problem opening is likely to be boring.

DEFINITION

Although a definition can be useful as an opening, do not define something with which the reader is familiar or provide a definition that is obviously a contrived opening (such as "Webster defines business as . . ."). A definition should be used as an opening only if it offers insight into what follows.

EXAMPLE *Risk* is a loosely defined term. It is used here in the sense of physical risk as a qualitative combination of the probability of an event and the severity of the consequences of that event. *Risk assessment* is the process of estimating the probabilities and consequences of events and of establishing the accuracy of these estimates. Another necessary term is *risk appraisal*. This goes far beyond assessment and involves judgment about people's perception of risk and their reactions to this perception; it extends to the final process of making decisions. Risk appraisal is thus an essential part of the work of a licensing and regulatory body in formulating safety policy and applying this policy to individual plants. It is part of the everyday work of inspectors. However, only in recent years have attempts been made to quantify the appraisal aspects of risk.

—H.J. Dunster and W. Vanek, "The Assessment of Risk—Its Values and Limitations," *Nuclear Engineering International* 24 (August 1979), p. 23.

INTERESTING DETAIL

Often an interesting detail of your subject can be used to gain the readers' attention and arouse their interest. This requires, of course, that you be aware of your readers' interests. The following opening, for example, might be especially interesting to members of a personnel department.

EXAMPLE The small number of graduates applying for jobs at Acme Corporation is disappointing, particularly after many years of steadily increasing numbers of applicants. There are, I believe, several reasons for this development. . . .

Sometimes it is possible to open with an interesting statistic.

EXAMPLE From asbestos sheeting to zinc castings, from a chemical analysis of the water in Lake Maracaibo (to determine its suitability for use in steam injection units) to pistol blanks (for use in testing power charges), the purchasing department attends to the company's material needs. Approximately 15,000 requisitions, each containing from one to fourteen separate items, are processed each year by this department. Every item or service that is bought . . .

ANECDOTE

An anecdote can also be used to catch your reader's attention and interest.

EXAMPLE In his poem "The Calf Path," Sam Walter Foss tells of a wandering, wobbly calf trying to find its way home at night through the lonesome woods. It made a crooked path, which was taken up the next day by a lone dog that passed that way. Then "a bellwether sheep pursued the trail over vale and steep, drawing behind him the flock, too, as all good bellwethers do." At last the path became a country road; then a lane that bent and turned and turned again. The lane became a village street, and at last the main highway of a flourishing city. The poet ends by saying, "A hundred thousand men were led by a calf, three centuries dead."

Many companies today follow a "calf path" because they react to events rather than planning. . . .

BACKGROUND

The background or history of a certain subject may be quite interesting and may even put the subject in perspective for your reader.

Consider the following example from a **newsletter article** describing the process of oil drilling:

> EXAMPLE From the bamboo poles the Chinese used when the pyramids were young to today's giant rigs drilling in a hundred feet of water, there has been a lot of progress in the search for oil. But whether four thousands years ago or today, in ancient China or a modern city, in twenty fathoms of water or on top of a mountain, the object of drilling is and has always been the same—to manufacture a hole in the ground, inch by inch. The hole may be either for a development well . . .

This type of opening is easily overdone, however; use it only if the background information is of some value to your reader. Never use it just as a way to get started.

QUOTATION

Occasionally, you can use a **quotation** to stimulate interest in your subject. To be effective, however, the quotation must be pertinent—not some loosely related remark selected from a book of quotations simply because it was listed under the subject heading that fits your writing project. Often a good quotation to use is one that predicts a new trend or development.

> EXAMPLE Richard Smith, president of P. R. Smith Corporation, recently said, "I believe that the Photon projector will revolutionize our industry." His statement represents a growing feeling among corporate . . .

OBJECTIVE

In reporting on a project or activity of some kind, you may wish to open with a statement of the project or activity's objective. Such an opening gives the reader a basis for judging the actual results as they are presented.

> EXAMPLE The primary objective of the project was to develop new techniques to measure heat transfer in a three-phase system. Our first step was to investigate . . .

SUMMARY

You can provide a summary opening by greatly compressing the results, conclusions, or recommendations of your article or report.

Do not start a summary, however, by writing "This report summarizes . . ."

CHANGE This report summarizes the advantages offered by the photon as a means of examining the structural features of the atom. The photon is a specially designed laser used for examining . . .

TO As a means of examining the structure of the atom, the photon offers several advantages. Since the photon is especially designed for examining . . .

FORECAST

Sometimes you can use a forecast of a new development or trend to arouse the reader's interest.

EXAMPLE In the very near future, we may be able to call our local library and have a videotape of *Hamlet* replayed on our wall television. This project and others are now being developed at Acme Industries. . . .

SCOPE

At times you may want to present the **scope** of your article or report in your opening. By providing the parameters of your material—the limitations of the subject or the amount of detail to be presented—you enable your readers to determine whether they want or need to read your article or report. Be on guard against letting this kind of opening become a narrated **table of contents** or an **abstract;** if either of those is appropriate, include it—but not as an artificial opening. The scope type of opening is often useful for small writing projects in which an abstract or table of contents would be unnecessary.

EXAMPLE This pamphlet provides a review of the requirements for obtaining an FAA pilot's license. It is not designed as a textbook to prepare you to take the examination itself; rather, it gives you an idea of the steps you need to take and the costs involved. . . .

oral presentations

Preparing your presentation is much like preparing to write; that is, you must determine your purpose, analyze your audience, develop your topic, and prepare an outline.

PURPOSE OF THE PRESENTATION

Summarize your specific message succinctly—in an single sentence if possible: "Tobacco kills!" or "We must reduce costs!" In order to summarize your message in a single sentence, you must know exactly what the purpose of your presentation is. Is it to tell the audience something they didn't know? Is it to update their knowledge? To ask them to take a specific action? To ask them for money?

AUDIENCE ANALYSIS

To communicate effectively with your listeners, you must first know who they are. Are they coworkers? Customers? Fellow members of a professional organization? You should consider their age, their sex, their occupation, their eduction—because who they are largely determines what they are interested in, and you can appeal effectively to an audience only by relating to their desires, needs, and anxieties.

Try to put yourself in your listeners' place. What do they already know about your topic? What do they need to know in order for your presentation to achieve its objective? What is your listeners' attitude toward your subject? Resistance? Apathy? Curiosity? Interest? Must you convince your listeners? Must you motivate them? Your listeners' attitude toward your topic should largely determine your approach. What is your audience's attitude toward *you*? The image you present and the prestige you bring with you will greatly influence their acceptance of you.

ORGANIZATION

Speaking well is mainly thinking logically and then organizing your talk in the same way. Logical reasoning is not only powerful and persuasive but is also a necessary ingredient for any speech. Your presentation will lack coherence without it.

Organizing your presentation consists of collecting the necessary data, developing the topic, and **outlining** the presentation.

Collecting the Data. You must be the master of your subject if you expect to be taken seriously. Therefore, it is a good idea to gather more information than you will use in your presentation, just to build your own confidence in your mastery of your subject.

In collecting data, use all the sources of information you can think of, including brainstorming, written material, interviewing others, and your own background (see **research**). Be sure of your facts. Do not be guilty of unfounded assumptions or unproved assertions.

Developing the Topic. Make certain that your **method of development** is appropriate to both your audience and your objective and is based on your listeners' needs.

For work-related presentations, one of two plans can be used to cover most situations:

1. State your facts.
   Argue from them.
   Appeal for action.
   or
2. Show that something is wrong.
   Show how to fix it.

Several other methods of development are also good for oral presentations.

The *problem–solution method of development* describes and analyzes the problem, then presents the criteria for evaluating possible solutions, and finally explains the advantages and disadvantages of each solution.

The **comparison method of development** sets your idea side by side with others and lets the audience compare them.

The **cause-and-effect method of development** explains why something happened or why you predict something is going to happen.

The **specific-to-general method of development** begins with specific details—such as statistics, expert opinions, and examples—and then uses them to build to a general conclusion.

The **general-to-specific method of development** presents a general statement and then follows it with such supporting details as statistics, expert opinions, and examples.

The **increasing-order-of-importance method of development** develops a topic from the least important to the most important point and thus leaves the audience with the most important points fresh in their minds at the end.

Structuring the Presentation. After selecting the most appropriate method of development, you must prepare an outline that reflects that method. If you used 3- × 5-inch note cards when gathering your data, lay them out on a desk top and group them by topic to organize your outline. When you have grouped and sequenced your note cards correctly, number them. However, don't consider this sequence final. You may find yourself renumbering your notes several times, even after you have typed your outline.

The Opening. Your opening is the most important part of your speech. It should not only catch the interest of your audience but also stimulate curiosity and impress your listeners with the importance of your topic.

For an interesting opening, use your personal experiences and your reading. Remember, however, that your opening must always be tied to your theme. The following are ways you might open a presentation:

- An unusual, startling statement

EXAMPLE "'Star Trek's' 'beam-me-up' technology may be the most common mode of transportation in twenty years."

- A rhetorical question

EXAMPLE "Will colonizing outer space ever be practical?"

- A dramatic story

EXAMPLE Terry Fox's attempt to run across Canada to raise money for cancer research.

- A personal experience
- A quotation from a famous person

EXAMPLE "We have nothing to fear but fear itself."
—Franklin D. Roosevelt

- A historical event

EXAMPLE "Do you remember where you were when Neil Armstrong first set foot on the moon?"

- A reference to a current news story
- A reference from literature or the Bible

EXAMPLE "We need the kind of frankness in this project that Samuel Pepys displayed in his diary back in the seventeenth century."

You may even use your opening to tell your audience why what you are about to say is important to them.

The Body. Begin the body of your presentation with a statement of your theme. (When delivering your presentation, pause both before and after this statement, to emphasize its importance.) Then relate all the evidence and proof necessary to support your theme statement. The strength of your proof will sell your ideas. Use analogies, stories, testimony, logic, statistics, and examples to support your theme.

Stories can sell your idea. Abraham Lincoln and Jesus Christ used stories to great effect. However, your story must relate directly to your theme, and it must be short. Avoid old stories that the audience has probably heard. Avoid off-color stories at all costs. And don't wait for a response when you have finished your story.

Testimony can provide authority and support for your position from people who have gone on record as agreeing with you. Testimony from an expert is especially helpful for questions of fact, and quotations from a prestigious book or person will help you gain acceptance from the audience. Be sure, however, that the person you are quoting has a right to speak on the subject.

Statistics, properly presented, can be interesting. But they should be presented in a frame of reference to which the reader can relate. "The ship was the length of three football fields" is much better, for example, than just the dry number of feet. You could say that the Astrodome is 210 feet high or that you could put a 20-story building inside it. Do not include endless statistics, however, no matter how interesting they may seem.

Analogies and *similes* can help you communicate by relating the unfamiliar to the familiar. They are comparisons of things that are not really alike but have one thing in common. "Pollution is to the environment what cancer is to the body."

The Closing. Plan your closing just as you planned your opening. Your closing is almost as important as your opening, because what you say in your closing is what your audience is most likely to remember. Therefore, make certain that your closing is tied directly to the purpose of your presentation. Your closing may also be designed to create a specific mood or stimulate a specific action.

If you want your closing to get action, tell your audience what you want them to do! If your objective is not to get your audience to do something, you might close with a summary or a recommendation, or by asking a rhetorical question or asking for their cooperation.

Whatever the objective of your closing, always try to end on a high note that emphasizes your central theme. And always make your closing as interesting as possible, since it is what is most likely to be remembered.

MAKING THE PRESENTATION

The key to effective delivery is practice. Public speaking isn't natural for anyone, but everyone can learn to do it well with enough practice.

Practice consists of two parts: (1) fixing your outline in your mind and (2) polishing your delivery. Memorize only your opening and your closing, plus perhaps a few important phrases. Read your outline over and over again until you are thoroughly familiar with it. Then try to make your presentation without notes. If you forget a part of the outline, go on as best you can until you have finished the whole talk. Repeat this process as many times as necessary. Rehearse with your outline until you know exactly how you want to move from one idea to another. Become so familiar with your outline that just a skeleton of it will guide you as you make the actual presentation.

To polish your delivery, rehearse on your feet and out loud. Don't be inhibited during rehearsal: force yourself to be dynamic, even though there is no live audience in front of you. Practice your gestures as well as your words and delivery.

Evaluate your voice for projection, inflection, and enthusiasm. Listen to yourself on a tape recorder. Practice a conversational style of presentation. Also listen for "uh," "y'know," and other nervous mannerisms.

In your practice sessions, vary your phraseology, gestures, and vocal inflections. Don't let your talk become canned. Use whatever words and gestures are natural. If you practice, you will have no problem with the gestures and words you want to use—they will come naturally.

If possible, arrange one practice session in the room where the speech is to be delivered so you can get the feel of the room and the equipment.

Last-minute rehearsal is generally not a good idea; in fact, it may actually work against you by making you more tense. It is generally better to try to relax immediately before your presentation.

EQUIPMENT AND ROOM

When you are to make your presentation, arrive well ahead of time to check out the room and equipment. Confirm in advance that the needed equipment is available. Will you need a chalkboard? A screen? A projector? A flipchart? If so, make sure they are available.

Consider the size of the room. Tailor everything to scale. Make sure that your visual aids can be seen from all parts of the room.

Check your slides (is it a rear or forward projector?). Check the location of wall plugs and light switches. Check the room for acoustics; if they are bad, you will know you must speak louder.

VISUAL AIDS

Well-planned visual aids can add interest and emphasis to your presentation. They can clarify and simplify your message because they communicate clearly, quickly, and vividly: we learn much better by seeing than by listening. Visual aids present an idea in a form that the audience can understand more quickly. But use visual aids *only* if they clarify or emphasize a point. Don't try to use them to flesh out a skimpy or weak presentation.

Keep your visual aids simple and uncluttered. Keep the amount of information on each to a minimum (usually one idea to a visual aid). When the visual aid is text, space it so the audience can easily read it.

Slide projectors and *overhead projectors* are easy to use. As you prepare your presentation, decide which slides you will use and where each one will go. Have them made up far enough in advance that you can rehearse with them. A line graph in a slide or overhead is good for showing growth and decline over a specified period, such as the growth of the U.S. population since 1790. Always put a title over a line graph, be sure to label both legs of the graph with horizontal labels, and make sure that the lines are dark enough to be seen from the back of the room. A bar graph in a slide or overhead can be either vertical or horizontal. Be sure to put a title over the bar graph. Label all bars horizontally (putting the label inside the bar helps make identification easy). Use different colors in the contrasting bars. Make sure everything can be read from the back of the room. (See **illustrations**.)

Flip charts are good for small- and medium-sized audiences. They give you good control of what you want the audience to see because you flip over the chart only when you need it. You can even draw illustrations on the flip chart ahead of time with light pencil lines that the audience can't see and then fill in the heavy lines with a felt-tip pen during your presentation (sketching it during the presentation helps hold the audience's attention).

A *chalkboard* is good with small- and medium-sized audiences, as it is easy to use and to control. It gives you complete flexibility to decide what goes on, what comes off, and when. It adds animation to your presentation because you must physically put the information on the board. If your information is too long to put on the board conveniently while you speak, put part of it on the board ahead of time. A disadvantage of using a chalkboard is that you must interrupt your presentation to write on it.

A *poster* can be used for medium- to large-sized audiences of up to 100 people. The poster must be large enough for your audience to see clearly what it contains; a common size is 2 × 3 feet. Prepare the poster ahead of time, and make your lettering large enough to be seen comfortably from the back of the room. Felt-tip markers are good writing instruments for posters. Put the poster on the easel only when you are ready to use it; otherwise, it may distract your audience.

Practice with your visual aids. During the practice sessions, point to the visual aid with both gestures and words. Practice talking to your audience, however, rather than to the visual aid. Be sure you do not block it with your body. Cover your visual aids both before and after discussing them so they won't distract your audience.

When using slides or overheads, give your audience time to absorb the information on the slide or overhead before you comment on it. Don't talk while you are changing overheads or doing something physical. Pause while you do it.

DELIVERY

The speaker's delivery can make the subject interesting or dull to the audience. Never open your presentation with an apology; open forcefully. Don't tell the audience you are a poor speaker; let *them* decide. Don't explain that you didn't have time to prepare; if you didn't, you should not have agreed to speak. If you stumble and apologize, your listeners will begin to question your knowledge of the subject.

Delivery is both audible and visual. Step briskly up to the lectern and take a deep breath. Look directly at your audience and remain silent until you have absolute quiet.

A strong voice, confident manner, and take-charge attitude will impress your audience favorably. If you seem reluctant to take charge, the audience will deny you the authority they would otherwise have granted you as the speaker.

Your first sentence should be forceful and attention getting.

> CHANGE "I would like to speak to you today about the problems we are experiencing with the new payroll management system."
>
> TO "If we don't meet our current schedule, the new payroll management system will be in jeopardy."

Talk with your audience in a conversational way—don't lecture to them. Make your message personal; don't be afraid of the pronoun *you.* Work at getting expressiveness and enthusiasm into your voice.

To emphasize a point, say it—then pause to let its significance register—and repeat the point. The "pregnant silence" is a great way to get your listeners' attention.

Let your notes guide you. If you lose your place, don't panic—just find it again. Remember that your audience is made up of human beings just like you.

Be sure not to pass out a long handout at the beginning of your presentation. If you do, the audience will read it instead of listening to you.

Nervousness. It is natural to be apprehensive, and that feeling will probably never go away. Stage fright shouldn't worry you. Indeed, perhaps the lack of it should, because it can be a helpful stimulant. But fear is a result of a lack of confidence, and the best defense against butterflies is to be as well prepared as possible. You can expect to be nervous, but if you know what you are going to say and how you are going to say it, you will probably relax as you start to speak.

You can gain strength and confidence simply by standing up and meeting the situation face to face. To avoid stage fright, try to relax both your body and your mind. Relax your body by whatever means works for you: meditation, isometric exercises, deep breathing, progressive relaxation techniques, and the like. Relax your mind by thinking positive thoughts. Avoid attempting to evaluate your ability to speak. Conjure up visual images of what you want to look like and do, and then assume that you will achieve the desired behavior. Give yourself a pep talk: "My subject is important. I am ready. My listeners have come here to hear what I have to say."

When you stand up to speak, take a long, deep breath. It will help calm you. Don't jingle coins in your pocket, fidget with notes or pencils, pull your ear, or run your fingers through your hair. If you can't keep your hands still, put them behind you and keep them there.

Don't mistake the audience's silence for hostility. The audience is on your side. They are eager for your information, not your scalp.

Enthusiasm. You must be convinced that you have something important to tell your listeners and then convey this conviction to them. If you are deeply interested in your subject and enthusiastic about communicating it to your listeners, you will automatically transmit that to your audience. Your presentation is most likely to be on a job-related topic, so you can speak with knowledge and conviction. Reach out and figuratively shake the hands of your listeners.

Eye Contact. It is important to establish rapport with the audience, and eye contact is the best way to do it. Pick out a few people in different parts of the audience and alternately speak directly to each one of them. Maintain eye contact with the good listeners—those who nod and react. In other words, hold a conversation with them. Look at your listeners as you would if you were actually talking to them. And look into the eyes of people—not over their heads. Maintaining eye contact creates an appearance of honesty, dependability, and security. Looking into someone's eyes is equivalent to a handshake; it tells your audience you are someone they can believe and trust.

If you memorize nothing else, memorize your opening and your closing so you can maintain eye contact with your audience during those important parts of your presentation.

Gestures. Use gestures to clarify, amplify, emphasize, demonstrate, and, finally, to draw attention to your visual aids. In other words, make your hands and body support your speaking. If you don't put your hands in your pockets, you will instinctively use them to gesture. But gesture just as naturally as you would in conversation.

Humor and Jokes. A discriminating sense of humor is a valuable asset to any speaker. It is an excellent device for gaining rapport with your audience, and it can offer comic relief when your subject is very serious.

Your humor should always relate directly to your message. Do not use humor from the current edition of a popular magazine or television program, because your audience probably saw it also. Above all, don't be vulgar.

QUESTION-AND-ANSWER SESSION

Try to anticipate what questions you will receive during any question-and-answer session, and be ready with your answers. Be sure you understand a question before attempting to answer it, and then answer it as completely as time allows or the question deserves. Be sure of the accuracy of your answers, and don't be afraid to say "I don't know."

Be polite and objective when responding to a hostile question, but be prepared to go on to the next one if it becomes evident that the questioner simply wants to argue. Do not be defensive, and do not play for laughs at the questioner's expense (you will offend not only him or her but the audience as well).

oral/verbal

Oral refers to what is spoken.

EXAMPLE He made an *oral* commitment to the policy.

Although it is sometimes used synonymously with *oral, verbal* literally means "in words" and can refer to what is spoken or written. To avoid possible confusion, do not use *verbal* if you can use *written* or *oral*.

CHANGE He made a *verbal* agreement to complete the work.
TO He made a *written* agreement to complete the work.
OR He made an *oral* agreement to complete the work.
CHANGE Avoid *verbal* orders.
TO Avoid *oral* orders.

When you must refer to something both written and spoken, you should use both *written* and *oral* to make your meaning clear.

EXAMPLE He demanded either a *written* or an *oral* agreement before he would continue the project.

organization

Organization is achieved, first, by developing your **topic** in a way that will enable your **reader** to understand your message and then by **outlining** your material on the basis of that **method of development.**

A logical method of development satisfies the reader's need for a controlling shape and structure for your subject. For example, if you were writing a proposal to your manager, trying to get him or her to fund a project you wanted to undertake, you would probably use an order-of-importance method of developing the topic. You would start with the most important reason that the project should be funded and end with the least important reason. On the other hand, if you were a Federal Aviation Administration agent reporting on an airplane crash, you would probably use a cause-and-effect method, starting with the cause of the crash and leading up to the crash (or starting with the crash and tracing back to the cause).

An appropriate method of development is the writer's tool for keeping things under control and the reader's means of following the writer's development of a theme. Many different methods of development are available to the writer; this book includes those that are likely to be used by technical people: **sequential, chronological, increasing-order-of-importance, decreasing-order-of-importance, division-and-classification, comparison, spatial, specific-to-general, general-to-specific,** and **cause-and-effect.** As the writer, you must choose the method of development that best suits your subject, your reader, and your **objective.**

Outlining provides structure to your writing by ensuring that it has a beginning, a middle, and an end. It gives proportion to your writing by making sure that one step flows smoothly to the next without omitting anything important, and it enables you to emphasize your key points by placing them in the positions of greatest importance. Using an outline makes larger and more difficult subjects easier to handle by breaking them into manageable parts. Finally, by forcing you to organize your subject and structure your thinking in the outline stage, creating a good outline allows you to concentrate exclusively on writing when you begin the rough draft.

organizational charts

An organizational chart shows how the various components of an organization are related to one another. It is useful when you want

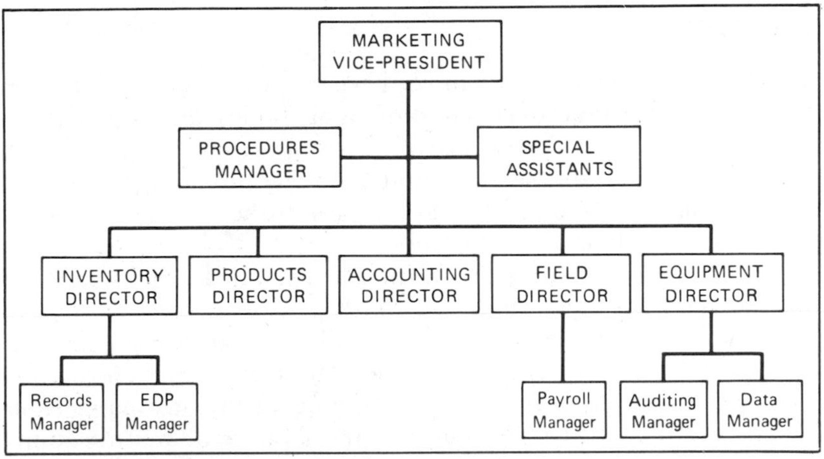

Figure 1 Organizational Chart Showing Positions

to give your **readers** an overview of an organization or to show them the lines of authority within it.

The title of each organizational component (office, section, division) is placed in a separate box. These boxes are then linked to a central authority. (See Figure 1.) If your readers need the information, include the name of the person occupying the position identified in each box. (See Figure 2.)

Figure 2 Organizational Box with Name of Person and Position

As with all **illustrations,** place the organizational chart as close as possible to the text that refers to it.

orient/orientate

Orientate is merely a **long variant** of *orient,* meaning "locate in relation to something else." The shorter form should be used because it is simpler.

CHANGE Let me *orientate* your group on our operation.
TO Let me *orient* your group on our operation.

outlining

Outlining provides structure to your writing by ensuring that it has a beginning (**introduction**), a middle (main body), and an end (**conclusion**). An outline gives your writing proportion so that one part flows smoothly to the next without omitting anything important. Outlining also enables you to emphasize your key points by placing them in the positions of greatest importance.

Like a road map, an outline indicates a starting point and keeps you moving logically so that you don't get lost before arriving at your conclusion. (Errors in logic are much easier to detect in an outline than in a draft.) Using an outline makes larger and more difficult subjects easier to handle by breaking them into manageable parts; therefore, the less certain you are about your writing ability or about your subject, the fuller your outline should be. The parts of an outline are easily moved about, so you can experiment to see what arrangement of your ideas is the most effective. Perhaps most important, creating a good outline frees you to concentrate on writing when you begin the rough draft (by forcing you to organize your subject and structure your thinking beforehand).

An outline is a tool that helps writers make their writing more logical and coherent. Do not feel that once you have created an outline you must follow it slavishly; as you write the draft, you may still make any necessary changes.

Two types of outlines are generally used: topic outlines and sentence outlines. The topic outline consists of short **phrases** that show the sequential order and relative importance of ideas; in this manner, a topic outline provides order and establishes the relationships of topics to one another. The topic outline alone is generally not sufficient for a large or complex writing job, but it may be used to structure the major and minor divisions of your topic in preparation for creating a sentence outline. (An outline for a small job is not as detailed as one for a larger job, but it is just as important; for example, a topic outline that lists your major and minor points can help greatly even in writing important letters.)

On a large writing project, it is wise to create a topic outline first and then use it as a basis for creating a sentence outline. In a sentence outline, the writer summarizes each idea in a single complete sentence that will become the **topic sentence** for a **paragraph** in the rough draft. The sentence outline begins with a main idea statement that establishes the subject and then follows with a complete sentence for each idea in the major and minor divisions. A sentence outline provides order and establishes the relationship of topics to one another to a considerably greater degree than does a topic outline. A well-developed sentence outline offers a sure test of the validity of your arrangement of your material. If most of your notes can be shaped into controlling topic sentences for paragraphs in your rough draft, you can be relatively sure before you begin its final composition that your paper will be well organized. A good sentence outline, when stripped of its numbering symbols, needs only to be expanded with full details to become a rough draft.

Some outlines begin with a thesis statement, which is a statement of the main idea that you intend to develop in your report. This statement gives you a goal that you can work toward as you create your outline. A thesis statement often suggests a particular **method of development.**

EXAMPLES The mission director must oversee a rigid chain of events that must occur prior to blast-off. (This thesis statement suggests a **sequential method of development.**)

Increased sales in the Chicago area appear to be a direct result of the new marketing techniques introduced there last year. (This thesis statement suggests a **cause-and-effect method of development.**)

The growth of the computer industry has closely paralleled the growth of conglomerates. (This thesis statement suggests a combination of the **chronological method of development** and the **comparison method of development.**)

Although the following outlining technique is mechanical, it is easy to master and well suited to unraveling the complexities of large and difficult subjects.

The first step is to group naturally related items and write them down on note cards and then to determine whether they are exactly what you need to meet the demands of your **objective,** your **reader,** and your **scope** of coverage. If they are—or when they are—arrange them in the proper order, and label them with roman numerals. For example, the major divisions for this discussion of outlining could be as follows:

I. Advantages of outlining
II. Types of outlining
III. Effective outlining
IV. Writing the draft from the outline

The second step is to establish your minor points by deciding on the minor divisions within each major head. Arrange them in the proper sequence under their major **heads,** and label them with capital letters. Keep major heads equal in importance and minor heads equal in importance; the purpose of this **parallel structure** is to keep your thinking in order. The major and minor heads for this discussion of outlining could be as follows:

I. Advantages of outlining
II. Types of outlining
  A. The topic outline
  B. The sentence outline
III. Effective outlining
  A. Establish major and minor heads
  B. Sort note cards by major and minor heads
  C. Complete the sentence outline
IV. Writing the draft from the outline

Of course, you will often need more than the two levels of heads illustrated here. If your subject is complicated, you may need four or five levels to keep all of your ideas straight and in proper relationship to one another. In that event, the following numbering scheme is recommended:

I. First-level head
  A. Second-level head
    1. Third-level head
      a. Fourth-level head
        (1) Fifth-level head

The third step is to mark each of your note cards with the appropriate roman numeral and capital letter. Sort the note cards by major and minor heads (roman numerals and capital letters). Then arrange the cards within each minor head, and mark each with the appropriate sequential arabic number. Transfer your notes to paper, converting them to complete sentences. You now have a complete rough sentence outline, and the most difficult part of the writing job is over.

You could also use a decimal numbering system with your outline, instead of roman numerals and **capital letters,** if you prefer.

EXAMPLE	1	FIRST-LEVEL HEAD
	1.1	Second-Level Head
	1.2	Second-Level Head
	1.2.1	Third-Level Head

1.2.2 Third-Level Head
1.2.2.1 Fourth-Level Head
1.2.2.2 Fourth-Level Head
1.3 Second-Level Head
1.3.1 Third-Level Head
1.3.2 Third-Level Head
2 FIRST-LEVEL HEAD

The final step is to polish your rough outline. Check to ensure that the **subordination** of minor heads to major heads is logical. Remember that all major heads should be parallel and all minor heads should be parallel.

CHANGE A. Major and minor heads should be established. (passive voice)
B. Note cards are sorted by major and minor heads. (passive voice)
C. Complete the sentence outline. (active voice)
TO A. Establish major and minor heads. (active voice)
B. Sort note cards by major and minor heads. (active voice)
C. Complete the sentence outline. (active voice)

Make certain that your outline follows your method of development. Check for **unity.** Does your outline stick to the subject, or does it stray into unrelated or only loosely related topics? Stay within your established scope of coverage. Resist the temptation to tell all you know unless it all is pertinent to your objective. It is much easier to correct this problem in the outline stage than in the rough draft. If some information is of importance to only a minority of your readers, consider including that information in an **appendix;** those few readers can then pursue the subject further, and the others will not be interrupted. Check your outline for completeness; scan it to see whether you need more information in any of your divisions, and insert any information that is needed.

You now have a final, polished sentence outline. It isn't sacred, however; change it if you need to as you write the draft; the outline should be your point of departure and return. Return to it to find your place and your direction as your work.

outside of

In the **phrase** *outside of,* the word *of* is redundant.

CHANGE Place the rack *outside of* the incubator.
TO Place the rack *outside* the incubator.

In addition, do not use *outside of* to mean ''aside from'' or ''except for.''

CHANGE *Outside of* his frequent absences, Jim has a good work record.
TO *Except for* his frequent absences, Jim has a good work record.

over with

In the expression *over with,* the word *with* is redundant; moreover, the word *completed* often better expresses the thought.

CHANGE You may enter the test chamber now that the experiment is *over with.*
TO You may enter the test chamber now that the experiment is *over.*
OR You may enter the test chamber now that the experiment has been *completed.*

P

pace

Pace is the speed at which the writer presents ideas to the **reader.** Your goal should be to achieve a pace that fits both your reader and your subject; at times you may need a fast pace, at other times a slow pace. The more knowledgeable the reader is, the faster your pace can be—but be careful not to lose control of it. In the first passage of the following example, facts are piled on top of each other at a rapid pace. In the second passage, the same facts are spread out over two sentences and presented in a more easily assimilated manner, even though the length is no greater; in addition, a different and more desirable **emphasis** is achieved.

> CHANGE The generator is powered by a 90-horsepower engine, is designed to operate under normal conditions of temperature and humidity, produces 110 volts at 60 Hertz, is designed for use under emergency conditions, and may be phased with other units of the same type to produce additional power when needed.
>
> TO The generator, which is powered by a 90-horsepower engine, produces 110 volts at 60 Hertz under normal conditions of temperature and humidity. Designed especially for use under emergency conditions, this generator may be phased with other units of the same type to produce additional power when needed.

Check your draft to determine the pace at which you are presenting your facts and ideas to the reader. If you find that you are jamming your ideas too closely together, you have a problem with pacing. There are at least three solutions to such problems: **subordination, sentence construction,** and **transition.**

The best way to solve a pacing problem is to search out minor thoughts and subordinate them to the major points, just as, in the example, " . . . is designed for use under emergency conditions . . ." was subordinated to ". . . this generator may be phased with other units . . ."

Another solution is to consider a different type of sentence structure. A complex idea is normally better stated in a simple sentence structure; a series of simple thoughts is better stated in a more complex sentence structure. Notice in the example above that the first passage is one long, complicated sentence, and that the same ideas are expressed in shorter sentences in the **revision**.

A final solution is to provide words and **phrases** of transition to alter the pace.

> CHANGE The report was not finished. It was 6:15. We all were tired. We left the office and went home.
>
> TO The report was *still* not finished at 6:15. *However*, we all were tired, *so* we left the office and went home.

(See also **telegraphic style.**)

paragraphs

A paragraph is a group of **sentences** that support and develop a single idea; it may be thought of as an essay in miniature, for its function is to expand the core idea stated in its **topic sentence.** A paragraph may use a particular **method of development** to expand this idea. The following paragraph uses the **general-to-specific method of development.** Its **topic sentence** is italicized.

> *The arithmetic of searching for oil is stark.* For all their scientific methods of detection, the only way oil drillers can actually know that there is oil in the ground is to drill a well. The average cost of drilling an oil well is over $300,000, and drilling a single well may cost over $8,000,000. And once the well is drilled, the odds against its containing any oil at all are 8 to 1!

The paragraph performs three functions: (1) it develops the unit of thought stated in the topic sentence; (2) it provides a logical break in the material; and (3) it creates a physical break on the page, which in turn provides visual assistance to the **reader.**

TOPIC SENTENCE

A topic sentence states the subject of a paragraph; the rest of the paragraph then supports and develops that statement with carefully related details. The topic sentence may appear anywhere in the paragraph—a fact that permits the writer to achieve **emphasis** and variety—and a topic statement may be more than one sentence if necessary.

The topic sentence is most often the first sentence of the paragraph because it states the subject that the paragraph is to develop. The topic sentence is effective in this position because the reader knows immediately what the paragraph is about. In the following paragraph, the first sentence is the topic sentence:

> *Spot size, critical in specifying resolution, should be as uniform as possible across the screen* [of the cathode ray tube, or CRT]. If spots are significantly larger in the corners than at the screen's center, the imaging in the corners will lack sharp definition. Although resolution depends largely on spot size or line width, a well-focused electron beam lacks a definite edge and thus has no absolute size. Light distribution has a tendency to decrease from its center in a bell-shaped or Gaussian curve. The spot size specified will depend on its application and attendant variables.
>
> —James A. Geissinger, "Cathode-Ray-Tube Display," *Chemical Engineering* 87 (1980), p. 145.

On rare occasions, the topic sentence logically falls in the middle of a paragraph. In the following example, the topic sentence, italicized, is in the middle of the paragraph:

> While the principles and techniques of system coordination can be taught in a one- to two-week training course, actual expertise cannot be developed until one has coordinated several systems. The reason for this is simple. *Coordination is at least 60% "art" and only 40% engineering.* For any given system, there are many different and acceptable solutions. The engineer performing coordination must have a depth of experience to select the optimum solution from the many choices.
>
> —Thad Brown, "Principles of Short-Circuit Analysis and Coordination," *Chemical Engineering* 87 (1980), p. 152.

Although the topic sentence is usually more effective early in the paragraph, when the reader's attention is greatest, a paragraph can lead up to the topic sentence, which is sometimes done to achieve emphasis. When a topic sentence concludes a paragraph, it can also serve as a summary or **conclusion,** based on the details that were designed to lead up to it. The following is a paragraph whose topic sentence is at the end:

> Energy does far more than simply make our daily lives more comfortable and convenient. Suppose you wanted to stop—and reverse—the economic progress of this nation. What

would be the surest and quickest way to do it? Find a way to cut off the nation's oil resources! Industrial plants would shut down; public utilities would stand idle; all forms of transportation would halt. The country would be paralyzed, and our economy would plummet into the abyss of national economic ruin. *Our economy, in short, is energy-based.*

—*The Baker World* (Los Angeles: Baker Oil Tools, Inc., 1964), p. 5.

Because multiple paragraphs are sometimes used to develop different aspects of an idea, not all paragraphs have topic sentences. In this situation, **transition** between paragraphs helps the reader know that the same idea is being developed through several paragraphs.

Topic sentence for all three paragraphs

To conserve valuable memory space, a large portion of the software package remains on disc; only the most frequently used portion resides in internal memory all of the time. The disc-resident software is organized into small modules that are called into memory as needed to perform specific functions.

Transition

The memory-resident portion of the operating system maintains strict control of processing. It consists of routines, subroutines, lists, and tables that are used to perform common program functions, such as processing input/output operations, calling other software routines from disc as needed, and processing errors.

Transition

The disc-resident portion of the operating system contains routines that are used less frequently in system operation, such as the peripheral-related software routines that are useful for correcting errors encountered on the various units, and the log and display routines that record unusual operating conditions in the system log. The disc-resident portion of the operating system also contains Monitor, the software program that supervises the loading of utility routines and the user's programs.

—*NCR Century Operating Systems Manual* (Dayton: The NCR Corporation, 1974), p. 17.

In this example, the reason for breaking the development of the idea expressed in the topic sentence into three paragraphs is to help the reader assimilate the fact that the main idea has two separate parts.

PARAGRAPH LENGTH

Paragraph length should be tailored to aid the reader's understanding of ideas. A series of short, undeveloped paragraphs can indicate poor **organization** of material, in which case you should look for a larger idea to which the ideas in the short paragraphs relate and then make the larger idea the topic sentence for a single paragraph. A series of short paragraphs can also sacrifice **unity** by breaking a single idea into several pieces. A series of long paragraphs, on the other hand, can fail to provide the reader with manageable subdivisions of thought. A good rule of thumb is that a paragraph should be just long enough to deal adequately with the subject of its topic sentence. A new paragraph should begin whenever the subject changes significantly. Occasionally, a one-sentence paragraph is acceptable if it is used as a **transition** between larger paragraphs or in letters and **memorandums** in which one-sentence openings and closings are sometimes appropriate.

A short paragraph sometimes immediately follows a long paragraph to emphasize the thought contained in the short paragraph. Also, short paragraphs increase the pace of your writing, whereas long paragraphs can slow down the pace.

> Because it has at times displaced some jobs, automation has become an ugly word in the American vocabulary. But the all-important fact that is so often overlooked is that it invariably creates many more jobs than it eliminates. (The vast number of people employed in the American automobile industry as compared with the number of people that had been employed in the harness- and carriage-making business is a classic example.) Almost always, the jobs that have been eliminated by automation have been menial, unskilled jobs, and those who have been displaced have been forced to increase their skills, which resulted in better and higher paying jobs for them.
>
> In view of these facts, is automation really bad? It has made our country the most wealthy and technologically advanced nation the world has ever known!

WRITING PARAGRAPHS

The paragraph is the basic building block of any writing effort. Careful paragraphing reflects the writer's accurate thinking and logical organization. Clear and orderly paragraphs help the reader follow the writer's thoughts more easily.

Outlining is the best guide to paragraphing. It is easy to group ideas into appropriate paragraphs when you follow a good working outline. Notice how the following outline plots the course of subsequent paragraphs:

Outline

I. Advantages of Chicago as location for new plant
   A. Transport facilities
      1. Rail
      2. Air
      3. Truck
      4. Sea (except in winter)
   B. Labor supply
      1. Engineering and scientific personnel
         a. Many similar companies in the area
         b. Several major universities
      2. Technical and manufacturing personnel
         a. Existing programs in community colleges
         b. Possible special programs designed for us

Resulting Paragraphs

Probably the greatest advantage of Chicago as a location for our new plant is its excellent transport facilities. The city is served by three major railroads. Both domestic and international air cargo service is available at O'Hare International Airport. Chicago is a major hub of the trucking industry, and most of the nation's large freight carriers have terminals there. Finally, except in the winter months when the Great Lakes are frozen, Chicago is a seaport, accessible through the St. Lawrence Seaway.

A second advantage of Chicago is that it offers an abundant labor force. An ample supply of engineering and scientific personnel is assured not only by the presence of many companies engaged in activities similar to ours but also by the presence of several major universities in the metropolitan area. Similarly, technicians and manufacturing personnel are in abundant supply. The seven colleges in the Chicago City College system, as well as half a dozen other two-year colleges in the outlying areas, produce graduates with associate degrees in a wide variety of technical specialties appropriate to our needs. Moreover, three of the outlying colleges have expressed an interest in establishing special courses attuned specifically to our requirements.

Consider not only the nature of the material you are developing but also the appearance of your page. An unbroken page looks forbidding.

PARAGRAPH COHERENCE AND UNITY

A good paragraph has unity, **coherence,** and adequate development. Unity is singleness of purpose, based on a topic sentence that states the core idea of the paragraph. When every sentence in the paragraph contributes to developing the core idea, the paragraph has unity. Coherence is holding to one **point of view,** one attitude, one **tense;** it is the joining of sentences into a logical pattern. Coherence is advanced by the careful choice of transitional words so that ideas are tied together as they are developed.

> Topic sentence *Any company which operates internationally today faces a host of difficulties.* Inflation is worldwide. Most countries are struggling Transition with other economic problems *as well. In addition,* there are many monetary uncertainties and growing economic nation- Transition alism directed against multinational companies. *Yet* there is ample business available in most developed countries if you have the right products, services, and marketing organization. To maintain the growth NCR has achieved overseas, we recently restructured our international operations into four ma- Transition jor trading areas. *This* will improve the services and support which the Corporation can provide to its subsidiaries around Transition the world. *At the same time* it established firm management con- Transition trol, ensuring consistent policies around the world. *So* you might say the problems of doing business abroad will be more difficult this year but we are better organized to meet those problems.
>
> —*1974 Annual Report* (Dayton: The NCR Corporation, 1974), p. 3.

Simple enumeration (*first, second, then, next,* and so on) can also provide effective transition within paragraphs. Notice how the italicized words and **phrases** give coherence to the following paragraph:

> Most adjustable office chairs have nylon hub tubes that hold metal spindle rods. To ensure trouble-free operation, lubricate these spindle rods occasionally. *First,* loosen the set screw in the adjustable bell. *Then* lift the chair from the base so that the entire spindle rod is accessible. *Next,* apply the lubricant to the spindle rod and the nylon washer, using the lubricant spar-

ingly to prevent dripping. *When you have finished,* replace the chair and tighten the set screw.

Good paragraphs often use details from the preceding paragraph, thereby preserving and advancing the thought being developed. Appropriate **conjunctions** and the **repetition** of key words and phrases can help provide unity and coherence among, as well as within, paragraphs, as the italicized words do in the following examples:

> Six high power thyristors connected in a three-phase bridge configuration form the basic armature module. When necessary, the modules can be arranged in parallel order to meet higher power requirements. The basic armature module is constructed as a convenient *pull-out tray*. All *armature trays* are interchangeable.
>
> For optimum shovel performance it is also desirable to use thyristors in the control of power fields. These smaller thyristors are also arranged in a *pull-out tray* arrangement called a *field tray*.
>
> Should a fault ever occur in a tray, an *indicator tray* is provided, which by means of pilot lights indicates which tray is the source of the trouble. Through the *pull-out tray* concept, the mine electrician can quickly replace the faulty tray and need not troubleshoot.

Sometimes a paragraph is used solely for transition, as in the following example:

> . . . that marred the progress of the company.
>
> There were two other setbacks to the company's fortunes that year which contributed to its present shaky condition: the loss of many skilled workers through the early retirement program and the intensification of the devastating rate of inflation.
>
> The early retirement program . . .

parallel structure

Parallel sentence structure requires that sentence elements that are alike in function be alike in construction as well, as in the following examples (in which similar actions are stated in similar **phrases**):

EXAMPLES The stream runs *under the culvert, behind the embankment,* and *into the pond.*

We need a supplementary work force *to handle* peak-hour activity, *to free* full-time employees from routine duties, *to relieve* operators during lunch breaks, and *to replace* vacationing employees.

Parallel structure achieves an economy of words, clarifies meaning, and pleases the **reader** aesthetically. In addition to adding a pleasing symmetry to a sentence, parallel structure expresses the equality of its ideas. This technique assists readers because they are able to anticipate the meaning of a sentence element on the basis of its parallel construction. When they recognize the similarity of word order or construction, readers know that the relationship between the new sentence element and the **subject** is the same as the relationship between the last sentence element and the subject. Because of this they can go from one idea to another more quickly and confidently.

EXAMPLES The computer instruction contains *fetch, initiate,* and *execute* stages. (parallel words)

The computer instruction contains *a fetch stage, an initiate stage,* and *an execute stage.* (parallel phrases)

The computer instruction contains a fetch stage, it contains an initiate stage, and *it contains an execute stage.* (parallel clauses)

Parallel structure is especially important in creating your **outline,** your **table of contents,** and your **heads** because it enables your reader to know the relative value of each item in your table of contents and each head in the body of your **report** (or other writing project).

In any type of writing, parallel structure channels the reader's attention and helps draw together related ideas or line up dissimilar ideas for comparison and contrast.

EXAMPLE Her book provoked much comment: *negative from her enemies, positive from her friends.*

The power of many of Lincoln's speeches comes from his use of parallel structure; in them, parallel structure harmonizes the elements within sentences and the relationships among sentences.

EXAMPLE But in a larger sense, *we cannot dedicate—we cannot consecrate—we cannot hallow*—this ground. The brave men, *living* and *dead,* who struggled here, have consecrated it far above our poor power to *add* or *detract. The world will little note nor long remember what we say here,* but *it can never forget what they did here.*

—Abraham Lincoln, Gettysburg Address

A century later, John F. Kennedy use the same technique with striking effect.

EXAMPLE We shall *pay any price, bear any burden, meet any hardship, support any friend, oppose any foe* to assure the survival and the success of liberty.

—John F. Kennedy, Inaugural Address

Although parallelism can be accomplished with words, phrases, or **clauses,** it is most frequently accomplished by the use of phrases.

EXAMPLES I was convinced of their competence *by their conduct, by their reputation,* and *by their survival in a competitive business.* (prepositional phrases)

Filling the gas tank, testing the windshield wipers, and *checking tire pressure* are essential to preparing for a long trip. (gerund phrases)

From childhood the artist had made it a habit *to observe people, to store up the impressions they made upon him,* and *to draw conclusions about mankind from them.* (infinitive phrases)

Correlative conjunctions (*either . . . or, neither . . . nor, not only . . . but also*) should always be followed by parallel structure. Both members of these pairs should be followed immediately by the same grammatical form: two words, two similar phrases, or two similar clauses.

EXAMPLES Viruses carry either *DNA* or *RNA*, never both. (words)

Clearly, neither *serologic tests* nor *virus isolation studies* alone would have been adequate. (phrases)

Either *we must increase our operational efficiency* or *we must decrease our production goals.* (clauses)

To make a parallel construction clear and effective, it is often best to repeat a **preposition,** an **article,** a **pronoun,** a **subordinating conjunction,** a **helping verb,** or the mark of an **infinitive.**

EXAMPLES The Babylonians had *a* rudimentary geometry and *a* rudimentary astronomy. (article)

My father and *my* teacher agreed that I was not really trying. (pronoun)

To run and be elected is better than *to* run and be defeated. (mark of the infinitive)

The driver *must* be careful to check the gauge and *must* move quickly when the light comes on. (helping verb)

New teams were being established *in* New York and *in* Hawaii. (preposition)

The history of factories shows both *the* benefits and *the* limits of standardization. (article)

When parallel constructions are closely related in thought and equal in length and construction, they balance one another. Such balance is useful for comparing and contrasting ideas.

EXAMPLE A qualified staff without a sufficient budget is helpless; a sufficient budget without a qualified staff is useless.

FAULTY PARALLELISM

Faulty parallelism results when joined elements are intended to serve equal grammatical functions but do not have equal grammatical form. Avoid this kind of partial parallelism. Make certain that each element in a series is similar in form and structure to all others in the same series.

CHANGE Before mowing the lawn, check the following items: the dipstick for the proper oil level, the gas tank for fuel, the spark plug wire attachment, and that no foreign objects are under or near the mower.

TO Before mowing the lawn, check the following items: the dipstick *for the proper oil level,* the gas tank *for fuel,* the spark plug wire *for attachment,* and the lawn *for foreign objects.*

In work-related writing, **lists** often cause problems with parallel structure. When you use a list that consists of phrases or clauses, each phrase or clause in the list should begin with the same **part of speech.**

CHANGE The following recommendations were made regarding the Cost Containment Committee's position statement:

1. *Stress* that this statement is for all departments.
2. *Start* the statement with "If the company continues to grow, the following steps should be taken."
3. *The statement* should emphasize that it applies both to department managers and staff.
4. *Such strong words* as *obligation, owe,* and *must* should be replaced with words that are less harsh.

TO The following recommendations were made regarding the Cost Containment Committee's position statement:

1. *Stress* that this statement is for all departments.
2. *Start* the statement with "If the company continues to grow, the following steps must be taken."
3. *Emphasize* that it applies both to department managers and staff.
4. *Replace* such strong words as *obligation, owe,* and *must* with words that are less harsh.

In the first list the items are not grammatically parallel in structure: items 1 and 2 begin with verbs, but items 3 and 4 do not. Notice how much more smoothly the corrected version reads, with all items beginning with verbs.

Faulty parallelism sometimes occurs because a writer tries to compare items that are not comparable.

CHANGE The company offers special training to help nonexempt employees move into professional and technical careers like data processing, bookkeeping, customer engineers, and sales trainees. (Notice that occupations—data processing and bookkeeping—are being compared to people—customer engineers and trainees.)

TO The company offers special training to help nonexempt employees move into professional and technical careers like data processing, bookkeeping, *customer service,* and *sales.*

Because faulty parallelism with correlative conjunctions (*either . . . or, neither . . . nor, not only . . . but also*) is one of the most common writing errors, the following examples are worth careful study:

CHANGE You may travel to the new plant either *by train* or *there is a plane.* (different grammatical forms: a prepositional phrase, *by train*; and a clause, *there is a plane*)

TO You may travel to the new plant either *by train* or *by plane.* (parallel grammatical forms: prepositional phrases)

OR You may travel to the new plant by either *train* or *plane.* (parallel grammatical forms: nouns)

CHANGE We are not only *responsible to our stockholders* but also *to our customers.* (different grammatical forms: an adjective with a modifying prepositional phrase, *responsible to our stockholders;* and a prepositional phrase, *to our customers*)

TO We are responsible not only *to our stockholders* but also *to our customers.* (parallel grammatical forms: prepositional phrases)

OR We are not only *responsible to our stockholders* but also *responsible to our customers*. (parallel grammatical forms: verb phrases)

CHANGE We either *will pay the bill* or *we will return the shipment.* (different grammatical forms: a verb with its object, *will pay the bill,* and a clause, *we will return the shipment*)

TO Either *we will pay the bill* or *we will return the shipment.* (parallel grammatical forms: clauses)

OR We will either *pay the bill* or *return the shipment.* (parallel grammatical forms: verbs with their objects)

Be careful not to throw your reader off balance. If you make a statement about two subjects and then follow with a further statement about one of them, complete the thought and make a balancing statement about the other subject.

CHANGE The Comanche Commander and the Cessna Hawk are both small aircraft; *the Commander seats four.*

TO The Comanche Commander and the Cessna Hawk are both small aircraft; *the Commander seats four* and *the Hawk seats six.*

paraphrasing

When you paraphrase a written passage, you rewrite it to state the essential ideas in your own words. Because you do not quote your source word for word when paraphrasing, it is unnecessary to enclose the paraphrased material in **quotation marks.** However, paraphrased material should be footnoted because the ideas are taken from someone else whether or not the words are identical.

Ordinarily, the majority of the notes you take during the **research** phase of writing your **report** will paraphrase the original material. (See also **note taking.**)

Original Material

One of the major visual cues used by pilots in maintaining precision ground reference during low-level flight is that of object blur. We are acquainted with the object-blur phenomenon experienced when driving an automobile. Objects in the foreground appear to be rushing toward us while objects in the background appear to recede slightly. There is a point in the observer's line of sight, however, at which objects appear to stand still for a moment, before once again rushing toward him with increasing angular velocity. The distance from the observer to this point where objects appear stationary is sometimes referred to as the ''blur threshold'' range.

—Wesley E. Woodson and Donald W. Conover, *Human Engineering Guide for Equipment Designers,* 2nd ed. (Berkeley and Los Angeles: University of California Press, 1964), p. 139.

Paraphrased

Object blur refers to the phenomenon by which observers in a moving vehicle look out and see the foreground objects appear to rush at them, while background objects appear to recede. But objects at some point appear temporarily stationary. The distance separating observers from this point is sometimes called the ''blur threshold'' range.

Note that the paraphrased version includes only the essential information from the original passage. Paraphrased material, consequently, will almost always be shorter than the original material. As with all summarized material, strive to put the original ideas into your words without distorting them. (See also **plagiarism** and **quotations.**)

parentheses

Parentheses ( ) are used to enclose words, **phrases,** or sentences. The material within parentheses can add **clarity** to a statement without altering its meaning. Parentheses de-emphasize (or play down) an inserted element. Parenthetical information may not be essential to a sentence, but it may be interesting or helpful to some **readers.**

EXAMPLE Aluminum is extracted from its ore (called bauxite) in three stages.

Parenthetical material applies to the word or phrase immediately preceding it.

EXAMPLE The development of International Business Machines (IBM) is a uniquely American success story.

Parentheses may be used to enclose figures or letters that indicate sequence. Enclose the figure or letter with two parentheses rather than using only one parenthesis.

EXAMPLE The following sections deal with (1) preparation, (2) research, (3) organization, (4) writing, and (5) revision.

Parenthetical material does not affect the **punctuation** of a sentence. If a parenthesis closes a sentence, the ending punctuation should appear after the parenthesis. Also, a **comma** following a parenthetical word, phrase, or **clause** appears outside the closing parenthesis.

EXAMPLE These oxygen-rich chemicals, as for instance potassium permanganate ($KMnO_4$) and potassium chromate ($KCrO_4$), were oxidizing agents (they added oxygen to a substance).

However, when a complete sentence within parentheses stands independently, the ending punctuation goes inside the final parenthesis.

EXAMPLE The new marketing approach appears to be a success; most of our regional managers report sales increases of 15 to 30 percent. (The only important exceptions are the Denver and Houston offices.) Therefore, we plan to continue . . .

Do not follow a spelled-out number with a figure in parentheses representing the same number.

CHANGE Send five (5) copies of the report.
TO Send five copies of the report.

In some footnote forms, parentheses enclose the publisher, place of publication, and date of publication.

EXAMPLE [1]J. Lamarsh, *Introduction to Nuclear Reactor Theory* (Reading, Mass.: Addison-Wesley, 1966), p. 9.

Use **brackets** to set off a parenthetical item that is already within parentheses.

EXAMPLE We should be sure to give Emanuel Foose (and his brother Emilio [1812–1882] as well) credit for his part in founding the institute.

Do not overuse parentheses. And guard against using parentheses when other marks of punctuation are more appropriate.

participial phrases

A participial phrase consists of a **participle** plus its **object** or **complement,** if any, and modifiers. Like the participle, a participial phrase functions as an **adjective** modifying a **noun** or **pronoun.**

EXAMPLES The division *having the largest sales increase* wins the trophy.
Finding the problem resolved, he went to the next item.
Having begun, we felt that we had to see the project through.

The relationship between a participial phrase and the rest of the sentence must be clear to the **reader.** For this reason, every sentence containing a participial phrase must have a noun or pronoun that the participial phrase modifies; if it does not, the result is a dangling participial phrase that is misplaced in the sentence and so appears to modify the wrong noun or pronoun. Both of these problems are discussed below.

DANGLING PARTICIPIAL PHRASES

A dangling participial phrase occurs when the noun or pronoun that the participial phrase is meant to modify is not stated but only implied in the sentence. (See also **dangling modifiers.**)

CHANGE *Being unhappy with the job,* his efficiency suffered. (His *efficiency* was not unhappy with the job; what the participial phrase really modifies—*he*—is not stated but merely implied.)
TO *Being unhappy with the job,* he grew less efficient. (Now what that participial phrase modifies—*he*—is explicitly stated.)

MISPLACED PARTICIPIAL PHRASES

A participial phrase is misplaced when it is too far from the noun or pronoun it is meant to modify and so appears to modify something else. This is an error that can sometimes make the writer look ridiculous indeed. (See also **modifiers.**)

CHANGE *Rolling around in the bottom of the vibration test chamber,* I found the missing bearings.
TO I found the missing bearings *rolling around in the bottom of the vibration test chamber.*
CHANGE We saw a large warehouse *driving down the highway.*
TO *Driving down the highway,* we saw a large warehouse.

participles

Participles are **verb** forms that function as **adjectives.**

EXAMPLES The *waiting* driver raced his engine.
Here are the *revised* estimates.
The *completed* report lay on his desk.
Rising costs reduced our profit margin.

Participles belong to a larger class of verb forms called **verbals** or nonfinite verbs. (The other two types of verbals are **gerunds,** which always function as nouns, and **infinitives,** which have several functions.) Because participles are formed from verbs, they share certain characteristics of verbs, even though they are adjectives: (1) they may take **objects** or **complements;** (2) they may be in the present, past, or perfect **tense;** and (3) they may be in the active or passive **voice.** But remember that a participle cannot be used as the verb of a sentence. Inexperienced writers sometimes make this mistake; the result is a **sentence fragment.**

CHANGE The committee chairman was responsible. His vote *being* the decisive one.

TO The committee chairman was responsible, his vote *being* the decisive one.

OR The committee chairman was responsible. His vote *was* the decisive one.

TENSE

The present participle ends in *ing.*

EXAMPLE *Declining* sales forced us to close one branch office.

Do not confuse present participles with gerunds, which are also verbals ending in *ing* but which always function as nouns, as in "*Running* is his favorite sport."

The past participle may end in *ed, t, en, n,* or *d.*

EXAMPLES Repair the *bent* lever.
What are the *estimated* costs?
Here is the *broken* calculator.
What are the metal's *known* properties?
The story, *told* many times before, was still interesting.

The perfect participle is formed with the present participle of the **helping verb** *have* plus the past participle of the main verb.

EXAMPLE *Having gotten* (perfect participle) a large raise, the *smiling* (present participle), *contented* (past participle) employee worked harder than ever.

VOICE

Participles formed from transitive verbs may be in either the active or the passive voice. Some form of the helping verb *be* is used to form the passive voice of the perfect participle.

EXAMPLES *Having finished* the job, we submitted our bill. (active)
The job *having been finished,* we submitted our bill. (passive)

parts of speech

Parts of speech is a term used to describe the class of words to which a particular word belongs, according to its function in the sentence; that is, each function in a sentence (naming, asserting, describing, joining, acting, modifying, exclaiming) is performed by a word belonging to a certain part of speech.

If a word's function is to *name* something, it is a **noun** or **pronoun.** If a word's function is to make an *assertion* about something, it is a **verb.** If its function is to *describe* or *modify* something, the word is an **adjective** or an **adverb.** If its function is to *join* or *link* one element of the sentence to another, it is a **conjunction** or a **preposition.** If its function is to express an exclamation, it is an **interjection.** (See also **functional shift.**)

party

In legal language, *party* refers to an individual, group, or organization.

EXAMPLE The injured *party* brought suit against my client.

The term is inappropriate to general writing; when an individual is being referred to, use *person.*

CHANGE The *party* whose file you requested is here now.
TO The *person* whose file you requested is here now.

Party is, of course, appropriate when it refers to a group.

EXAMPLE Arrangements were made for the members of our *party* to have lunch after the tour.

per

Per is a common business term that means "by means of," "through," or "on account of," and in these senses it is appropriate.

EXAMPLES *per* annum, *per* capita, *per* diem, *per* head

When used to mean "according to" (*per* your request, *per* your order), the expression is business **jargon** at its worst and should be avoided. Equally annoying is the phrase *as per.*

CHANGE *Per your request,* I am enclosing the production report.
TO *As you requested,* I am enclosing the production report.

CHANGE *As per our discussion,* I will send revised instructions.
TO *As we agreed,* I will send revised instructions.

per cent/percent/percentage

Percent, which is replacing the two-word *per cent,* is used instead of the symbol (%) except in tables.

EXAMPLE Only 25 *percent* of the members attended the meeting.

Percentage, which is never used with numbers, indicates a general size.

EXAMPLE Only a small *percentage* of the managers attended the meeting.

periods

A period (also called a full stop or end stop) usually indicates the end of a declarative sentence. Periods also link (when used as leaders) and indicate omissions (when used as **ellipses**).

Although the primary function of periods is to end declarative sentences, periods also end imperative sentences that are not emphatic enough for an **exclamation mark.**

EXAMPLE Send me any information you may have on the subject.

Periods may also end questions that are really polite requests and questions to which an affirmative response is assumed.

EXAMPLE Will you please send me the specifications.

Periods end incomplete sentences when the meaning is clear from the context. These sentences are common in advertising. (See also **sentence faults.**)

EXAMPLE Bell and Howell's new Double-Feature Cassette Projector will change your mind about home movies. *Because if you can press a button, now you can show movies. Instantly. Easily.*

IN QUOTATIONS

Do not use a period after a declarative sentence that is quoted in the context of another sentence.

CHANGE "There is every chance of success." she stated.
TO "There is every chance of success," she stated.

A period is conventionally placed inside **quotation marks.**

EXAMPLES He liked to think of himself as a "tycoon."
He stated clearly, "My vote is yes."

WITH PARENTHESES

A sentence that ends in a **parenthesis** requires a period *after* the parenthesis.

EXAMPLE The institute was founded by Harry Denman (1902–1972).

If a whole sentence (beginning with an initial **capital letter**) is in parentheses, the period (or any other end mark) should be placed inside the final parenthesis.

EXAMPLE The project director listed the problems facing her staff. (This was the third time she had complained to the board.)

CONVENTIONAL USES OF PERIODS

Use periods after initials in names.

EXAMPLES W. T. Grant, J. P. Morgan

Use periods as decimal points with numbers.

EXAMPLES 109.2, $540.26, 6.9%

Use periods to indicate **abbreviations.**

EXAMPLES Ms., Dr., Inc.

When a sentence ends with an abbreviation that ends with a period, do not add another period.

EXAMPLE Please meet me at 3:30 p.m.

Use periods following the **numbers** in numbered **lists.**

EXAMPLE 1.
2.
3.

AS ELLIPSES

When you omit words in quoted material, use a series of three spaced periods—called **ellipsis** dots—to indicate the omission (do not use ellipses to indicate a pause). Such an omission must not detract from or alter the essential meaning of the passage.

ORIGINAL "Technical material distributed for promotional use is sometimes charged for, particularly in high-volume distribution to educational institutions, although prices for these publications are not uniformly based on the cost of developing them."

WITH OMISSION "Technical material distributed for promotional use is sometimes charged for . . . although prices for these publications are not uniformly based on the cost of developing them."

When introducing a **quotation** that does not begin with the first word of a sentence (that is, when your quotation starts somewhere in the middle of a sentence of the material from which you are quoting), you do not need ellipsis dots, as the lowercase letter with which you begin the quotation already indicates an omission.

ORIGINAL "When the programmer has determined a system of runs, he must create a systems flowchart to provide a picture of the data flow through the system."

WITH OMISSION The booklet states that the programmer "must create a systems flowchart to provide a picture of the data flow through the system."

If an ellipsis follows the end of a sentence, retain the period at the end of the sentence and add the three ellipsis dots.

ORIGINAL "During the year, every department participated in the development of a centralized computer system. The basic plan centered on the use of the computer as a cost reduction tool. At the beginning of the year, each department received a booklet explaining the purpose of the system."

WITH OMISSION "During the year, every department participated in the development of a centralized computer system. . . . At the beginning of the year, each department received a booklet explaining the purpose of the system."

AS LEADERS

When spaced periods are used in a **table** to connect one item to another, they are called *leaders*. The purpose of leaders is to help the reader align the data.

150 lbs	1.7 psi
175 lbs	2.8 psi
200 lbs	3.9 psi

The most common use of leaders is in **tables of contents**.

PERIOD FAULTS

The incorrect use of a period is sometimes referred to as a *period fault.* When a period is inserted prematurely, the result is a **sentence fragment.**

CHANGE After a long day at the office in which we finished the report. We left hurriedly for home.
TO After a long day at the office in which we finished the report, we left hurriedly for home.

When a period is left out, the result is a "fused," or run-on, sentence. Be careful never to leave out necessary periods.

CHANGE Bill was late for ten days in a row Ms. Sturgess had to fire him.
TO Bill was late for ten days in a row. Ms. Sturgess had to fire him.

(See also **sentencc faults.**)

person

Person refers to the form of a **personal pronoun** that indicates whether the **pronoun** represents the speaker, the person spoken to, or the person (or thing) spoken about. A pronoun representing the speaker is in the first person.

EXAMPLE *I* could not find the answer in the manual.

If the pronoun represents the person or persons spoken to, the pronoun is in the second person.

EXAMPLE *You* are going to be a good supervisor.

If the pronoun represents the person or persons spoken about, the pronoun is in the third person.

EXAMPLE *They* received the news quietly.

The accompanying table shows first, second, and third person pronouns. Identifying pronouns by person helps the writer avoid illogi-

Person	Singular	Plural
First	I, me, my	we, ours, us
Second	you, your	you, your
Third	he, him, his she, her, hers it, its	they, them, their

cal shifts from one person to another. A common error is to shift from the third person to the second person.

CHANGE *Managers* should spend the morning hours on work requiring mental effort, for *your* mind is freshest in the morning.
TO *Managers* should spend the morning hours on work requiring mental effort, for *their* minds are freshest in the morning.
OR *You* should spend the morning hours on work requiring mental effort, for *your* mind is freshest in the morning.

(See also **number, case,** and **one.**)

personal/personnel

Personal is an **adjective** meaning "of or pertaining to an individual person."

EXAMPLE He left work early because of a *personal* problem.

Personnel is a **noun** meaning a "group of people engaged in a common job."

EXAMPLE All *personnel* should pick up their paychecks on Thursday.

Be careful not to use *personnel* when the word you need is *persons* or *people.*

CHANGE The remaining two *personnel* will be moved next Thursday.
TO The remaining two *persons* will be moved next Thursday.

personal pronouns

The personal pronouns are *I, me, my, mine; you, your, yours; he, him, his; she, her, hers; it, its; we, us, our, ours;* and *they, them, their, theirs.*

Don't attempt to avoid using the personal pronoun *I* when it is natural. The use of unnatural devices to avoid using *I* is more likely to call attention to the writer than would use of the pronoun *I.*

CHANGE *The writer* wishes to point out that the proposed solution has several weaknesses.
TO *I* wish to point out that the proposed solution has several weaknesses.
OR The proposed solution has several weaknesses.

Also, avoid substituting the **reflexive pronouns** *myself* for *I* or *me,* or *ourselves* for *we* or *us*.

CHANGE Joe and *myself* worked all day on it.
TO Joe and *I* worked all day on it.
CHANGE He gave it to my assistant and *myself.*
TO He gave it to my assistant and *me.*
CHANGE This decision cannot be made by managers like *myself.*
TO This decision cannot be made by managers like *me.*
CHANGE Nobody can solve this problem but *ourselves.*
TO Nobody can solve this problem but *us.*

The use of *we* for general reference ("*We* are living in a time of high inflation") is acceptable. But using the editorial *we* to avoid *I* is pompous and stiff. It may also assume, presumptuously, that "we" all agree.

CHANGE *We* must state unequivocally that *we* disapprove of such practices.
TO *I* must state unequivocally that *I* disapprove of such practices.

Bear in mind that the use of *we* can be legally construed to commit the writer's company as well as the writer personally. If you are speaking for yourself—even in your role as an employee—it is better to use *I*.

One is also acceptable for general reference. Be aware, however, that it is impersonal and somewhat stiff and that overdoing it makes your writing awkward.

CHANGE *One* must be careful about *one's* work habits lest *one* become sloppy.
TO *People* must be careful about *their* work habits lest *they* become sloppy.

(See also **one** and **point of view.**)

persons/people

When we use *persons,* we are usually referring to individual people thought of separately.

EXAMPLE We need to find three qualified *persons* to fill the vacant positions.

When we say *people,* we are identifying a large or anonymous group.

EXAMPLE Many *people* have never even heard of our product.

persuasion

Persuasion in writing attempts to convince the **reader** to adopt the writer's point of view. The range of persuasive writing required on the job varies widely. Technical writers may use persuasive techniques when writing **memorandums,** pleading for safer working conditions in a shop or factory, justifying the expense of a new program, or writing a **proposal** for a multimillion-dollar contract. Persuasive writing is also found in many types of **correspondence,** especially when the **"you" viewpoint** is used.

In persuasive writing, the way that you present your ideas is as important as the ideas themselves. You must support your appeal with **logic**—a sound presentation of facts, statistics, and examples. You must also acknowledge any real or possible conflicting opinions. Only then can you demonstrate and argue for the merit of your point of view. But remember your audience, and take into account your readers' feelings. Always maintain a positive **tone.** And be careful not to wander from your main point—avoid ambiguity and never make trivial, irrelevant, or false claims. But if a reader may be skeptical of your main idea, you may need to build up to it with strong, specific points or arguments that support it. (This *inductive* method is illustrated in **specific-to-general method of development.**)

The passage presented in Figure 1, on page 498, is taken from a guide on the selection and proper use of chainsaws. The intended readers, loggers, use chainsaws daily in their work. The author explains the necessity of maintaining the proper tension on the saw's chain. In explaining some of the technical problems that had to be overcome so that the proper tension could be maintained under all conditions, the author convincingly makes the case that chainsaws of a particular design are the most effective.

Notice that the author acknowledges that this design is not entirely free of potentially harmful side effects: a saw with this bar will kick back harder than others will, and the bar itself is more fragile. By acknowledging negative details or opposing views, a writer also

For maximum chain life, the chain should be properly tensioned. Maintaining the proper tension has always been difficult with solid-nosed bars because the chain develops a great deal of heat from friction as it slides over the nose of the bar. The heat causes the chain to expand and hence become too slack. If the chain is properly tensioned when cold, it will be too slack when running. If the chain is tensioned after it warms up, then stopping the saw long enough to refuel it might result in the chain shrinking to the bar so tight that it will not turn.

The first attempt at solving these problems was the roller-nosed bar. This bar cuts down on the friction to a considerable extent, but is somewhat fragile, and since the chain is unsupported during part of its travel, it is not good practice to cut with the tip of the bar. The sprocket-nosed bar seems to have solved these problems. Here a thin sheet-metal sprocket, roughly the thickness of the drive links, is all but hidden inside the nose of the bar. As the chain approaches the end of the bar, the sprocket lifts the drive link and therefore keeps the saw from running on the rails of the bar as it travels around the nose. This cuts friction while supporting the chain well enough so that the nose of the bar can be used for cutting. It is true that a saw with a sprocket-nosed bar will kick back harder than other saws because there is so little friction between the chain and the bar, and the sprocket nose may be somewhat more fragile, particularly if it is pinched in a cut. On the other hand, its advantages often outweigh its disadvantages. It gives faster cutting, better chain life, and better chain tension, which in itself is a safety factor. Most sprockets and bearings for bars can be replaced by a dealer, although if the nose portion of the bar is bent, the bar is ruined. Some sprocket-nosed bars are available with the entire tip of the bar made as a section that may be easily replaced by the user.

—R. P. Sarna, *Chain Saw Manual* (Danville, Ill.: The Interstate Printers & Publishers, Inc., 1979), pp. 8-9.

Figure 1 Example of Persuasive Writing

gains credibility through the readers' impression of him or her as a person. That's why the appearance of a **letter of application** and a résumé, for example, is important. (See also **methods of development** and **forms of discourse.**)

phenomenon/phenomena

A *phenomenon* is an observable thing, fact, or occurrence. Its plural form is *phenomena.*

EXAMPLES The natural *phenomenon* of earth tremors *is* a problem we must anticipate in designing the California installation.

The *phenomena* associated with atomic fission *were* only recently understood.

photographs

Photographs are vital to many publications. Pictures are the best way to show the surface appearance of an object or to record an event or the development of a phenomenon over a period of time. Not all representations, however, call for photographs. They cannot depict the internal workings of a mechanism or below-the-surface details of objects or structures. Such details are better represented in **drawings** or **schematic diagrams**.

HIGHLIGHTING PHOTOGRAPHIC SUBJECTS

Stand close enough to the object so that it fills your picture frame. To get precise and clear photographs, choose camera angles carefully. A camera will photograph only what it is aimed at; accordingly, select the important details and the camera angles that will record them. To show relative size, place a familiar object—such as a ruler, a book, or a person—near the object being photographed.

USING COLOR

The preparation and printing of color photographs are complex technical tasks performed by graphics and printing experts. If you are planning to use color photographs in your publication, discuss with these experts the type and quality of photographs required. Generally, they prefer color transparencies (slides) and color negatives (negatives of color prints) for reproduction.

The advantages of using color illustrations are obvious: color is a good way of communicating crucial information. In medical, chemical, geological, and botanical publications, to name but a few, readers often need to know exactly what an object or phenomenon looks like. In these circumstances, color reproduction is the only legitimate option available.

Be aware, however, that color reproduction is significantly more expensive than black and white. Color can also be tricky to reproduce accurately without losing contrast and vividness. For this reason, the original must be sharply focused and rich in contrasts.

TIPS FOR USING PHOTOGRAPHS

Like all illustrative materials, photographs must be handled carefully. When preparing photographs for printing, observe the following guidelines:

1. Mount photographs on white bond paper with rubber cement or another adhesive and allow ample margins.
2. If the photograph is the same size as the paper, type the caption, figure and page numbers, and any other important information on labels and fasten them to the photograph with rubber cement or another adhesive. The labels that identify key features in the photograph are referred to as *callouts*. (Photographs are given figure numbers in sequence with **illustrations** in a publication; see Figure 1.)
3. Position the figure number and caption so that the reader can view them and the photograph from the same orientation.
4. Do not draw crop marks (lines showing where the photo should be trimmed for reproduction) directly across a photograph. Draw them at the very edge of the photograph.

Figure 1 Example of a Photograph

5. Do not write on a photograph, front or back. Tape a tissue-paper overlay over the face of the photograph, and then write very lightly on the overlay with a soft-lead pencil. Never write on the overlay with a ball-point pen.
6. Do not use paper clips or staples directly on photographs.
7. Do not fold or crease photographs.
8. If only a color photograph or slide is available for black-and-white reproduction, have a photographer produce a black-and-white glossy copy or slide for printing. Otherwise, the printed image will not have an accurate tone.

phrases

Below the level of the sentence, there are two ways to combine words into groups: by forming **clauses,** which combine **subjects** and **verbs,** and by forming phrases, which are based on **nouns,** nonfinite verb forms, or verb combinations without subjects. A phrase is the most basic meaningful group of words; it cannot make a full statement as a clause can, however, because it does not contain a subject and a verb.

EXAMPLE He encouraged his staff (clause) *by his calm confidence.* (phrase)

A phrase may function as an **adjective,** an **adverb,** a noun, or a verb.

EXAMPLES The subjects *on the agenda* were all discussed. (adjective)
We discussed the project *with great enthusiasm.* (adverb)
Working hard is her way of life. (noun)
The chief engineer *should have been notified.* (verb)

Even though phrases function as adjectives, adverbs, nouns, or verbs, they are normally named for the kind of word around which they are constructed—**preposition, participle, infinitive, gerund,** verb, or noun.

PREPOSITIONAL PHRASES

A preposition is a word that shows the relationship of its **object** to another word; that is, it combines with its object (a noun or a **pronoun**) to form a modifying phrase. A **prepositional phrase,** then, consists of a preposition plus its object (the noun or pronoun) and its **modifiers.**

EXAMPLE *After the meeting,* the regional managers adjourned *to the executive dining room.*

PARTICIPIAL PHRASES

A participle is any form of a verb that is used as an adjective. A **participial phrase** consists of a participle plus its object and its modifiers.

EXAMPLE *Looking very pleased with himself,* the sales manager reported on the success of the policies he had introduced.

INFINITIVE PHRASES

An infinitive is the bare form of a verb (*go, run, talk*) without the restrictions imposed by **person** and **number;** an infinitive is generally preceded by the word *to* (which is usually a preposition but in this use is called the sign, or mark, of the infinitive). An **infinitive phrase** consists of the word *to* plus an infinitive and any objects or modifiers.

EXAMPLE *To succeed in this field,* you must be willing *to assume responsibility.*

GERUND PHRASES

A gerund phrase, which consists of a gerund plus any objects or modifiers, always functions as a noun.

EXAMPLES *Preparing an annual report* is a difficult task. (subject)
She liked *running the department.* (direct object)

VERB PHRASES

A **verb phrase** consists of a main verb and its **helping verb.**

EXAMPLE Company officials discovered that a computer *was emitting* more data than it *had been asked* for. After investigation, police suggested that an unauthorized program *had been run* through the computer at the coded command of another computer belonging to a different company. The police obtained a warrant to search for electronic impulses in the memory of the suspect machine. The case (involving trade secrets) and its outcome *should make* legal history.

NOUN PHRASES

A noun phrase consists of a noun and its modifiers.

EXAMPLES *Many large companies* use computers.
Have *the two new employees* fill out *these forms.*

plagiarism

To use someone else's exact words without **quotation marks** and appropriate credit, or to use the unique ideas of someone else without acknowledgment, is known as plagiarism. In publishing, plagiarism is illegal; in other circumstances it is, at the least, unethical. (For detailed guidance on quoting correctly, see **quotations.**)

You may quote or **paraphrase** the words and ideas of another if you document your source. (See **documenting sources.**) However, if you intend to publish or reproduce and distribute material in which you have included quotations from published works, you may have to obtain written permission to do so from the **copyright** holder. Although you need not enclose paraphrased material in quotation marks, you must document the source. Paraphrased ideas are taken from someone else whether or not the words are identical. Paraphrasing a passage without citing the source is permissible only when the information paraphrased is common knowledge in a field. (Common knowledge refers to historical, scientific, geographical, technical, and other types of information on a topic readily available in handbooks, manuals, atlases, and other references.)

plurals

Nouns, pronouns, and **verbs** may be either singular or plural.

NOUNS

Nouns normally form the plural by simply adding *s* to their singular forms.

EXAMPLES Many new business *ventures* have failed in the past ten *years*.
Partners are not always personal *friends*.

Nouns ending in *s, z, x, ch,* and *sh* form the plural by adding *es*.

EXAMPLES Partners in successful *businesses* are not always friends.
How many size *sixes* did we produce last month?
The letter was sent to all area *churches*.
Technology should not inhibit our individuality; it should fulfill our *wishes*.
The young engineers were known throughout the company as *whizzes*.
The instructor gave three *quizzes* during the course.

Nouns ending in a consonant plus *y* form the plural by changing *y* to *ies*.

EXAMPLE The store advertises prompt delivery but limits the number of *deliveries* it will make in one day.

Some nouns ending in *o* add *es* to form the plural; others add only *s*.

EXAMPLES One tomato plant produced twelve *tomatoes*.
We installed two *dynamos* in the plant.

Some nouns ending in *f* or *fe* add *s* to form the plural; others change the *f* or *fe* to *ves*.

EXAMPLES cliff/cliffs, fife/fifes, knife/knives

Some nouns require an internal change to form the plural.

EXAMPLES woman/women, man/men, mouse/mice, goose/geese

Some nouns do not change in the plural.

EXAMPLE *Fish* swam lazily in the brook while *deer* mingled with *sheep* in a nearby meadow.

Compound nouns form the plural in the main word.

EXAMPLES sons-in-law, high schools

Compound nouns written as one word add *s* to the end.

EXAMPLE Use seven *spoonfuls* of freshly ground coffee to make seven cups of coffee.

PRONOUNS

All pronouns except *you* change to form the plural.

EXAMPLES I/we, he/they, she/they, it/they

A problem sometimes occurs because the masculine pronoun has traditionally been used to refer to both sexes.

EXAMPLE *Each* may stay or go as *he* chooses.

To avoid the sexual bias implied in such usage, use *he or she* or the plural form of the pronoun, *they*.

CHANGE *Each* may stay or go as *he* chooses.
TO *Each* may stay or go as *he or she* chooses.
OR *All* may stay or go as *they* choose.

If you use the plural form of the pronoun, be sure to change the indefinite pronoun *each* to its plural form, *all*.

VERBS

Verbs may also be singular or plural.

EXAMPLES The company *pays* high taxes. (singular)
The companies *pay* high taxes. (plural)

Most verbs show the singular of the third **person,** present **tense,** indicative **mood** by adding an *s* or *es.*

EXAMPLES he *stands,* she *works,* it *goes*

The verb *to be* normally changes its form to indicate the plural.

EXAMPLES I *am* ready to begin work. (singular)
We *are* ready to begin work. (plural)

If in doubt about the plural form of a word, look it up in a **dictionary.** Most dictionaries give the plural if it is formed in any way other than by adding *s* or *es.*

point of view

Point of view indicates the writer's relation to the information presented, as reflected in the use of **person.** The writer usually expresses point of view in first-, second-, or third-person **personal pronouns.** The use of the first person indicates that the narrator is a participant or observer ("This happened to me," "I saw that"). The second and third person indicate that the narrator is writing about other people or something impersonal or is giving directions, instructions, or advice ("This happened to her, to them, to it"; "Enter the data after pressing the ENTER key once"). The writer may also avoid using personal pronouns, thus adopting an impersonal point of view.

EXAMPLE It is regrettable that the material shipped on March 12 is defective.

Consider the same sentence written from the personal, first-person point of view.

EXAMPLE I regret that we cannot accept your shipment of March 12.

Although the meaning of both sentences is the same, the sentence with the personal point of view indicates that two people are involved in the communication. Years ago technical people preferred the impersonal point of view because they thought it made writing sound more objective and professional. Unfortunately, however, the impersonal point of view often prevents clear and direct communication and can cause misunderstanding. Today, good technical writers adopt a more personal point of view whenever it is appropriate.

Writers should not avoid *I* by using *one* when they are really talking about themselves. They do not increase objectivity but merely make the statement impersonal.

CHANGE *One* can only conclude that the absorption rate is too fast.
TO *I* can only conclude that the absorption rate is too fast.

Writers should never use *the writer* to replace *I* in a mistaken attempt to sound formal or dignified.

CHANGE *The writer* believes that this project will be completed by the end of June.
TO *I* believe that this project will be completed by the end of June.

However, the personal point of view should not be used when an impersonal point of view would be more appropriate or more effective.

CHANGE I am inclined to think that each manager should attend the final committee meeting to hear the committee's recommendations.
TO Each manager should attend the final committee meeting to hear the committee's recommendations.

The preceding examples make clear that when the impersonal point of view is used, the focus shifts from the writer to the person or thing being discussed. For the subject matter to receive more emphasis than either the writer or the reader does, an impersonal point of view should be used.

EXAMPLE The evidence suggests that the absorption rate is too fast.

In a letter on company stationery, use of the pronoun *we* may be interpreted as reflecting company policy, whereas *I* clearly reflects personal opinion. Which pronoun to use should be decided according to whether the matter discussed in the letter is a corporate or an individual concern.

EXAMPLES *I* appreciate your suggestion regarding our need for more community activities.
We appreciate your suggestion regarding our need for more community activities.

The pronoun *we* may commit an organization to what the writer says. If the following statement accurately reflects an organization's opinion, with its implication that the proposal being singled out will be accepted, it is justified.

EXAMPLE *We* believe that your proposal was the best submitted.

If, however, the writer can speak only personally instead of for the organization, the statement should be hedged so that the reader understands that it represents the writer's opinion rather than the organization's official position.

EXAMPLE I *personally* believe that your proposal was the best submitted, but the policy commitee will make the final decision.

Whether the writer adopts a personal or an impersonal point of view should depend on the objective and the reader. For example, in a business letter to an associate, the writer will most likely adopt a personal point of view. But in a report to a large group, the writer will probably choose to emphasize the subject by adopting an impersonal point of view.

In descriptive writing, point of view refers to the physical point or vantage of the writer's presentation. The writer can present a scene from a fixed or moving point of view. Once a vantage has been taken, however, the writer should not shift it without indicating to the reader that the perspective has changed. (See also **correspondence**.)

positive writing

Presenting positive information as though it were negative is a trap that technical writers fall into quite easily because of the complexity of the information they must write about. It is a practice that confuses the **reader,** however, and one that should be avoided.

CHANGE If the error does *not* involve data transmission, the special function will *not* bc used.

TO The special function is used only if the error involves data transmission.

In the first sentence, the reader must reverse two negatives to understand the exception that is being stated; the second sentence presents an exception to a rule in a straightforward manner.

On the other hand, negative facts or conclusions should be stated negatively; stating a negative fact or conclusion positively can mislead the reader.

CHANGE For the first quarter of this year, employee exposure to airborne lead has been maintained to within 10 percent of acceptable state health standards.

TO For the first quarter of this year, employee exposure to airborne lead continues to be 10 percent above acceptable state health standards.

Even if what you are saying is negative, do not use any more negative words than are necessary or words that are more negative than necessary.

CHANGE We are withholding your shipment until we receive your payment.

TO We will forward your shipment as soon as we receive your payment.

possessive adjectives (see **possessive case**)

possessive case

A **noun** or **pronoun** is in the possessive case when it represents a person or thing owning or possessing something.

EXAMPLE A recent scientific analysis of *New York City's* atmosphere concluded that New Yorkers on the street took into *their* lungs the equivalent in toxic materials of 38 cigarettes per person daily.

Singular nouns show the possessive by adding an **apostrophe** and *s*.

EXAMPLE company/company's

Nouns that form their **plurals** by adding an *s* show the possessive by placing an apostrophe after the *s* that forms the plural.

EXAMPLES a managers' meeting/the technicians' handbook

The following **list** shows the relationships among singular, plural, and possessive nouns that form their plurals by adding *s* or changing *y* to *ies*.

singular	company	employee
singular possessive	company's	employee's
plural	companies	employees
plural possessive	companies'	employees'

Nouns that do not add *s* to form their plurals add an apostrophe and an *s* to show possession in both the plural and the singular forms.

singular	child	man
singular possessive	child's	man's
plural	children	men
plural possessive	children's	men's

Singular nouns that end in *s* form the possessive either by adding only an apostrophe or by adding both an apostrophe and an *s*.

EXAMPLES a *waitress'* uniform/an *actress'* career
a *waitress's* uniform/an actress's career

Singular nouns of one syllable always form the possessive by adding both an apostrophe and an *s*.

EXAMPLE The *boss's* desk was cluttered.

Although exceptions are relatively common, it is a good rule of thumb to use an apostrophe and an *s* with nouns referring to persons and living things and to use an *of* phrase for possessive nouns referring to inanimate objects.

EXAMPLES The *chairman's* address was well received.
The keys *of the typewriter* were sticking.

If this rule leads to awkwardness or wordiness, however, be flexible.

EXAMPLES The *company's* plants are doing well.
The *plane's* landing gear failed.

Several indefinite pronouns (*any, each, few, most, none,* and *some*) form the possessive only in *of* phrases (*of any, of some*). Others, however, use an apostrophe (*anyone's*).

The established **idiom** calls for the possessive case in many stock phrases.

EXAMPLES a *day's* journey, a *day's* work, a *moment's* notice, at his *wit's* end, the *law's* delay

In a few cases, the idiom even calls for a double possessive using both the *of* and the *'s* forms.

EXAMPLE That colleague *of* George*'s* was at the conference.

When several words make up a single term, add *'s* to the last word only.

EXAMPLES The *chairman of the board's* statement was brief.
The *Department of Energy's* new budget shows increased revenues of $192 million for uranium enrichment.

With coordinate nouns, the last noun takes the possessive form to show joint possession.

EXAMPLE *Michelson and Morely's* famous experiment on the velocity of light was made in 1887.

To show individual possession with coordinate nouns, each noun should take the possessive form.

EXAMPLE The difference between *Thomasson's* and *Silson's* test results were statistically insignificant.

To form the possessive of a **compound word,** add *'s*.

EXAMPLES the *vice-president's* car, the *pipeline's* diameter, the *antibody's* reaction.

When a noun ends in multiple consecutive *s* sounds, form the possessive by adding only an apostrophe.

EXAMPLES Jesus' disciples, Moses' journey

Do not use an apostrophe with possessive pronouns.

EXAMPLES *yours, its, his, ours, whose, theirs*

It's is a contraction of *it is,* not the possessive form of *it.*

In the names of places and institutions, the apostrophe is often omitted.

EXAMPLES Harpers Ferry, Writers Book Club

The use of possessive pronouns does not normally cause problems except with **gerunds** and indefinite pronouns. Several indefinite pronouns (*all, any, each, few, most, none,* and *some*) require *of* phrases to form the possessive case.

EXAMPLE Both dies were stored in the warehouse, but rust had ruined the surface *of each.*

Others, however, use an apostrophe.

EXAMPLE *Everyone's* contribution is welcome.

Only the possessive form of a pronoun should be used with a gerund.

EXAMPLES The safety officer insisted in *my* wearing a respirator.
Our monitoring was not affected by changing weather conditions.

practicable/practical

Practicable means that something is possible or feasible. *Practical* means that something is both possible and useful.

EXAMPLE The program is *practical,* but considering the company's recent financial problems, is it *practicable?*

Practical, not *practicable,* is used to describe a person, implying common sense, a commitment to what works, rather than to theory.

predicate adjectives (see **adjectives**)

predicate nominatives (see **predicates**)

predicates

The predicate is that part of a sentence that contains the main **verb** and any other words used to complete the thought of the sentence (the verb's **modifiers** and **complements**). The principal part of the predicate is the verb, just as a **noun** (or noun substitute) is the principal part of the **subject.**

EXAMPLE Bill *piloted the airplane.*

The simple **predicate** is the verb (or **verb phrase**) alone; the complete predicate is the verb and its modifiers and complements.

A compound predicate consists of two or more verbs with the same subject. It is an important device for economical writing.

EXAMPLE The company *tried* but *did not succeed* in that field.

A predicate nominative is a noun construction that follows a **linking verb** and renames the **subject.**

EXAMPLES He is my *lawyer.* (noun)
His excuse was *that he had been sick.* (noun clause)

The predicate nominative is one kind of **subjective complement;** the other is the predicate adjective.

prefixes

A prefix is a letter, or letters, placed in front of a root word that changes its meaning, often causing the new word to mean the opposite of the root word. When a prefix ends with a vowel and the root word begins with the same vowel, the prefix may be separated from the root word with a **hyphen** (re-enter, co-operate, re-elect); the second vowel may be marked with a dieresis (reënter, coöperate, reëlect), although this practice is rare now; or the word may be written with neither hyphen nor dieresis (reenter, cooperate, reelect). The hyphen and the dieresis are visual aids that help the **reader** recognize that the two vowels are pronounced differently. Since typewriters do not have the dieresis mark, the hyphen is the more practical way to help the reader.

Except between identical vowels, the hyphen rarely appears between a prefix and its root word. At times, however, a hyphen is necessary for clarity of meaning; for example, *reform* means "correct" or "improve," and *re-form* means "change the shape of."

Root Word	*With Prefix*
enter	*re*enter (enter again)
acceptable	*un*acceptable (not acceptable)
operate	*co*operate (operate together)
symmetrical	*a*symmetrical (not symmetrical)
honest	*dis*honest (not honest)
science	*meta*science (beyond science)

preparation

To write clearly and effectively does not require innate creative talent. It does require a systematic approach, however, which should logically begin with the proper preparation.

Preparation consists of determining the **objective** of your writing project, its **readers,** and the **scope** within which you will cover your subject. Although these fundamentals of the writing process are often overlooked, you must resolve them if you hope to write well. They are the foundations on which all the other steps of the writing process are built.

OBJECTIVE

What *exactly* do you want your readers to know or be able to do when they have finished reading what you have written? When you can answer this question—again, *exactly*—you will have determined the objective of your writing. A good test of whether you have formulated your objective adequately is to state it in a single sentence. This statement may also function as the thesis statement in your **outline,** although such a function should not be a requirement of your stated objective. (For further discussion, see **objective**.)

READERS

Know who your readers are and learn certain key facts about them, such as their educational level, technical knowledge, and needs relative to your subject. Their level of technical knowledge, for example, should determine whether you need to cover the fundamentals of your subject and which terms you must define. (For more detail, see **readers**.)

SCOPE

If you know the objective of your writing project and your readers' needs, you will know the type and amount of detail you will need to include in your writing. This is called *scope,* which may be defined as the depth and breadth of your coverage of the subject. If you do not determine your scope of coverage, you will not know how much or what kind of information to include, and therefore you are likely to invest unnecessary work during the **research** phase of your writing project.

One of the most common—and damaging—mistakes that technical writers make is to begin writing too soon. The preparation stage of the writing process is analogous to focusing a camera before taking a picture. By determining who your reader is, what your objective is, and what your scope of coverage should be, you are bringing your whole writing effort into focus before beginning to write the draft. As a result, your **topic** will be much more clearly focused for your reader. (See also the Checklist of the Writing Process.)

prepositional phrases

A **preposition** is a word that shows relationship and combines with a **noun** or **pronoun** (its **object**) to form a modifying **phrase.** A prepositional phrase, then, consists of a preposition plus its object and the object's **modifiers.**

EXAMPLE *After the meeting,* the district managers adjourned *to the executive dining room.*

Prepositional phrases, because they normally modify nouns or **verbs,** usually function as **adjectives** or **adverbs.**

A prepositional phrase may function as an adverb of motion.

EXAMPLE Turn the dial four degrees *to the left.*

A prepositional phrase may function as an adverb of manner.

EXAMPLE Answer customers' questions *in a courteous fashion.*

A prepositional phrase may function as an adverb of placc.

EXAMPLE We ate lunch *in the company cafeteria.*

A prepositional phrase may function as an adjective.

EXAMPLE Garbage *with a high paper content* has been turned into protein-rich animal food.

When functioning as adjectives, prepositional phrases follow the nouns they modify.

EXAMPLE Here are the results *of our study.*

When functioning as adverbs, prepositional phrases may appear in different places.

EXAMPLES *In residential and farm wiring,* one of the wires must always be grounded.

One of the wires must always be grounded *in residential and farm wiring.*

Be careful when using prepositional phrases, as separating a prepositional phrase from the noun it modifies can cause **ambiguity**.

CHANGE *The man* standing by the drinking fountain *in the gray suit* is our president.

TO *The man in the gray suit* who is standing by the drinking fountain is our president.

(See also **misplaced modifiers**.)

prepositions

A preposition is a word that links a **noun** or **pronoun** (its **object**) to another sentence element, by expressing such relationships as direction (*to, into, across, toward*), location (*at, in, on, under, over, beside, among, by, between, through*), time (*before, after, during, until, since*), or figurative location (*for, against, with*). Although only about seventy prepositions exist in the English language, they occur frequently. Together, the preposition, its object, and the object's modifiers form a **prepositional phrase,** which acts as a **modifier.**

The object of a preposition, the word or phrase following the preposition, is always in the objective **case.** This situation gives rise to a problem in such constructions as "between you and *me*," a phrase that is frequently and incorrectly written as "between you and *I*." *Me* is the objective form of the pronoun, and *I* is the subjective form.

Many words that function as prepositions also function as **adverbs.** If the word takes an object and functions as a connective, it is a preposition; if it has no object and functions as a modifier, it is an adverb.

EXAMPLES The manager sat *behind* the desk *in* his office. (prepositions)

The customer lagged *behind*; then she came *in* and sat *down*. (adverbs)

AVOIDING ERRORS WITH PREPOSITIONS

Do not use redundant prepositions, such as "off *of*," "in back *of*," and "inside *of*."

CHANGE The client arrived *at about* four o'clock.

TO The client arrived *at* four o'clock. (to be exact)

OR The client arrived *about* four o'clock. (to be approximate)

Do not omit necessary prepositions.

CHANGE He was oblivious and not distracted by the view from his office window.

TO He was oblivious *to* and not distracted *by* the view from his office window.

Avoid unnecessarily adding the preposition *up* to **verbs.**

CHANGE Call *up* and see if he is in his office.

TO Call and see if he is in his office.

USING A PREPOSITION AT THE END OF A SENTENCE

If a preposition falls naturally at the end of a sentence, leave it there.

EXAMPLE I don't remember which file I put it *in*.

Be aware, however, that a preposition at the end of a sentence can be an indication that the sentence is awkwardly constructed.

CHANGE The branch office is where he was *at*.

TO He was at the branch office.

COMMON USES OF PREPOSITIONS

Certain verbs, adverbs, and **adjectives** are used with certain prepositions. For example, we say "interested *in*," "aware *of*," "devoted *to*," "equated *with*," "abstain *from*," "adhere *to*," "conform *to*," "capable *of*," "comply *with*," "object *to*," "find fault *with*," "inconsistent *with*," "independent *of*," "infer *from*," and "interfere *with*." (See also **idioms.**)

USING PREPOSITIONS IN TITLES

When a preposition appears in a title, it is capitalized if it is the first word in the title or if it has four letters or more. (See also **capital letters.**)

EXAMPLES *Composition for the Business and Technical World* was reviewed recently by the *Journal of Technical Communications*.

The book *Managing Like Mad* should be taken seriously in spite of its title.

principal/principle

Principal, meaning an "amount of money on which interest is earned or paid" or a "chief official in a school or court proceeding," is sometimes confused with *principle,* meaning a "basic truth or belief."

EXAMPLES The bank will pay 6.5 percent per month on the *principal.*
She is a person of unwavering *principles.*
He sent a letter to the *principal* of the high school.

Principal is also an adjective, meaning "main" or "primary."

EXAMPLE My *principal* objection is that it will be too expensive.

process explanation

Many kinds of technical writing explain a process, an operation, or a procedure. The explanation involves putting in the appropriate order the steps that a specific mechanism or system uses to accomplish a certain result. The process itself might range from the legal steps necessary to form a corporation, to the steps necessary to develop a roll of film. At the least, you should divide the process into distinct steps and present them in their normal order.

In your opening **paragraph,** tell your **reader** why it is important to become familiar with the process you are explaining. Before you explain the steps necessary to form a corporation, for example, you could cite the tax savings that incorporation would permit. To provide your reader with a framework for the details that will follow, you might present a brief overview of the process. Finally, you might describe how the process works in relation to a larger whole. In explaining the air brake system of a large dump truck, you might note that the braking system is one part of the vehicle's air system, which also controls the throttle and transmission-shifting mechanisms.

A process explanation can be long or short, depending on how much detail is necessary. The following description of the way in which a camera controls light to expose a photographic film fits into one paragraph:

EXAMPLE The camera is the basic tool for recording light images. It is simply a box from which all light is excluded except that which passes through a small opening at the front. Cameras are equipped with various devices for controlling the light rays as they enter this opening. At the press of a button, a mechanical blade or curtain, called a shutter, opens and closes automatically. During the fraction of a second that the shutter is open, the light reflected from the subject toward which the camera is

Surface Mining of Coal

The process of removing the earth, rock, and other strata (called *overburden*) to uncover an underlying mineral deposit is generally referred to as surface mining. Strip mining is a specific kind of surface mining in which all the overburden is removed in strips, one cut at a time. Three types of strip-mining methods are used to mine coal: *area, contour,* and *mountain-top removal.* Which method is used depends upon the topography of the area to be mined.

AREA STRIP MINING

Area strip mining is used in regions of flat to gently rolling terrain, like those found in the Midwest and West. Depending on applicable reclamation laws, the topsoil may be removed from the area to be mined, stored, and later reapplied as surface material during reclamation of the mined land. Following removal of the topsoil, a trench is cut through the overburden to expose the upper surface of the coal to be mined. The length of the cut generally corresponds to the length of the property or of the deposit. The overburden from the first cut is placed on unmined land adjacent to the cut. After the first cut is completed, the coal is removed and a second cut is made parallel to the first. The overburden (now referred to as spoil) from each of the succeeding cuts is deposited in the adjacent pit from which the coal was just removed. The final cut leaves an open trench equal in depth to the thickness of the overburden plus the coal bed, bounded on one side by the last spoil pile and on the other side by the undisturbed soil. The final cut may be as far as a mile from the first cut. The overburden from all the cuts, unless graded and leveled, resembles the ridges of a giant washboard.

CONTOUR STRIP MINING

In areas of rolling or very steep terrain, such as in the eastern United States, contour strip mining is used. In this method, the overburden is removed from the mineral seam in a pattern that follows the contour line around the hillside. The overburden is then deposited on the downslope side of the cut until the depth of the overburden becomes too great for economical recovery of the coal. This method leaves a bench, or shelf, on the side of the hill, bordered on the inside by a highwall (30 to 100 feet high) and on the other side by a high ridge of spoil.

A method of mining that is often used in conjunction with contour mining is *auger mining.* This method is employed when the overburden becomes too thick and renders contour mining

Figure 1 Explaining a Process

uneconomical and when extraction by underground mining would be too costly or unsafe. In auger mining an instrument bores holes horizontally into the coal seam. The coal can then be removed like the shavings produced by a drill bit. The exposed coal seam in the highwall is left with a continous series of bore holes from which the coal was removed.

MOUNTAIN-TOP REMOVAL

In areas of rolling or steep terrain an adaptation of area mining to conventional contour mining is used; it is called the mountain-top removal method. With this method entire mountain tops are removed down to the coal seam by a series of parallel cuts. This method is economical when the coal lies near the tops of mountains, ridges, or knobs. If there is excess overburden that cannot be stored on the mined land, it may be transported elsewhere.

aimed passes into the camera through a piece of optical glass called the *lens*. The lens focuses, or projects, the light rays onto the wall at the back of the camera. These light reflections are captured on a sheet of film attached to the back wall.

Many process explanations require more details and will, of course, be longer than a paragraph. The example presented in Figure 1 discusses several methods of surface mining for coal. The writer begins with an overview of the elements common to all the processes, defines the terms important to the explanation, and then describes each separately. Transitional words and **phrases** serve to achieve **unity** within paragraphs, and headings mark the **transition** from one process to the next. **Illustrations** often help convey your message.

progress and activity reports

A progress report provides information about a project—its current status, whether it is on schedule, whether it is within the budget, and so on. Progress reports are often submitted by a contracting company to a client company. They are issued at regular intervals throughout the life of a project and state what has been done in a specified interval and what has yet to be done before the project can be completed. The progress report is used primarily with projects that involve many steps over a period of weeks, months, or even years. Progress reports help keep projects running smoothly by allowing management to assign workers, adjust schedules, allocate budgets, or schedule supplies and equipment, as necessary.

All reports in a series of progress reports should have the same **format.** Sincc progress reports are normally sent outside the company, they are often written as letters. (See also **correspondence.**)

The **introduction** to the first progress report should identify the project, any methods and necessary materials, and the date by which the project is to be completed. Subsequent reports should then summarize the progress since the first report.

The body of the progress report should describe the project's present status, including such details as schedules and costs.

The report should end with conclusions and recommendations about changes in the schedule, materials, techniques, and so on. The **conclusion** may also include a statement of the work done and an estimate of future progress.

The example shown in Figure 1 is the first progress report submitted by an electrical contractor (REMCON) to a client (the Arena Committee).

Within an organization, professional employees often submit reports on the progress and current status of all ongoing projects assigned to them by their managers. The managers combine these *activity reports* (also called *status reports*) into larger activity reports that they, in turn, submit to their managers. Activity reports help keep all levels of management aware of the progress and status of projects within an organization.

An activity report includes information about the status of all current projects, including any problems related to their completion, the action currently being taken to resolve any problems, and the writer's plans for the coming month. A manager's activity report should also indicate the current number of employees.

Because the activity report is issued periodically, often monthly, and contains material familiar to its readers, it normally needs no introduction or conclusion. The format of an activity report may vary from company to company, or even among different parts of the same company. The following sections are typical and would be adequate for most situations:

Current Projects
This section lists every project assigned to the employee or manager and summarizes its current status.

REMCON ELECTRIC
5099 Seventh Street, St. Paul, Minnesota 55101 (612) 555-1212

August 20, 19--

Arena Committee
Minnesota Sports
708 N. Case St.
St. Paul, MN 55101

Subject: Arena Rewiring Progress Report

This report, as agreed to in our contract, covers the progress on the rewiring program at the Sports Arena from May 5 to August 15 of 19--. Although the costs of certain equipment are higher than our original bid indicated, we expect to complete the project without going over cost; the speed with which the project is being completed will save labor costs.

Work Completed

On August 15, we finished installing the circuit-breaker panels and meters of Level I service outlets and of all subfloor rewiring. Lighting fixture replacement, Level II service outlets, and the upgrading of stage lighting equipment are in the preliminary stages (meaning that the wiring has been completed but installation of the fixtures has not yet begun).

Costs

Equipment used up to this point has cost $10,800, and labor has cost $31,500 (including some subcontracted plumbing). My estimate for the rest of the equipment, based on discussions with your lighting consultant, is $11,500; additional labor costs should not exceed $25,000.

Work Schedule

I have scheduled the upgrading of stage lighting equipment from August 16 to October 5, the completion of Level II service outlets from October 6 to November 12, and the replacement of lighting fixtures from November 15 to December 17.

Conclusion

Although my original estimate on equipment ($20,000) has been exceeded by $2,300, my original labor estimate ($60,000) has been reduced by $3,500; so I will easily stay within the limits of my original bid. In addition, I see no difficulty in having the arena finished for the Christmas program on December 23.

Sincerely,
John Remcon
John Remcon
President and Owner

ics

Figure 1 Progress Report

> *Current Problems*
> This section details any problems confronting the employee or manager and explains the steps being taken to resolve them.
> *Plans for the Next Period*
> This section projects what the writer expects to achieve on each project during the next month or reporting period.
> *Current Staffing Level*
> This section, included by managers and project leaders, lists the number of subordinates assigned to the writer and correlates that number with the staffing level considered necessary for the projects assigned.

The activity report shown in Figure 2, using a **memorandum** form, was submitted by a manager of software development (Wayne Tribinski) who supervises 11 employees. The reader of the report (Kathryn Hunter), Tribinski's supervisor, is the director of engineering.

pronoun reference

Avoid vague and uncertain references between a **pronoun** and its antecedents.

CHANGE Studs and thick treads make snow tires effective. *They* are implanted with an air gun.
TO Studs and thick treads make snow tires effective. *The studs* are implanted with an air gun.

CHANGE We made the sale and delivered the product. *It* was a big one.
TO We made the sale, which was a big one, and delivered the product.

The **noun** to which a pronoun refers must be clear. Three problems are encountered in regard to pronoun references. One is an ambiguous reference, or one that can be interpreted in more than one way.

CHANGE Jim worked with Tom on the report, but *he* wrote most of it. (Who wrote most of it, Jim or Tom?)
TO Jim worked with Tom on the report, but *Tom* wrote most of it.

A general (or broad) reference, or one that has no real antecedent, is another problem that often occurs when the word *this* is used by itself.

INTEROFFICE MEMORANDUM

Date: June 5, 19--
To: Kathryn Hunter, Director of Engineering KH
From: Wayne Tribinski, Manager, Software Development

Subject: Activity Report for May, 19--

Projects

1. The Problem-Tracking System now contains both software and hardware problem-tracking capabilities. The system upgrade took place over the weekend of May 11 and 12 and was placed online on the 13th.

2. For the Software Training Mailing Campaign, we anticipate producing a set of labels for mailing software-training information to customers by June 10.

3. The Search Project is on hold until the PL/1 training has been completed, probably by the end of June.

4. The project to provide a data base for the Information Management System has been expanded in scope to provide a data base for all training activities. We are in the process of rescheduling the project to take the new scope into account.

5. The Metering Reports project is part of a larger project called "Reporting Upgrade." We have completed the Final Project Requirements and sent it out for review. The Resource Requirements Estimate has also been completed, and Phase Three is scheduled for completion by June 17.

Problems

The Information Management System has been delayed. The original schedule was based on the assumption that a systems analyst who was familiar with the system would work on this project. Instead, the project was assigned to a newly hired systems analyst who was inexperienced and required much more learning time than expected.

Bill Michaels, whose activity report is attached, is correcting a problem in the CNG Software. This correction may take a week.

The Beta Project was delayed for approximately one week because of two problems: interfacing and link handling. The interfacing problem was resolved rather easily. The link-handling problem, however, was more severe, and Debra Mann has gone to the customer's site in France to resolve it.

Plans for Next Month

- Complete the Software Training Mailing Campaign.
- Resume the Search Project.
- Restart the project to provide a data base on information management with a schedule that reflects its new scope.
- Complete the Phase Three project.
- Write a report to justify the addition of two software engineers to my department.
- Congratulate publicly the recipients of Meritorious Achievement Awards: Bill Thomasson and Nancy O'Rourke.

Current Staffing Level

Current staff: 11
Open requisitions: 0

ja
Enclosure

Figure 2 Activity Report

CHANGE He deals with personnel problems in his work. *This* helps him in his personal life.
TO He deals with personnel problems in his work. This experience helps him in his personal life.

A hidden reference, or one that has only an implied antecedent, is the third problem.

CHANGE A high lipid, low carbohydrate diet is "ketogenic" because it favors *their* formation.
TO A high lipid, low carbohydrate diet is "ketogenic" because it favors the formation of *ketone bodies*.

For the sake of **coherence,** pronouns should be placed as close as possible to their antecedents. The danger of creating an ambiguous reference increases with distance. Don't force your **reader** to go back too far to find out what a pronoun stands for.

CHANGE The *house* in the meadow at the base of the mountain range was resplendent in *its* coat of new paint.
TO The house, resplendent in *its* coat of new paint, was nestled in a meadow at the base of the mountain range.

Do not repeat an antecedent in **parentheses** following the pronoun. If you feel that you must identify the pronoun's antecedent in this way, you need to rewrite the sentence.

CHANGE The senior partner first met Bob Evans when he (Evans) was a trainee.
TO Bob Evans was a trainee when the senior partner first met him.

When referring to a noun that includes both sexes (student, teacher, clerk, everyone), the pronouns *he* and *his* are traditionally used.

EXAMPLE Each employee is to have *his* annual X-ray taken by Friday.

However, it is now generally accepted that this conventional use of the masculine **personal pronoun** implies sexual bias. If there is danger of offending, it is often best to avoid the problem by substituting an **article** or changing the statement from singular to plural. (See also **he/she.**)

EXAMPLES Each employee is to have *the* annual X-ray taken by Friday.
All employees are to have *their* annual X-rays taken by Friday.

pronouns

A pronoun is a word that is used as a substitute for a **noun** (the noun for which a pronoun substitutes is called its *antecedent*). Using pronouns in place of nouns relieves the monotony of repeating the same noun over and over. Pronouns fall into several different categories: personal, demonstrative, relative, interrogative, indefinite, reflexive, intensive, and reciprocal. (See also **pronoun reference**.)

PERSONAL PRONOUNS

Personal pronouns refer to the person or persons speaking (*I, me, my, mine; we, us, our, ours*), the person or persons spoken to (*you, your, yours*), or the person or thing spoken of (*he, him, his; she, her, hers; it, its; they, them, their, theirs*). (See also **person.**)

EXAMPLES *I* wish *you* had told *me* that *she* was coming with *us*.
If *their* figures are correct, *ours* must be in error.

DEMONSTRATIVE PRONOUNS

Demonstrative pronouns (*this, these; that, those*) indicate or point out the thing being referred to.

EXAMPLES *This* is my desk.
These are my coworkers.
That will be a difficult job.
Those are incorrect figures.

RELATIVE PRONOUNS

The **relative pronouns** are *who* (or *whom*), *which,* and *that.* A relative pronoun performs a dual function: (1) it takes the place of a noun, and (2) it connects, and establishes the relationship between, a **dependent clause** and its main **clause.**

EXAMPLE The personnel manager decided *who* would be hired.

The pronoun *that* is normally used with restrictive clauses and the pronoun *which* with nonrestrictive clauses. (See also **restrictive and nonrestrictive elements.**)

EXAMPLES Companies *that adopt the plan* nearly always show profit increases. (restrictive clause)
The annual report, *which was distributed yesterday,* shows that sales increased 20 percent last year. (nonrestrictive clause)

INTERROGATIVE PRONOUNS

Interrogative pronouns (*who* or *whom, what,* and *which*) ask questions. They differ from relative pronouns in two ways: They are used only to ask questions, and they may introduce independent interrogative sentences, whereas relative pronouns connect or show relationship and introduce only dependent clauses.

Notice that, unlike relative pronouns, interrogative pronouns normally lack expressed antecedents; the antecedent is actually in the expected answer.

EXAMPLES *Who* said that?
What is the trouble?
Which of these is best?

INDEFINITE PRONOUNS

Indefinite pronouns specify a class or group of persons or things rather than a particular person or thing. *All, any, another, anyone, anything, both, each, either, everybody, few, many, most, much, neither, nobody, none, several, some,* and *such* are indefinite pronouns.

EXAMPLE Not *everyone* liked the new procedures; *some* even refused to follow them.

REFLEXIVE PRONOUNS

A **reflexive pronoun,** which always ends with the **suffix** *-self* or *-selves*, indicates that the **subject** of the sentence acts upon itself.

EXAMPLE The electrician accidentally shocked *herself.*

The reflexive and intensive pronouns are *myself, yourself, himself, herself, itself, oneself, ourselves, yourselves,* and *themselves.*

Myself is not a substitute for *I* or *me* as a personal pronoun.

CHANGE John and *myself* completed the report on time.
TO John and *I* completed the report on time.

CHANGE The assignment was given to Wally and *myself.*
TO The assignment was given to Wally and *me.*

INTENSIVE PRONOUNS

Intensive pronouns are identical in form with the reflexive pronouns, but they perform a different function: to give **emphasis** to their antecedents.

EXAMPLE I *myself* asked the same question.

RECIPROCAL PRONOUNS

Reciprocal pronouns (*one another* and *each other*) indicate the relationship of one item to another. *Each other* is commonly used when referring to two persons or things and *one another* when referring to more than two.

EXAMPLES They work well with *each other.*
The crew members work well with *one another.*

GRAMMATICAL PROPERTIES OF PRONOUNS

Case. Pronouns have forms to show the subjective, objective, or possessive **case,** as the following chart shows:

	Subjective	*Objective*	*Possessive*
1st person singular	I	me	my, mine
2nd person singular	you	you	your, yours
3rd person singular	he	him	his
	she	her	her, hers
	it	it	its
1st person plural	we	us	our, ours
2nd person plural	you	you	your, yours
3rd person plural	they	them	their, theirs

A pronoun that is used as the subject of a clause or sentence is in the subjective case (*I, we, he, she, it, you, they, who*). The subjective case is also used when the pronoun follows a **linking verb.**

EXAMPLES *He* is my boss.
My boss is *he.*

A pronoun that is used as the **object** of a **verb** or **preposition** is in the objective case (*me, us, him, her, it, you, them, whom*).

EXAMPLES Mr. Davis hired Tom and *me.* (object of verb)
Between *you* and *me,* he's wrong. (object of preposition)

A pronoun that is used to express ownership is in the possessive case (*my, mine, our, ours, his, hers, its, your, yours, their, theirs, whose*).

EXAMPLE He took *his* notes with him on the business trip.

Only the possessive form of a pronoun should ever be used with a **gerund.**

EXAMPLE The boss objected to *my* arriving late every morning.

Several indefinite pronouns (*any, each, few, most, none,* and *some*) form the possessive only in "of" phrases (*of any, of some*). Others, however, use an **apostrophe** (*anyone's*).

CHANGE *Each's* opinion was solicited.
TO The opinion *of each* was solicited.

A pronoun **appositive** takes the case of its antecedents.

EXAMPLES Two systems analysts, Joe and *I,* were selected to represent the company. (*Joe and I* is in apposition to the subject, *systems analysts,* and must therefore be in the subjective case.)
The systems analysts selected two members—Joe and *me.* (*Joe and me* is in apposition to *two members,* which is the object of the verb *selected,* and therefore must be in the objective case.)

How to Determine the Case of a Pronoun. To determine the case of a pronoun, try it with a transitive verb, such as *resembled.* If the pronoun can precede the verb, it is in the subjective case; if it must follow the verb, it is in the objective case.

EXAMPLES *She* resembled her father. (subjective)
Her father resembled *her.* (objective)

If compound pronouns cause problems in determining case, try using them singly.

EXAMPLES In his letter, John mentioned *you* and *me.*
In his letter, John mentioned *you.*
In his letter, John mentioned *me.*

They and *we* must discuss the terms of the merger.
They must discuss the terms of the merger.
We must discuss the terms of the merger.

When a pronoun modifies a noun, try it without the noun to determine its case.

EXAMPLES (*We/Us*) pilots fly our own planes.
We fly our own planes. (You would not write "*Us* fly our own planes.")
He addressed his remarks directly to (*we/us*) technicians.
He addressed his remarks directly to *us.* (You would not write "He addressed his remarks directly to *we.*")

To determine the case of a pronoun that follows *as* or *than,* try mentally adding the words that are normally omitted.

EXAMPLES The director does not have as much formal education *as he* [does]. (You would not write "*him* does.")

His friend was taller *than he* [was]. (You would not write "*him* was.")

Gender. A pronoun must agree in **gender** with its antecedent. A problem sometimes occurs because the masculine pronoun has traditionally been used to refer to both sexes.

EXAMPLE *Each* may stay or go as *he* chooses.

To avoid the sexual bias implied in such usage, use *he or she* or the plural form of the pronoun, *they.*

CHANGE *Each* may stay or go as *he* chooses.

TO *Each* may stay or go as *he or she* chooses.

OR *All* may stay or go as *they* choose.

If you use the plural form of the pronoun, be sure to change the indefinite pronoun *each* to its plural form, *all.*

Number. **Number** is a frequent problem only with a few indefinite pronouns (*each, either, neither;* and those ending with *-body,* or *-one,* such as *anybody, anyone, everybody, everyone, nobody, no one, somebody, someone*), which are normally singular and so require singular verbs and are referred to by singular pronouns.

EXAMPLE As *each arrives* for the meeting and takes *his* seat, please hand *him* a copy of the confidential report. *Everyone is* to return *his* copy before *he* leaves. *No one* should be offended by these precautions when the importance of secrecy has been explained to *him.* I think *everybody* on the committee *understands* that *neither* of our major competitors *is* aware of the new process that we have developed.

Person. Third-person personal pronouns usually have antecedents.

EXAMPLE John presented the report to the members of the board of directors. *He* (John) first read *it* (the report) to *them* (the directors) and then asked for questions.

First- and second-person personal pronouns do not normally require antecedents.

EXAMPLES *I* like my job.

You were there at the time.

We all worked hard on the project.

proofreaders' marks

Publishers have established **symbols,** called *proofreaders' marks,* which writers and editors use to communicate with printers in the production of publications. A familiarity with these symbols makes it easy for you to communicate your changes to others. (See Figure 1.)

proofreading

Proofreading should be done in several stages: read through the material three times, looking for specific things each time, and then read it a final time for content.

During the first time through, ask yourself, "Does it look right?" Then look for:

- accurate and clean typing,
- aesthetic page placement of material,
- acceptable format,
- correct spelling of names and places,
- accuracy of numbers.

During the second time, ask, "Am I following the rules?" Then check for:

- typographical errors,
- capitalization,
- punctuation,
- spelling,
- grammar.

During the third time, ask, "Is it complete?" Then look for:

- omissions,
- deletions,
- doubly typed words.

When you read the material for content, make certain that everything is there, is correct, and is in the right place.

To proofread effectively, you must be critical and alert. Much like revising, proofreading requires that you look at your writing objectively. In her article "36 Aids to Successful Proofreading," Ruth H. Turner offered useful tips for proofreading three types of material: short narratives (letters or memorandums), long narratives (man-

Mark in Margin	Instruction	Mark on Manuscript	Corrected Type
ℯ	Delete	the ~~lawyer's~~ bible	the bible
lawyer's	Insert	The ^ bible	The lawyer's bible
(stet)	Let stand	the ~~lawyer's~~ bible	the lawyer's bible
(cap)	Capitalize	the bible	the Bible
(lc)	Make lower case	the /Law	the law
(ital)	Italicize	the lawyer's bible	the *lawyer's* bible
(tr)	Transpose	the bible lawyer's	the lawyer's bible
⁀	Close space	the Bi ble	the Bible
(sp)	Spell out	(2) bibles	two bibles
#	Insert space	TheBible	The Bible
¶	Start paragraph	¶ The lawyer's...	The lawyer's...
(run in)	No paragraph	...marks. Below is a....	....marks. Below is a....
(sc)	Set in small capitals	The bible	The BIBLE
(rom)	Set in roman type	The *bible*	The bible
(bf)	Set in boldface	The bible	The **bible**
(lf)	Set in lightface	The **bible**	The bible
⊙	Insert period	The lawyers have their own bible ^	The lawyers have their own bible.
^,^	Insert comma	However ^ we cannot....	However, we cannot....
=/=	Insert hyphens	half ^ and ^ half	half-and-half
⊙:	Insert colon	We need the following ^	We need the following:
^;^	Insert semicolon	Use the law ^ don't....	Use the law; don't....
'	Insert apostrophe	Johns law book	John's law book
" / "	Insert quotation marks	The law is law.	The "law" is law.
(/)	Insert parentheses	John's ^ law ^ book	John's (law) book
[/]	Insert brackets	John ^ 1920-1962 ^ went....	John [1920-1962] went....
1/N	Insert en dash	1920 ^ 1962	1920-1962
1/M	Insert em dash	Our goal ^ victory	Our goal—victory
2	Insert superior type	3 = 9	$3^2 = 9$
2	Insert inferior type	HSO4	H_2SO_4
*	Insert asterisk	The law	The law*
†	Insert dagger	The law ^	The law†
‡	Insert double dagger	The bible ^	The bible‡
§	Insert section symbol	^ Research	§Research

Figure 1 Proofreader's Marks

uals or **reports** of ten or more pages), and technical material and statistical tables.*

SHORT NARRATIVES

1. To concentrate on typographical errors, read backwards (from right to left). Of course, since this method will not reveal omissions or duplications, you must read again for content.
2. Arrange with another person to proofread each other's material. However, do not allow this procedure to delay your progress.
3. Concentrate as you slowly read for errors. Speed reading does not help.
4. Pay attention to dates; do not assume they are correct. Check the **spelling** of months and even the correctness of the year.
5. Do not overlook names, addresses, subject or reference lines, signature lines, or even the copies list. Errors can sneak into these items as well as into the body of the letter or memorandum.
6. Examine the end and beginning of lines in the body of the letter or memorandum to make sure that little words like *that* or *and* have not been unnecessarily repeated.
7. If you have duplicated typed material and have ended each line at the same place, check your copy by laying it over the original and holding them both up to a strong light. This procedure will reveal errors without the need for reading.
8. Proofread personalized letters or memorandums that are identical except for names, addresses, or short insertions by carefully examining the personalized portions. Then hold them both up to a light, as described in the previous suggestion.

LONG NARRATIVES

1. Make a style sheet to ensure consistency throughout a manuscript regarding the spelling of specialized terminology or **jargon,** capitalization and **punctuation, heads, formats** for **outlines** or **tables,** and names and titles. Any items that are unique to your project can be jotted down on this style sheet. Keep this sheet for proofreading later revisions or additional manuals or reports in a series.

*Adapted, with permission, from *The Secretary,* Vol. 38, no. 2, p. 12, Feb. 1978 by the National Secretaries Association (International), Kansas City, MO 64108.

2. Proofread in steps. For example, check all the heads and titles first. Be sure that you have consistently followed the style sheet.
3. Scan page numbers. Have you missed or duplicated any?
4. Read the body of the material against the original, sentence by sentence.
5. Make sure any references to other parts of the work are correct. For example, "See page 6 for a similar list." Possibly, in the final typing, the list moved to page 7.
6. Verify the title and page numbers of other reference materials if you have that responsibility. Look up each item if at all possible.
7. When proofreading from a draft that has handwritten inserts, make sure that none have been overlooked in the final draft.

TECHNICAL MATERIAL AND STATISTICAL TABLES

1. If possible, use two people—one reading aloud, the other checking copy—to verify itemized data, codes, figures, and so on.
2. When the columns of a table have been typed by tabulating across the page, proofread your copy down the columns, folding the original from top to bottom along the column and laying it next to the corresponding column of the copy.
3. If the draft is typewritten, lay a ruler under each line. This technique will help you keep your place on both the original and the copy. It also helps when the inevitable interruptions occur.
4. Check for both vertical and horizontal alignment of table columns. Proofread across the page, folding the original beneath each line and laying it under the corresponding lines of typed entries so that you can easily read the original against your copy.
5. Count the number of entries in each table column, and compare the total with the number in your copy. If there is a difference, hunt down the culprit.
6. Proofread outlines by breaking the task into components: check headings, then check the mechanics (**numbers**, letters, **indentations**), and finally perform a continuous reading of each item against the original.

A CHECKLIST FOR PROOFREADING

1. Make the **dictionary** your best friend. Is it *privilege, privelege,* or *priviledge; gauge* or *guage; discression* or *discretion; chief* or *cheif?*

Never be embarrassed to look up a word, for you might be more embarrassed at an error in the completed work. Watch closely for an omitted *-ed* or final *-s*.

2. Study related words until you understand the differences in their meaning and use: **affect/effect, imply/infer, diagnosis/prognosis, advice/advise, insure/ensure/assure.**
3. Mentally repeat each syllable of long words with many vowels, such as *evacuation, responsibilities, continuously, individual.* It is easy to omit one.
4. Be careful with double letters. They are hard to remember, especially in words like *accommodation* and *occurrence.*
5. Remember that *supersede* is the only word in English ending in *-sede* and that *succeed, proceed,* and *exceed* are the only ones ending in *-ceed.* All others (*cede, recede, precede,* and the like) end in *-cede.*
6. Be careful with silent letters: *rhythm, rendezvous, malign.*
7. Check all punctuation marks after completing all other proofreading. Be careful not to omit a closing **parenthesis** or **quotation mark.**
8. Refer to the dictionary to determine whether to spell **compound words** as separate words, a hyphenated word, or one word. If the dictionary does not list the word as hyphenated or separate words, write it as one word.
9. Be sure that **capital letters** are used consistently. Follow any established style for the type of document or subject matter. If an established style does not exist, adopt one of your own.
10. One last suggestion—proofread tomorrow what you worked on today.

For further information on proofreading, you may wish to consult the following books:

Lasky, Joseph, *Proofreading and Copy-Preparation*
A Manual of Style. University of Chicago Press
Skillin, M.E., and Gay, R.M., eds. *Words into Type*

proper nouns

A proper noun names a specific person, place, thing, concept, action, or quality and therefore is always capitalized.

Proper Nouns	Common Nouns
Jane Jones	person
Chicago	city
General Electric	company
Tuesday	day
Declaration of Independence	document

(For the **case** and **number** of proper nouns, see **nouns.**)

proposals

A proposal is a document written to persuade someone to follow a plan or course of action. It may be internal to an organization or be sent outside the organization to a potential client.

An **internal proposal** usually recommends a change or improvement within an organization. The recommendation might be to expand the cafeteria service, to combine multiple manufacturing operations, or to introduce new working procedures. A capital appropriations proposal, for example, is an internal proposal submitted for management approval to spend large sums of money. An internal proposal is normally prepared by an employee or department and then sent to a higher-ranking person in the organization for approval.

External proposals include both **sales proposals** and **government proposals**. Both represent a company's offer to provide goods or services to a potential buyer within a certain amount of time and at a specified cost. The purpose of the proposal is to present a product or service in the best possible light and to explain why a buyer should choose it over the competitors. Remember that the survival of some companies depends on their ability to produce effective proposals.

Since proposals offer plans to fill a need, your **readers** will evaluate your plan according to how well your written presentation answers their questions about what you are proposing to do, how you plan to do it, when you plan to do it, and how much it is going to cost.

To answer these questions satisfactorily, make certain that your proposal is written at your reader's level of knowledge. If you have more than one reader—and proposals often require more than one level of approval—take into account all your readers. For example, if your immediate reader is an expert on your subject but the next higher level of management (which must also approve the proposal) is not, provide an **executive summary** written in language that is as nontechnical as possible. You might also include a **glossary** of technical terms used in the body of the proposal, or an appendix that explains the technical information in nontechnical language. On the other hand, if your immediate reader is not an expert but the person at the next level is, write the proposal with the nonexpert in mind and include an appendix that contains the technical details.

Proposals usually consist of an **introduction,** a body, and a **conclusion.** The introduction should summarize the problem you are proposing to solve and your solution. It may also indicate the benefits that your reader will receive from your solution and its total cost. The body should explain in detail (1) how the job will be done, (2) what methods will be used to do it (and if applicable, the materials to be used and any other pertinent information), (3) when work will begin, (4) when the job will be completed, and (5) a cost breakdown for the entire job. The conclusion should emphasize the benefits for the reader and should urge him or her to take action. Your conclusion should have an encouraging, confident, and reasonably assertive **tone.**

proved/proven

Both *proved* and *proven* are acceptable past **participles** of *prove,* although *proved* is currently in wider use.

EXAMPLES They had *proved* more obstinate than expected.
They had *proven* more obstinate than expected.

Proven is more commonly used as an **adjective.**

EXAMPLE She was hired because of her *proven* competence as a manager.

pseudo/quasi

As a **prefix,** *pseudo,* meaning "false or counterfeit," is joined to a word without a **hyphen** unless the word begins with a **capital letter.**

EXAMPLES *pseudo*science (false science)
pseudo-Americanism (pretended Americanism)

Pseudo is sometimes confused with quasi, meaning "somewhat" or "partial." Unlike, *semi, quasi* does not mean half. *Quasi* is usually hyphenated in combinations.

EXAMPLE *quasi*-scientific literature

(See also **bi-/semi-.**)

punctuation

Punctuation is a system of **symbols** that helps the **reader** understand the structural relationship within (and the intention of) a sentence. Marks of punctuation may link, separate, enclose, indicate omissions, terminate, and classify sentences. Most of the thirteen punctuation marks can perform more than one function. The use of punctuation is determined by grammatical conventions and the writer's intention—in fact, punctuation often substitutes for the writer's facial expressions. Misuse of punctuation can cause your reader to misunderstand your meaning. Detailed information on each mark of punctuation is given in its own entry. The following are the thirteen marks of punctuation:

apostrophe	'
brackets	[ ]
colon	:
comma	,
dash	—
exclamation mark	!
hyphen	-
parentheses	( )
period	.
question mark	?
quotation marks	" "
semicolon	;
slash	/

Q

question marks

The question mark (?) has the following uses:
Use a question mark to end a sentence that is a direct question.

EXAMPLE Where did you put the specifications?

Use a question mark to end any statement with an interrogative meaning (a statement that is declarative in form but asks a question).

EXAMPLE The report is finished?

Use a question mark to end an interrogative **clause** within a declarative sentence.

EXAMPLE It was not until July (or was it August?) that we submitted the report.

Retain the question mark in a title that is being cited, even though the sentence in which it appears has not ended.

EXAMPLE *Should Engineers Be Writers?* is the title of her book.

When used with **quotations,** the question mark indicates whether the writer who is doing the quoting or the person being quoted is asking the question. When the writer doing the quoting asks the question, the question mark is outside the **quotation marks.**

EXAMPLE Did she say, "I don't think the project should continue"?

If, on the other hand, the quotation itself is a question, the question mark goes inside the quotation marks.

EXAMPLE She asked, "When will we go?"

If the writer doing the quoting and the person being quoted both ask questions, use a single question mark inside the quotation marks.

EXAMPLE Did she ask, "Will you go in my place?"

Question marks may follow a series of separate items within an interrogative sentence.

EXAMPLE Do you remember the date of the contract? its terms? whether you signed it?

A question mark should never be used at the end of an indirect question.

CHANGE He asked me whether sales had increased this year?
TO He asked me whether sales had increased this year.

When a directive or command is phrased as a question, a question mark is usually not used. However, a request (to a customer or a superior, for instance) almost always requires a question mark.

EXAMPLES Will you make sure that the machinery is operational by August 15.
Will you please telephone me collect if your entire shipment does not arrive by June 10?

questionnaires

A questionnaire—a series of questions on a particular topic, sent out to a number of people—is a sort of interview on paper. It offers several advantages over the personal interview but has some disadvantages. A questionnaire allows you to sample many more people than personal interviews can. It enables you to obtain responses from people in different parts of the country. Even people who live near you may be easier to reach by mail than in person. Those responding to a questionnaire do not face the constant pressure posed by someone jotting down their every word—a fact that can produce more thoughtful answers from questionnaire respondents. And the questionnaire reduces the possibility that the interviewer might influence an answer by tone of voice or facial expression. Finally, the cost of a questionnaire is lower than the cost of numerous personal interviews.

Questionnaires have drawbacks too. People who have strong opinions on a subject are more likely to respond to a questionnaire than those who do not. This can skew the results. An interviewer can follow up on an answer with a pertinent question; at best, a questionnaire can be designed to let one question lead logically to another. Furthermore, mailing a batch of questionnaires and waiting for replies take considerably longer than a personal interview does.

A questionnaire must be properly designed. Your goal should be to obtain as much information as possible from your respondents with as little effort on their part as possible. The first rule to follow is to keep the questionnaire brief. The longer the questionnaire is, the less likely the recipient will be to complete and return it. Next, the questions should be easy to understand. A confusing question will yield confusing results, whereas a carefully worded question will be easy to answer. Ideally, questions should be answerable with a yes or no.

> Would you be willing to work a four-day work week, ten hours a day, with every Friday off?
>
> Yes ________
>
> No ________
>
> No opinion ________

When it is not possible to phrase questions in such a straightforward way, offer an appropriate range of answers. Questions must be phrased neutrally; their wording must not lead respondents to a particular answer.

CHANGE How many hours of overtime would you be willing to work each week?

4 hours ________ 10 hours ________

6 hours ________ More than 10 hours ________

8 hours ________ No overtime ________

TO Would you be willing to work overtime?

Yes ________

No ________

If yes, how many hours of overtime would you be willing to work each week?

4 hours ________ 10 hours ________

6 hours ________ 12 hours ________

8 hours ________ More than 12 hours ________

When preparing your questions, remember that you must eventually tabulate the answers; therefore, try to formulate questions whose answers can be readily computed. The easiest questions to tabulate are those for which the recipient does not have to compose an answer. Any questions that require a comment for an answer take time to think about and write and thus lessen your chances of obtaining a response. They are also difficult to interpret. Questionnaires should include a section for additional comments where recipients may clarify their overall attitude toward the subject. If the information will help interpret the answers, include questions about the recipient's age, education, occupation, and so on. Include your name, your address, the purpose of the questionnaire, and the date by which an answer is needed.

A questionnaire sent by mail must be accompanied by a letter explaining who you are, the purpose of the questionnaire, how it will be used, and the date by which you would like to receive a reply. If the information provided will be kept confidential or if the recipient's identity will not be disclosed, say so in the letter.

The sample questionnaire (pp. 542–3) was sent to employees in a large organization who had participated in a six-month program of flexible working hours. Under the program, employees worked a forty-hour, five-day week, with flexible starting and quitting times. Employees could start work between 7 and 9 a.m. and leave between 3:30 and 6:30 p.m., provided that they worked a total of eight hours each day and took a one-half-hour lunch period midway through the day.

Select the recipients for your questionnaire carefully. If you want to survey the opinions of all the employees in a small laboratory, simply send each worker a questionnaire. To survey the members of a professional society, mail questionnaires to everyone on the membership list. But to survey the opinions of large groups in the general population—for example, all medical technologists working in private laboratories—is not so easy. Since you cannot include everybody in your survey, you have to choose a representative cross-section.

Each of the following books contains advice on the design and interpretation of sampling techniques:

Arkin, Herbert, and Raymond R. Colton. *Statistical Methods.* 5th ed. New York: Barnes & Noble Books, 1967.

May 18, 19--

To: All Company Employees

From: Nelson Barrett, Director of Personnel *nb*

Subject: Review of Flexible Working Hours Program

Please complete this questionnaire regarding Luxwear Corporation's trial program of flexible working hours. Your answers will help us to decide whether the program should continue. Return the questionnaire to the Personnel Department by May 28.

If you want to discuss any item in more detail, call Tania Peters in Personnel at extension 8812.

1. What kind of position do you hold?

   ____________ supervisory ____________ nonsupervisory

2. Indicate your exact starting time under flexitime.

   ____

3. Where do you live?

   Talbot County ________ Greene County ________

   Montgomery County ____ Other, specify _______

4. How do you usually travel to work?

   Drive alone ______ Walk _________

   Car pool _________ Bus __________

   Train ____________ Motorcycle __________

   Bicycle __________ Other, specify __________

5. Has flexitime affected your commuting time? If so, please indicate the approximate number of minutes.

   ________ increase ________ decrease ________ no change

6. If you drive, has flexitime affected the amount of time it takes to find a parking space?

   ________ increase ________ decrease ________ no change

Figure 1 Sample Questionnaire

-2-

7. Do you think that flexitime has affected your productivity?

   __________ increase   __________ decrease   __________ no change

8. Have you had difficulty getting in touch with colleagues whose work schedules are different from yours?

   Yes__________   No__________

9. Have you had trouble scheduling meetings?

   Yes__________   No__________

10. Has flexitime affected the way you feel about your job?

   feel better   feel worse   no change

11. How important is it for you to have flexibility in your working hours?

   Very__________   Not very__________

   Somewhat__________   Not at all__________

12. If you have children, has flexitime made it easier or more difficult for you to obtain babysitting or day-care services?

   __________ easier   __________ more difficult   __________ no change

13. Do you recommend that the flexitime program be made permanent?

   Yes__________   No__________

14. Do you have suggestions for any changes in the program? If so, please specify.

   ____

   ____

mo

Beardsley, Monroe C. *Thinking Straight: Principles of Reasoning for Readers and Writers.* 4th ed. Englewood Cliffs, N.J.: Prentice-Hall, 1975.

Raja, Des. *The Design of Sample Surveys.* New York: McGraw-Hill, 1972.

Weddle, Perry. *Argument: A Guide to Critical Thinking.* New York: McGraw-Hill, 1978.

quid pro quo

Quid pro quo, which is Latin for "one thing for another," can suggest mutual cooperation or "tit for tat" in a relationship between two groups or individuals. The term may be appropriate to business and legal contexts if you are sure your **reader** understands its meaning.

EXAMPLE It would be unwise to grant their request without insisting on some *quid pro quo.*

quotation marks

Quotation marks (" ") are used to enclose direct **repetition** of spoken or written words. They should not be used to emphasize. There are a variety of guidelines for using quotation marks.

Enclose in quotation marks anything that is quoted word for word (direct quotation) from speech.

EXAMPLE She said clearly, "I want the progress report by three o'clock."

Do not enclose indirect **quotations**—usually introduced by *that*—in quotation marks. Indirect quotations are **paraphrases** of a speaker's words or ideas.

EXAMPLE She said that she wanted the progress report by three o'clock.

Handle quotations from written material the same way: place direct quotations, but not indirect quotations, within quotation marks.

EXAMPLES The report stated, "The potential in Florida for our franchise is as great as in California."

The report indicated that the potential for our franchise is as great in Florida as in California.

If you use quotation marks to indicate that you are quoting, you may not make any changes in the quoted material unless you clearly indicate what you have done. (See **quotations** and **brackets.**)

Quotations longer than four typed lines (at least fifty characters per line) are normally indented (*all* the lines) five spaces from the left margin, single-spaced, and *not* enclosed in quotation marks.

Unless it is indented as just described, a quotation of more than one **paragraph** is given quotation marks at the beginning of each new paragraph, but at the end of only the last paragraph.

Use single quotation marks (on a typewriter use the **apostrophe** key) to enclose a quotation that appears within a quotation.

EXAMPLE John said, "Jane told me that she was going to 'hang tough' until the deadline is past."

Use quotation marks to set off special words or technical terms only to point out that the term is used in context for a unique or special purpose (used, that is, in the sense of *the so-called*).

EXAMPLE Typical of deductive analyses in real life are accident investigations: What chain of events caused the sinking of an "unsinkable" ship such as the *Titanic* on its maiden voyage?

However, slang, colloquial expressions, and attempts at humor, although infrequent in technical writing, seldom rate being set off by quotation marks.

CHANGE Our first six months in the new office amounted to little more than a "shakedown cruise" for what lay ahead.

TO Our first six months in the new office amounted to little more than a shakedown cruise for what lay ahead.

Use quotation marks to enclose titles of reports, short stories, articles, essays, radio and television programs, short musical works, paintings, and other art works.

EXAMPLE Did you see the article, "No-Fault Insurance and Your Motorcycle" in last Sunday's *Journal*?

Titles of books and periodicals are underlined (to be typeset in **italics**).

EXAMPLE Articles in the *Business Education Forum* and *Scientific American* quoted the same passage.

Some titles, by convention, are neither set off by quotation marks nor underlined, although they are capitalized.

EXAMPLES the Bible, the Constitution, Lincoln's Gettysburg Address, the Montgomery Ward Catalog

Commas and **periods** always go inside closing quotation marks.

EXAMPLE "Reading *Space Technology* gives me the insider's view," he says, adding "it's like having all the top officials sitting in my office for a bull session."

Semicolons and **colons** always go outside the closing quotation marks.

EXAMPLES He said, "I will pay the full amount"; this certainly surprised us.
The following are her favorite "sports": eating and sleeping.

All other **punctuation** follows the logic of the context: if the punctuation is a part of the material quoted, it goes inside the quotation marks; if the punctuation is not part of the material quoted, it goes outside the quotation marks.

Quotation marks may be used as ditto marks, instead of repeating a line of words or numbers directly beneath an identical set. In formal writing, this use is confined to **tables** and **lists.**

EXAMPLE

A	is	at	a	point	equally	distant	from	L and M.
B	"	"	"	"	"	"	"	S and T.
C	"	"	"	"	"	"	"	R and Q.

quotations

When you have borrowed words, facts, or ideas of any kind from someone else's work, acknowledge your debt by giving your source credit in a footnote. Otherwise, you will be guilty of **plagiarism.** Also be sure that you have represented the original material honestly and accurately.

DIRECT QUOTATIONS

Direct word-for-word quotations are enclosed in **quotation marks.** They are usually, although not always, separated from the rest of the sentence in which they occur by either a **comma** or a **colon.**

EXAMPLES The noted economist says, "If monopolies could be made to behave as if they were perfectly competitive, we would be able to enjoy the benefits both of large-scale efficiency and of the perfectly working price mechanism."

The noted economist pointed out: "There are three options available when technical conditions make a monopoly the natural outcome of competitive market forces: private monopoly, public monopoly, or public regulation."

INDIRECT QUOTATIONS

Indirect quotations, which are essentially **paraphrases** and are usually introduced by *that,* are not set off from the rest of the sentence by **punctuation** marks. (See also **footnotes/end notes.**)

DIRECT He said in a recent interview, "Regulation cannot supply the dynamic stimulus that in other industries is supplied by competition." (direct quotation)

INDIRECT In a recent interview he said that regulation does not stimulate the industry as well as competition does. (indirect quotation)

When a quotation is divided, the material that interrupts the quotation is set off, before and after, by commas, and quotation marks are used around each part of the quotation.

EXAMPLE "Regulation," he said in a recent interview, "cannot supply the dynamic stimulus that in other industries is supplied by competition."

At the end of a quoted passage, commas and **periods** go inside the quotation marks, and colons and **semicolons** go outside the quotation marks.

DELETIONS OR OMISSIONS

Deletions or omissions from quoted material are indicated by three **ellipsis** dots within a sentence and four ellipsis dots at the end of a sentence, the first of the four dots being the period that ends the sentence.

EXAMPLE "If monopolies could be made to behave . . . we would be able to enjoy the benefits of . . . large-scale efficiency. . . ."

When a quoted passage begins in the middle of a sentence rather than at the beginning, however, ellipsis dots are not necessary; the fact that the first letter of the quoted material is not capitalized tells the **readcr** that the quotation begins in midsentence.

CHANGE He goes on to conclude that ". . . coordination may lessen competition within a region."

TO He goes on to conclude that "coordination may lessen competition within a region."

Such quotations should be worked into one of your sentences rather than left standing alone. When quoted material is worked into a sentence, be sure that it is related logically, grammatically, and syntactically to the rest of the sentence.

INSERTING MATERIAL INTO QUOTATIONS

When it is necessary to insert a clarifying comment within quoted material, use **brackets.**

EXAMPLE "The industry is organized as a relatively large, integrated system serving an extensive [geographic] area, with smaller systems existing as islands within the larger system's sphere of influence."

When quoted material contains an obvious error, or might in some other way be questioned, the expression *sic,* enclosed in brackets, follows the questionable material to indicate that the writer has quoted the material exactly as it appeared in the original. (See also **sic.**)

EXAMPLE In *Basic Astronomy,* Professor Jones noted that the "earth does not revolve around the son [*sic*] at a constant rate."

INCORPORATING QUOTATIONS INTO TEXT

Depending on the length, there are two mechanical methods of incorporating quotations into your text. Quotations of four or fewer lines are incorporated into your text and enclosed in quotation marks. (Many writers use this method regardless of length. When this method is used with multiple **paragraphs,** quotation marks appear at the beginning of each new paragraph, but at the end of only the last paragraph.) Material longer than five lines is now usually inset; that is, it is set off from the body of the text by being indented five spaces from the left margin and by triple-spacing above and below the quotation. The quoted passage is single-spaced and not enclosed in quotation marks.

OVERQUOTING

Do not rely too heavily on the use of quotations in the final version of your **report** or paper. If you do, your work will appear merely derivative. The temptation to overquote during the **note-taking** phase of your **research** can be avoided if you concentrate on summarizing what you read. Quote word for word only when your source concisely sums up a great deal of information or points to a trend or development important to your subject. As a rule of thumb, avoid quoting anything over one paragraph.

R

raise/rise

Both *raise* and *rise* mean "move to a higher position." However, *raise* is a transitive **verb** and always takes an **object** (*raise* crops), whereas *rise* is an intransitive verb and never takes an object (heat *rises*).

EXAMPLES *Raise* the lever to the ON position.
Turn off the motor if the internal temperature *rises* above 110°F.

re

Re (and its variant form *in re*) is business and legal **jargon** meaning "in reference to" or "in the case of." *Re* is sometimes used in **memorandums.**

EXAMPLE To: Elaine Barr
From: Edgar Roden
Re: Revised Office Procedures

Re, however, is now giving way to *subject.*

EXAMPLE To: Elaine Barr
From: Edgar Roden
Subject: Revised Office Procedures

readers

The first rule of effective writing is to help the reader. The responsibility that technical writers have to their readers is nicely expressed in the Society of Technical Communications's *Code for Communicators*, which calls for technical writers to hold themselves responsible for how well their audience understands their message and to satisfy the readers' need for information rather than the writer's need for self-expression. This code reminds writers to concentrate on writing *for* the needs of specific readers rather than merely *about* certain subjects. Too often, writers are blinded by their own familiarity with the technical subject and therefore tend to overlook their readers' lack of knowledge.

As a technical writer, you must usually assume that your readers are less familiar than you are with the subject. You have to be careful, for example, in writing **instructions** for operators of equipment that you designed or in writing a **report** about a technical system for executives whose training is in such nontechnical areas as management, accounting, law, and so on. Such readers are unlikely to have had training in your technical specialty and thus need definitions of technical terms as well as clear explanations of principles that you, as a specialist, take for granted. Even if you write a **journal article** for others in the field. you must remember to explain new or special uses of standard terms and principles.

Before starting to write, you must determine who your readers are and then adjust the amount of detail and technical vocabulary to their training and experience. For example, to describe a new technique for repairing a vibration dampener to an experienced auto mechanic, you could use standard auto repair terms and procedures. To describe the same technique in a training manual, however, the terms and procedures would require a much fuller explanation.

As the previous example implies, when you write for many readers (hundreds of auto mechanic trainees, for example), try to visualize a single, typical member of that group and write for that reader. You might also make a list of characteristics (experience, training, and work habits for example) of that reader. This technique, used widely by professional writers, enables you to decide what should or should not be explained according to that reader's needs. When you are writing to a group, it is usually best to aim at those with the least training and experience. The extra information and explanation will not insult those who know more; they can simply read or follow the instructions more quickly, perhaps skipping sections with which they are familiar.

When your readership includes people with widely varied backgrounds, however, such as in a report or **proposal**, consider aiming various sections of the document at different sets of readers. Recommendations, **executive summaries**, and **abstracts** can be aimed at executives, who will be reading to understand the general implications of projects or technical systems. **Appendixes** containing **tables**, **graphs**, and raw technical data can be aimed at specialists who wish to examine or use such supporting data. A **glossary**, of course, must be written for readers who are not familiar with standard technical terminology. The body of a report or proposal should be aimed at those readers with the most serious interest or who need to make decisions based on the details in the contents.

Routine **memorandums** and short reports that are written for one individual reader do not require such elaborate design. Simply remember that person's exact needs as you write.

Letters written to readers outside your organization, however, require a special **tone**. In such letters, you must concern yourself not only with the readers' understanding of your **topic** but with their reaction to it as well. You must be as careful in representing your organization as you are in explaining your subject matter. (See **correspondence** for a discussion of the extra courtesy required in such business letters.)

No matter what your topic or format, never forget that your readers are the most important consideration, for the most eloquent prose is meaningless unless it reaches its audience.

reason is because

The **phrase** *reason is because* is a colloquial expression to be avoided in writing. *Because,* which in this phrase only repeats the notion of cause, should be replaced by *that.*

CHANGE Sales have increased more than 20 percent. The *reason is because* our sales force has been more aggressive this year.

TO Sales have increased more than 20 percent. The *reason is that* our sales force has been more aggressive this year.

OR Sales have increased more than 20 percent this year *because* our sales force has been more aggressive.

reciprocal pronouns

The reciprocal pronouns *one another* and *each other* indicate a relationship between two or more persons or things. *Each other* is normally used when referring to two persons or things.

EXAMPLE The two partners often disagreed with *each other.*

One another is normally used when referring to more than two persons or things.

EXAMPLE The three vice-presidents talked to *one another* for an hour.

reference books

The following list provides many basic sources for **library research** and general reference:

ABSTRACTS

Agricultural Engineering Abstracts, 1976–. (monthly)

American Statistics Index (ASI), 1974–. (monthly)
(index and abstracts of government publications containing statistics)

Astronomy and Astrophysics Abstracts, 1969–. (semianually)

Biological Abstracts, 1926–. (every two weeks)

Biological Abstracts/RRM, 1962–. (monthly)
(formerly *Bioresearch Index)*

Chemical Abstracts (CA), 1907–. (every two weeks)

Computer Abstracts, 1967–. (monthly)
(Section C of *Science Abstracts.)*

Dissertation Abstracts International, 1938–. (monthly)
Part B contains science and engineering; Part C contains European abstracts.

Ecological Abstracts, 1974–. (every other month)

Electrical and Electronic Abstracts, 1898–. (monthly)
(Section B of *Science Abstracts.)*

Energy Information Abstracts, 1976–. (monthly)

Energy Research Abstracts, 1976–. (every two weeks)

Engineering Index, 1906–. (monthly)

Environment Abstracts, 1971–. (monthly)

INIS Atomindex, 1970–. (every two weeks)
(formerly *Nuclear Science Abstracts*)

International Aerospace Abstracts, 1961–. (every two weeks)

Key Abstracts: Electrical Measurement and Instrumentation, 1976–. (monthly)

Key Abstracts: Physical Measurement and Instrumentation, 1976–. (monthly)

Key Abstracts: Solid State Devices, 1978–. (monthly)

Mathematical Reviews, 1940–. (monthly)

Metals Abstracts, 1968–. (monthly)

Oceanic Abstracts, 1966–. (every other month)

Physics Abstracts (PA), 1898–. (every two weeks) (Section A of *Science Abstracts.*)

Pollution Abstracts, 1970–. (every two months)

Review of Plant Pathology, 1922–. (monthly) (Formerly *Review of Applied Mycology.*)

ATLASES

Moore, Patrick. *The Atlas of the Universe.*

National Geographic Atlas of the World.

Rand McNally Commercial Atlas and Marketing Guide. (updated annually)

The Rand McNally Atlas of the Oceans.

Times Atlas of the World: Comprehensive Edition.

United States Geological Survey. *National Atlas of the United States of America.*

BIBLIOGRAPHIES

Alred, Gerald J., Diana C. Reep, and Mohan R. Limaye. *Business and Technical Writing: An Annotated Bibliography of Books, 1880–1980.*

Besterman, Theodore. *A World Bibliography of Bibliographies.*

Bibliographic Index: A Cumulative Bibliography of Bibliographies, 1937–. (updated three times annually)

Hernes, Helga. *The Multinational Corporation: A Guide to Information Sources.*

Philler, Theresa Ammennito, et al., eds. *An Annotated Bibliography on Technical Writing, Editing, Graphics, and Publishing 1950–1965.*

Sheehy, Eugene P. *Guide to Reference Books.*

Walford, A. J. *Walford's Guides to Reference Material.* (Volume 1 contains science and technology.)

BIOGRAPHIES

American Men and Women of Science.

Biography Index, 1946–1973.

Current Biography, 1940–. (monthly)

Dictionary of Scientific Biography, 1970–1980.

Ireland, Norma Olin. *Index to Scientists of the World from Ancient to Modern Times: Biographies and Portraits.*

Who's Who in America, 1899–. (updated every two years)

BOOK GUIDES

Books in Print. (issued annually) (lists author, title, publication data, edition, and price)

Cumulative Book Index, 1928–. (lists all books published in English, with author, title, and subject listings; issued annually with monthly supplements)

New Technical Books, 1915–. (annotated selections, arranged by subject and updated every two months)

Proceedings in Print, 1964–. (announcements of conference proceedings with citation, subject, agency, and price; updated every two months)

BOOK REVIEWS

Book Review Digest, 1905–. (monthly, with annual accumulation)

Current Book Review Citations, 1976–.

Technical Book Review Index, 1935–. (guide to reviews in scientific, technical, and trade journals; issued monthly except for July and August)

COMMERCIAL GUIDES AND PROFESSIONAL DIRECTORIES

Encyclopedia of Associations. (annual guide to professional groups and publications with a quarterly supplement, *New Associations and Projects*)

MacRae's Blue Book (annually updated buying directory for industrial material and equipment)

National Trade and Professional Associations of the United States and Canada and Labor Unions, 1966–. (annual)

Thomas Register of American Manufacturers, 1905–. (monthly)

CURRENT AWARENESS SERVICES

Chemical Titles, 1960–. (lists current journal articles, indexed by key-word and author, issued every two weeks)

Conference Papers Index, 1973–. (monthly)

Current Contents, 1961–. (weekly listing of current journals' tables of contents; sections include agriculture, biology and environment services, engineering and technology; physical, chemical and earth sciences; life sciences)

Current Papers in Electrical and Electronics Engineering, 1969–. (monthly)

Current Papers in Physics, 1966–. (every two weeks)

Current Papers on Computer and Control, 1969–. (monthly)

Environmental Periodicals Bibliography, 1972–. (monthly)

Index to Scientific and Technical Proceedings, 1978–. (monthly)

DICTIONARIES

For unabridged dictionaries and desk dictionaries, see **dictionary.** For selected specialized subject dictionaries, see **library research.**

DISSERTATIONS

Dissertation Abstracts International, 1938–. (monthly)
(since 1966, Part B has been devoted to science and engineering)

Masters Abstracts, 1962–. (quarterly)

ENCYCLOPEDIAS

Encyclopedia Americana.

Encyclopedia of Computer Science and Technology.

Gray, Peter, ed. *The Encyclopedia of the Biological Sciences.*

International Encyclopedia of Statistics.

McGraw-Hill Encyclopedia of Science and Technology.

The New Encyclopaedia Britannica.

Newman, James R., ed. *Harper Encyclopedia of Science.*

Sills, David L., ed. *International Encyclopedia of the Social Sciences.*

Van Nostrand's Scientific Encyclopedia.

FACTBOOKS AND ALMANACS

The Budget in Brief. (a condensed version of the annual U.S. federal budget)

Demographic Yearbook, 1979–. (annual U.N. collection of information on world economics and trade)

Facts on File, 1974–. (weekly digest of world news, collected annually)

Johnson, Otto T., et al. *Information Please Almanac; Atlas Yearbook.*

Statistical Abstract of the United States (annual publication of the U.S. Bureau of the Census that summarizes data on industrial, political, social, and economic organizations in the United States)

GOVERNMENT PUBLICATIONS

Government Reports Announcements (GRA), 1946–. (twice-monthly listing of federally sponsored research reports arranged by subject with abstracts)

Government Reports Index, 1965–. (semimonthly index of reports abstracted in *GRA* and arranged by subject, author, and report number)

Index to U.S. Government Periodicals, 1970–. (quarterly)

Monthly Catalog of U.S. Government Publications, 1895–. (lists nonclassified publications of all federal agencies by subject, author, title, and report number; also gives ordering instructions)

Monthly Checklist of State Publications, 1940–. (state government publications arranged by state with an annual subject and title index)

Schmeckebier, Laurance F., and Roy B. Eastin. *Government Publications and Their Use.* (guide to government publications)

GRAPHICS

American Society for Testing and Materials. *Illustrations for Publication and Projection.*

Arkin, Herbert, and Raymond R. Colton. *Statistical Methods.*

Cardamone, Tom. *Chart and Graph Preparation Skills.*

Enrick, Norbert Lloyd. *Handbook of Effective Graphic and Tabular Communication.*

Field, Ron M. *A Guide to Micropublishing.*

Holmes, Nigel. *Designer's Guide to Creating Charts and Diagrams.*

Kepler, Harold B. *Basic Graphical Kinematics.*

LANGUAGE USAGE

Bernstein, Theodore M. *The Careful Writer: A Guide to English Usage.*

Copperud, Roy H. *American Usage and Style: The Consensus.*

Morris, William, and Mary Morris. *Harper Dictionary of Contemporary Usage.*

Nicholson, Margaret. *A Dictionary of American English Usage.*

LOGIC

Beardsley, Monroe C. *Thinking Straight: Principles of Reasoning for Readers and Writers.*

Brennan, Joseph G. *A Handbook of Logic.*

Copi, Irving M. *Introduction to Logic.*

Weddle, Perry. *Argument: A Guide to Critical Thinking.*

PARLIAMENTARY PROCEDURE

Deschler, Lewis. *Deschler's Rules of Order.*

Robert, Henry M., et al. *Robert's Rules of Order.*

PERIODICAL INDEXES AND DIRECTORIES

Applied Science and Technology Index, 1958–. (alphabetical subject list of more than 50,000 periodicals; issued monthly)

Bibliography of Agriculture, 1942–. (every two months)

General Science Index, 1978–. (monthly)

Index Medicus, 1880–. (monthly)

New York Times Index, 1851–. (every two weeks)

Readers' Guide to Periodical Literature, 1900–. (every two weeks)

Science Citation Index, 1962–. (quarterly)

Ulrich's International Periodicals Directory: A Classified Guide to Current Periodicals, Foreign and Domestic (annually)

Wildlife Review, 1935–. (quarterly)

Zoological Record, 1864–. (annually)

SPEECH AND GROUP DISCUSSION

Capp, Glen R., and G. Capp, Jr., *Basic Oral Communication.*

Harnack, R. Victor, et al. *Group Discussion: Theory and Technique.*

Maier, Norman. *Problem-solving Discussions and Conferences: Leadership Methods and Skills.*

Wilcox, Roger P. *Oral Reporting in Business and Industry.*

(See also **oral presentations.**)

STYLE MANUALS AND GUIDES

For a list of style manuals and guides, see **documenting sources.**

SUBJECT GUIDES

Jenkins, Frances B. *Science Reference Sources.*

Lasworth, Earl J. *Reference Sources in Science and Technology.*

Malinowsky, Harold R. *Science and Engineering Literature: A Guide to Reference Sources.*

Mount, Ellis. *Guide to Basic Information Sources in Engineering.*

Owen, Dolores B. *Abstracts and Indexes in Science and Technology: A Descriptive Guide.*

SYNONYMS

Rodale, J. L. *The Synonym Finder.*

Roget, Peter M. *Roget's International Thesaurus.*

Webster's Collegiate Thesaurus.

reference letters

Almost everyone is called upon at some time to provide a recommendation or reference for a colleague, an acquaintance, or an employee. Reference letters may range from completing an admission form for a prospective student to composing a detailed description of professional accomplishments and personal characteristics for someone seeking employment.

In order to write an effective letter of recommendation, you must be familiar enough with the applicant's abilities or actual performance to offer an evaluation. Then you must *truthfully,* without embellishment, communicate that evaluation to the inquirer. For the reference letter to serve as a valid selection device, you must address specifically the applicant's skills, abilities, knowledge, and personal characteristics in relation to the requested objective. You may also be asked to say what you know about the person's former employment and education or even his or her military discharge and personnel records.

In a reference letter, always respond directly to the inquiry, being careful to address the specific questions asked. For the record, you must identify yourself: name, title or position, employer, and address. You will also have to state how long you have known the applicant and the circumstances of your acquaintance. You should mention with as much substantiation as possible, one or two outstanding characteristics of the applicant. Organize the details in your letter in order of importance; put the most important details first. Conclude with a statement of recommendation and a brief summary of the applicant's qualifications.

The Privacy Act of 1974 has had a major impact on reference letters. Under the act, applicants have a legal right to examine the materials in an organization's files that concern them—unless they sign a waiver of their right to do so. Organizations seeking to avoid the privacy issue may offer applicants the option of signing a waiver against reading letters of reference written about them. When you are requested to serve as a reference or asked to supply a letter of reference, you should thus inquire about the confidentiality status of your letter—whether the applicant's file will be open or closed. Letter 1 is a typical reference letter.

references (see **documenting sources**)

reflexive pronouns

A reflexive pronoun, which always ends with the **suffix** *-self* or *-selves,* indicates that the **subject** of the sentence acts upon itself.

IVY COLLEGE
DEPARTMENT OF BUSINESS
WEST LAFAYETTE, IN 47906

Mr. Phillip Lester
Personnel Director
Thompson Enterprises
201 State Street
Springfield, IL 62705

Dear Mr. Lester:

As her employer and her former professor, I am happy to have the opportunity to recommend Kerry Hawkins. I've known Kerry for the last four years, first as a student in my class and for the last year as a research assistant.

I have found Kerry to be an excellent student, with above average grades in our program. On the basis of a GPA of 5.6 (A=6), Kerry was offered a research assistantship to work on a grant under my supervision. In every instance, Kerry completed her library search assignment within the time agreed upon. The material provided in the reports Kerry submitted met the requirements for my work and more. These reports were always well written. While working 15 hours a week on this project, Kerry has maintained a class load of 12 hours per semester.

I strongly recommend Kerry for her ability to work independently, to organize her time efficiently, and for her ability to write clearly and articulately. Please let me know if I can be of further service.

Sincerely yours,

Michael Paul

Michael Paul
Professor of Business

MP:mt

Letter 1 Reference Letter

EXAMPLE The lathe operator cut *himself.*

The reflexive form of the **pronoun** may also be used as an intensive to give **emphasis** to its antecedent. The reflexive and intensive pronouns are *myself, yourself, himself, herself, itself, oneself, ourselves, yourselves, themselves.*

EXAMPLE I collected the data *myself.*

Never use a reflexive pronoun as a subject or as a substitute for a subject.

CHANGE Joe and *myself* worked all day on it.
TO Joe and *I* worked all day on it.

A common mistake is substituting a reflexive pronoun for *I, me,* or *us.*

CHANGE John and *myself* completed the report.
TO John and *I* completed the report.

CHANGE He gave it to my assistant and *myself.*
TO He gave it to my assistant and *me.*

CHANGE Nobody can solve the problem but *ourselves.*
TO Nobody can solve the problem but *us.*

refusal letters

When you receive a **complaint letter** or an **inquiry letter** to which you must give a negative reply, you may need to write a refusal letter. For example, a professional organization may have asked you to speak at their regional meeting, but they require your firm to pay your expenses. Because your firm has declined the request, you must write a refusal letter turning down the invitation. The refusal letter is difficult to write because it contains bad news; however, you can tactfully and courteously convey the bad news. In this kind of **correspondence,** that should be your goal.

In your letter you should lead up to the refusal. To state the bad news in your **opening** would certainly affect you **reader** negatively. The ideal refusal letter says no in such a way that you not only avoid antagonizing your reader but keep his or her goodwill. To achieve such an **objective,** you must convince your reader of the justness of your refusal *before refusing.* The following pattern is an effective way to deal with this problem:

1. A buffer beginning.
2. A review of the facts.
3. The bad news, based on the facts.
4. A positive and pleasant closing.

The primary purpose of a buffer beginning is to establish a pleasant and positive **tone.** You want to convince your reader that you are a reasonable person. One way is to indicate some form of agreement with, or approval of, the reader or the project. If your refusal is in response to a complaint letter, do not begin by recalling the reader's disappointment ("We regret your dissatisfaction . . ."). Keep your buffer **paragraph** positive and pleasant. You can express appreciation for your reader's time and effort, if appropriate. Stating such appreciation will soften the disappointment and pave the way for future good relations.

EXAMPLE Thank you for your cooperation, your many hours of extra work answering our questions, and your patience with us as we struggled to make a very difficult decision. We believe our long involvement with your company certainly indicates our confidence in your products.

Next you should analyze the circumstances of the situation sympathetically. Place yourself in your reader's position and try to see things from his or her point of view. Establish clearly the reasons you cannot do what the reader wants—even though you have not yet said you cannot do it. A good explanation of the reasons should so thoroughly justify your refusal that the reader will accept your refusal as a logical conclusion.

EXAMPLE The Winton Check Sorter has all the features that your Abbott Check sorter offers and, in fact, offers two additional features that your sorter does not. The more important of the two is a backup feature that retains totals in memory, even if the power fails, so that processing doesn't have to start again from scratch following a power failure. The second additional feature that the Winton Sorter offers is stacked pockets, which take less space than do the linear pockets on your sorter.

State your refusal quickly, clearly, and as positively as possible.

EXAMPLE After much deliberation, we have decided to purchase Winton Check Sorters because of the extra features they offer.

Close your letter with a friendly remark, whether to assure the reader of your high opinion of his product or merely to wish him success.

EXAMPLE We have enjoyed our discussions with your company and feel we have learned a great deal about check processing. Although we did not select your sorter, we were very favorably impressed with your systems, your people, and your company. Perhaps we will work together in the future on another project.

The example shown in Letter 1 refuses an invitation to speak at a professional organization's regional meeting. Letter 2 (page 566) refuses a company credit for establishing an open account. Letter 3 (p. 567) refuses a job offer, and Letter 4 (p. 568) refuses a claim.

(See also **resignation letter or memorandum**.)

relative adjectives (see **adjectives**)

relative pronouns

Relative **pronouns** (the most common are *who, whom, which, what,* and *that*) perform a dual function: They substitute for a **noun** and link the **dependent clause** in which they appear to a main **clause.** (Relative pronouns differ in use from relative **adjectives,** which are pronouns in form and have a similar linking function but which *modify* [describe] nouns.) In each of the following examples, two words are italicized: the relative pronoun and the word for which it is substituting *(this entry continues on page 569)*:

EXAMPLES The specifications were sent to the *manager, who* immediately approved them.

The project was assigned to *Ed Jones, whom* everyone respects for his skill as a designer.

The customer's main *objection, which* can probably be overcome, is to the high cost of installation.

The *memo* from Martha Goldstein *that* you received yesterday sets forth the most important points.

Global Insurance Company
301 Decatur Street
Decatur, Illinois 62525
May 2, 19--

George M. Johnson
Good Deal Insurance Company
9001 Cummings Drive
St. Louis, Missouri 63129

Dear Mr. Johnson:

I am honored to have been invited to address your regional meeting in St. Louis on May 17. To be considered one who might make a meaningful contribution to such a gathering of experts is indeed flattering.

On checking with those who control the purse strings here at Global Insurance Company, however, I learned that it would be contrary to company policy for Global to pay my expenses for such a trip. Therefore, as much as I would enjoy attending your meeting, I must decline.

I have been very favorably impressed over the years with your organization's contribution to the insurance industry, and I am proud to have received your invitation.

Sincerely,

Ralph P. Morgan

Ralph P. Morgan
Adviser/GIC

RPM/mo

Letter 1 Refusal Letter

Titus Packaging, Inc.
2063 Eldorado Dr.
Billings, Montana 59102

April 4, 19--

Ms. Edna Kohls
Graphic Arts Services, Inc.
936 Grand Avenue
Billings, Montana 59102

Dear Ms. Kohls:

We appreciate your interest in establishing an open account at Titus Packaging, Inc. In the two years since you began operations, your firm has earned an excellent reputation in the business community.

As you know, interest rates have been rising sharply this past year, while sales in general have declined. With the current negative economic climate, and considering the relatively recent establishiment of your company, we believe that an open account would not be appropriate at this time.

We will be happy to have you renew your request around the first of next year, when the economic climate is expected to improve and when your company will be even more firmly established. In the meantime, we will be happy to continue our present cash relationship, with a 2-percent discount for payment made in ten days.

Sincerely,

Conrad C. Atkins

Conrad C. Atkins
Manager, Credit Department

CCA/mo

Letter 2 Refusal Letter

127 Idlewild Road
Boston, MA 02173
October 17, 19--

Ms. Juanita Perez
Director, Personnel Department
Manchester Iron Works, Inc.
1258 Halsey Avenue
Boston, MA 02181

Dear Ms. Perez:

I was pleased to receive your offer of employment as an Administrative Assistant in the Executive Office. I was very favorably impressed with your company and with the position as you described it.

Since interviewing with you, however, I have been offered another position that is even more in line with my long-range goals, and I have accepted that position.

Thank you for your time and consideration. I enjoyed meeting with your staff.

Sincerely yours,

Jason L. Wytosh

Jason L. Wytosh

Letter 3 Refusal Letter

General Television, Inc.
5521 West 23rd Street
New York, N.Y. 10062

September 28, 19--

Mr. Fred J. Swesky
7821 Ranchero Drive
Tucson, AZ 85761

Dear Mr. Swesky:

Thank you for your letter regarding the replacement of your KL-71 television set.

You stated in your letter that you used the set on an uncovered patio. As our local service representative pointed out, this model is not designed to operate in extreme heat conditions. As the instruction manual that accompanied your television set cautioned, such exposure can seriously damage this model. Since your set was used in such extreme heat conditions, we cannot honor the two-year replacement warranty.

We trust you will understand our position and we hope to provide for your future television needs.

Sincerely yours,

Susan Siegel

Susan Siegel
Assistant Director

SS/ar

Letter 4 Refusal Letter

Sometimes a relative pronoun does not stand for another specific noun but does act as a noun in serving as the subject of a clause. In the following example, *what* acts as a subject in the dependent clause while also linking that clause to the main clause:

EXAMPLE We invited the committee to come to see *what* was being done about the poor lighting.

The relative pronoun *who* (*whom*) normally refers to a person; *that* and *which* normally refer to a thing; and *that* can refer to either. The pronoun *that* is normally used with restrictive clauses and the pronoun *which* with nonrestrictive clauses. (See also **restrictive and nonrestrictive elements.**)

EXAMPLES The annual report, *which was distributed yesterday,* shows that sales increased 20 percent last year. (nonrestrictive)
Companies *that adopt the plan* nearly always show profit increases. (restrictive)

Relative pronouns can often be omitted.

EXAMPLE I received the handcrafted ornament (that) I ordered.

Do not omit the word *that,* however, when it is used as a conjunction and its omission could cause the **reader** to misread a sentence.

CHANGE This time delay ensures the operator has time to load the card hopper.
TO This time delay ensures *that* the operator has time to load the card hopper.

(See also **who/whom**.)

repetition

The deliberate use of repetition to build a sustained effort or emphasize a feeling or idea can be quite powerful.

EXAMPLE Similarly, atoms *come and go* in a molecule, but the molecule *remains;* molecules *come and go* in a cell, but the cell *remains;* cells *come and go* in a body, but the body *remains;* persons *come and go* in an organization, but the organization *remains.*

—Kenneth Boulding, *Beyond Economics* (Ann Arbor: University of Michigan Press, 1968), p. 131.

Repeating key words from a previous sentence or **paragraph** can also be used effectively to achieve **transition.**

EXAMPLE For many years, *oil* has been a major industrial energy source. However, *oil* supplies are limited, and other sources of energy must be developed.

Be consistent in the word or phrase you use to refer to something. In technical writing, it is generally better to repeat a word (so there will be no question in the reader's mind that you mean the same thing) than to use a synonym in order to avoid repeating it. Your primary goal in technical writing is effective and precise communication rather than elegance.

CHANGE Several recent *analyses* support our conclusion. These *studies* cast doubts on the feasibility of long-range forecasting. The *reports,* however, are strictly theoretical.

TO Several recent theoretical *studies* support our conclusion. These *studies* cast doubts on the feasibility of long-range forecasting. They are, however, strictly theoretical.

Purposeless repetition, however, makes a sentence awkward and hides its key ideas.

CHANGE He *said that* the customer *said that* the order was canceled.

TO He *said that* the customer canceled the order.

The harm caused to your writing by careless repetition is not limited to words and **phrases;** the needless repetition of ideas can be equally damaging.

CHANGE In this modern world of ours today, the well-informed, knowledgeable executive will be well ahead of the competition.

TO To succeed, the contemporary executive must be well informed.

(See also **elegant variation.**)

reports

A report is an organized presentation of factual information that answers a request by supplying the results of an investigation, a trip, a test, a research project, and the like. All reports can be divided into two broad categories: formal and informal.

Formal reports present the results of projects that may require months of work and involve large sums of money. These projects may be done either for one's own organization or as a contractual requirement for an outside organization. Formal reports generally follow a stringent **format** and include some or all of the report elements discussed in **formal reports.**

Informal reports normally run from a few paragraphs to a few pages and include only the essential elements of a report: introduction, body, conclusions, and recommendations. Because of their brevity, informal reports are customarily written as a letter (if the report is to be sent to someone outside the organization) or as a **memorandum** (if it is to be sent to someone inside the organization).

The *introduction* serves several functions: it announces the subject of the report, states its purpose, and, when appropriate, gives essential background information. The **introduction** should also summarize any conclusions, findings, or recommendations made in the report. Managers, supervisors, and clients find a concise summary useful because it gives them the essential information at a glance and helps focus their thinking as they read the rest of the report. The *body* of the report should present a clearly organized account of the report's subject—the results of a test carried out, the status of a construction project, and so on. The amount of detail to include depends on the complexity of the subject and on your reader's familiarity with it. The *conclusion* should summarize your findings and tell the reader what you think their significance may be. In some reports a final section gives *recommendations.* (Sometimes the **conclusions** and recommendations sections are combined.) In this section, you make suggestions for a course of action based on the data you have presented.

Many different types of informal reports are described in separate entries in this handbook: **feasibility reports, investigative reports, progress and activity reports, trip reports, laboratory reports, test reports,** and **trouble reports.** Many of these can be formal or informal, depending on their length and complexity.

research

Research is the process of investigation, the discovery of facts. As part of the writing process, research must be preceded by **preparation** (that is, you must first determine your **reader**, **objective**, and

scope). During the preparation phase, you establish the extent of the coverage needed for your work, and then during the research stage you can logically determine what information you will need and in how much detail. Without this preparation, the research will not be adequately defined or focused.

Researchers frequently distinguish between primary and secondary information, and depending on the goal of the research, the distinction can be important. *Primary information* refers to the "raw" data compiled from observations, surveys, experiments, questionnaires, and the like. This information is frequently gathered or located in notebooks, computer printouts, telephone logs, transcripts of speeches and interviews, tape recordings, and unedited film. *Secondary information* refers to primary information that has been analyzed, assessed, evaluated, compiled, or otherwise organized into accessible form. The forms include books, articles, reports, dissertations, operating and procedure manuals, brochures, and so forth. Use the information most appropriate to your research needs, recognizing that some projects will require both types.

Many projects require information from a variety of sources. The most common ones include written materials, interviews with experts, and firsthand observations. However, before investigating any of them, do not overlook the invaluable resource that *you* may represent. Before going elsewhere, interview yourself to tap into your own knowledge and experience. You may find that you already have enough information to get started. This technique, commonly known as *brainstorming*, may also stimulate your thinking about additional places to seek information.

First, write down as many ideas as you can think of about the subject, as well as the objective and scope of your project, and jot them down in random order *as they occur to you.* (This may be done by a group as well as alone.) Put down what you know and, if possible, where you learned of it. For every idea noted, ask what, when, who, where, how and why. List the details these additional questions bring to mind. You can write down these ideas on a chalkboard, pad of paper, or note cards. (Note cards are preferable if you are working alone because they can easily be shuffled and rearranged.) Once your memory runs dry, group the items in the most logical order, based on your objective and your reader's needs.

The end result of this process will be a tentative outline of your project. The outline will have sketchy or missing sections, thus helping to show where additional research is needed, and it will also help integrate the various details of the additional research.

If you must conduct a formal, structured search of published sources, see **library research.** Scan the written material quickly before reading it in order to get an idea of the kind of information it includes. Then if it seems useful, read it carefully, taking notes of any information that falls within the scope of your research. Be sure to note down where you found the information so you can find it again if necessary or compile it for a list of works cited. You may think of additional questions about the topic as you read; jot these down and identify them with a question mark. Some of them may eventually be answered in other research sources; those that remain unanswered can guide you to further research. (For detailed guidance on taking notes, see **note taking** and **paraphrasing.**)

If after researching these other sources you still need more information, consider talking to an expert. Not only can someone skilled in a field answer many of your questions, but he or she can also suggest other sources of information. (To make the most of such discussions, see **interviewing for information** and **listening.**)

Finally, you can use your own observations or hands-on experience. In fact, this is the only way to obtain certain kinds of information: the behavior of people and animals, certain natural phenomena, mechanical processes, the operation of tools and equipment, and the like. When planning research that involves observation, choose your sites and times carefully. Also, be sure to obtain permission in advance when necessary. Such observations may be illegal, resented, or contrary to an organization's policy. Keep accurate, complete records, indicating date, time of day, and the like. Save interpretations of your observations for future analysis. Be aware that research involving observations may be time-consuming, complicated, and expensive and that your presence may inadvertently influence what you observe.

If you have to write an operating manual for equipment or a procedure for some task, you must first acquire hands-on experience by operating the equipment or performing the task yourself. Afterwards, you will in effect be interviewing yourself based on your experience. To do so, make a rough outline of your experience using the brainstorming technique described in this entry. After writing the draft, have someone unfamiliar with the equipment operate it or someone unfamiliar with the procedure perform it. When he or she has a problem, rewrite the passage until it is clear and easy to follow.

As with observations, obtain permission to operate the equipment or carry out a procedure. Comply with all safety rules, and make them a part of the manual or procedures you write. (See also **questionnaires.**)

resignation letter or memorandum

Start your letter or memorandum of resignation on a positive note, regardless of the circumstances under which you are leaving. You might, for example, point out how you have benefited from working for the company. Or you might say something complimentary about how well the company is run. Or you might say something positive about the people with whom you have been associated.

Then explain why you are leaving. Make your explanation objective and factual, and avoid recriminations. Your letter or memorandum will become part of your permanent file with that company, and if it is angry and accusing, it could haunt you in the future: you may need references from the company.

Your letter or memorandum should give enough notice to allow your employer time to find a replacement. It may be no more than two weeks, or it may be enough time to enable you to train your replacement. Some organizations ask for a notice that is equivalent to the number of weeks of vacation you receive.

The sample memorandum of resignation in Figure 1 is from an employee who is leaving to take a job offering greater opportunities. The memorandum of resignation presented in Figure 2 is from an employee who is leaving under unhappy circumstances. Notice that it opens and closes positively and that the reason for the resignation is stated without apparent anger or bitterness.

respective/respectively

Respective is an **adjective** that means "pertaining to two or more things regarded individually."

EXAMPLE The committee members prepared their *respective* reports.

To: W.R. Johnson, Director of Purchasing
From: J.L. Washburn, Purchasing Agent JLW
Date: January 7, 19--
Subject: Resignation from Barnside Appliances, effective February 7, 19--

My three years at Barnside Appliances has been an invaluable period of learning and professional development. I arrived as a novice, and I believe that today I am a professional--primarily as a result of the personal attention and tutoring I have received from my superiors and the fine example set both by my superiors and my peers.

I believe the time has come for me to move on, however, to a larger company that can give me an opportunity to continue my professional development. Therefore, I have accepted a position with General Electric, where I am scheduled to begin on February 12 Thus, my last day at Barnside will be February 7. However, if you need more time to hire and train my replacement, I can make arrangements to work longer.

Many thanks for the experiences I have gained, and best wishes for the future.

Figure 1 Sample Resignation Memo (Resignation to Accept a Better Position)

To: T.W. Haney, Vice-President, Administration
From: L.R. Rupp, Executive Assistant *LR*
Date: February 12, 19--
Subject: Resignation from Winterhaven, effective March 1, 19--

My ten-year stay with the Winterhaven Company has been a very pleasant experience, and I am sure it has been mutually beneficial.

Because the recent realignment of my job leaves no career path open to me, however, I have accepted a position with another company that I feel will give me a better future. I am, therefore, submitting my resignation, to be effective on March 1, 19--.

I have enjoyed working for Winterhaven and wish the company success in the future.

Figure 2 Sample Resignation Memo (Resignation Under Negative Conditions)

Respectively is the **adverb** form of *respective,* meaning "singly, in the order designated."

EXAMPLE The first, second, and third prizes in the sales contest were awarded to Maria Juarez, Gloria Hinds, and Margot Luce, *respectively.*

Respective and *respectively* are often unnecessary because the meaning of individuality is already clear.

CHANGE The committee members prepared their *respective* reports.
TO Each committee member prepared a report.

restrictive and nonrestrictive elements

Modifying **phrases** and **clauses** may be either restrictive or nonrestrictive. A nonrestrictive phrase or clause provides additional information about what it modifies, but it does not limit, or restrict, the meaning of what it modifies. Therefore, the nonrestrictive phrase or clause can be removed without changing the essential meaning of the sentence. It is, in effect, a parenthetical element, and so it is set off by **commas** to show its loose relationship with the rest of the sentence.

EXAMPLES This instrument, *called a backscatter gauge,* fires beta particles at an object and counts the particles that bounce back. (nonrestrictive phrase)
The annual report, *which was distributed yesterday,* shows that sales increased 20 percent last year. (nonrestrictive clause)

A restrictive phrase or clause limits, or restricts, the meaning of what it modifies. If it were removed, the essential meaning of the sentence would be changed. Because a restrictive phrase or clause is essential to the meaning of the sentence, it is never set off by commas.

EXAMPLES All employees *wishing to donate blood* may take Thursday afternoon off. (restrictive phrase)
Companies *that adopt the plan* nearly always show profit increases. (restrictive clause)

It is important for writers to distinguish between nonrestrictive and restrictive elements. The same sentence can take on two entirely different meanings depending on whether a modifying element is set off by commas (because nonrestrictive) or not set off (because restrictive). The results of a slip by the writer can be not only misleading but also downright embarrassing.

> CHANGE I think you will be impressed by our systems engineers who are thoroughly experienced in projects like yours.
>
> TO I think you will be impressed by our systems engineers, who are thoroughly experienced in projects like yours.

The problem with the first sentence is that it suggests that "you may not be so impressed by our other, less experienced, systems engineers."

résumés

A résumé* is an essential element in any **job search** because it is, in effect, a catalog of what you have to offer to prospective employers. Sent out with your **application letter,** it is the basis for their decision to invite you for a job interview. (See **interviewing for a job**.) It tells them who you are, what you know, what you can do, what you have done, and what your job objectives are. It may be the basis for the questions you are asked in the interview, and it can help your prospective employer evaluate the interview, as an organized reminder of what you covered orally. The résumé, then, is an important document that deserves careful planning and execution.

Three steps will help you prepare your résumé:

1. Analyze yourself and your background.
2. Identify those to whom you will submit your résumé (your **readers**).
3. Organize and prepare the résumé.

GATHERING INFORMATION

The starting point in preparing a résumé is a thorough analysis of your experience, education and training, professional and personal skills, and personal traits.

List the major points for each category before actually beginning to organize and write your résumé. (It helps to write each item on a

*When typeset, *résumé* is set with accent marks (´) to distinguish it from the verb *resume.* When *résumé* is typewritten, the accent marks are not required.

3- × 5-inch card.) The following questions can serve as an inventory of what you should consider in each category.

Experience. Note all your employment—full-time, part-time, vacation jobs, and free-lance work—and analyze each job on the basis of the following list of questions:

1. What was the job title?
2. What did you do (in reasonable detail)?
3. What experience did you gain that you can apply to another job?
4. Why were you hired for the job?
5. What special skills did you learn on that job?
6. Were you promoted or given a job with more responsibility?
7. Why did you leave it?
8. When did you start and when did you leave the job?
9. Would your former employer give you a reference?
10. What special traits were required of you on the job (initiative, leadership, ability to work with details, ability to work with people, imagination, ability to organize, and so on)?

Education. For the applicant with little work experience, such as a recent graduate, the education category is of primary importance. For the applicant with extensive work experience, education is still quite important, even though it is now secondary to work experience. List the following information about your education:

1. Colleges attended and the inclusive dates.
2. Degrees and the dates they were awarded.
3. Major and minor subjects.
4. Courses taken or skills acquired that might be important to the job you are applying for.
5. Internships, work-study programs, co-op positions.
6. Cumulative grade averages and academic honors.
7. Extracurricular activities.
8. Scholarships and awards.
9. Special training courses (academic or industrial).

ORGANIZING AND WRITING THE RÉSUMÉ

After listing all the items in the preceding two categories, you should analyze them in terms of the job you are seeking, evaluating the items, rejecting some of them, and finally selecting those that you will include in your résumé. Base your decision on the following questions:

1. What exact job are you seeking?
2. Who are the prospective employers?
3. What information is most pertinent to that job and those employers?
4. What details should be included and in what order?
5. How can you present your qualifications most effectively to get an interview?

Résumés should be brief—preferably one or two pages, depending on your level of experience—and should include the following information:

1. Your name, address, and phone number.
2. Your immediate and long-range job objectives.
3. Your professional training.
4. Your professional experience, including the firms where you have worked, and your responsibilities.
5. Your specific skills.
6. Pertinent personal information.

FORMAT OF THE RÉSUMÉ

A number of different formats can be used. Most important is to make sure that your résumé is attractive, well organized, easy to read, and free of errors. A common format uses the following **heads:**

Employment Objective
Education
Employment Experience
Special Skills and Activities
References (optional)

Underline or capitalize the heads to make them stand out on the page. Whether you list education or experience first depends on which is stronger in your background. If you are a recent graduate, list education first; if you have substantial job experience, list your most recent experience first, your next most recent experience second, and so on.

The Heading. Center your name in all capital letters (and boldface if you're using a word processor) at the top of the page. Follow it with your address and telephone number, also usually centered.

Employment Objective. State both your immediate and long-range employment objectives. However, if an objective is not appropriate for a particular position (a temporary job, for example), you may exclude it.

Education. List the college or colleges you attended, the dates you attended each one, any degree or degrees received, your major field of study, and any academic honors you earned. First-time applicants frequently give the name of their high school, its location (city, state), and the dates they attended.

Employment Experience. List all your full-time jobs, from the most recent to the earliest that is appropriate. If you have had little full-time work experience, list your part-time and temporary jobs, too. Give the details of your employment, including the job title, dates of employment, and the name and address of your employer. Provide a concise description of your duties only for those jobs whose duties were similar to those of the job you are seeking; otherwise, give only a job title and a brief description. Specify any promotions or pay increases you received. Do not, however, list your present salary. (Salary depends on your experience and your potential value to an employer.) If you have been with one company for a number of years, highlight your accomplishments during that time. List military service as a job. Give the dates you served, your duty specialty, and your rank at discharge. Discuss the duties only if they relate to the job you are applying for.

Some résumés arrange employment experience by function rather than by chronological time. Instead of listing your jobs first, starting with the most recent, you list the functions you've performed ("Management," "Project Development," "Training," "Sales," and the like). The functional arrangement is useful for persons who wish to stress their skills or who have been employed at one job and wish to demonstrate their diversity of experience at that single position. It is also used by those who have gaps in time on their résumé caused by unemployment or illness. Although the functional arrangement can be effective, employment directors know that it may be used to cover weaknesses, and so you should use it carefully and be prepared to explain any gaps.

Special Skills and Activities. This category usually comes near the end and may replace the Personal Data category. You may include such skills as knowledge of foreign languages, writing and editing abilities, specialized knowledge (such as experience with a computer language or a background in electronics), hobbies, student or community activities, professional or club memberships, and published works. Be selective: do not overload this category, do not duplicate information given in other categories, and do not include any items that do not support your employment objective. Other titles, such as "Abilities" and "Professional Affiliations," may be used for this category.

Personal Data. Federal legislation limits the inquiries an employer can make before hiring an applicant—especially requests for such personal information as age, sex, marital status, race, and religion. Consequently, some job seekers exclude this category because they feel their personal data could have a negative impact on the employer. You may also eliminate this category simply because you need the space for more significant information about your qualifications. However, if you believe the Personal Data category will help you land a particular job, then include your date of birth and your marital status (it's better to give your date of birth than your age, so that your résumé won't become out of date after your next birthday). In any case, you should see yourself through the eyes of your prospective employers and provide whatever information would be most useful to them.

References. You can state on your résumé that "references will be furnished upon request" or include references as part of the résumé (only if you need to fill space or if including names is traditional to your profession). Either way, do not list anyone as a reference without first obtaining his or her permission to do so.

Style. As you write your résumé, use action **verbs**, and state your ideas concisely. You can avoid the pronoun *I* since the résumé is only about you.

> CHANGE I was promoted to office manager in June 19--.
> TO Promoted to office manager in June 19--.

The following is a list of typical verbs used in résumés:

administered	conducted
analyzed	coordinated
completed	created

designed	oriented
developed	prepared
directed	programmed
evaluated	recruited
increased	started
managed	supervised
operated	trained
organized	wrote

Be truthful in your résumé. If you give false data and are found out, the consequences could be serious. At the very least, you will have seriously damaged your credibility with your employer.

Your résumé should be flawless. If you are not a skilled typist, you may want to have it professionally done. Printing usually produces more professional-looking copies than does photocopying. A word processor is ideal for creating a résumé, since it allows you to customize and update it easily. (See **word processing.**)

The sample résumés presented on pages 584–8 are all for the same person. Figure 1 is a student's first résumé, and Figures 2 and 3 show her work experience later in her career (one is organized by job, the other by function). Examine as many résumés as possible, and select the format that best suits you and your goals.

revision

The more natural a work of writing seems to the **reader,** the more effort the writer has probably put into revision. Anyone who has ever had to say of his or her own writing, "Now I wonder what I meant by that," can testify to the importance of revision. The time you invest in review will make the difference between clear writing and unclear writing.

If you have followed the appropriate steps of the writing process, you will have a very rough draft that can hardly be considered finished writing. Revision is the obvious next step.

Allow a couple of days, if possible, to go by without looking at the draft before beginning to revise it. Without a cooling period, you are too close to the draft to be objective in evaluating it. Only when enough time has elapsed so that you can say, "How did I ever write that?" or "I wonder why I said that so awkwardly?" are you ready to start revising. *(This entry continues on page 589.)*

CAROL ANN WALKER
273 East Sixth Street
Bloomington, IN 47401
(913) 555-1212

Employment Objective	Position as financial research assistant, leading to a management position in corporate finance.
Education	B.B.A., Indiana University (expected June 19--) Major: Finance Minor: Computer Science Honors: Dean's List (G.P.A.: 3.88/4.00), Senior Honor Society, 19--
Employment Experience	First Bank, Research Department (Bloomington, IN) Intern research assistant: Assisted manager of corporate planning and developed computer model for long-range financial planning. (Summer and Fall semesters 19--) Martin Financial Research Services (Bloomington, IN) Intern editorial assistant: Provided research assistance to staff. Developed a design concept for in-house financial audits. Various part-time jobs to finance education (60%)(from 19-- to 19--)
Special Skills and Activities	Associate editor, Business School Alumni Newsletter Edited submissions, surveyed business periodicals for potential stories, wrote two stories on financial planning with computer models. President of Women's Transit program Coordinated efforts to provide safe nighttime transportation to and from dormitories and campus buildings. Amateur Photographer Won several local awards for nature photos.
References	Available upon request.

Figure 1 Student Résumé

CAROL ANN WALKER
1436 W. Schantz Avenue
Dayton, Ohio 45401
(513) 555-1212

Employment Objective Senior Research Financial Analyst in corporate offices of growth and research-oriented major manufacturing company.

Employment Experience
(19-- to Present)

January 19-- to Present KERFHEIMER CORPORATION (Dayton, Ohio)

Senior Financial Analyst (September 19-- to October 19--)
Report to the Senior Vice-President for Corporate Financial Planning of this $1-billion manufacturer of heavy mining and construction equipment, which is also purchased by various branches of the federal government and the Department of Defense (DOD). Work with government procurement officers and engineers in developing manufacturing cost estimates for complex and massive machinery, based on prototypes developed in our research lab. Develop program funding requirements and determine best available sources, some of which have been in the $100- to $300-million range. Recipient of the "Financial Planner of the Year" award from the Association of Financial Planners.

Financial Analyst (January 19-- to September 19--)
Reported to Senior Financial Analyst and assisted in the development of funding for major DOD contracts for troop carriers, digging and earth-moving machines, and extension booms. Researched funding options and recommended those with the most favorable rates and terms. Recommended by Senior Financial Analyst as the person to replace him on his retirement.

June 19-- to December 19-- FIRST BANK, INC. (Bloomington, IN)

Planning Analyst (September 19-- to December 19--)
Reported to Manager of Corporate Planning and developed computer models for short- and long-range planning that reduced by 65% percent the time required to compute costs.

Figure 2 Advanced Résumé: Organized by Job

CAROL ANN WALKER
Page 2

Education

Wharton School, University of Pensylvania (19-- to Present)
Graduate Seminars Taken
Advanced Corporate Planning and Demographics
Computer Analysis of Corporate Planning
Theory of Corporate Policies
Regulation and Corporate Financial Policy

University of Wisconsin--Milwaukee (June 19--)
Degree: M.B.A.
"Special Executive Curriculum": A special fast-track program for M.B.A. students who are identified as promising and who are sponsored by their employer.

Indiana University (June 19--)
Degree: B.B.A. (magna cum laude)
Major: Finance
Minor: Computer Science

Special Skills and Activities

Article published in Midwest Finance Journal: "Developing Computer Models for Financial Planning" (Vol. 34, No. 2, 19--), pp. 122-136.

Guest Lecturer at Indiana University (June 29, 19--)

Senior Member, Association for Corporate Financial Planning

Member of Audubon Society, Photographer, and Runner

References and a portfolio of computer financial models on request.

CAROL ANN WALKER
1436 W. Schantz Avenue
Dayton, Ohio 45401
(513) 555-1212

Employment Objective Senior Research Financial Analyst in corporate offices of growth- and research-oriented major manufacturing company.

Major Accomplishments

Financial Planning

- Provided research on funding options that enabled company to achieve a 23-percent return on investment.
- Developed long-range funding requirements to respond to government and military contracts that totaled over $1 billion.
- Developed computer model for long- and short-range planning, which saved 65 percent in preparation time.

Capital Acquisition

- Developed strategies enabling employers to acquire over $1 billion at 3 percent below market rate.
- Responsible for securing over $100 million through private and government grants for research.
- Developed computer models for capital acquisition that enabled company to increase its long-term debt during several major building expansions.

Research and Analysis

- Developed sophisticated computer models based on cutting-edge research applied to practical problems of corporate finance.
- Served primarily as a researcher for 11 years at two different types of firms.
- Recipient of the "Financial Planner of the Year" award from the Association of Financial Planners, a group composed of both academics and practitioners.
- Published academic article and am taking doctoral-level seminars at the Wharton School, which are described in the education section that follows.

Figure 3 Advanced Résumé: Organized by Function

CAROL ANN WALKER
Page 2

Education

Wharton School, University of Pennsylvania (19-- to Present)
Graduate Seminars Taken
Advanced Corporate Planning and Demographics
Computer Analysis of Corporate Planning
Theory of Corporate Policies
Regulation and Corporate Financial Policy
University of Wisconsin--Milwaukee (June 19--)
Degree: M.B.A.
"Special Executive Curriculum": A special fast-track program for M.B.A. students who are identified as promising and who are sponsored by their employer.
Indiana University (June 19--)
Degree: B.B.A. (magna cum laude)
Major: Finance
Minor: Computer Science

Work History

KERFHEIMER CORPORATION (Dayton, Ohio): January 19-- to Present
Senior Financial Analyst (19-- to 19--)
Financial Analyst (19-- to 19--)
FIRST BANK, INC. (Bloomington, IN): June 19-- to December 19--
Planning Analyst (19-- to 19--)
Student Intern (19-- to 19--)

Special Skills and Activities

Article published in Midwest Finance Journal: "Developing Computer Models for Financial Planning" (Vol. 34, No. 2, 19--), pp. 122-136.
Guest Lecturer at Indiana University (June 29, 19--)
Senior Member, Association for Corporate Financial Planning
Member of Audubon Society, Photographer, and Runner

References and a portfolio of computer financial models on request.

A different frame of mind is required for revising than for **writing the draft.** Read and evaluate the draft, which should have been written quickly, with cool deliberation and objectivity—from the point of view of a reader or critic. Be anxious to find and correct faults, and be honest. In revising, consider your reader first.

Do not try to do all your revision at once. Read through your rough draft several times, each time searching for and correcting a different set of problems.

Check your draft for completeness. Your writing should give readers exactly what they need, but it should not burden them with unnecessary information or sidetrack them into insignificant or only loosely related subjects. Check your draft against your **outline** to make certain that you did, in fact, follow your plan. If not, now is the time to insert any missing points into your draft. This may mean inserting a line here and there; it may mean writing a **paragraph** or passage on a separate sheet and attaching it to the appropriate page; or it may mean major additions to your rough draft, depending on how carefully you followed your outline.

Take time to examine the facts in your draft for accuracy. Keep in mind that no matter how careful and painstaking you may have been in conducting your **research,** compiling your notes, and creating your outline, you could easily have made errors when transferring your thoughts from the outline to the rough draft. You can quickly and easily check the facts in your draft for accuracy if you took complete notes, and this can be critical in catching errors. Be certain that contradictory facts have not crept into your draft.

Be consistent with names. Make sure, for example, that you have not called the same item a "routine" on one page and a "program" on another.

Your **introduction,** if you use one, should state your **objective** and provide a frame in which your reader can place the detailed information that follows in the body of your draft. Check your introduction to see that it does, in fact, give the reader such assistance. If you do not use an introduction, check your **opening** to see that it is useful and interesting to the reader.

To determine whether you have good **transition,** look for **unity** and **coherence.** If a paragraph has unity, all its sentences and ideas are closely tied together and contribute directly to the main idea expressed in the **topic sentence** of the paragraph. Writing that is coherent flows smoothly from one point to another, from one sentence to another, and from one paragraph to another. Where transition is missing, provide it; where transition is weak, strengthen it. Also look for unity in the entire report, and provide any transition that is needed between paragraphs or sections.

Check your draft for **pace.** If you find that you are jamming your ideas too closely together, space them out and slow down the pace.

Check your use of **jargon.** If you have any doubt that all your readers will understand any jargon you may have used, eliminate it.

Check your draft for **conciseness**. Tighten your writing so that it says exactly what you mean by pruning unnecessary words, **phrases**, sentences, and even paragraphs.

Replace **clichés** with fresh **figures of speech** or direct statements.

Watch for words that could carry an undesired **connotation.** Also delete or replace vague or pretentious words, coined words, and unnecessary intensifiers. (See also **word choice.**)

Check your draft for **awkwardness,** determine the reason for any you may find, and correct it.

Check your draft for possible grammatical errors. Any such errors should be caught in your check for awkwardness, but a final review may save you embarrassment.

Finally, check your final draft for typographical errors by **proofreading** it.

rhetorical questions

A rhetorical question is a question that a writer asks the **reader**—a question to which a specific answer is neither needed nor expected. The question is often intended to make the reader think about the subject from a different perspective; the writer then answers the question in the article or essay.

EXAMPLES Is methanol the answer to the energy shortage?
Does advertising lower consumer prices?

The answer to a rhetorical question may not be a yes or no; in the above examples, it might be a detailed explanation of the pros and cons of the value of methanol as a source of energy or the effect of advertising on consumer prices.

The rhetorical question can be an effective **opening**, and it is often used for a **title.** By its nature, however, it is somewhat informal and should therefore be used judiciously in technical writing. For example, a rhetorical question would not be an appropriate opening for a **report** or **memorandum** addressed to a busy superior in your company. When you do use a rhetorical question, be sure that it is not trivial, obvious, or forced. More than any other writing device, the rhetorical question requires that you know your reader.

run-on sentences

A run-on sentence, sometimes called a fused sentence, is two or more sentences without **punctuation** to separate them. The term is also sometimes applied to a pair of **independent clauses** separated by only a **comma**, although this variation is usually called a **comma fault** or comma splice. Run-on sentences can be corrected by (1) making two sentences, (2) joining the two clauses with a **semicolon** (if they are closely related), (3) joining the two clauses with a comma and a **coordinating conjunction,** or (4) subordinating one clause to the other.

CHANGE The new manager instituted several new procedures some were impractical. (run-on sentence)

TO The new manager instituted several new procedures. Some were impractical. (period)

OR The new manager instituted several new procedures; some were impractical. (semicolon)

OR The new manager instituted several new procedures, but some were impractical. (comma plus coordinating conjunction)

OR The new manager instituted several new procedures, some of which were impractical. (one clause subordinated to the other)

(See also **sentence faults.**)

S

sales proposals

The *sales proposal,* one of the major marketing tools in use in business and industry today, is a company's offer to provide specific goods or services to a potential buyer within a specified period of time and for a specified price. The primary purpose of a sales proposal is to demonstrate that the prospective customer's purchase of the seller's products or services would be a wise investment, because doing so will solve a problem, improve operations, or offer benefits. Indeed, the sales proposal should sell solutions to problems rather than equipment or services.

Sales proposals vary greatly in size and relative sophistication. Some may be a page or two long and be written by one person; others may be many pages long and be written by several people; still others may take hundreds of pages and be written by a team of professional proposal writers. A short sales proposal might bid for the construction of a single home; a sales proposal of moderate length might bid for the installation of a network of computer systems; and a very long sales proposal might be used to bid for the construction of a multibillion-dollar aircraft carrier.

Your first task in writing a sales proposal is to find out exactly what your prospective customer needs. The information in a proposal should be gathered in a survey of the potential customer's business and needs. You must then determine whether your organization can satisfy the customer's needs. Before preparing a sales proposal, you should try to find out who your principal competitors are. Then compare your company's strengths with those of the competing firms, determine your advantages over your competitors, and emphasize them in your proposal. For example, a small software company bidding for an Air Force contract at a local base would be familiar with its competitors. If the proposal writer believes that his or her company has better-qualified personnel than do its competitors, he or she might include the **résumés** of the key people who would be involved in the project, as a way of emphasizing that advantage.

SOLICITED AND UNSOLICITED PROPOSALS

Sales proposals may be either solicited or unsolicited. The latter are not so unusual as they may sound: companies often operate for years with a problem they have never recognized (unnecessarily high maintenance costs, for example, or poor inventory-control methods). If you learn of such a company, you might prepare an unsolicited proposal if you were convinced that the potential customer could realize substantial benefits by adopting your solution to the problem. You would, of course, need to persuade the customer of the excellence of your idea and of his or her need for what you are proposing.

To ensure that you don't waste time and effort, you might precede an unsolicited proposal with an inquiry to determine any potential interest. If you receive a positive response, you might then need to conduct a detailed study of the prospective customer's needs and requirements in order to determine whether you can be of help and, if so, exactly how. You would then prepare your proposal on the basis of your study.

An unsolicited proposal should clearly identify the potential customer's problem, but at the same time it should be careful not to overstate it. Furthermore, the proposal must convince the customer that the problem needs to be solved. One way to do this is to emphasize the benefits the customer will realize from the solution being proposed.

The other type of sales proposal—the solicited sales proposal—is a response to a request for goods or services. Procuring organizations that would like competing companies to bid for a job commonly issue a *Request for Proposals.* The Request for Proposals is the means used by many companies and government agencies to find the best method of doing a job and the most qualified company to do it. A Request for Proposals may be rigid in its specifications governing how the proposal should be organized and what it should contain, but it is normally quite flexible about the approaches that the bidding firms may propose. Normally, the Request for Proposals simply defines the basic work that the firm needs.

The procuring organization generally publishes its Request for Proposals in one or more journals, in addition to sending it to certain companies that have good reputations for doing the kind of work needed. Some companies and government agencies even hold a conference for the competing firms at which they provide all pertinent information about the job being bid for.

Managers interested in responding to Requests for Proposals regularly scan the appropriate publications. Upon finding a project of interest, an executive in the sales department of such a company obtains all available information from the procuring company or agency. The data is then presented to management for a decision on whether the company is interested in the project. If the decision is positive, the technical staff is assigned to develop an approach to the work described in the Request for Proposals. The technical staff normally considers several alternatives, selecting one that combines feasibility and a price that offers a profit. The staff's concept is presented to higher management for a decision on whether the company wishes to present a proposal to the requesting organization. Assuming that the decision is to proceed, preparation of the proposal is the next step.

When you respond to a Request for Proposals, pay close attention to any specifications in the request governing the preparation of the proposal, and follow them carefully.

WRITING A SIMPLE SALES PROPOSAL

Even a short and uncomplicated sales proposal should be carefully planned. The **introduction** should state the purpose and **scope** of the proposal. It should indicate the dates on which you propose to begin and complete work on the project, any special benefits of your proposed approach, and the total cost of the project. The introduction could also refer to any previous association your company may have had with the potential customer, assuming that the association was positive and mutually beneficial.

The body of a sales proposal should itemize the products and services you are offering. It should include, if applicable, a discussion of the procedures you would use to perform the work and any materials to be used. It should also present a time schedule indicating when each stage of the project would be completed. Finally, the body should include a breakdown of the costs of the project.

The **conclusion** should express your appreciation for the opportunity to submit the proposal and your confidence in your company's ability to do the job. You might add that you look forward to establishing good working relations with the customer and that you would be glad to provide any additional information that might be needed. Your conclusion could also review any advantages your company may have over its competitors. It should specify the time period during which your proposal can still be considered a valid offer. If any supplemental materials, such as blueprints or price sheets, accompany the proposal, include a list of them at the end of the proposal. Figure 1, on page 596, shows a typical short sales proposal.

WRITING A LONG SALES PROPOSAL

Wherever possible in a sales proposal, use eye-catching graphics instead of sentences and paragraphs. **Lists** and **flowcharts** are particularly effective. Customize your proposal whenever possible.

EXAMPLE A PROPOSAL PREPARED EXPRESSLY FOR

Mr. William A. Kurtz
Vice-President
The Cambridge Company
Tucson, Arizona

Also, whenever possible, include testimonial letters from other users, stressing specific benefits. Be sure to include the appropriate advertising literature, a copy of your company's **annual report**, and the contract.

A large sales proposal should contain the following elements:

- Transmittal letter
- Title page
- Executive summary
- Description of method currently being used
- Description of proposed method
- Comparison of present and proposed methods
- Equipment recommendations
- Cost analysis
- Delivery schedule
- Summary of advantages
- Statement of responsibilities
- Description of vendor
- Advertising literature
- Contract

PROPOSAL
TO LANDSCAPE THE NEW CORPORATE HEADQUARTERS
OF THE
WATFORD VALVE CORPORATION

Submitted to: Ms. Tricia Olivera, Vice-President
Submitted by: Jerwalted Nursery, Inc.
Date Submitted: February 1, 19--

Introduction states purpose and scope of proposal, indicates when project can be started and completed.

Jerwalted Nursery, Inc., proposes to landscape the new corporate headquarters of the Watford Valve Corporation, on 1600 Swason Avenue, at a total cost of $8,000. The lot to be landscaped is approximately 600 feet wide and 700 feet deep. Landscaping will begin no later than April 30, 19-- and will be completed by May 31.

The following trees and plants will be planted, in the quantities and sizes given and at the prices specified.

Body lists products to be provided, cost per item.

4 maple trees (not less than 7 ft.) @ $40 each—$160
41 birch trees (not less than 7 ft.) @ $65 each—$2,665
2 spruce trees (not less than 7 ft.) @ $105 each—$210
20 juniper plants (not less than 18 in.) @ $15 each—$300
60 hedges (not less than 18 in.) @ $7 each—$420
200 potted plants (various kinds) @ $2 each—$400

Total Cost of Plants	=	$4,155
Labor	=	$3,845
Total Cost	=	$8,000

Conclusion specifies time limit of proposal, expresses confidence, and looks forward to working with prospective customer.

All trees and plants will be guaranteed against defect or disease for a period of 90 days, the warranty period to begin June 1, 19--.

The prices quoted in this proposal will be valid until June 30, 19--.

Thank you for the opportunity to submit this proposal. Jerwalted Nursery, Inc., has been in the landscaping and nursery business in the St. Louis area for thirty years, and our landscaping has won several awards and commendations, including a citation from the National Association of Architects. We are eager to put our skills and knowledge to work for you, and we are confident that you will be pleased with our work. If we can provide any additional information or assistance, please call.

Figure 1 Short Sales Proposal

The **transmittal letter** expresses your appreciation for the opportunity to submit your proposal and for any assistance you may have received in studying the customer's requirements. The letter should acknowledge any previous association with the customer, assuming that it was a positive experience. Then it should summarize the recommendations offered in the proposal and express your confidence that they will satisfy the customer's needs. Figure 2, on page 598, shows the cover letter for the proposal illustrated in Figures 3 through 11, a proposal that the Waters Corporation of Tampa provide a computer system for the Cookson chain of retail stores.

An **executive summary** follows the transmittal letter. Written to the executive who will ultimately accept or reject the proposal, it should summarize in nontechnical language how you plan to approach the work. Figure 3 (p. 599) shows the executive summary of the Waters Corporation proposal.

If your proposal offers products as well as services, it should include a *general description* of the products, as in Figure 4 (p. 600).

After the executive summary and the general description, explain exactly how you plan to do what you are proposing. Because this section will be read by specialists who can understand and evaluate your plan, you can feel free to use technical language and discuss complicated concepts. Figure 5 (pages 601–2) shows one part of the detailed solution appearing in the Waters Corporation proposal, which included several other applications in addition to the payroll application. Notice that this discussion, like that found in an unsolicited sales proposal, begins with a statement of the customer's problem, follows with a statement of the solution, and concludes with a statement of the benefits to the customer.

Essential to any sales proposal are a *cost analysis* and a *delivery schedule.* The cost analysis itemizes the estimated cost of all the products and services that you are offering; the delivery schedule commits you to a specific timetable for providing those products and services. Figure 6 (pp. 603–4) shows the cost analysis, and Figure 7 (p. 605) shows the delivery schedule of the Waters Corp. proposal.

Also essential if your recommendations include modifying the customer's physical facilities is a *site preparation description* that details the required modifications. Figure 8, on page 606, shows the site preparation section of the Waters proposal.

If the products you are proposing require training the customer's

THE WATERS CORPORATION
17 North Waterloo Blvd., Tampa, Florida 33607

September 1, 19--

Mr. John Yeung, General Manager
Cookson's Retail Stores, Inc.
Savannah, Georgia 31499

Dear Mr. Yeung:

Express your appreciation for the opportunity to make the proposal.

The Waters Corporation appreciates the opportunity to respond to Cookson's Request for Proposals, dated July 26, 19--. We would like to thank Mr. Becklight, Director of your Management Information Systems Department, for his invaluable contributions to the study of your operations that we conducted before preparing our proposal.

Summarize the proposal's recommendations.

Our proposal describes a Waters Interactive Terminal/Retail Processor System designed to meet Cookson's network and processing needs. It will provide all of your required capabilities, from the point-of-sale operational requirements at the store terminals to the host processor. The system uses the proven Retail III modular software, with its point-of-sale applications, and the superior Interactive Terminal with its advanced capabilities and design. This system is easily installed without massive customer reprogramming.

Make a soft sales pitch.

The Waters Interactive Terminal/Retail Processor System, which is compatible with much of Cookson's present equipment, not only will answer your present requirements but also will provide the flexibility to add new features and products in the future. The system's unique hardware modularity, efficient microprocessor design, and flexible programming capability will greatly reduce the risk of obsolescence.

Thank you for the opportunity to present this proposal.

Sincerely yours,

Janet A. Curtain

Janet A. Curtain
Account Manager
General Merchandise Systems

JA/mo

Figure 2 Cover Letter for a Proposal

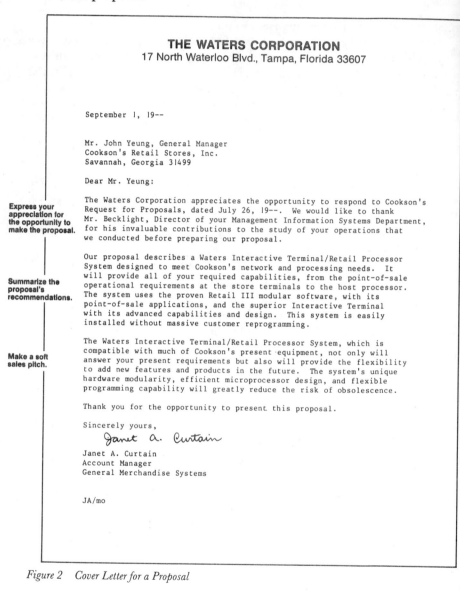

EXECUTIVE SUMMARY

The Waters 319 Interactive Terminal/615 Retail Processor System will give your management the tools necessary to manage people and equipment more profitably with procedures that will yield more cost-effective business controls for Cookson's.

The equipment and applications proposed for Cookson's will respond to your current requirements and allow for a logical expansion in the future.

The features and hardware in the system were determined from data acquired through the comprehensive survey we conducted at your stores in February of this year. The total of 71 Interactive Terminals proposed to service your four stores is based on the number of terminals currently in use and on the average number of transactions processed during normal and peak periods. The planned remodeling of all four stores was also considered, and the suggested terminal placement has been incorporated into the working floor plan. The proposed equipment configuration and software applications have been simulated to determine system performance based on the volumes and anticipated growth rates of the Cookson stores.

The information from the survey was also used in the cost justification, which was checked and verified by your controller, Mr. Deitering. The cost effectiveness of the Waters Interactive Terminal/Retail Processor System is apparent. Expected savings, such as the projected 46-percent reduction in sales audit expenses, are realistic projections based on Waters's experience with other installations of this type.

Figure 3 Executive Summary

GENERAL SYSTEM DESCRIPTION

The point-of-sale system that Waters is proposing for Cookson's includes two primary Waters products. These are the 319 Interactive Terminal and the 615 Retail Processor.

Waters 319 Interactive Terminal

The primary component in the proposed retail system is the Interactive Terminal. It contains a full microprocessor, which gives it the flexibility that Cookson's has been seeking.

The 319 Interactive Terminal gives you freedom in sequencing a transaction. You are not limited to a preset list of available steps or transactions. The terminal program can be adapted to provide unique transaction sets, each designed with a logical sequence of entry and processing to accomplish required tasks. In addition to sales transactions recorded on the selling floor, specialized transactions, such as theater tickets sales and payments, can be designed for your customer service area.

The 319 Interactive Terminal also functions as a credit authorization device, either by using its own floor limits or by transmitting a credit inquiry to the 615 Retail Processor for authorization.

Data-collection formats have been simplified so that transaction editing and formatting are easier. Mr. Sier has already been given the documentation on these formats and has outlined all data-processing efforts that will be necessary to adapt the data to your current systems. These projections have been considered in the cost justification.

Waters 615 Retail Processor

The Waters 615 Retail Processor is a minicomputer system designed to support the Waters family of retail terminals. The processor will reside in the computer room in your data center in Buffalo. Operators already on your staff will be trained to initiate and monitor its activities.

The 615 will collect data transmitted from the retail terminals, process credit, check authorization inquiries, maintain files to be accessed by the retail terminals, accumulate totals, maintain a message-routing network, and control the printing of various reports. The functions and level of control performed at the processor depend on the peripherals and software selected.

Software

The Retail III software used with the system has been thoroughly tested and is operational in many Waters customer installations.

The software provides the complete processing of the transaction, from the interaction with the operator on the sales floor through the data capture on cassette or disk in stores and your data center.

Retail III provides a menu of modular applications for your selection. Parameters adapt each of them to your hardware environment and operation requirements. The selection of hardware will be closely related to the selection of the software applications.

Figure 4 General Description of Products

PAYROLL APPLICATION

Current Procedure

Your current system of reporting time requires each hourly employee to sign a time sheet; the time sheet is reviewed by the department manager and sent to the Payroll Department on Friday evening. Since the week ends on Saturday, the employee must show the scheduled hours for Saturday and not the actual hours; therefore, the department manager must adjust the reported hours on the time sheet for employees who do not report on the scheduled Saturday or who do not work the number of hours scheduled.

The Payroll Department employs a supervisor and three full-time clerks. To meet deadlines caused by an unbalanced work flow, an additional part-time clerk is used for 20 to 30 hours per week. The average wage for this clerk is $4.25 per hour.

Advantage of Waters's System

The 319 Interactive Terminal can be programmed for entry of payroll data for each employee on Monday mornings by department managers, with the data reflecting actual hours worked. This system would eliminate the need for manual batching, controlling, and keypunching. The Payroll Department estimates conservatively that this work consumes 40 hours per week.

Hours per week	40
Average wage	×4.25
Weekly payroll cost	$170.00
Annual savings	$8,840.00

Elimination of the manual tasks of tabulating, batching, and controlling can save .25 units. Improved work flow resulting from timely data in the system without keypunch processing will allow more efficient use of clerical hours. This would reduce payroll by the .50 units currently required to meet weekly check disbursement.

Eliminate manual tasks	.25
Improve work flow	.75
40-hour unit reduction	1.00
Hours per week	40
Average wage	4.75
Savings per week	$190.00
Annual savings	$9,880.00

TOTAL SAVINGS: $18,720.00

Figure 5 Detailed Solution

personnel, your proposal should specify the *required training* and its cost. Fig. 9 (p. 607) shows the training section of the Waters proposal.

To prevent misunderstandings about what you and your customer's responsibilities will be, you should draw up a *statement of responsibilities,* as shown in Figure 10, on page 608. It usually appears toward the end of the proposal. Also toward the end of the proposal is a *description of the vendor* (Figure 11, p. 609), which should give a factual description of your company, its background, and its present position in the industry. After this, many proposals add what is known as an *institutional sales pitch* (Figure 12, p. 610). Up to this point, the proposal has attempted to sell specific goods and services. The sales pitch, striking a somewhat different chord, is designed to sell the company and its general capability in the field. Less factual than the vendor description, the sales pitch promotes the company and concludes the proposal on an upbeat note.

same

When used as a **pronoun,** *same* is awkward and outmoded.

CHANGE Your order has been received, and we will respond to *same* next week.

TO Your order has been received, and we will respond to *it* next week.

OR We received your order, and we will comply with *it* next week.

COST ANALYSIS

This section of our proposal provides detailed cost information for the Waters 319 Interactive Terminal and the 615 Waters Retail Processor. It then extends these major elements by the quantities required at each of your four locations.

319 Interactive Terminal

	Price	Maint. (1yr)
Terminal	$2,895	$167
Journal Printer	425	38
Receipt Printer	425	38
Forms Printer	525	38
Software	220	--
TOTALS	$4,490	$281

615 Retail Processor

	Price	Maint. (1Yr)
Processor	$57,115	$5,787
CRT I/O Writer	2,000	324
Matrix Printer	4,245	568
Software	12,480	---
TOTALS	$75,840	$6,679

The following breakdown itemizes the cost per store.

Store No. 1

Description	Qty	Price	Maint.
Terminals	16	$68,400	$4,496
Digital Cassette	1	1,300	147
Thermal Printer	1	2,490	332
Software	16	3,520	---
TOTALS		$75,710	$4,975

Figure 6 Cost Analysis

Store No. 2

Description	Qty	Price	Maint.
Terminals	20	$85,400	$5,620
Digital Cassette	1	1,300	147
Thermal Printer	1	2,490	332
Software	20	4,400	---
TOTALS		$93,590	$6,099

Store No. 3

Description	Qty	Price	Maint.
Terminals	17	$72,590	$4,777
Digital Cassette	1	1,300	147
Thermal Printer	1	2,490	332
Software	17	3,740	---
TOTALS		$80,120	$5,256

Store No. 4

Description	Qty	Price	Maint.
Terminals	18	$76,860	$5,058
Digital Cassette	1	1,300	147
Thermal Printer	1	2,490	332
Software	18	3,960	---
TOTALS		$84,610	$5,537

Data Center at Buffalo

Description	Qty	Price	Maint.
Processor	1	$57,115	$5,787
CRT I/O Writer	1	2,000	324
Matrix Printer	1	4,245	568
Software	1	12,480	---
TOTALS		$75,840	$6,679

The following summarizes all costs.

Location	Hardware	Maint.	Software
Store No. 1	$72,190	$4,975	$3,520
Store No. 2	89,190	6,099	4,400
Store No. 3	76,380	5,256	3,740
Store No. 4	80,650	5,537	3,960
Data Center	63,360	6,679	12,480
Subtotals	$381,770	$28,546	$28,100

TOTAL $438,416

DELIVERY SCHEDULE

Waters is normally able to deliver 319 Interactive Terminals and 615 Retail Processors within 120 days from the date of the contract. This can vary depending on the rate and size of incoming orders.

All the software recommended in this proposal is available for immediate delivery. We do not anticipate any difficulty in meeting your tentative delivery schedule.

Figure 7 Delivery Schedule

schematic diagrams

The schematic diagram, which is used primarily in electronics and chemistry and in electrical and mechanical engineering, attempts to show the operation of its subject with lines and **symbols** rather than by a physical likeness to it. The schematic diagram is usually a highly abstract representation of its subject because it emphasizes the relationships among the parts at the expense of precise proportions. Because a schematic diagram is a symbolic representation of the subject, as opposed to a realistic representation of it, the schematic relies heavily on the symbols and **abbreviations** that are common to the subject. (See Figure 1 on page 611.)

For guidelines on how to use illustrations and incorporate them into your texts, see **illustrations**. *(Please turn to page 609.)*

SITE PREPARATION

Waters will work closely with Cookson's to ensure that each site is properly prepared prior to system installation. You will receive a copy of Waters's installation and wiring procedures manual, which lists the physical dimensions, service clearance, and weight of the system components in addition to the power, logic, communication-cable, and environmental requirements. Cookson's is responsible for all building alterations and electrical facility changes, including the purchase and installation of communication cables, connecting blocks, and receptacles.

Wiring

For the purpose of future site considerations, Waters's in-house wiring specifications for the system call for two twisted pair wires and twenty-two shielded gauges. The length of communications wires must not exceed 2,500 feet.

As a guide for the power supply, we suggest that Cookson's consider the following:

1. The branch circuit (limited to 20 amps) should service no equipment other than 319 Interactive Terminals.
2. Each 20-amp branch circuit should support a maximum of three Interactive Terminals.
3. Each branch circuit must have three equal-sized conductors--one hot leg, one neutral, and one insulated isolated ground.
4. Hubbell IG 5362 duplex outlets or the equivalent should be used to supply power to each terminal.
5. Computer-room wiring will have to be upgraded to support the 615 Retail Processor.

Figure 8 Site Preparation Section

TRAINING

To ensure a successful installation, Waters offers the following training course for your operators.

Interactive Terminal/Retail Processor Operations

Course number: 8256
Length: three days
Tuition: $500.00 per person

This course provides the student with the skills, knowledge, and practice required to operate an Interactive Terminal/Retail Processor System. On-line, clustered, and stand-alone environments are covered.

We recommend that students have department store background and that they have some knowledge of the system configuration with which they will be working.

Figure 9 Training Section

RESPONSIBILITIES

On the basis of its years of experience in installing information-processing systems, Waters believes that a successful installation requires a clear understanding of certain responsibilities.

Generally, it is Waters's responsibility to provide its users with needed assistance during the installation so that live processing can begin as soon thereafter as is practical.

Waters's Responsibilities

--Provide operations documentation for each application that you acquire from Waters.

--Provide forms and other supplies as ordered.

--Provide specifications and technical guidance for proper site planning and installation.

--Provide advisory assistance in the conversion from your present system to the new system.

Customer's Responsibilities

--Identify an installation coordinator and system operator.

--Provide supervisors and clerical personnel to perform conversion to the system.

--Establish reasonable time schedules for implementation.

--Ensure that the physical site requirements are met.

--Provide competent personnel to be trained as operators and ensure that other employees are trained as necessary.

--Assume the responsibility for implementing and operating the system.

Figure 10 Statement of Responsibilities

DESCRIPTION OF VENDOR

The Waters Corporation develops, manufactures, markets, installs, and services total business information-processing systems for selected markets. These markets are primarily in the retail, financial, commercial, industrial, health-care, education, and government sectors.

The Waters total system concept encompasses one of the broadest hardware and software product lines in the industry. Waters computers range from small business systems to powerful general-purpose processors. Waters computers are supported by a complete spectrum of terminals, peripherals, data-communication networks, and an extensive library of software products. Supplemental services and products include data centers, field service, systems engineering, and educational centers.

The Waters Corporation was founded in 1934 and presently has approximately 26,500 employees. The Waters headquarters is located at 17 North Waterloo Boulevard, Tampa, Florida, with district offices throughout the United States and Canada.

Figure 11 Description of Vendor

scope

If you know your **reader** and the **objective** of your writing project, you will know the type and amount of detail to include in your writing. This is *scope,* which may be defined as the depth and breadth to which you need to cover your subject. If you do not determine your scope of coverage in the planning stage of your writing project, you will not know how much or what kind of information to include, and as a result, you are likely to invest unnecessary work during the **research** phase of your project.

Your scope should be designed to satisfy the needs of your objective and your reader. By keeping your objective and your reader's

WHY WATERS?

Corporate Commitment to the Retail Industry

Waters's commitment to the retail industry is stronger than ever. We are continually striving to provide leadership in the design and implementation of new retail systems and applications that will ensure our users of a logical growth pattern.

Research and Development

Over the years, Waters has spent increasingly large sums on research and development efforts to ensure the availability of products and systems for the future. In 19--, our research and development expenditure for advanced systems design and technological innovations reached the $70 million level.

Leading Point-of-Sale Vendor

Waters is a leading point-of-sale vendor, having installed over 150,000 units. The knowledge and experience that Waters has gained over the years from these installations ensure well-coordinated and effective systems implementations.

Figure 12 Institutional Sales Pitch

profile in mind as you work, you can determine those items that should and should not be included in your writing.

semicolons

The semicolon links **independent clauses** or other sentence elements of equal weight and grammatical rank, especially **phrases** in a series that have **commas** within them. The semicolon indicates a greater pause between clauses than a comma would, but not as great a pause as a **period** would.

When the independent clauses of a **compound sentence** are not joined by a comma and a **conjunction,** they are linked by a semicolon.

EXAMPLE No one applied for the position; the job was too difficult.

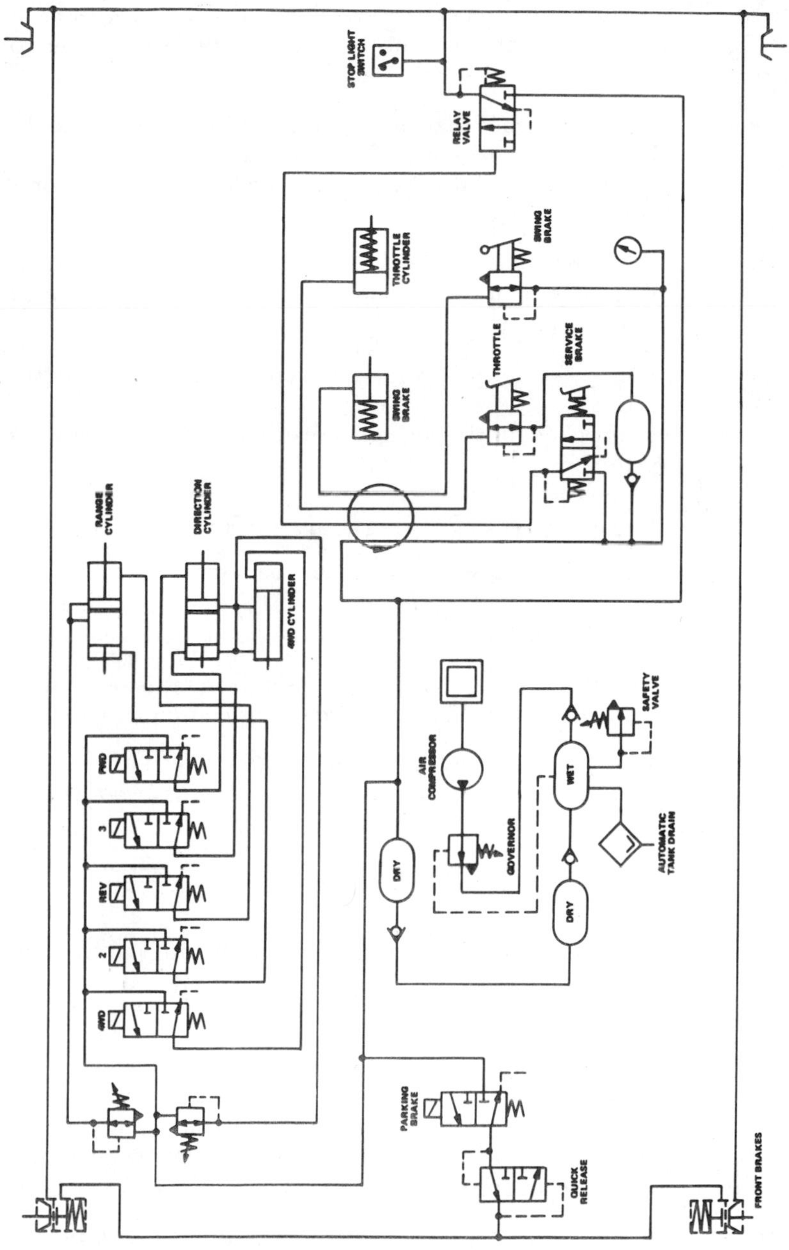

Figure 1 Air System Schematic for a Self-Propelled Hydraulic Crane

Source: Harnischfeger Corporation

Make sure, however, that such **clauses** balance or contrast with each other. The relationship between the two statements should be so clear that further explanation is not necessary.

EXAMPLE It is a curious fact that there is little similarity between the chemical composition of river water and that of sea water; the various elements are present in entirely different proportions.

Do not use a semicolon between a **dependent clause** and its main clause. Remember that elements joined by semicolons must be of equal grammatical rank or weight.

CHANGE No one applied for the position; even though it was heavily advertised.

TO No one applied for the position, even though it was heavily advertised.

USING A SEMICOLON WITH STRONG CONNECTIVES

In complicated sentences, a semicolon may be used before transitional words or phrases (*that is, for example, namely*) that introduce examples or further explanation.

EXAMPLE The study group was aware of his position on the issue; that is, federal funds should not be used for the housing project.

A semicolon should also be used before **conjunctive adverbs** (such as *therefore, moreover, consequently, furthermore, indeed, in fact, however*) that connect independent clauses.

EXAMPLE I won't finish today; moreover, I doubt that I will finish this week.

The semicolon in this example shows that *moreover* belongs to the second clause.

USING A SEMICOLON FOR CLARITY IN LONG AND COMPLICATED SENTENCES

Use a semicolon between two main clauses connected by a **coordinating conjunction** (*and, but, for, or, nor, yet*) if the clauses are long and contain other **punctuation.**

EXAMPLE In most cases these individuals are corporate executives, bankers, Wall Street lawyers; but they do not, as the economic determinists seem to believe, simply push the button of their economic power to affect fields remote from economics.

—Robert Lubar, "The Prime Movers," *Fortune* (February 1960), p. 98.

A semicolon may also be used if items in a series contain commas within them.

EXAMPLE Among those present were John Howard, president of the Omega Paper Company; Carol Martin, president of Alpha Corporation; and Larry Stanley, president of Stanley Papers.

Do not, however, use semicolons to enclose a parenthetical element that contains commas. Use **parentheses** or **dashes** for this purpose.

CHANGE All affected job classifications; typists, secretaries, clerk-stenographers, and word processors; will be upgraded this month.

TO All affected job classifications (typists, secretaries, clerk-stenographers, and word processors) will be upgraded this month.

OR All affected job classifications—typists, secretaries, clerk-stenographers, and word processors—will be upgraded this month.

Do not use a semicolon as a mark of anticipation or enumeration. Use a **colon** for this purpose.

CHANGE Three decontamination methods are under consideration; a zeolite-resin system, an evaporation and resin system, and a filtration and storage system.

TO Three decontamination methods are under consideration: a zeolite-resin system, an evaporation and resin system, and a filtration and storage system.

The semicolon always appears outside closing **quotation marks.**

EXAMPLE The attorney said, "You must be accurate"; the client said, "I will."

sentence construction

A sentence is the most basic and versatile tool available to the writer. Consider the many different ways the same idea can be expressed to achieve different **emphasis.**

EXAMPLES Every person in the department, from manager to typists, must work overtime this week to ensure that we meet the deadline.

To ensure that we meet the deadline, every person in the department, from manager to typists, must work overtime this week.

From manager to typists, every person in the department must work overtime this week to ensure that we meet the deadline.

To ensure that we meet the deadline, every person in the department must work overtime this week, from manager to typists.

This week, every person in the department, from manager to typists, must work overtime to ensure that we meet the deadline.

When shifting word order for emphasis, however, be aware that word order can make a great difference in the meaning of a sentence.

EXAMPLES He was *only* the engineer.

He was the *only* engineer.

Except for **expletives,** all words or word groups in a sentence function as a **subject,** a **verb,** a **complement,** a **modifier,** a connective, an **appositive,** or an **absolute.** Subjects, verbs, and complements are the main elements of the sentence. Everything else is subordinate to them in one way or another. A sentence that progresses quickly from subject to verb to complement is clear and easy to understand. The problem is to write sentences that contain more information, and in more complicated form, with the same **clarity** and directness.

The basic sentence patterns in English are the following:

EXAMPLES The cable snapped. (subject-verb)

Generators produce electricity. (subject-verb-direct object)

The test results gave us confidence. (subject-verb-indirect object-direct object)

Repairs made the equipment operational. (subject-verb-direct object-objective complement)

The metal was aluminum. (subject-linking verb-subjective complement)

Most sentences follow the subject-verb-object pattern. In "The company dismissed Joe," we know the subject and the object by their positions relative to the verb. The knowledge that the usual sentence order is subject-verb-object helps **readers** interpret what they read. The fact that readers tend to expect this order explains why the writer's departures from it can be effective if used sparingly for **emphasis** and variety, but annoying if overdone.

An inverted sentence places the elements in other than normal order.

EXAMPLES A better job I never had. (direct object-subject-verb)
More optimistic I have never been. (subjective complement-subject-linking verb)

Inverted sentence order may be used in questions and exclamations and also to achieve emphasis.

EXAMPLES Have you a pencil? (verb-subject-complement)
How heavy your book feels! (complement-subject-verb)
A sorry sight we presented! (complement-subject-verb)

In sentences introduced by expletives (*there, it*), the subject generally follows the verb because the expletive precedes the verb.

EXAMPLES *There* (expletive) *are* (verb) certain *principles* (subject) of drafting that must not be ignored.
It (expletive) *is* (verb) *difficult* (complement) *to work* (subject) in a noisy office.

CHANGE *There are* certain principles of drafting that must be followed.
TO Certain principles of drafting must be followed.

Beginning sentences with expletives is often wordy.

CONSTRUCTING CLEAR SENTENCES

Use uncomplicated sentences to state complex ideas. If readers must cope with a complicated sentence in addition to a complex idea, they are likely to become confused.

CHANGE When you are purchasing parts, remember that although an increase in the cost of aluminum forces all the vendors to increase their prices, some vendors will have a supply of aluminum purchased at the old price, and they may be willing to sell parts to you at the old price in order to get your business.

TO Although an increase in the cost of aluminum forces all vendors to increase their prices, some vendors will have a supply of aluminum purchased at the old price. When you are purchasing aluminum parts, remember that these vendors may be willing to sell you the parts at the old price in order to get your business.

Just as simpler sentences make complex ideas more digestible, a complex sentence construction makes a series of simple ideas more palatable.

CHANGE The computer is a calculating device. It was once known as a mechanical brain. It has revolutionized industry.

TO The computer, a calculating device once known as a mechanical brain, has revolutionized industry.

Do not string together in a series a number of thoughts that should be written as separate sentences or some of which should be subordinated to others. Sentences carelessly tacked together this way are monotonous and hard to read because all ideas seem to be of equal importance.

CHANGE We started the program three years ago, there were only three members on the staff, and each member was responsible for a separate state, but it was not an efficient operation.

TO When we started the program three years ago, there were only three members on the staff, each having responsibility for a separate state; however, that arrangement was not efficient.

Sentences can often be improved by eliminating trailing constructions and ineffective **repetition.**

CHANGE We conducted a new experiment last month and learned much from it.

TO We learned much from a new experiment last month.

CONSTRUCTING PARALLEL SENTENCES

Express coordinate ideas in similar form. The very construction of the sentence helps the reader grasp the similarity of its components. (See also **parallel structure.**)

EXAMPLE Similarly, atoms come and go in a molecule, but the molecule remains; molecules come and go in a cell, but the cell remains; cells come and go in a body, but the body remains; persons come and go in an organization, but the organization remains.

—Kenneth Boulding, *Beyond Economics* (Ann Arbor: University of Michigan Press, 1968), p. 131.

CONSTRUCTING SENTENCES TO ACHIEVE EMPHASIS

Subordinate your minor ideas to emphasize your more important ideas. (See also **subordination.**)

CHANGE We all had arrived, and we began the meeting early.
TO Since we all had arrived, we began the meeting early.

The most emphatic positions within a sentence are at the beginning and the end. Do not waste them by tacking on **phrases** and **clauses** almost as an afterthought or by burying the main point in the middle of a sentence between less important points. For example, consider the following original and revised versions of a statement written for a company's **annual report** to its stockholders:

CHANGE Sales declined by 3 percent in 19--, but nevertheless the Company had the most profitable year in its history, thanks to cost savings that resulted from design improvements in several of our major products; and we expect 19-- to be even better, since further design improvements are being made.

TO Cost savings from design improvements in several major products not only offset a 3-percent sales decline but made 19-- the most profitable year in the Company's history. Further design improvements now in progress promise to make 19-- even more profitable.

Reversing the normal word order is also used to achieve emphasis.

EXAMPLES I will never agree to that.
That I will never agree to.
Never will I agree to that.

BEGINNING A SENTENCE WITH A COORDINATING CONJUNCTION

There is no rule against beginning a sentence with a **coordinating conjunction;** in fact, coordinating conjunctions can be strong transitional words and at times provide emphasis.

EXAMPLE I realize the project was more difficult than expected and that you have also encountered personnel problems. *But* we must meet our deadline.

Starting sentences with **conjunctions** is acceptable in all but the most formal English. But like any other writing device, this one should be used sparingly lest it become ineffective and even annoying. (See also **sentence variety, run-on sentences, sentence fragments,** and **sentence faults.**)

sentence faults

A number of problems can create sentence faults.

FAULTY SUBORDINATION

Faulty **subordination,** one of the most common sentence faults, occurs (1) when a grammatically subordinate element, such as a **dependent clause,** actually contains the main idea of the sentence or (2) when a subordinate element is so long or detailed that it dominates or obscures the main idea. Avoiding the first problem (main idea expressed in a subordinate element) depends on the writer's knowing which idea is *meant* to be the main idea. Note that both of the following sentences appear logical; which one really is logical depends on which of two ideas the writer means to emphasize.

EXAMPLES Although the new filing system saves money, many of the staff are unhappy with it.

The new filing system saves money, although many of the staff are unhappy with it.

In this example, if the writer's main point is that *the new filing system saves money,* the second sentence is correct. If the main point is that *many of the staff are unhappy,* then the first sentence is correct. It is easier than you may think to slip and put your main idea into a grammatically subordinate element (especially since people commonly do so in conversation, when they can raise their voices and make gestures to stress the main point).

The other major problem with subordination is the loading of so much detail into a subordinate element that it "crushes" the main point by its sheer size and weight.

CHANGE Because the noise level on a typical street in New York City on a weekday is as loud as an alarm clock ringing three feet away, New Yorkers often have hearing problems.

TO Because the noise level in New York City is so high, New Yorkers often have hearing problems.

CLAUSES WITH NO SUBJECTS

Writers sometimes inappropriately assume a subject that is not stated in a clause.

CHANGE Your application program can request to end the session after the next command. (Request *who* or *what* to end the session? The reader doesn't know.)

TO Your application program can request *the host program* to end the session after the next command.

CHANGE This command enables sending the entire message again if an incomplete message transfer occurs. (The sentence contains no subject for the verb *sending*. The reader doesn't know *who* or *what* is doing the sending.)

TO This command enables *you to send* the entire message again if an incomplete message occurs.

(See also **telegraphic style.**)

RAMBLING SENTENCES

Sentences that contain more information than the reader can comfortably absorb in one reading are known as *rambling sentences.* The obvious remedy for a rambling sentence is to divide it into two or more sentences. In doing so, however, put the main message of the rambling sentence into the first of the revised sentences.

CHANGE The payment to which a subcontractor is entitled should be made promptly in order that in the event of a subsequent contractual dispute we, as general contractors, may not be held in default of our contract by virtue of nonpayment.

TO Pay subcontractors promptly. Then if a contractual dispute should occur, we cannot be held in default of our contract because of nonpayment.

MISCELLANEOUS SENTENCE FAULTS

The assertion made by a sentence's **predicate** about its **subject** must be logical. "Mr. Wilson's *job* is a salesman" is not logical, but "*Mr. Wilson* is a salesman" is. "Jim's *height* is six feet tall" is not logical, but "*Jim* is six feet tall" is.

Do not omit a required **verb.**

CHANGE The floor is swept and the lights out.

TO The floor is swept and the lights *are* out.

CHANGE I never have and probably never will write the annual report.

TO I never have *written* and probably never will write the annual report.

Do not omit a subject.

CHANGE He regarded price fixing as wrong, but until abolished by law, he engaged in it, as did everyone else.

TO He regarded price fixing as wrong, but until *it was* abolished by law, he engaged in it, as did everyone else.

Avoid **compound sentences** containing **clauses** that have little or no logical relationship to one another.

CHANGE The reactor oxidizes the harmful exhaust ingredients into harmless water vapor and carbon dioxide, and it is housed in a double-walled metal shell.

TO The reactor oxidizes the harmful exhaust ingredients into harmless water vapor and carbon dioxide. It is housed in a double-walled metal shell.

(See also **run-on sentences** and **sentence fragments.**)

sentence fragments

A sentence fragment is an incomplete grammatical unit that is punctuated as a sentence.

EXAMPLES He quit his job. (sentence)
And quit his job. (fragment)

Sentence fragments are often introduced by **relative pronouns** (*who, which, that*) or **subordinating conjunctions** (such as *although, because, if, when,* and *while*). This knowledge can tip you off that what follows is a dependent clause and must be combined with a main clause.

CHANGE The new manager instituted several new procedures. *Many of which are impractical.* (The last is an adjective clause modifying *procedures* and linked to it by the relative pronoun *which.*)

TO The new manager instituted several new procedures, many of which are impractical.

A sentence must contain a finite verb; **verbals** will not do the job. The following examples are sentence fragments because their verbals (*providing, to work, waiting*) cannot perform the function of a finite verb.

EXAMPLES *Providing* all employees with hospitalization insurance.
To work a forty-hour week.
The customer *waiting* to see you.

Fragments usually reflect incomplete and sometimes confused thinking. The most common type of fragment occurs because of the careless addition of an afterthought. Such fragments should either be a part of the preceding sentence or converted into their own sentences.

CHANGE These are my coworkers. *A fine group of people.*
TO These are my coworkers, a fine group of people.

The following examples are common types of sentence fragments:

CHANGE Some of our customers prefer to pay at the time of purchase. *While others find installment payments preferable.* (adverbial clause)
TO Some of our customers prefer to pay at the time of purchase, while others find installment payments preferable.

CHANGE The board approved the project. *After much discussion.* (prepositional phrase)
TO The board approved the project after much discussion.

CHANGE We reorganized the department. *Distributing the work load more evenly.* (participial phrase)
TO We reorganized the department, distributing the work load more evenly.

CHANGE The staff decided to take a break. *It being midafternoon.* (absolute phrase)
TO The staff decided to take a break, it being midafternoon.

CHANGE The field tests showed the prototype to be extremely rugged. *The most durable we've tested this year.* (appositive phrase)
TO The field tests showed the prototype to be extremely rugged, the most durable we've tested this year.

CHANGE We have one major goal this month. *To increase the strength of the alloy without reducing its flexibility.* (infinitive phrase)
TO We have one major goal this month: to increase the strength of the alloy without reducing its flexibility.

Explanatory **phrases** beginning with *such as, for example,* and similar terms often lead writers to create sentence fragments.

CHANGE The staff wants additional benefits. For example, the use of company automobiles. (fragment)

TO The staff wants additional benefits, such as the use of company automobiles. (one sentence)

OR The staff wants additional benefits. For example, one possible benefit is the use of company automobiles. (two sentences)

A hopelessly snarled fragment simply has to be rewritten. This rewriting is best done by pulling the main points out of the fragment, listing them in the proper sentence, and then rewriting the sentence. (See also **garbled sentences.**)

CHANGE Removing the protection cap and the piston secured in the housing by means of the spring placed between the piston and the housing.

MAIN POINTS
1. Remove the protection cap.
2. The piston is held in the housing by a spring.
3. The spring is connected to the piston at one end and the housing at the other.
4. To remove the piston, disconnect the spring.

TO Removing the protection cap enables you to remove the piston by disconnecting a spring that connects to the piston at one end and the housing at the other.

(See also **sentence construction, sentence faults, subjects of sentences,** and **run-on sentences.**)

sentence types

Sentences may be classified according to *structure* (simple, compound, complex, compound-complex); *intention* (declarative, interrogative, imperative, exclamatory); and *stylistic use* (loose, periodic, minor).

BY STRUCTURE

A **simple sentence** consists of one **independent clause.** At its most basic, the simple sentence contains only a **subject** and a **predicate.**

EXAMPLES Profits (subject) rose (predicate).
The strike (subject) finally ended (predicate).

A **compound sentence** consists of two or more independent clauses connected by a **comma** and a **coordinating conjunction,** by a **semicolon,** or by a semicolon and a **conjunctive adverb.**

EXAMPLES Drilling is the only way to collect samples of the layers of sediment below the ocean floor, but it is by no means the only way to gather information about these strata. (comma and coordinating conjunction)

—Bruce C. Heezen and Ian D. MacGregor, "The Evolution of the Pacific," *Scientific American* (November 1973), p. 103.

There is little similarity between the chemical composition of sea water and that of river water; the various elements are present in entirely different proportions. (semicolon)

It was 500 miles to the site; therefore, we made arrangements to fly. (semicolon and conjunctive adverb)

The **complex sentence** provides a means of subordinating one thought to another. A complex sentence contains one independent clause and at least one **dependent clause** that expresses a subordinate idea.

EXAMPLE The generator will shut off automatically (independent clause) if the temperature rises above a specified point (dependent clause).

A **compound-complex sentence** consists of two or more independent clauses plus at least one dependent clause.

EXAMPLE Productivity is central to controlling inflation (independent clause), for when productivity rises (dependent clause), employers can raise wages without raising prices (independent clause).

BY INTENTION

By intention, a sentence may be declarative, interrogative, imperative, or exclamatory.

A declarative sentence conveys information or makes a factual statement.

EXAMPLE This motor powers the conveyor belt.

An interrogative sentence asks a direct question.

EXAMPLE Does the conveyor belt run constantly?

An imperative sentence issues a command.

EXAMPLE Start the generator.

An exclamatory sentence is an emphatic expression of feeling, fact, or opinion. It is a declarative sentence that is stated with great feeling.

EXAMPLE The heater exploded!

BY STYLISTIC USE

A *loose* sentence is one that makes its major point at the beginning and then adds subordinate **phrases** and **clauses** that develop the major point. It is the pattern in which we express ourselves most naturally and easily. A loose sentence could be ended at one or more points before it actually does.

EXAMPLE It went up (.), a great ball of fire about a mile in diameter (.), changing colors as it kept shooting upward (.), an elemental force freed from its bonds (.) after being chained for billions of years.

Compound sentences are generally classed as loose, since the sentence could end after the first independent clause.

EXAMPLE Copernicus is frequently called the first modern astronomer; he was the first to develop a complete astronomical system based on the motion of the earth.

Complex sentences are loose if their subordinate clauses follow their main clause.

EXAMPLE The installation will not be completed on schedule (.) because heavy spring rains delayed construction.

A *periodic* sentence delays its main idea until the end by presenting subordinate ideas or modifiers first, thus holding the reader's interest until the end. If skillfully handled, a periodic sentence lends force, or **emphasis,** to the main point by arousing the reader's anticipation and then presenting the main point as a climax.

EXAMPLE During the last decade or so, the attitude of the American citizen toward automation has undergone a profound change.

Do not use periodic sentences too frequently, however, for overuse results in a **style** that is irritating to the **reader.**

A *minor* sentence is an incomplete sentence. It makes sense in its context because the missing element is clearly implied by the preceding sentence.

EXAMPLE In view of these facts, is automation really useful? *Or economical?* There is no question that it has made our country the most wealthy and technologically advanced nation the world has ever known.

In short, minor sentences are elliptical expressions that are equivalent to complete sentences because the missing words are clearly understood without being stated.

EXAMPLES Why not?
How much?
Ten dollars.
At last!
This way, please.
So much for that idea.

Minor sentences are common in advertising copy and fictional dialogue; they are not normally appropriate to technical writing.

CHANGE You can use the one-minute long-distance rate any time after eleven at night all the way until eight in the morning. *Any night of the week.*

TO You can use the one-minute long-distance rate any time after eleven at night all the way until eight in the morning, any night of the week.

(See also **sentence construction, sentence variety,** and **sentence faults.**)

sentence variety

Sentences may be long or short; loose or periodic; **simple, compound, complex,** or **compound-complex;** declarative, interrogative, exclamatory, or imperative—even elliptical. There is never a legitimate excuse for letting your sentences become tiresomely alike. However, sentence variety is best achieved during **revision;** do not let it concern you when you are **writing the draft.**

SENTENCE LENGTH

Varying sentence length makes writing more interesting to the **reader** because a long series of sentences of the same length is monotonous. For example, avoid stringing together a number of short **independent clauses.** Either connect them with subordinating **connectives,** thereby making some dependent, or make some **clauses** into separate sentences.

CHANGE The river is 60 miles long, *and* it averages 50 yards in width, *and* its depth averages 8 feet.
TO This river, *which* is 60 miles long and averages 50 yards in width, has an average depth of 8 feet.
OR This river is 60 miles long. It averages 50 yards in width and 8 feet in depth.

Short sentences can often be effectively combined by converting **verbs** into **adjectives.**

CHANGE The steeplejack *fainted.* He collapsed on the scaffolding.
TO The *fainting* steeplejack collapsed on the scaffolding.

Although too many short sentences make your writing sound choppy and immature, a short sentence can be effective at the end of a passage of long ones.

EXAMPLE During the past two decades, many changes have occurred in American life, the extent, durability, and significance of which no one has yet measured. *No one can.*

In general terms, short sentences are good for emphatic, memorable statements. Long sentences are good for detailed explanations and support. There is nothing inherently wrong with a long sentence, or even with a complicated one, as long as its meaning is clear and direct. Sentence length becomes an element of **style** when varied for **emphasis** or contrast; a conspicuously short or long sentence can be used to good effect.

WORD ORDER

When a series of sentences all begin in exactly the same way, the result is likely to be monotonous. You can make your sentences more interesting by occasionally starting with a modifying word, **phrase,** or clause. This could be a single adjective, **adverb, participle,** or **infinitive;** a **prepositional phrase,** a **participial phrase,** or an **infinitive phrase;** or even a subordinate clause. But overdoing this technique can be monotonous, so use it in moderation.

EXAMPLES *Exhausted,* the project director slumped into a chair. (adjective)
Recently, sales have been good. (adverb)
Smiling, he extended his hand. (participle)
To advance, one must work hard. (infinitive)
In the morning, we will finish the report. (prepositional phrase)
Reading the report, she found several errors. (participial phrase)
To reach the top job, she presented constructive alternatives when current policies failed to produce results. (infinitive phrase)
Because we now know the result of the survey, we may proceed with certainty. (adverb clause)

Inverted sentence order can be an effective way to achieve variety.

EXAMPLES Then occurred the event that gained us the contract.
Never have sales been so good.

Too many inverted sentences, however, can lead to the kind of criticism that was aimed at *Time* magazine's writing style many years ago: "Backwards ran the sentences until reeled the mind." Use inverted sentence order sparingly.

Be careful in your sentences to avoid the unnecessary separation of **subject** and verb, **preposition** and **object,** and parts of a **verb phrase.**

CHANGE Electrical equipment can, if not carefully handled, cause serious accidents. (parts of verb phrase separated)
TO Electrical equipment can cause serious accidents if not carefully handled.

This advice, however, does not mean that a subject and verb should never be separated by a modifying phrase or clause.

EXAMPLE John Stoddard, who founded the firm in 1943, is still an active partner.

Vary the position of **modifiers** in your sentences to achieve variety as well as precision. The following examples illustrate four different ways that the same sentence could be written by varying the position of its modifiers:

EXAMPLES Gently, with the square end up, slip the blasting cap down over the time fuse.

With the square end up, gently slip the blasting cap down over the time fuse.
With the square end up, slip the blasting cap gently down over the time fuse.
With the square end up, slip the blasting cap down over the time fuse gently.

LOOSE/PERIODIC/INSERTION SENTENCES

A loose sentence makes its major point at the beginning and then adds subordinate phrases and clauses that develop or modify the point. A periodic sentence delays its main idea until the end by presenting modifiers or subordinate ideas first, thus holding the reader's interest until the end.

EXAMPLES The attitude of the American citizen toward automation has undergone a profound change during the last decade or so. (loose)
During the last decade or so, the attitude of the American citizen toward automation has undergone a profound change. (periodic)

Experiment in your own writing, especially during revision, with shifts from loose sentences to periodic sentences. Avoid the sing-song monotony of a long series of loose sentences, particularly a series containing coordinate clauses joined by **conjunctions.** Subordinating some thoughts to others makes your sentences more interesting.

CHANGE The auditorium was filled to capacity, *and* the chairman of the board came onto the stage. The meeting started at eight o'clock, *and* the president made his report of the company's operations during the past year. The audience of stockholders was obviously unhappy, *but* the members of the board of directors all were reelected.

TO By eight o'clock, *when* the chairman of the board came onto the stage and the meeting began, the auditorium was filled to capacity. *Although* the audience of stockholders was obviously unhappy with the president's report of the company's operations during the past year, the members of the board of directors all were reelected.

For variety, you may also alter the normal sentence order with an inserted phrase or clause.

EXAMPLE Titanium fills the gap, both in weight and strength, between aluminum and steel.

The technique of inserting such a phrase or clause is good for **emphasis,** for providing detail, for breaking monotony, and for regulating **pace.** (See also **sentence construction,** and **sentence faults.**)

sequential method of development

The sequential, or step-by-step, method of development is especially effective for explaining a process or describing a mechanism in operation. It would also be the logical method for writing **instructions** (Figure 1).

PROCESSING FILM

Developing. In total darkness, load the film on the spindle and enclose it in the developing tank. Be careful not to allow the film to touch the tank walls or other film. Add the developing solution, turn the lights on, and set the timer for seven minutes. Agitate for five seconds initially and then every half minute.

Stopping. When the timer sounds, drain the developing solution from the tank and add the stop bath. Agitate continually for 30 seconds.

Fixing. Drain thc stop bath and add the fixing solution. Allow the film to remain in the fixing solution for two to four minutes. Agitate for five seconds initially and then every half minute.

Washing. Remove the tank top and wash the film for at least 30 minutes under running water.

Drying. Suspend the film from a hanger to dry. It is generally advisable to place a drip pan below the rack. Sponge the film gently to remove excess water. Allow the film to dry completely.

Figure 1 Sequential Method of Development

The main advantage of the sequential method of development is that it is easy to follow because the steps correspond to the elements of the process or operation being described.

The disadvantages of the sequential method are that it can become monotonous and does not lend itself very well to achieving **emphasis.**

Practically all **methods of development** have elements of sequence within them. The **chronological method of development,** for example, is also sequential: to describe a trip chronologically, from beginning to end, is also to describe it sequentially.

service

When used as a **verb,** *service* means "keep up or maintain" as well as "repair."

EXAMPLE Our company will *service* your equipment.

To mean providing a more general benefit, use *serves.*

CHANGE Our company *services* the northwest area of the state.
TO Our company *serves* the northwest area of the state.

set/sit

Sit is an intransitive **verb**; it does not, therefore, require an **object.** Its past tense is *sat.*

EXAMPLES I *sit* by a window in the office.
We *sat* around the conference table.

Set is usually a transitive **verb**, meaning "put or place," "establish," or "harden." Its past tense is *set.*

EXAMPLES Please *set* the trophy on the shelf.
The jeweler *set* the stone beautifully.
Can we *set* a date for the tests?
The high temperature *sets* the epoxy quickly.

Set is occasionally intransitive.

EXAMPLES The sun *sets* a little earlier each day.
The glue *sets* in 45 minutes.

shall/will

It was at one time fairly common to use *shall* for first-person constructions and *will* for second- and third-person constructions.

EXAMPLES I *shall* go.
You *will* go.
He *will* go.

This distinction is unnecessary, however, because no one could be confused by *I will go* to express an action in the near future. *Shall* is sometimes used in all **persons,** nonetheless, to emphasize determination.

EXAMPLE Employees *shall* submit a written reason for a leave of absence.

sic

Latin for "thus," *sic* is used in **quotations** to indicate that the writer has quoted the material exactly as it appeared in the original source. It is most often used when the original material contains an obvious error or might in some other way be questioned. *Sic* is placed within **brackets.**

EXAMPLE In *Basic Astronomy,* Professor Jones noted that the "earth does not revolve around the son [*sic*] at a constant rate."

simile

A simile is a direct **comparison** of two essentially unlike things, making the comparison with the word *like* or *as.* Similes state that A is *like* B.

EXAMPLE Constructing the frame is *like piecing together the borders of a jigsaw puzzle.*

Like **metaphors,** similes can help illuminate difficult or obscure ideas.

EXAMPLE The odds against having your plant or business destroyed by fire are at least as great *as the odds against hitting the jackpot on a Las Vegas slot machine.* Because of the long odds, top management may consider sophisticated fire protection and alarm systems more costly than the risk merits and refuse to approve the expenditure. The plant fire protection chief is free to recommend whatever system he likes, but unless the system is required by fire codes or ordered by the insurance company, chances are it will never win top management's approval.

And yet, plants are destroyed by fire, *just as gamblers sometimes hit the jackpot on a one-armed bandit.* Losing big if fire strikes the plant, or winning big on a slot machine both depend on the chance alignment of three events or factors. . . .

—Jay R. Asher and Daniel T. Williams, Jr., "How Not to Lose the Fire Protection Gamble," *Occupational Hazards* (September 1974), p. 130.

simple sentences

A simple sentence has one **clause.** At its most basic, the simple sentence contains only a **subject** and a **predicate.**

EXAMPLES Profits rose.
The strike ended.

Both the subject and the predicate may be compounded without changing the basic structure of the simple sentence.

EXAMPLES *Bulldozers and road graders* have blades. (compound subject)
Bulldozers *strip, ditch, and backfill.* (compound predicate)

Although **modifiers** may lengthen a simple sentence, they do not alter its basic structure.

EXAMPLE The *recently introduced* procedure works *very well.*

Inverting the subject and predicate does not alter the basic structure of the simple sentence.

EXAMPLE A better job I never had.

A simple sentence may contain noun phrases and **prepositional phrases** in either the subject or the predicate.

EXAMPLES The man *in the blue suit* is my boss. (prepositional phrase in the subject)
I wrote the report *in a day.* (prepositional phrase in the predicate)

A simple sentence may include modifying **phrases** in addition to the **independent clause.** This fact causes most of the confusion about simple sentences. The following sentence, for example, is a simple sentence because the introductory phrase is an adjective phrase and not a **dependent clause:**

EXAMPLE *Hard at work in my office,* I did not realize how late it was.

-size/-sized

As **modifiers,** the **suffixes** *-size* and *-sized* are more common to advertising copy than to general writing.

EXAMPLES a *king-sized* bed, an *economy-sized* carton

However, they are usually redundant unless they are part of the name of a product and should generally not be used.

CHANGE It was a *small-size* desk.
TO It was a *small* desk.

slashes

Although not always considered a mark of **punctuation,** the slash performs punctuating duties by separating and showing omission. The slash is called a variety of names, including slant line, virgule, bar, solidus, shilling sign.

The slash is often used to separate parts of addresses in continuous writing.

EXAMPLE The return address on the envelope was Ms. Rose Howard/62 W. Pacific Court/Dalton/Ontario/Canada.

The slash can indicate alternative items.

EXAMPLE David's telephone number is 549–2278/2335.

The slash often indicates omitted words and letters.

EXAMPLES miles/hour for "miles per hour"
c/o for "in care of"
w/o for "without"

In fractions the slash separates the numerator from the denominator.

EXAMPLES 2/3 (2 of 3 parts), 3/4 (3 of 4 parts), 27/32 (27 of 32 parts)

The slash is sometimes used to indicate **brackets** on a typewriter that has no bracket key.

EXAMPLE Dr. Smith wrote that the earth "does not revolve around the son /*sic*/ at a constant rate."

In informal writing, the slash is also used to separate day from month and month from year in **dates.**

EXAMPLE 12/29/87

(See also **he/she** and **and/or.**)

so/such

So is often vague and should be avoided if another word would be more precise.

CHANGE She writes faster, *so* she finished before I did.
TO *Because* she writes faster, she finished before I did.

Another problem occurs with the **phrase** *so that,* which should never be replaced with *so* or *such that.*

CHANGE The report should be written *such that* it can be copied.
TO The report should be written *so that* it can be copied.

Such, an **adjective** meaning "of this or that kind," should never be used as a **pronoun.**

CHANGE Our company does not need computers, and we do not anticipate using *such.*
TO Our company does not need computers, and we do not anticipate using *any.*

some

When *some* functions as an **indefinite pronoun** for plural count **nouns**, or as an **indefinite adjective** modifying plural count nouns, use a plural **verb.**

EXAMPLES *Some* people *are* kinder than others.
Some of us *are* prepared to leave.

Some is singular, however, when used with mass nouns.

EXAMPLES *Some* sand *has* trickled through the crack.
Some oil *was* spilled on the highway.
Some stationery *is* missing from the supply cabinet.
Most of the water evaporated, but *some remains.*

some/somewhat

Some, an **adjective** or **pronoun** meaning "an undetermined quantity" or "certain unspecified persons," should not replace the **adverb** *somewhat,* which means "to some extent."

CHANGE His writing has improved *some.*
TO His writing has improved *somewhat.*
OR His writing is *somewhat* improved.

some time/sometime/sometimes

Some time refers to a duration of time.

EXAMPLE We waited for *some time* before calling the customer.

Sometime refers to an unknown or unspecified time.

EXAMPLE We will visit with you *sometime.*

Sometimes refers to occasional occurrences at unspecified times.

EXAMPLE He *sometimes* visits the branch offices.

spatial method of development

In a spatial sequence, you describe an object or a process according to the physical arrangement of its features. Depending on the subject, you may describe the features from top to bottom, side to side, east to west (or west to east), inside to outside, and so on. Descriptions of this kind rely mainly on dimension (height, width, length), direction (up, down, north, south), shape (rectangular, square, semicircular), and proportion (one-half, two-thirds). Features are described in relation to one another.

EXAMPLE One end is raised six to eight inches higher than the other end to permit the rain to run off.

Features are also described in relation to their surroundings.

EXAMPLE The lot is located on the east bank of the Kingman River.

The spatial method of development is commonly used in **descriptions** of laboratory equipment, **proposals** for landscape work, construction-site **progress and activity reports,** and, in combination with a step-by-step sequence, in many types of **instructions.**

The following instructions (Figure 1), which explain how a two-

Conducting a Methodical Room Search

First, look around the room to decide how the room should be divided for the search and to what height the first searching sweep should extend. The first sweep should include all items resting on the floor up to the selected height.

As nearly as possible, divide the room into two equal parts. Base the division on the number and type of objects in the room, not on the size of the room. Divide the room with an imaginary line that extends from one object to another—for example, the window on the north wall to the floor lamp next to the south wall.

Next, select a search height for the first sweep. Base this height on the average height of the majority of objects resting on the floor. In a typical room, this height is established by such objects as table and desk tops, chair backs, and so on. As a rule, the first sweep will be made hip high and below.

After dividing the room and establishing the first search height, go to one end of the agreed upon room division. This point will be the starting point for the first and all subsequent search sweeps. Beginning back to back, work your way around the room along the walls toward the other team member at the other end of the imaginary dividing line. Check all items resting on the floor adjacent to the walls; be sure to check the floor under the rugs. When you meet your partner at the other end of the room, return to the starting point and search all items in the middle of the room up to the first search height. Include all items mounted in or on the walls in the first search sweep, such as air conditioning ducts, baseboards, heaters, built-in storage units, and so on. During this and all subsequent searches, use an electronic or a medical stethoscope.

Then determine the search height for the second search sweep. This height is usually set at the chin or at the top of the head of the searchers. Return to the starting point and repeat the searching technique up to the second search height. This sweep typically covers objects hanging on walls, built-in storage units, tall standing items on the floor, and the like.

Next, establish the third search height. This area usually includes everything in the room from the top of the searcher's head to the ceiling. In this sweep, features like hanging light fixtures and mounted air conditioning units are examined.

Finally, if the room has a false or suspended ceiling, perform a fourth sweep. Check flush or ceiling light fixtures, air conditioning or ventilation ducts, speaker systems, structural frames, and so forth.

Figure 1 Spatial Sequencing Method of Development

person security team should conduct a methodical room search, use spatial sequencing.

specie/species

Specie means "coined money" or "in coin."

EXAMPLE Paper currency was virtually worthless, and creditors began to demand payment in *specie.*

Species means a category of animals, plants, or things having some of the same characteristics or qualities. *Species* is the correct **spelling** for both singular and plural.

EXAMPLES The wolf is a member of the canine *species.*
Many animal *species* are represented in the Arctic.

specific-to-general method of development

The specific-to-general **method of development** begins with a specific statement and builds to a general conclusion. It is somewhat like the **increasing-order-of-importance method of development** in that it carefully builds its case, often with examples and **analogies** in addition to facts or statistics, and does not actually make its point until the end. For example, if your subject were highway safety, you might begin with a specific highway accident and then go on to generalize about how details of the accident were common enough to many similar accidents that recommendations could be made to reduce the probability of such accidents. Figure 1, on page 638, is an example of the specific-to-general method of development.

specifications

By definition, a specification is "a detailed and exact statement of particulars; especially a statement prescribing materials, dimensions, and workmanship for something to be built, installed, or manufactured." The most significant words in this definition are *detailed* and *exact.* Although there are two broad categories of specifications—government specifications and industrial specifications—both require the writer of the specification to be both accurate and exact. A specification must be written so clearly that no one could misinterpret any statement contained in it; therefore, do not imply

The Facts About Seat Belts

Recently the Highway Safety Foundation studied the use of seat
Statistic belts in 4,500 accidents involving nearly 13,000 people. Nearly all these accidents occurred on routes which had a speed limit of at least 40 mph. Only 20 percent of all the vehicle occupants were
Statistic wearing any kind of seat belt.

The shoulder-type belts were even more unpopular than the lap belts, and only 4 percent of the occupants who had shoulder belts were wearing them.

In this study, as in other studies, it was found that vehicle
Statistic passengers not using seat belts were more than four times as likely to be killed as those using them. The driver in some cases escaped a more serious injury by being thrown against the steering wheel.

A conservative estimate is that 40 percent of the front-seat
General Conclusion passenger car deaths could be prevented if everyone used the seat belts, which the law requires the manufacturer to put into each automobile. If you are in an accident, your chances of survival are far greater if you are using your seat belt.

—*The Safe Driving Handbook* (New York: Grosset & Dunlap, 1970), pp. 84–85.

Figure 1 Specific-to-General Method of Development

or suggest—state explicity what is needed. **Ambiguity** in a specification not only can waste money but can result in a lawsuit. Because of the stringent requirements for completeness and exactness of detail in writing specifications, careful **research** and **preparation** are especially important before you begin to write, as is careful **revision** after you have completed the draft of your specification.

GOVERNMENT SPECIFICATIONS

Government agencies are required by law to contract for equipment strictly according to definitions provided in formal specifications. Government specifications are contractual documents that protect both the procuring government agency and the contractor.

A government specification is a precise definition of exactly what the contractor is to provide for the money he or she is paid. In addition to a technical description of the device to be purchased, the specification normally includes an estimated cost, an estimated delivery date, and the standards for the design, manufacture, workmanship, testing, training of government personnel, governing codes, inspection, and delivery of the item to be purchased. Government specifications are often used to prescribe the content and deadline for a **government proposal** submitted by a vendor or a company that wishes to bid on a project.

Government specifications are engineering and contractual documents with rules, **formats**, and peculiarities often known only to people trained in this specialty. What follows is a general description of the contents of a government specification. Detailed information on military and nonmilitary government specifications can be found in the following publications:

MILITARY *Specifications, Types and Forms.* Mil-S-83490. October 1968.
Standardization Policies, Procedures and Instructions. DOD-4120.3-M. January 1972.

Both are available from the Naval Publications and Forms Center, 5801 Tabor Avenue, Philadelphia, Pennsylvania 19120.

NONMILITARY *Index of Federal Specifications, Standards, and Commercial Descriptions.* FPMR101-29. (updated annually)

This publication is available from the U.S. General Services Administration, Specification and Consumer Information Distribution Section, Building 197, Navy Yard Annex, Washington, D.C. 20407.

Government specifications contain details on (1) the scope of the project, (2) the documents that the contractor is required to furnish with the purchased device, (3) the required product characteristics and functional performance of the purchased device, (4) the required tests, test equipment, and test procedures, (5) the required preparations for delivery, (6) notes, and (7) an **appendix**.

The "characteristics and performance" section of the specification (item 3 in the previous paragraph) must precisely define every product characteristic not covered by the engineering drawings, and it must precisely define every functional performance requirement the device must meet when it is operational. The "test" section of the specification (item 4 in the previous paragraph) must specify how the device is to be tested to verify that it meets all requirements. The test section includes the precise tests that are to be performed, the procedure to be used in conducting the tests, and the test equipment to be used.

The example shown in Figure 1, on pages 640–2, is the beginning of a very long government specification for a computer installation.

Specification for JC/80 Computerized Building Automation System

1700. PREFACE

1700.00.01 The building automation system as herein specified shall be provided in its entirety by the building automation contractor. The building automation contractor shall base his bid on the system as specified. Alternate techniques, modifications, or changes to any aspect of these specifications shall be submitted as a voluntary alternate.

Identification of this specification
Identification of a major section within this specification
Identification of a minor section within this major section
Identification of a paragraph within this minor section

1700.01. VOLUNTARY ALTERNATES

1700.01.01 Voluntary alternates shall be fully documented and submitted to the consulting engineer ten working days prior to bid date for permission to bid the voluntary alternate. Permission to bid a voluntary alternate in no way implies that the alternate will be acceptable. Each voluntary alternate will be itemized as an add or deduct to base bid.

1700.01.02 No voluntary alternate shall be considered from any bidder who has failed to submit a base quotation in full compliance with the building automation system specified herein.

1701. General Requirements

1701.01. GENERAL

1701.01.01 The "computerized building automation system" herein specified shall be fully integrated and installed as a complete package by the building automation contractor. The system shall include all computer software and hardware, sensors, transmission equipment, required wiring, piping, preassembled control consoles, local panels, and labor supervision. Adjustment and calibration shall be provided as a prerequisite in the service contract specified hereafter.

1701.01.02 All input/output for central processing unit operation shall be ASCII (American Standard Code for Information Interchange) coded with standard EIA (Electronic Industries Association) interface hardware. Future incorporation of marketplace equipment prohibits any deviations from the standard ASCII-coded input/output or the standard EIA interface hardware.

1701.01.03 The building automation system operator shall have the capability to make several on-line adjustments to various system para-

Figure 1 Government Specifications

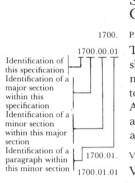

meters and response to his requests shall occur immediately. The system shall be a combination of hardware and software to permit simultaneous data processing, output printing, and operator communications.

1701.02. CONTRACTOR

1701.02.01 The building automation contractor shall have a local office within a 75-mile radius of the job site, staffed with factory-trained engineers fully capable of providing instructions and routine emergency maintenance service on all system components.

1701.03. EXPERIENCE RECORD

1701.03.01 The building automation contractor shall have a five-year successful history of the design and installation of fully computerized building systems similar in performance to that specified herein and shall be prepared to evidence this history as condition of acceptance and approval prior to bidding.

1701.04. INSTALLATION

1701.04.01 The installation shall include computer programming, drawings, supervision, adjusting, validating, and checkout necessary for an operational system.

1701.04.02 The building automation contractor shall provide all other wiring necessary for system operation, including tie-ins from building automation system relays into motor starting circuits.

1701.04.03 All wiring performed by the building automation contractor shall be installed in accordance with all applicable local and national electrical codes.

1701.04.04 It shall be the responsibility of the building automation contractor to make a complete job survey prior to bid date to ascertain job conditions.

1702. WORK PROGRESS

1702.00.01 To ensure on-time completion of the project, the building automation contractor shall use a computerized quantitative weekly reporting system. Each week a progress report will be submitted to the owner's representative. The report shall be of actual construction progress, within plus or minus 1-percent accuracy, and shall be submitted within one week after work has been performed at the job site. Subjective reports and estimates of project completion are not acceptable.

1702.00.02 All weekly reports shall relate to manpower analysis of the project, giving dates the work is planned for completion and the number of calendar days actual job progress varies from the original plan.

1703. CONTRACT COMPLETION, GUARANTEE, AND SERVICE

1703.00.01 All components, parts, and assemblies shall be guaranteed against defects in workmanship and materials for a period of one year after acceptance. In addition, the building automation contractor shall provide operator instruction and, if desired by the owner, system maintenance training as described hereinafter for the primary system as well as the subsystems.

1703.00.02 The following procedures shall govern the guarantee period. Within thirty days after the owner is receiving beneficial use of the building automation system, the contractor shall initiate the guarantee period by formally transmitting to the owner a 12-month service and maintenance contract marked "Paid in full." The contract shall become effective upon being dated with countersignature by the owner or his authorized representative. This service contract shall be a formal service agreement of the contractor, signed by an authorized employee, and shall include the monthly cost of the services to be provided.

1703.01. SERVICE CONTRACT

1703.01.01 The service contract shall include these minimum provisions:

01. Provide regularly scheduled maintenance and service of at least one man-day per month by factory-trained service representatives of the building automation contractor.
02. Replace all defective parts and components as required.
03. Make available, upon request, emergency maintenance service.
04. Unless canceled by the owner 45 days prior to termination of the service agreement, the building automation contractor shall agree to continue the service on a monthly basis for an additional one-year period at the maintenance contract rates set forth in the agreement. . . .

INDUSTRIAL SPECIFICATIONS

The industrial specification is used in areas like computer software in which there are no engineering drawings or other means of documentation. It is a permanent document, whose purpose is twofold: (1) to document the item being implemented so that it can be maintained if the person who designed it is promoted, transferred, or leaves the company and (2) to provide detailed technical information on the item being implemented to all those in the company who need it (this includes other engineers and technicians, technical writers, technical instructors, and possibly salespeople and purchasing agents).

The industrial specification describes (1) a planned project, (2) a newly completed project, or (3) an old project. The specification for a planned project describes how it is going to be implemented; the specification for a newly completed project describes how the newly completed project was implemented; and the specification for an old project describes the project as it finally exists after it has been operational long enough for all the problems to have been discovered and corrected. All three types of industrial specifications contain a detailed technical description of all aspects of the item being described, including what was done, how it was done, what is required to use the item, how it is used, what its function is, who would use it, and so on. In addition, the specification for a planned project often estimates the time to complete the project (in man-hours or man-months in addition to calendar months) and estimated costs.

The industrial specification differs from the **technical manual** that is given to the customer for the same project in that the specification contains detailed information that the customer does not need.

Shown in Figure 2 on pages 644–6 is a specification for a computer utility routine that reads a deck of punched cards and prints in English the contents of each card. The name of the routine is CDPRINT. GAC is the name of another utility routine, and NEAT/3 is the name of a computer programming language.

spelling

Learning to spell requires a systematic effort. The following system will help you learn to spell correctly *(this entry continues on p. 646)*:

1. Keep your **dictionary** handy, and use it regularly. If you are unsure about the spelling of a word, don't rely on memory or

2.8.8 CARD TO PRINTER

The purpose of this routine is to enable the user quickly and easily to verify and list a Hollerith Code punched card deck.

2.8.8.1 *Scope*

Identification of this specification.
Identification of a major section within this specification.
Identification of a minor section within this major section.
Identification of a paragraph within this minor section.

The CDPRNT routine will be valuable to any site that uses cards either as Monitor Control String records, NEAT/3 source lines, and/or data records. It also can be used in conjunction with other utility routines. For example, a deck of cards that has just been punched may be printed for easy visual validation. As another example, if (during a GAC run, using a card deck as the source file) an error or errors occur on the reading and translating of the cards, a simple run of the deck through the CDPRNT routine will list the cards and also verify the punch configurations against the set being used. If any of the punches are in error, the columns containing the errors will be printed behind the card image in the listing.

2.8.8.2 *User Options*

The user is provided with a few simple but useful options. These are as follows:

1. *SET TYPE* The user may choose either the extended A-set, H-set, or Site Code for translation of his cards.
2. *SLEW* The user may slew either one or two lines between each card image on the printer.
3. *FILENAME* An optional filename of ten characters may be entered, which will be printed at the top of each page.
4. *SPREAD* The spread function, also optional, will differentiate between NEAT/3 Source Card types (C,D,F, etc.) and list them in a format similar to the compiler. If no entry is made on the parameter for the spread option, the cards will be printed in a straight 80-character format.

2.8.8.3 *Parameter Card*

The parameter card is designed as follows:

Figure 2 Industrial Specifications

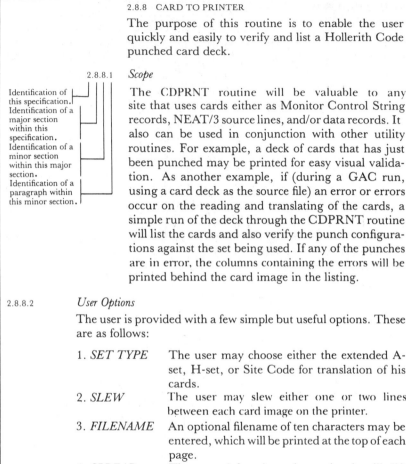

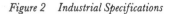

PAGE	LINE	7	REFERENCE	OPERATION	OPERANDS
1 2 3	4 5 6	7	8 9 10 11 12 13 14 15 16 17	18 19 20 21 22 23	24 25 26 27 28 29 30 31 32 33 34 35 36 37 38 39 40 41 42 43 44
		C	SPECSUTIL	CDPRNT	SUD, S, S, FILENAMEDO, S,
		C			E (28) L (30) P (43)
		C			T (28) E (30) R (43)
		C			W (30) E (43)
		C			A (43)
		C			D (43)

The explanation of the parameter is:

Column 7-C indicates a control card.
8-16-SPEC$UTIL, identifies a utility routine.
18-21-CDPRNT, routine name.
24-26-SUD, symbolic unit designator of the card reader to be used as source.
28-SET, card set to be used.
H = Extended H-set
A = Extended A-set
S = SITECODE of Alternate Systems Disc
30-SLEW, signals printer number of lines to slew.
1 = Single space
2 = Double space
32-41-FILENAME, optional entry, allows the user to give a name to card file being printed. Ten-character name and version number.
43-SPREAD, an *S* entered in this column will cause the cards to be printed in NEAT/3 compiler type format. No entry in this column gives 80-character format.
S = SPREAD
□ = 80-character image

The parameter card is entered in the normal utility fashion.

There should be only one parameter card for each run.

2.8.8.4 *Card Verification*

As the card file is being read and translated, the CDPRNT routine detects any characters that cannot be translated and notes the column of the card that is in error. The column or columns in error are then printed out immediately following the card image. The routine checks up to ten errors in any single card. Upon reaching the tenth, the routine moves a message "BAD CARD" under the CHECK COLUMN heading on the print page. If such a message appears, it should inform the user that this particular card needs to be completely re-created as it has at least ten errors in it.

2.8.8.5 *Translation Using Site Code*

By using the Side Code, the CDPRNT routine will be able to handle a nonstandard card code. The routine will expect the user to have the pack with the nonstandard code (which he has built by using SITECD routine) on the Alternate Systems Disc unit. CDPRNT will access the Site Code from the Alternate Systems Disc. The reason for specifying the alternate disc is that the program may be loaded in the standard A-set from the current systems disc, while the card file can still be translated in any code desired by the user.

guesswork—consult the dictionary. When you look up a word, focus on both its spelling and its meaning.

2. After you have looked in the dictionary for the spelling of the word, write the word from memory several times. Then check the accuracy of your spelling. If you have misspelled the word, repeat this step. If you do not follow through by writing the word from memory, you lose the chance of retaining it for future use. Practice is essential.
3. Keep a list of the words you commonly misspell, and work regularly at whittling it down. Do not load the list with exotic words; many of us would stumble over *asphyxiation* or *pterodactyl.* Concentrate instead on words like *calendar, maintenance,* and *unnecessary.* These and other frequently used words should remain on your list until you have learned to spell them.
4. Check all your writing for misspellings by **proofreading.**

Any of the following dictionaries should serve you well:

American Heritage Dictionary of the English Language
Twenty-Thousand Words Spelled and Divided for Quick Reference
Webster's New Collegiate Dictionary
Webster's New World Dictionary of the American Language
Webster's Third International Dictionary of the English Language
Random House Dictionary of the English Language

spin-off

In technical usage, *spin-off* usually refers to benefits that come about in one area (for example, housing materials) as the result of achievements in another area (for example, space technology research). Because it is **jargon,** do not use the term unless you are certain that all your **readers** understand it.

EXAMPLE The Teflon coating on cookware is a *spin-off* from the space program.

standard English (see **English, varieties of**)

strata

Strata is the plural form of *stratum,* meaning a "layer of material."

EXAMPLES The land's *strata are* exposed by erosion.
Each *stratum is* clearly visible in the cliff.

style

A **dictionary** definition of style is "the way in which something is said or done, as distinguished from its substance." Writers' styles are determined by the way they think and transfer their thoughts to paper—the way they use words, sentences, images, **figures of speech,** and so on.

A writer's style is the way his or her language functions in particular situations. For example, a letter to a friend would be relaxed, even chatty in **tone,** whereas a job **application letter** would be more restrained and deliberate. Obviously, the style appropriate to the one letter would not be appropriate to the other. In both letters, the reader and situation determine the manner or style the writer adopts.

Standard English can be divided into two broad catagories of style—formal and informal—according to how it functions in certain situations. Understanding the distinction between formal and informal writing styles helps writers use the appropriate style in the appropriate place. We must recognize, however, that no clear-cut line divides the two categories and that some writing may call for a combination of the two.

A formal writing style can perhaps best be defined by pointing to certain material that is clearly formal, such as scholarly and scientific articles in professional journals, lectures read at meetings of professional societies, and legal documents.

Material written in a formal style is usually the work of a specialist writing to other specialists, or writing that embodies laws or regulations. As a result, the vocabulary is specialized and precise. The writer's tone is impersonal and objective because the subject matter looms larger in the writing than does the author's personality. (See **point of view.**) Unlike an informal writing style, a formal writing style does not use **contractions,** slang, or dialect. (See **English, varieties of.**) Sentences may be elaborate because complex ideas are generally being examined.

Formal writing need not be dull and lifeless, however. By using such techniques as the active **voice** whenever possible, **sentence variety,** and **subordination,** a writer can make formal writing lively and even interesting, especially if the subject matter is inherently interesting to the **reader.**

EXAMPLE Although a knowledge of the morphological chemical constitution of cells is necessary to the proper understanding of living things, in the final analysis it is the activities of their cells that distinguish organisms from all other objects in the world. Many of these activities differ greatly among the various types of living things, but some of the basic sorts are shared by all, at least in their essentials. It is these fundamental actions with which we are concerned here. They fall into two major groups—those that are characteristic of the cell in the *steady state*, that is, in the normally functioning cell not engaged in reproducing itself, and those that occur during the process of *cellular reproduction*.

—Lawrence S. Dillon, *The Principles of Life Sciences* (New York: Macmillan, 1964), p. 32.

Whether you should use a formal style in a particular instance depends on your reader and **objective.** When writers attempt to force a formal style when it should not be used, they are likely to fall into **affectation, awkwardness,** and **gobbledygook.**

An informal writing style is a relaxed and colloquial way of writing standard English. It is the style found in most private letters and in some **correspondence, memorandums,** nonfiction books of general interest, and mass-circulation magazines. There is less distance between the writer and reader because the tone is more personal than in a formal writing style. Contractions and elliptical constructions are commonplace. Consider the following passage, written in an informal style, from a nonfiction book:

EXAMPLE All you need is a talent for spotting the idiocies now built into the system. But you'll have to give up being an administrator who loves to run others and become a manager who carries water for his people so they can get on with the job. And you'll have to keep a suspicious eye on the phonies who cater to your uncertainties or feed your trembling ego on press releases, office perquisites, and optimistic financial reports. You'll have to give substance to such tired rituals as the office party. And you'll certainly have to recognize, once you get a hunk of your company's stock, that you aren't the last man who might enjoy the benefits of shareholding. These elegant simplicities require a sense of justice that won't be easy to hang on to.

—Robert Townsend, *Up the Organization* (New York: Knopf, 1970), p. 11.

As this example illustrates, the vocabulary of an informal writing style is made up of generally familiar rather than unfamiliar words and expressions, although slang and dialect are usually avoided. An informal style approximates the cadence and structure of spoken English, while conforming to the grammatical conventions of written English.

When we consciously attempt to create a "style," we usually defeat our purpose. One writer may attempt to impress the reader with a flashy writing style, which can lead to affectation. Another may attempt to impress the reader with scientific objectivity and produce a style that is dull and lifeless. Technical writing need be neither affected nor dull. It can and should be simple, clear, direct, and even interesting—the key is to master the basic writing skills and always to keep your reader in mind. What will be both informative and interesting to your reader? When this question is uppermost in your mind as you apply the steps of the writing process, you will achieve an interesting and informative writing style. (See the Checklist of the Writing Process, at the beginning of this book.)

The following guidelines will help you produce a brisk, interesting style. Concentrate on them as you revise your rough draft.

1. Use the active **voice**—not exclusively but as much as possible without becoming awkward or illogical.
2. Use **parallel structure** whenever a sentence presents two or more thoughts that are of equal importance.

3. Avoid the monotony of a sing-song style by using a variety of sentence structures. (See **sentence variety**.)
4. Avoid stating positive thoughts in negative terms (write *40 percent responded* instead of *60 percent failed to respond*). (See also **positive writing.**)
5. Concentrate on achieving the proper balance between **emphasis** and subordination.

Beyond an individual's personal style, there are various kinds of writing that have distinct stylistic traits, such as **technical writing style.**

subjective complements

A subjective complement is a **noun** or **adjective** in the **predicate** of a sentence; it completes the meaning of a **linking verb** by describing or renaming the subject of the **verb.**

EXAMPLES The project director seems *confident.* (adjective)
Acme Corp. is *our major competitor*. (noun phrase)
His excuse was *that he had been sick.* (noun clause)
He is an *engineer.* (noun)

The subjective complement is also known as a **predicate nominative** (noun) or a **predicate adjective** (adjective).

subjects of sentences

The subject of a sentence is a word or group of words about which the sentence or **clause** makes a statement. It indicates the topic of the sentence, telling what the **predicate** is about. It may appear anywhere in a sentence but most often appears at the beginning.

EXAMPLES *To increase sales* is our goal.
The wiring is defective.
We often work late.
That he will be fired is now doubtful.

The simple subject is a **substantive;** the complete subject is the simple subject and its **modifiers.** In the following sentence, the simple subject is *procedures;* the complete subject is *the procedures that you instituted.*

EXAMPLE The procedures that you instituted have increased efficiency.

Grammatically, a subject must agree with its **verb** in **number.**

EXAMPLES The *departments have* much in common.
The *department has* several advantages.

The subject is the actor in active-**voice** sentences.

EXAMPLE The *aerosol bomb* propels the liquid as a mist.

A compound subject has two or more elements as the subject of one verb.

EXAMPLE *The president* and *the treasurer* agreed to withhold the information.

Alternative subjects are joined by *or* and *nor.*

EXAMPLE Either *cash* or a *check* is acceptable.

Be careful not to shift subjects in a sentence, as doing so may confuse your **reader.**

CHANGE *Radio amateurs* stay on duty during emergencies, and *sending messages* to and from disaster areas is their particular job.
TO *Radio amateurs,* who stay on duty during emergencies, are responsible for sending messages to and from disaster areas.
OR *Radio amateurs* stay on duty during emergencies, sending messages to and from disaster areas.

subordinating conjunctions

Subordinating conjunctions connect sentence elements of varying importance, normally **independent clauses** and **dependent clauses;** they usually introduce the dependent clause. The most frequently used subordinating conjunctions are *so, although, after, because, if, where, than, since, as, unless, before, though, when,* and *whereas.* They are distinguished from **coordinating conjunctions,** which connect elements of equal importance.

EXAMPLES He finished his report, *and* he left the office. (coordinating conjunction)
He left the office *after* he finished his report. (subordinating conjunction)

A clause introduced with a subordinating conjunction is a dependent clause.

EXAMPLE *Because we did not oil the crankcase,* the engine was ruined.

The word *because* is a subordinating conjunction. Do not use the coordinating conjunction *and* as a substitute for it.

CHANGE We didn't add oil to the crankcase, *and* it ruined the engine.
TO *Because* we didn't add oil to the crankcase, the engine was ruined.

When the subordinating conjunction begins the sentence, a comma should follow the dependent clause in which it appears.

EXAMPLE *Because* we were late, the client was angry.

When the dependent clause beginning with a subordinating conjunction comes at the end of the sentence, do not separate the independent clause and the dependent clause with a comma.

EXAMPLE The client was angry *because* we were late.

Be aware that words may serve multiple functions. *When, where, how,* and *why* serve as both interrogative **adverbs** and subordinating conjunctions.

EXAMPLES *When* will we go? (interrogative adverb)
We will go *when* he arrives. (subordinating conjunction)

Since, until, before, and *after* are used as both subordinating conjunctions and **prepositions.**

EXAMPLES *Since* we all had arrived, we decided to begin the meeting early. (subordinating conjunction)
I have worked on this project *since* May. (preposition)

subordination

Subordination is a technique used by writers to show, in the structure of a sentence, the appropriate relationship between ideas of unequal importance by subordinating the less important ideas to the more important ideas. The skillful use of subordination is a mark of mature writing.

CHANGE Beta Corp. now employs 500 people. It was founded just three years ago.
TO Beta Corp., *which now employs 500 people,* was founded just three years ago.
OR Beta Corp., *which was founded just three years ago*, now employs 500 people.

Effective subordination can be used to achieve **sentence variety, conciseness,** and **emphasis.** For example, consider the following sentence. "The city manager's report was carefully illustrated, and it covered five typed pages." See how it might be rewritten, using subordination, in any of the following ways:

EXAMPLES The city manager's report, *which covered five typed pages,* was carefully illustrated. (adjectival clause)

The city manager's report, *covering five typed pages,* was carefully illustrated. (participial phrase)

The carefully illustrated report of the city manager covered five typed pages. (participial phrase)

The city manager's *five-page* report was carefully illustrated. (single modifier)

The city manager's report, *five typed pages,* was carefully illustrated. (appositive phrase)

We sometimes use a **coordinating conjunction** to concede that an opposite or balancing fact is true; however, a subordinating connective can often make the point more smoothly.

CHANGE Their bank has a lower interest rate on loans, *but* ours provides a fuller range of essential services.

TO *Although* their bank has a lower interest rate on loans, ours provides a fuller range of essential services.

The relationship between a conditional statement and a statement of consequences will be clearer if the condition is expressed as a subordinate **clause.**

CHANGE The bill was incorrect, *and* the customer was angry.

TO The customer was angry *because* the bill was incorrect.

Subordinating connectives (such as *because, if, while, when, though*) and **relative pronouns** (*who, whom, which, that*) achieve subordination effectively when the main clause states a major point and the **dependent clause** establishes a relationship of time, place, or logic with the main clause.

EXAMPLE A buildup of deposits is impossible *because* the apex seals are constantly sweeping the inside chrome surface of the rotor housing.

Relative pronouns (*who, whom, which, that*) can be used effectively to combine related ideas that would be stated less smoothly as independent clauses or sentences.

CHANGE The generator is the most common source of electric current. It uses mechanical energy to produce electricity.

TO The generator, *which* is the most common source of electric current, uses mechanical energy to produce electricity.

Avoid overlapping subordinate constructions, with each depending on the last. Often the relationship between a relative pronoun and its antecedent will not be clear in such a construction.

CHANGE Shock, *which* often accompanies severe injuries, severe infections, hemorrhages, burns, heat exhaustion, heart attacks, food or chemical poisoning, and some strokes, is a failure of the circulation, *which* is marked by a fall in blood pressure *that* initially affects the skin (*which* explains pallor) and later the vital organs of kidneys and brain; there is a marked fall in blood pressure.

TO Shock often accompanies severe injuries, severe infections, hemorrhages, burns, heat exhaustion, heart attacks, food or chemical poisoning, and some strokes. It is a failure of the circulation, initially to the skin (this explains pallor) and later to the vital organs of kidneys and brain; there is a marked fall in blood pressure.

substantives

A substantive is a word, or a group of words, that functions in its sentence as a **noun.** It may be a noun, a **pronoun,** or a **verbal** (**gerund** or **infinitive**), or it may be a **phrase** or even a **clause** that is used as a noun.

EXAMPLES The *report* is due today. (noun)
We must finish the project on schedule. (pronoun)
Several college graduates applied for the job. (noun phrase)
Drilling is expensive. (gerund)
To succeed will require hard work. (infinitive)
What I think is unimportant. (noun clause)

suffixes

A suffix is a letter or letters added to the end of a word to change its meaning in some way. Suffixes can change the **part of speech** of a word.

EXAMPLES The *wire* is on the truck. (noun)
The *wirelike* tubing is on the truck. (adjective)

What is the *length* of the unit? (noun)
Move the unit *lengthwise* through the conveyor. (adverb)

sweeping generalizations

When the scope of an opinion is unlimited, the opinion is called a *sweeping generalization.* Such statements, though at times tempting to make, should be qualified during **revision.** Consider the following:

EXAMPLES Anyone who succeeds in business today is dishonest.
Engineers are poor writers.

These statements ignore any possibility that someone may have succeeded in business without being dishonest or that there might be an engineer who writes superbly. Moreover, one person's definition of "success" or "dishonesty" or "poor writing" may be different from another's. No matter how certain you are of the general applicability of an opinion, use such all-inclusive terms as *anyone, everyone, no one, all, always, never,* and *in all cases* with caution. Otherwise, your opinions are likely to be judged irresponsible. (See also **logic.**)

syllabication

When a word is too long to fit on a line, the proper place to divide it is often a troublesome question. The most general rule is to divide words between syllables. The following guidelines may be useful:

1. One-syllable words should not be divided.
2. Words with less than six letters should normally not be divided.
3. Fewer than three letters should not be left at the end of a line or carried over to begin a new line.
4. Words with **suffixes** or **prefixes** should be divided at, rather than within, the suffix or prefix.

   CHANGE su-permarket
   TO super-market

5. A hyphenated word should not be divided at the **hyphen** if the hyphen is essential to the meaning of the word. For example, *re-form* (to change the shape of something) and *reform* (to improve something) have different meanings. Thus it could be confusing to divide *re-form* at the hyphen.
6. **Abbreviations** and **contractions** should never be divided.
7. Proper names and company titles should not be divided.

8. The last word of a **paragraph** should not be divided.
9. The last word on a page of a typewritten manuscript should not be divided.

Secretarial handbooks with **spellings** and word divisions and **dictionaries** that indicate syllable breaks can be helpful. A handy guide is *Webster's Instant Word Guide* (Springfield, Mass.: G. & C. Merriam Company, 1980).

symbols

From highway signs to mathematical equations, people communicate in written symbols. When a symbol seems appropriate in your writing, either be certain that your **reader** understands its meaning or place an explanation in **parentheses** following the symbol the first time it appears. However, never use a symbol when your reader would more readily understand the full term. The following is a list of symbols and their appropriate uses.

Symbol	*Meaning and Use*
£	pound (basic unit of currency in the United Kingdom)
$	dollar (basic unit of currency in the United States)
O	oxygen (For a listing of all symbols for chemical elements, see a periodic table of elements in a **dictionary** or handbook.)
$+$	plus
$-$	minus
$\pm$	plus or minus
$\mp$	minus or plus
$\times$	multiplied by
$\div$	divided by
$=$	equal to
$\neq$ or $\not=$	not equal to
$\approx$ or $\doteq$	approximately (or nearly equal to)
$\equiv$	identical with
$\not\equiv$	not identical with
$>$	greater than
$\not>$	not greater than
$<$	less than
$\not<$	not less than

Symbol	*Meaning and Use*
:	is to (or ratio)
≐	approaches (but does not reach equality with)
‖	parallel
⊥	perpendicular
√	square root
∛	cube root
∞	infinity
π	*pi*
∴	therefore (in **mathematical equations**)
∵	**because** (in mathematical equations)
( )	parentheses (See also **punctuation.**)
[ ]	brackets (See also **punctuation.**)
{ }	braces (used to group two or more lines of writing, to group figures in **tables** and to enclose figures in mathematical equations)
°F or °C	degree (Fahrenheit or Celsius)
′	minute *or* foot
″	second *or* inch
#	number
*	asterisk (used to indicate a **footnote** when there are very few)
&	**ampersand**
♂	**male**
♀	**female**
©	**copyright**
%	**percent**
℅	in care of
a/o	account of
@	at (used in **tables,** but never in writing)
´	acute (accent mark in French and other languages)
`	grave (accent mark in French and other languages)
ˆ	circumflex (accent mark in French and other languages)
˜	tilde (**diacritical mark** identifying the palatal nasal in Spanish and Portuguese)
¯	macron (marks a long phonetic sound, as in *cāke*)
˘	breve (marks a short phonetic sound, as in *brăcket*)
¨	dieresis or umlaut (mark placed over the second of two

Symbol	*Meaning and Use*
	consecutive vowels indicating that the sound is to be pronounced—*coöperate*)
	cedilla (mark placed beneath the letter *c* in French, Portuguese, and Spanish to indicate the letter is pronounced as *s*—*garçon*)
^	caret (a proofreader's mark used to indicate inserted material)
FR	franc (basic unit of currency in France)
Mex $	peso (basic unit of currency in Mexico)
$	peso (Philippine peso)
R	ruble (basic monetary unit of the U.S.S.R.)
¥	yen (basic unit of currency in Japan)

(See also **abbreviations** and **proofreader's marks.**)

synonyms

A synonym is a word that means nearly the same thing as another word does.

EXAMPLES purchase, acquire, buy;
seller, vendor, supplier

The **dictionary** definitions of synonyms are usually identical; the **connotations** of such words, however, may differ. (A *seller* may be the same thing as a *supplier,* but the term *supplier* does not suggest a commercial transaction as strongly as *seller* does.)

Do not try to impress your **reader** by finding fancy or obscure synonyms in a **thesaurus;** the result is likely to be **affectation.** (See also **connotation/denotation** and **antonyms.**)

syntax

Syntax refers to the way that words, **phrases,** and **clauses** are combined to form sentences. In English, the most common structure is the **subject-verb-object** pattern. (For more information about the word order of sentences, see **sentence construction, sentence faults, sentence fragments, sentence types,** and **sentence variety.**)

T

table of contents

A table of contents is a list of **heads** in a **report** or chapters in a book. It lists them in their order of appearance and cites their page numbers. Appearing at the front of a work, a table of contents permits **readers** to preview what is in the work and assess the work's value to them. It also aids readers who may want to read only certain sections.

The length of your report should determine whether it needs a table of contents. If it does, use the major **heads** and subheads of your **outline** to create your table of contents, as shown in Figure 1 (p. 660).

To punctuate a table of contents, see "leaders" in **periods**, and for guidance on the placement of the table of contents in a report, see **formal reports.**

tables

A table is useful for showing large numbers of specific, related facts or statistics in a small space. A table can present **data** more concisely than the text can, and it is more accurate than graphic presentations are because it provides numerous facts that a **graph** cannot convey. A table facilitates **comparisons** among figures because of the arrangement of the figures into rows and columns. Overall trends about the information, however, are more easily seen in charts and graphs. But do not rely on a table (or any **illustration**) as the only method of conveying significant information. (See Figure 1, on page 661.)

GUIDELINES FOR CREATING TABLES

Table Number. If you are using several tables, assign each a number, and then center the number and title above the table. The numbers are usually arabic, and they should be assigned sequentially to the tables throughout the text. Tables should be referred to in the text by table number rather than by direction ("Table 4" rather than "the above table"). If there are more than five tables in your **report** or paper, give them a separate heading ("List of Tables"), and list them by title and page number, together with any figure

TABLE OF CONTENTS

iii

Figure 1 Sample Table of Contents

Table number — Caption — Boxhead — Column captions — Stub — Body — Rule — Footnote — Source line

Table 1 Recreational fresh-water angling by water-body type and geographical region*

Geographical Regions	Reservoirs	Man-Made Ponds	Natural Lakes & Ponds	Rivers & Streams	Farm Ponds
New England	130	40	570	410	410
Middle Atlantic	710	290	780	1200	630
East North Central	1200	760	3100	1600	1300
West North Central	810	550	1200	970	980
South Atlantic	1100	760	640	1500	1600
East South Central	890	630	190	670	1200
West South Central	1700	610	430	880	1300
Mountain	820	50	280	600	230
Pacific	950	200	820	1400	470
Totals	8300	3900	8000	9200	7800

*In thousands of anglers. Anglers who fished in more than one water body or region are represented in more than one category.

Source: U.S. Department of the Interior

Figure 1 Table of Data

numbers, on a separate page immediately after the **table of contents.** The first text reference to the table should *precede* the table. If a document contains several chapters or sections, tables can be numbered by chapter or section (Table 1-1, 1-2, . . . 3-1, 3-2).

Title. The title, which is placed just above the table, should describe concisely what the table represents.

Boxhead. The boxhead carries the column headings. These should be kept concise but descriptive. Units of measurement, where necessary, should be specified either as part of the heading or enclosed in **parentheses** beneath the heading. Avoid vertical lettering if possible.

Stub. The left-hand vertical column of a table is the stub. It lists the items about which information is given in the body of the table.

Body. The body comprises the data below the boxhead and to the right of the stub. Within the body, columns should be arranged so that the terms to be compared appear in adjacent rows and columns. Leaders are sometimes used between figures to aid the eye in following data from column to column. Where no information exists for a specific item, leave an empty space to acknowledge the gap.

Rules. These are the lines that separate the table into its various parts. Horizontal lines are placed below the title, below the body of the table, and between the column headings and the body of the table. They should not be closed at the sides. The columns within the table may be separated by vertical lines if they aid clarity.

Footnotes. Footnotes are used for explanations of individual items in the table. Symbols (*, #) or lowercase letters, rather than numbers, are ordinarily used to key table footnotes, because numbers might be mistaken for the data in a numerical table.

Source Line. The source line, which identifies where the data were obtained, appears below any footnotes (when a source line is appropriate).

Continued Lines. When a table must be divided so that it can be continued on another page, repeat the boxhead and give the table number at the head of each new page with a "continued" label ("Table 3, continued").

technical information letters

The letter of technical information is, in effect, a small technical **report**. It may take the form of a **memorandum** for use within your firm, or it may be a letter or a report to another firm.

Since this letter or memo is an abbreviated report, follow the appropriate steps of the writing process as you compose it. It is most helpful to your **reader** if you restate the problem in the letter or memo. Above all, be as clear and to the point as possible.

Consider Letter 1 (p. 664) which addresses an updating problem peculiar to a specific computer. The writer spells out the problem in the introduction, gives the background of the problem, and then goes on to suggest two possible solutions to it.

technical manuals

Technical manuals help technical specialists and customers use and maintain products, from word processors purchased by general consumers to multimillion-dollar aircraft or sophisticated mainframe computer systems purchased by manufacturers, businesses, and governments. Technical manuals are normally written by professional technical writers, although in smaller companies, engineers and technicians may write them.

Today, many manufacturers consider good technical manuals to be an important marketing tool; if a manual makes equipment easy to operate or repair, consumers are more likely to buy it. The following list describes typical technical manuals and the **readers** and **objectives** they serve.

User Manuals

User manuals are aimed at skilled or unskilled consumers of such equipment as word processors. For example, in the case of word processors, these manuals enable users to set up, operate, and care for both the hardware and software of their systems.

Tutorials

Tutorials are self-study, workbook-like teaching guides for users of a product or system. Either packaged with user manuals or published separately, tutorials walk the novice user through the operation of the product or system. Both user manuals and tutorials are written for home and office applications.

Training Manuals

Training manuals are textbooks used in the classroom training of individuals in some procedure or skill, such as flying an airplane or processing an insurance claim. In many technical and vocational

Parkside Office Machines Co.
123 Oceanview Drive
Seattle, WA 98002

April 2, 19--

Ms. Judith Sparks
Technical Director
Components Division
Parkside Office Machines Co.
Pines, NJ 04113

Dear Judy:

We've discovered a problem with our current block size restriction for the Decade 2000 computer. Many of our internal and customer publications on utility routines state that the Decade 2000 computer is normally restricted to a maximum block length of 4023 characters, but that this restriction can be lifted by setting Flag 20. This statement is misleading, and I am beginning to wonder if we should take a different approach. As I understand it, the problem is a hardware restriction in that the processor was wired to accommodate a maximum block length of 4023 characters. This restriction created a problem for FRAN users because a FRAN track could accept a block of 4623 characters. The solution was to make a wiring change to make use of the first bit in the second "TA" character of control words, thereby doubling the maximum block length to 8046. This change was made about two years ago, but existing systems were not modified. Consequently, if a user who has one of the old systems sets Flag 20, the restriction is not going to be lifted as we say it will.

A couple of solutions quickly come to mind. We could instruct the user to set the Flag only if his system can accommodate an 8046-character block (if we assume that he has this information). We could give him the change number of the update to the processor and instruct him to use Flag 20 only if his processor has that number or greater. I propose that we use the second solution because the user is more likely to know his change number than whether he can accommodate an 8046-character block.

Since the customer is directly involved, the solution to the problem should probably be approved by your department. Please let me know as quickly as possible whether you approve of this solution because some of our customer manuals are presently being revised.

Sincerely yours,

Gerald Stein

Gerald Stein

GS/bb

Letter 1 Technical Information Letter

fields, they are the primary teaching device. Training manuals are often accompanied by slides, tapes, or other audiovisual material.

Operators' Manuals

Written for skilled operators of construction, manufacturing, computer, or military equipment, operators' manuals contain essential instructions and safety warnings.

Service Manuals

Service manuals help trained technicians repair equipment or systems, usually at the customer's location. Such manuals often contain elaborate troubleshooting guides for locating technical problems.

Maintenance Manuals

Maintenance manuals are written for semiskilled operators or technicians who must keep such equipment as aircraft and manufacturing machinery operating efficiently.

Repair Manuals

Repair manuals guide skilled technicians as they perform extensive repairs or rebuild equipment, usually at the manufacturing site. Such manuals contain detailed illustrations and sophisticated technical information, sometimes including the theory or operation for highly technical electronic and computer equipment.

Special-Purpose Manuals

Programmer reference manuals provide computer programmers with definitions, syntax rules, and other technical information about specific programming languages, such as FORTRAN and COBOL.

Overhaul manuals are guides for rebuilding items at the factory. Overhaul manuals are often used in the heavy-equipment industry, in which rebuilding a huge crane, for example, is much less costly than manufacturing a new one.

Handling and setup manuals are guides for skilled and unskilled employees in the safe handling, installation, or setup of various kinds of equipment.

Safety manuals are guides for operators of potentially dangerous equipment, illustrating safe operating practices.

Some manuals, of course, are combinations or variations of those described.

Technical manuals often use extensive **illustrations**, such as exploded-view **drawings, flowcharts, photographs,** and **tables**. They include such sections as parts lists, troubleshooting guides, warning statements, the standard symbols for potential dangers, and **indexes**.

The packaging of manuals is important. Although they may be either bound or looseleaf, manuals are more often looseleaf to enable easy revision and updating. They are sized and packaged for easy use. A word-processing manual, for example, may use a small (7″ × 9″) three-ring binder that can be easily held or that lays flat open on a desk while the user views the screen. Or an aircraft training manual may be spiral bound at the top to allow it to fit neatly in the cockpit of the aircraft.

Writing a technical manual requires careful adherence to the steps outlined in the Checklist of the Writing Process (p. xviii). **Preparation** and **organization** are especially important because of the complexity of the products and systems for which the manuals are written. Before beginning to write a technical manual, give particular care to **outlining**, and review the entries on **format, instructions, process explanation**, and **technical writing style**. These elements are central to writing a manual, regardless of its purpose and **scope**. Above all, pay particular attention to the readers' needs and their level of technical knowledge of the product or system you are documenting.

To ensure that manuals are helpful and accurate, you should submit them to technical, legal, and peer reviews. Some companies test the manuals by using prototypes with typical manual readers. (See also **proofreading**.)

technical writing style

Technical writing style is standard **exposition** in which the **tone** is objective, with the author's voice taking a back seat to the subject matter. Because the focus is on helping the **reader** use a device or understand a process or concept, the language is utilitarian—emphasizing exactness rather than elegance for its own sake. Thus the writing is usually not adorned with figurative language, except when a **figure of speech** would facilitate understanding. Technical writing requires an effective **introduction** or **opening**, good **organization**, and **sentence variety** and benefits greatly from **heads** and **illustrations**.

Good technical writing avoids overusing the passive **voice**. Its vocabulary is appropriately technical, although the general word is preferable to the technical word. **Gobbledygook** and **jargon** are poor substitutes for clear and direct writing. Do not use a big or technical term merely because you know it—make sure that your readers also understand it. (See also **affectation**.)

Technical writing is direct and often is aimed at readers who are not experts in the subject—such as consumers needing to learn how to operate unfamiliar equipment or managers trained in business rather than technical methods. Figure 1 (p. 668) is an excerpt from a manual that teaches customers how to operate a word-processing system. As this opening for a **technical manual** illustrates, its **format** (or the page design) is important because it keeps the reader oriented and it helps make the writing readable. (See also **style** and **formal writing style**.)

telegraphic style

Telegraphic style condenses writing by omitting **articles, pronouns, conjunctions,** and transitional expressions. Although **conciseness** is important in writing, writers sometimes make their sentences too brief by omitting these words. Telegraphic style forces the reader to mentally supply the missing words. Compare the following two passages, and notice how much easier the revised version reads (the added words are italicized).

CHANGE Take following action when treating serious burn. Remove loose clothing on or near burn. Cover injury with clean dressing and wash area around burn. Secure dressing with tape. Separate fingers/toes with gauze/cloth to prevent sticking. Do not apply medication unless doctor prescribes.

TO Take *the* following action when treating *a* serious burn. Remove *any* loose clothing on or near *the* burn. Cover *the* injury with *a* clean dressing, and wash *the* area around *the* burn. *Then* secure *the* dressing with tape. Separate fingers *or* toes with gauze *or* cloth to prevent *them from* sticking *together*. Do not apply medication unless *a* doctor prescribes *it*.

Telegraphic style can also produce **ambiguity,** as the following example demonstrates:

CHANGE Grasp knob and adjust lever before raising boom.

Does this sentence mean that the reader should *adjust the lever* or *grasp an adjust lever?*

INTRODUCTION

You will soon see that the Alpha operating with the word-processing software, Alphaword, is an easy system to operate. By following the instructions, you will learn to create, edit, save, and print a document in a very short time.

As you read the instructions, perform each step as it is presented.

GETTING STARTED

You are ready to start word processing. In this section you will learn to:

Load the FirstWord software.
Initialize a disk.
Copy a disk.
Create your first document.
Edit your first document.
Store your first document on disk.
Print your first document.

LOADING FIRSTWORD

1. Remove the paper envelope from the FirstWord disk. Note that disk still remains mostly covered except for a small slot on each side. Avoid touching this exposed area.
2. Turn the disk so that the label is up and toward you and the exposed area of the disk is away from you. Gently slide the disk into Drive A (left side) and snap down the latch, as shown below.

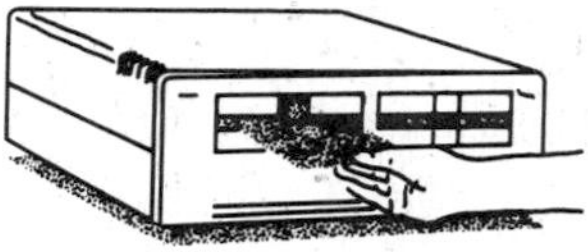

3. Turn on the screen unit.
4. Turn on the disk unit.

After a short delay while internal diagnostics verify the proper operation of the hardware, the FirstWord software is loaded into memory, and a request for the date and time is displayed.

Figure 1 Example of Technical Writing

TO Grasp the knob and the adjust lever before raising the boom.
OR Grasp the knob and adjust the lever before raising the boom.

Some writers excuse telegraphic style with the claim that their readers will "understand." Maybe so, but they will probably have to work hard to do so—and may therefore even misunderstand. Though telegraphic writing can save space, it never really saves time. As a writer, remember that although you may save yourself work by writing telegraphically, your readers will have to work that much harder to read—or decipher—your writing. (See also **transitions.**)

tenant/tenet

A *tenant* is one who holds or temporarily occupies a property owned by another person.

EXAMPLE The *tenants* were upset by the increased rent.

A *tenet* is an opinion or belief held by a person or an organization.

EXAMPLE The idea that competition will produce adequate goods and services for a society is a central *tenet* of capitalism.

tense

Tense is the grammatical term for **verb** forms that indicate time distinctions. There are six tenses in English: past, past perfect, present, present perfect, future, and future perfect. Each of these has a corresponding progressive form.

Tense	*Basic*	*Progressive*
Past	I began	I was beginning
Past Perfect	I had begun	I had been beginning
Present	I begin	I am beginning
Present Perfect	I have begun	I have been beginning
Future	I will begin	I will be beginning
Future Perfect	I will have begun	I will have been beginning

Perfect tenses allow you to express a prior action or condition that continues in a present, past, or future time.

EXAMPLES I *have begun* to write the annual report, and I will work on it for the rest of the month. (present perfect)

I *had begun* to read the manual when the lights went out. (past perfect)
I *will have begun* this project by the time funds are allocated. (future perfect)

Note from the table that the progressive forms are created by combining the **helping verb** *be,* in the appropriate tenses, with the present **participle** (*-ing*) form of the main verb.

PAST TENSE

The simple past tense indicates that an action took place in its entirety in the past. The past tense is usually formed by adding *-d* or *-ed* to the root form of the verb.

EXAMPLE We *closed* the office early yesterday.

PAST PERFECT TENSE

The past perfect tense indicates that one past event preceded another. It is formed by combining the helping verb *had* with the past participle form of the main verb.

EXAMPLE He *had finished* by the time I arrived.

PRESENT TENSE

The simple present tense represents action occurring in the present, without any indication of time duration.

EXAMPLE I *use* the beaker.

A general truth is always expressed in the present tense.

EXAMPLE He learned that the saying "time *heals* all wounds" is true.

The present tense can be used to present actions or conditions that have no time restrictions.

EXAMPLE Water *boils* at 212°F.

The present tense can be used to indicate habitual action.

EXAMPLE I *pass* the paint shop on the way to my department every day.

The present tense can be used as the "historical present" to make things that occurred in the past more vivid.

EXAMPLE He *asks* for more information on production statistics and *receives* a detailed report on every product manufactured by the company. Then he *asks,* "Is each department manned at full strength?" In his office, surrounded by his staff, he *goes* over the figures and *plans* for the coming year.

PRESENT PERFECT TENSE

The present perfect tense describes something from the recent past that has a bearing on the present—a period of time before the present but after the simple past. The present perfect tense is formed by combining a form of the helping verb *have* with the past principle form of the main verb.

EXAMPLES He *has retired,* but he visits the office frequently.
We *have finished* the draft and are ready to begin revising it.

FUTURE TENSE

The simple future tense indicates a time that will occur after the present. It uses the helping verb *will* (or *shall*) plus the main verb.

EXAMPLE I *will finish* the job tomorrow.

Do not use the future tense needlessly.

CHANGE This system *will be* explained on page 3.
TO This system *is* explained on page 3.

CHANGE When you press this button, the card *will be* moved under the read station.
TO When you press this button, the card *is* moved under the read station.

FUTURE PERFECT TENSE

The future perfect tense indicates action that will have been completed at a future time. It is formed by linking the helping verbs *will have* to the past participle form of the main verb.

EXAMPLE He *will have driven* the test car 40 miles by the time he returns.

TENSE AGREEMENT OF VERBS

The verb of a subordinate **clause** should usually agree in tense with the verb of the main clause. (See also **agreement.**)

EXAMPLES When the supervisor *presses* the starter button, the assembly line *begins* to move.

When the supervisor *pressed* the starter button, the assembly line *began* to move.

Shift in Tense. Be consistent. The only legitimate shift in tense records a real change in time. When you choose a tense in telling a story or discussing an idea, stay with that tense. Illogical shifts in tense will only confuse your **reader.**

CHANGE Before he *installed* the printed circuit, the technician *cleans* the contacts.

TO Before he *installed* the printed circuit, the technician *cleaned* the contacts.

test reports

The *test report* differs from the more formal **laboratory report** in both size and **scope**. Considerably smaller and less formal than the laboratory report, the test report can be a **memorandum** or a formal business letter, depending on its recipient. Either way, the report should have a subject line at the beginning to identify the test being discussed.

The **opening** of a test report should state the test's purpose, unless it is obvious. The body of the report presents the data. A report on the tensile strength of metal, for example, includes the readings from the test equipment. If the procedure used to conduct the test would interest the **reader**, it should be described. The results of the test should be stated and, if necessary, interpreted. Often there is reason to discuss their significance. The report should conclude with any recommendations made as a result of the test.

The test report in Figure 1 does not explain how the tests were performed because such an explanation is unnecessary. The example in Figure 2 (p. 674), however, does explain this.

that

Avoid the unnecessary repetition of *that.*

CHANGE I think *that* when this project is finished *that* you should write the results.

TO I think *that* you should write the results when the project is finished.

OR Write the results when the project is finished.

Biospherics, Inc.
4928 Wyaconda Road
Rockville, MD 20852

March 14, 19--

Mr. John Sebastiani, General Manager
Midtown Development Corporation
114 West Jefferson Street
Milwaukee, WI 53201

Subject: Results of Analysis of Soil Samples for Arsenic

Dear Mr. Sebastiani:

Following are the results of the analysis of 22 soil samples for arsenic. The arsenic values listed are based on a wet-weight determination. The moisture content of the soil is also given to allow conversion of the results to a dry-weight basis if desired.

Hole Number	Depth	Percent of Moisture	Arsenic Total As ppm
1	12"	19.0	312.0
2	Surface	11.8	737.0
3	12"	12.7	9.5
4	12"	10.8	865.0
5	12"	17.1	4.1
6	12"	14.2	6.1
7	12"	24.2	2540.0
8	Surface	13.6	460.0

I noticed that some of the samples contained large amounts of metallic iron coated with rust. Arsenic tends to be absorbed into soils high in iron, aluminum, and calcium oxides. The large amount of iron present in some of these soil samples is probably responsible for retaining high levels of arsenic. The soils highest in iron, aluminum, and calcium oxides should also show the highest levels of arsenic, provided the soils have had approximately equal levels of arsenic exposure.

If I can be of further assistance, please do not hesitate to contact me.

Yours truly,

Gunther Gottfried

Gunther Gottfried
Chemist

GG/jrm

Figure 1 *Test Report*

Biospherics, Inc.
4928 Wyaconda Road
Rockville, MD 20852

September 9, 19--

Mr. Leon Hite, Administrator
The Angle Company, Inc.
1869 Slauson Boulevard
Waynesville, VA 23927

Dear Mr. Hite:

On Tuesday, 30 August, Biospherics, Inc., performed asbestos-in-air monitoring at your Route 66 construction site, near Front Royal, Virginia. Six persons and three construction areas were monitored.

All monitoring and analyses were performed in accordance with "Occupational Exposure to Asbestos," U.S. Department of Health, Education and Welfare, Public Health Service, National Institute for Occupational Safety and Health, 1972. Each worker or area was fitted with a battery-powered personal sampler pump operating at a flow rate of approximately one liter per minute. We collected the airborne asbestos on a 37-mm Millipore type AA filter mounted in an open-face filter holder over an 8-hour period.

We mounted a wedge-shaped piece of each filter on a microscope slide with a drop of 1:1 solution of dimethyle phthalate and diethyl oxalate and then covered it with a cover clip. We counted samples within 24 hours after mounting, using a microscope with phase contrast option.

In all cases, the workers and areas monitored were exposed to levels of asbestos fibers well below the NIOSH standard. The highest exposure we found was that of a driller who was exposed to 0.21 fibers per cubic centimeter. We analyzed the driller's sample by scanning electron microscopy followed by energy dispersive X-ray techniques which identify the chemical nature of each fiber, thereby, verifying the fibers as asbestos or identifying them as other fiber types. Results from these analyses show that the fibers present are tremolite asbestos. We found no nonasbestos fibers.

If you have any questions about the tests, please call or write me.

Yours truly,

Gary Willis

Gary Willis
Chemist

GW/jrm

Figure 2 Test Report

CHANGE You will note *that* as you assume greater responsibility and as your years of service with the company increase, *that* your benefits will increase accordingly.

TO You will note *that* as you assume greater responsibility and as your years of service with the company increase, your benefits will increase accordingly.

OR You will note *that* your benefits will increase as you assume greater responsibility and as your years of service with the company increase.

However, do not delete *that* from a sentence in which it is necessary for the **reader's** understanding.

CHANGE Quarreling means trying to show the other person is in the wrong.

TO Quarreling means trying to show *that* the other person is in the wrong.

CHANGE Some engineers fail to recognize sufficiently the human beings who operate the equipment constitute an important safety system.

TO Some engineers fail to recognize sufficiently *that* the human beings who operate the equipment constitute an important safety system.

(See also **conciseness/wordiness.**)

that/which/who

Who refers to persons, whereas *that* and *which* refer to animals and things.

EXAMPLES John Brown, *who* is retiring tomorrow, has worked for the company for twenty years.

The test animal *that* was given the injection responded exactly as anticipated.

The jet stream, *which* is approximately 8 miles above the earth, blows at an average of 64 miles per hour from the west.

That is often overused. (See **that.**) However, do not eliminate it if to do so would cause **ambiguity** or problems with **pace.**

CHANGE On the file specifications input to the compiler for any chained file, the user must ensure the number of sectors per main file section is a multiple of the number of sectors per bucket.

TO On the file specifications input to the compiler for any chained file, the user must ensure *that* the number of sectors per main file section is a multiple of the number of sectors per bucket.

Which, rather than *that,* should be used with nonrestrictive **clauses** (clauses that do not change the meaning of the basic sentence).

EXAMPLES After John left the restaurant, *which* is one of the best in New York, he came directly to my office. (nonrestrictive)
A company *that* diversifies often succeeds. (restrictive)

(See also **who/whom, relative pronoun,** and **restrictive and nonrestrictive elements.**)

there/their/they're

There, their, and *they're* are often confused because they sound alike. *There* is an **expletive** or an **adverb.**

EXAMPLES *There* were more than 1,500 people at the conference. (expletive)
More than 1,500 people were *there.* (adverb)

Their is the **possessive** form of *they.*

EXAMPLE Our employees are expected to keep *their* desks neat.

They're is a **contraction** of *they are.*

EXAMPLE If *they're* right, we should change the design.

thesaurus

A thesaurus is a book of words with their **synonyms** and **antonyms**, arranged by categories. Thoughtfully used, it can help you with **word choice** during the **revision** phase of the writing process. However, this variety of words may tempt you to choose inappropriate or obscure synonyms just because they are available. Use a thesaurus only to clarify your meaning, not to impress your **reader.** Never use a word unless you are sure of its meanings; **connotations** of the word that might be unknown to you could mislead your reader. (See "synonyms" in **reference books** for the names of thesauruses, and also see **affection** and **defining terms**.)

thus/thusly

Thus is an **adverb** meaning "in this manner" or "therefore." The *-ly* in *thusly* is superfluous and should be omitted.

CHANGE The committee's work is done. *Thusly* we should have the report by the end of the week.
TO The committee's work is done. *Thus* we should have the report by the end of the week.

'til/until

'Til is a nonstandard **spelling** of *until,* meaning "up to the time of." Use *until.*

CHANGE We worked *'til* eight o'clock.
TO We worked *until* eight o'clock.

titles

The title of a document should indicate its topic and announce its **scope** and **objective.** Often the title is the only basis on which readers can decide whether they should read something. Titles are also increasingly used by information specialists as a source of key terms by which documents can be indexed for automated information storage and retrieval systems. Imprecise titles thus defeat human and automated information retrieval.

To create accurate titles, keep the following guidelines in mind:

1. Be specific. Do not use general or vague terms when specific ones will pinpoint the topic and its scope. Although the title "Electric Fields and Living Organisms" announces the topic of a report, it leaves the reader with important unanswered questions: What is the relationship between electric fields and living organisms? What kind of organisms does it refer to? What is the intensity of the electric field? What stage in the organism's life cycle does it discuss? The addition of details identifies not only the topic but its scope and objective as well: *Effects of 60-Hz Electric Fields on Embryo and Chick Development, Growth, and Behavior.*

2. Be concise, but don't make your title short just for the sake of being short. As the example above indicates, making a title accurate may mean adding a few terms. Conciseness, instead, means eliminating words that do not contribute to accuracy. Avoid titles that begin "Notes on . . .," "Studies on . . .," "A Report on . . .," or "Observations on. . . ." These notions are self-evident to the reader. On the other hand, certain works, like **annual reports** or **feasibility studies** should be identified as such in the title because this information helps define the purpose and scope.
3. Do not indicate dates in the title of a periodic or progress report, but put them in a subtitle:

   The Effects of Acid Rain on Red Spruce
   in the White Mountains
   Quarterly Report
   January–March 1987

4. Avoid using **abbreviations, acronyms and initialisms,** chemical formulas, and the like *unless* the work is addressed exclusively to specialists in the field.
5. Do not put titles in sentence form.

   CHANGE How Residential Passive Solar Heating Could Affect Seven Electric Utilities
   TO Potential Effects of Residential Passive Solar Heating on Seven Utilities

6. For multivolume publications, repeat the title on each volume. The distinction among the volumes is made by the volume number (Volume 1, 2, 3, and so on), and sometimes each volume will have a different subtitle. For example, Volume 1 could be an **executive summary**; Volume 2 could be the main report; and any subsequent volumes could be lengthy **appendixes.**

(For guidelines on how to capitalize titles and when to use italics and quotation marks, see **capital letters, italics,** and **quotation marks.**)

to/too/two

To, too, and *two* are confused only because they sound alike. *To* is used as a **preposition** or to mark an **infinitive.**

EXAMPLES Send the report *to* the district manager. (preposition)
I wish *to* go. (mark of the infinitive)

Too is an **adverb** meaning "excessively" or "also."

EXAMPLES The price was *too* high. ("excessively")
I, *too,* thought it was high. ("also")

Two is a number.

EXAMPLE Only *two* buildings have been built this fiscal year.

tone

In writing, tone is the writer's attitude toward the subject and his or her **readers.** The tone may be casual or serious, enthusiastic or skeptical, friendly or hostile. In technical writing, the tone in **correspondence, memorandums,** and **formal reports** may range widely—depending on the **objective,** situation, and context. For example, in a memo read only by an associate who is also a friend, your tone might be casual and friendly.

EXAMPLE I think your proposal to Smith and Sons is great. If we get the contract, I owe you a lunch! I've marked a couple of places where we could cover ourselves on the schedule. See what you think.

In a memo to a superior, however, your tone might be quite different.

EXAMPLE I think your proposal to Smith and Sons is excellent. I have marked a couple of places for your consideration where we could ensure that we are not committing ourselves to a schedule we might not be able to keep. If I can help in any other way, please let me know.

In a memo that serves as a **report** to numerous readers, the tone would again be different.

EXAMPLE The Smith and Sons proposal appears complete and thorough, based on our department's evaluation. Several small revisions, however, would ensure that Acme is not committing itself to an unrealistic schedule. These are marked on the copy of the report being circulated.

Your tone will be set by many factors in your writing. As illustrated above, a formal writing **style** will normally have a different tone than will an informal writing style. (See also **technical writing style.**) **Word choice,** the **introduction** or **opening,** and even your **title** all contribute to the overall tone. For instance, a title such as "Some Observations on the Diminishing Oil Reserves in Wyoming" clearly sets a tone quite different from that of "What Happens When We've Pumped Wyoming Dry?" The first title would most likely be appropriate for a report; the second title would be more appropriate for a **newsletter article.** The important thing is to make sure that your tone is the one best suited to your objective, by always keeping your reader in mind.

In correspondence, tone is particularly important because the letter represents a direct communication between two people. Furthermore, good business letters establish a rapport between your organization and the public, and a positive and considerate tone is essential.

topics

On the job, the topic of a writing project is usually determined by need. In a college writing course, on the other hand, you may have to select your own topic. If you do, keep the following points in mind:

1. Select a topic that interests you.
2. Select a topic that you can **research** adequately with the facilities available to you.
3. Limit your topic so that its **scope** is small enough to handle within the time you are given. A topic like "Air Pollution," for example, would be too broad. On the other hand, keep the topic broad enough so that you will have enough to write about.
4. Select a topic for which you can make adequate **preparation** to ensure a good final **report.**

topic sentences

A topic sentence states the controlling idea of a **paragraph;** the rest of the paragraph supports and develops that statement with carefully related details. The topic sentence may appear anywhere in the paragraph—which permits the writer variety in **style**—and a topic statement may also be more than one sentence if necessary.

The topic sentence is often the first sentence because it states the subject.

> *The arithmetic of searching for oil is stark.* For all his scientific methods of detection, the only way the oil driller can actually know for sure that there is oil in the ground is to drill a well. The average cost of drilling an oil well is over $300,000, and drilling a single well may cost over $8,000,000! And once the well is drilled, the odds against its containing any oil at all are 8 to 1! Even after a field has been discovered, one out of every four holes drilled in developing the field is a dry hole because of the uncertainty of defining the limits of the producing formation. The oil driller can never know what Mark Twain once called "the calm confidence of a Christian with four aces in his hand."

On rare occasions, the topic sentence logically falls in the middle of a paragraph.

> It is perhaps natural that psychologists should awaken only slowly to the possibility that behavioral processes may be directly observed, or that they should only gradually put the older statistical and theoretical techniques in their proper perspective. But it is time to insist that science does not progress by carefully designed steps called "experiments," each of which has a well-defined beginning and end. *Science is a continuous and often a disorderly and accidental process.* We shall not do the young psychologist any favor if we agree to reconstruct our practices to fit the pattern demanded by current scientific methodology. What the statistician means by the design of experiments is design which yields the kind of data to which *his* techniques are applicable. He does not mean the behavior of the scientist in his laboratory devising research for his own immediate and possibly inscrutable purposes.
>
> —B. F. Skinner, "A Case History in Scientific Method," *The American Psychologist* 2 (May 1956), p. 232.

Although the topic sentence is usually most effective early in the paragraph, a paragraph can lead up to the topic sentence; this is sometimes done to achieve **emphasis.** When a topic sentence concludes a paragraph, it can also serve as a summary or **conclusion,** based on the details that were designed to lead up to it.

> Energy does far more than simply make our daily lives more comfortable and convenient. Suppose you wanted to stop—and reverse—the economic progress of this nation. What would be the surest and quickest way to do it? Find a way to cut off the nation's oil resources! Industrial plants would shut down; public utilities would stand idle; all forms of transpor-

tation would halt. The economy would plummet into the abyss of national economic ruin. *Our economy, in short, is energy-based.*
—*The Baker World* (Los Angeles: Baker Oil Tools, 1964), p. 5.

Occasionally, the controlling idea of a paragraph—what would normally be its topic sentence—is not explicitly stated but only implied. Nevertheless, if the paragraph has **unity,** a sentence stating the controlling idea *could* be written, and all the sentences in the paragraph would be found to be related directly to it. However, people writing in business and industry are well advised to make their topic sentences *explicit.*

tortuous/torturous

Tortuous means "marked by twisting and winding."

EXAMPLE The highway through the mountain was *tortuous.*

Torturous means "causing pain, as to punish."

EXAMPLE The prisoner's experience was *torturous.*

toward/towards

Both *toward* and *towards* are acceptable varient spellings of the **preposition** meaning "in the direction of." *Toward* is more common in the United States, and *towards* is more common in Great Britain.

EXAMPLES We walked *toward* the Golden Gate Bridge.
They were headed *towards* London Bridge.

transition

Transition is the means of achieving a smooth flow of ideas from sentence to sentence, **paragraph** to paragraph, and subject to subject. Transition is a two-way indicator of what has been said and what will be said; that is, it provides a means of linking ideas to clarify the relationship between them. This may be done with a word, a **phrase,** a sentence, or even a paragraph. Without the guideposts of transition, **readers** can lose their way. Transition can be quite obvious.

EXAMPLE Having considered the economic feasibility of this alloy as a transformer core, *we turn now* to the problem of inadequate supply.

Or it can be more subtle.

EXAMPLE Even if this alloy is economically feasible as a transformer core, there still remains the problem of inadequate supply.

Either way, you now have your reader's attention fastened on the problem of inadequate supply, exactly what you set out to do.

Certain words and phrases are inherently transitional. Consider the following terms and their functions:

Result: *therefore, as a result, consequently, thus, hence*
Example: *for example, for instance, specifically, as an illustration*
Comparison: *similarly, likewise*
Contrast: *but, yet, still, however, nevertheless, on the other hand*
Addition: *moreover, furthermore, also, too, besides, in addition*
Time: *now, later, meanwhile, since then, after that, before that time*
Sequence: *first, second, third, then, next, finally*

Within a paragraph, such transitional expressions clarify and smooth the movement from idea to idea. Conversely, the lack of transitional devices can make the going bumpy for the reader. Consider first the following passage, which lacks adequate transition:

EXAMPLE People had always hoped to fly. Until 1903 it was only a dream. It was thought by some that human beings were not meant to fly. The Wright brothers launched the world's first heavier-than-air flying machine. The airplane has become a part of our everyday life.

Now read the same passage with words and phrases of transition added (in **italics**), and notice how much more smoothly the thoughts flow:

EXAMPLE People had always hoped to fly, *but* until 1903 it was only a dream. *Before,* it was thought by some that human beings were not meant to fly. *In 1903* the Wright brothers launched the world's first heavier-than-air flying machine. *Now* the airplane has become a part of our everyday life.

And finally, read the same passage with stronger transition provided:

EXAMPLE People had always hoped to fly, but until 1903 it was only a dream. *Before that time,* it was thought by some that human beings were not meant to fly. *However,* in 1903 the Wright brothers launched the world's first heavier-than-air flying machine. *Since then* the airplane has become a part of our everyday life.

If your **organization** and **outline** are good, your transitional needs will be less difficult to satisfy (although they must nonetheless *be* satisfied). Just as the outline is a road map for the writer, transition is a road map for the reader.

TRANSITION BETWEEN SENTENCES

In addition to using transitional words and phrases such as those shown above, the writer may achieve effective transition between sentences by repeating key words or ideas from preceding sentences, by using **pronouns** that refer to antecedents in previous sentences, and by using **parallel structure**—that is, by repeating the pattern of a phrase or **clause.** Consider the following short paragraph, in which all these means are employed. (A few examples are indicated by italics; however, to mark them all would be confusing.)

EXAMPLE Representative of many American university towns is Millville. *This midwestern town,* formerly a *sleepy farming community,* is today the home of a large and bustling *academic community.* Attracting students from all over the Midwest, *this university* has grown very rapidly in the last ten years. *This same decade* has seen a physical expansion of *the campus.* The state, recognizing *this expansion,* has provided additional funds for the acquisition of land adjacent to the university. The *university* has become *Millville's* major industry, generating most of *the town's* income—and, of course, many of *its* problems, too.

Another device for achieving transition is enumeration:

EXAMPLE The recommendation rests upon *three conditions. First,* the department staff must be expanded to a sufficient size to handle the increased work load. *Second,* sufficient time must be provided for the training of the new members of the staff. *Third,* a sufficient number of qualified applicants must be available.

TRANSITION BETWEEN PARAGRAPHS

All the means discussed above for achieving transition between sentences—and especially the repetition of key words or ideas—may also be effective for transition between paragraphs. For paragraphs, however, longer transitional elements are often required. One technique is to use an opening sentence that summarizes the preceding paragraph and then to move ahead to the business of the new paragraph.

EXAMPLE One property of material considered for manufacturing processes is hardness. Hardness is the internal resistance of the material to the forcing apart or closing together of its molecules. Another property is ductility, the characteristic of material that permits it to be drawn into a wire. The smaller the diameter of the wire into which the material can be drawn, the greater the ductility. Material also may possess malleability, the property that makes it capable of being rolled or hammered into thin sheets of various shapes. Engineers, in selecting materials to employ in manufacturing, must consider these properties before deciding on the most desirable for use in production.

The requirements of hardness, ductility, and malleability account for the high cost of such materials. . . .

Ask a question at the end of one paragraph and answer it at the beginning of the next.

EXAMPLE Automation has become an ugly word in the American vocabulary because it has at times displaced some jobs. But the all-important fact that is often overlooked is that it invariably creates many more jobs than it eliminates. (The vast number of people employed in the great American automobile industry as compared with the number of people that had been employed in the harness-and-carriage-making business is a classic example.) Almost always, the jobs that have been eliminated by automation have been menial, unskilled jobs, and those who have been displaced have been forced to increase their skills, which resulted in better and higher-paying jobs for them. *In view of these facts, is automation really bad?*

Certainly automation has made our country the most wealthy and technologically advanced nation the world has ever known. . . .

A purely transitional paragraph may be inserted to aid readability.

EXAMPLE . . . that marred the progress of the company.

There were two other setbacks to the company's fortunes that year that also marked the turning of the tide: the loss of many skilled workers

> *through the Early Retirement Program and the intensification of the devastating rate of inflation.*
> The Early Retirement Program . . .

CHECKING FOR TRANSITION DURING REVISION

Check for **unity** and **coherence** to determine whether the transition is effective. If a paragraph has unity and coherence, the sentences and ideas will be tied together and contribute directly to the subject of the paragraph. Look for places where transition is missing, and add it. Look for places where it is weak, and strengthen it.

transitive verbs (see **verbs**)

transmittal letters

The transmittal letter, also known as a cover letter, identifies the item being sent, the person to whom it is being sent, and the reason for sending it. A transmittal letter provides a permanent record of the transmittal for both the writer and the **reader.**

Keep your remarks brief in a transmittal letter. Open with a **paragraph** explaining what is being sent and why. In an optional second paragraph, you might include a summary of any information you're sending. A letter accompanying a **proposal,** for example, might point out any sections in the proposal of particular interest to the reader. The letter could then go on to present evidence that the writer's firm is the best one to do the job. This paragraph could also mention the conditions under which the material was prepared, such as limitations of time or budget. The closing paragraph should contain acknowledgments, offer additional assistance, or express the hope that the material will fulfill its purpose.

The first of the following examples (Letter 1) is brief and to the point. The second example (page 688) is a bit more detailed, for it touches on the manner in which the information was gathered.

trip reports

Many companies require or encourage reports of the business trips their employees take. A trip report both provides a permanent record of a business trip and its accomplishments and enables many

WATERFORD PAPER PRODUCTS
P.O. BOX 413
WATERFORD, WI 53474

(414) 738-2191

November 12, 19--

Ms. Nancy Melcher
2021 State Street
Racine, WI 53307

Dear Ms. Melcher:

Thank you for your interest in Waterford Paper Products. The enclosed brochure describes our product line and our current prices.

If I can be of future help, please call me at extension 2349.

Sincerely,

Lawrence Smith

Lawrence Smith
Customer Relations

LS/ik
Enclosure

Letter 1 Brief Transmittal Letter

WATERFORD PAPER PRODUCTS
P.O. BOX 413
WATERFORD, WI 53474

(414) 738-2191

January 16, 19--

Mr. Roger Hammersmith
Ecology Systems, Inc.
1015 Clarke Street
Chicago, IL 60615

Dear Mr. Hammersmith:

Enclosed is the report estimating our power consumption for the year as requested by John Brenan, Vice President, on September 4.

The report is a result of several meetings with the Manager of Plant Operations and her staff and an extensive survey of all our employees. The survey was delayed by the temporary layoff of key personnel in Building "A" from October 1 to December 5. We believe, however, that the report will provide the information you need to furnish us with a cost estimate for the installation of your Mark II Energy Saving System.

We would like to thank Diana Biel of ESI for her assistance in preparing the survey. If you need any more information, please let me know.

Sincerely,

James G. Evans
New Projects Office

JGE/fst
Enclosure

Letter 2 Transmittal Letter

employees to benefit from the information that one employee has gained.

A trip report should normally be in the **format** of a **memorandum**, addressed to your immediate superior. On the subject line give the destination and dates of the trip. The body of the **report** will explain why you made the trip, whom you visited, and what you accomplished. The report should devote a brief section to each major event and many include a **head** for each section (you needn't give equal space to each event, but instead, elaborate on the more important events). Follow the body of the report with any appropriate **conclusions** and recommendations. The trip report shown in Figure 1, on pages 690–691, is typical.

trite language

Trite language is made up of words, **phrases**, or ideas that have been used so often that they are stale.

> CHANGE *It may interest you to know* that all the folks in the branch office are *hale and hearty.* I should finish my report *quick as a wink,* and we should *clean up* on it.
>
> TO Everyone here at the branch office is well. I should finish my project within a week, and I'm sure it will prove profitable for us.

Trite language shows that the writer is thoughtless in his or her **word choice** or is not thinking carefully about the **topic** but is relying instead on what others have thought and said. (See also **affectation.**)

trouble reports

The trouble report is used to report an accident, an equipment failure, an unplanned work stoppage, and so on. The **report** enables the management of an organization to determine the cause of the problem and to make any changes necessary to prevent its recurrence. The trouble report normally follows a simple **format** (and is often a **memorandum**), since it is an internal document and is not large enough in either size or **scope** to require the format of a **formal report.**

In the subject line of the memorandum, state the precise problem you are reporting. Then begin your description of the problem in the body of your report. What happened? Where did it occur?

MEMORANDUM

Date: February 6, 19--
TO: J.L. Watson
FROM: T.R. Santow TRS
SUBJECT: Trip to Milwaukee Division the Week of January 17, 19--

The purpose of this trip was to gather information on (1) operating the new disc in a single-spindle environment, (2) system scheduling, (3) data base, (4) indexed sequential, (5) the small inquiry system, (6) random filing system, and (7) the roll-in/roll-out software.

<u>Disc</u>

I talked with Kevin Hutch, Janet Martin, and Gerald MacDougal about the need for a separate card control string to bring the new disc up in a single-spindle mode. I also discussed the possibility of altering the software so that it will search for a program on the current system disc and then on the library disc in the single-spindle mode, and the possibility of identifying a library disc on the PAL printout.

I observed the disc in operation when Ralph Stevens ran some test routines. It appeared to run faster than the same routines run on the old disc. It soon became apparent, as I watched the tests being run, that the speed of the actuator and the rotational speed of the disc make the operating environment considerably different from what anyone had anticipated. Timing tests will have to be run on the relationship between file positions and throughput before we can write our file placement document.

<u>System Scheduling</u>

I talked to Bill Wilson about S7 System Scheduling. At the conclusion of our discussion, he agreed to send someone to Dayton to work with us when all the necessary information is available. I also talked to Bill about the interim publication/pilot site relationship; he felt that we should have our interim publications, as well as the writer, at the key pilot sites so that documentation can be tested as well as the software and the hardware.

<u>Data Base</u>

I discussed Data Base with Wayne Sewell and Ellen Golden to get a feel for the current status of the project, as well as for its progress, to determine when we might expect to become involved. However, the project does not seem to be far enough along yet for us to be concerned.

Figure 1 Trip Report

Indexed Sequential

I discussed Indexed Sequential with Margaret Miller; she had the information I needed, but informed me that Indexed Sequential will not be her responsibility in the future. Therefore, I tried to determine new assignments of responsibility in Milwaukee.

Small Inquiry System and Random Filing System

I discussed the Small Inquiry System with Wally Bevins. He feels that the software is firm and that we should begin work on it at our earliest opportunity. I also discussed the Random Filing System with him; we must add the new disc to this publication and do some additional updating.

Roll-In/Roll-Out Software

I talked to Janet Martin about Roll-In/Roll-Out specifications and found that they are as complete as we can expect them to be. She feels that the specifications are firm and that we should begin work on the project soon.

bb

MEMORANDUM

To: James K. Arburg, Safety Officer
From: Lawrence T. Baker, Foreman of Section A-40 210
Date: November 30, 19--

Subject: Personal-Injury Accident in Section A-40
October 10, 19--

On October 10, 19--, at 10:15 p.m., Jim Hollander, operating punch press #16, accidentally brushed the knee switch of his punch press with his right knee as he swung a metal sheet over the punching surface. The switch activated the punching unit, which severed Hollander's left thumb between the first and second joints as his hand passed through the punch station. While an ambulance was being summoned, Margaret Wilson, R.N., administered first aid at the plant dispensary. There were no witnesses to the accident.

The ambulance arrived from Mercy Hospital at 10:45 p.m., and Hollander was admitted to the emergency room at the hospital at 11:00 p.m. He was treated and kept overnight for observation, then released the next morning.

Hollander returned to work one week later, on October 17. He has been given temporary duties in the tool room until his injury heals.

Conclusions About the Cause of the Accident

The Maxwell punch press on which Hollander was working has two switches, a hand switch and a knee switch, and both must be pressed to activate the punch mechanism. The hand switch must be pressed first, and then the knee switch, to trip the punch mechanism. The purpose of the knee switch is to leave the operator's hands free to hold the panel being punched. The hand switch, in contrast, is a safety feature. Because the knee switch cannot activate the press until the hand switch has been pressed, the operator cannot trip the punching mechanism by touching the knee switch accidentally.

Inspection of the punch press that Hollander was operating at the time of the accident made it clear that Hollander had taped the hand switch of his machine in the ON position, effectively eliminating its safety function. He could then pick up a panel, swing it onto the machine's punching surface, press the knee switch, stack the newly punched panel, and grab the next unpunched panel, all in one continuous motion--eliminating the need to let go of the panel, after placing it on the punching surface, in order to press the hand switch.

To prevent a recurrence of this accident, I have conducted a brief safety session with all punch press operators, at which I described Hollander's experience and cautioned them against tampering with the safety features of their machines.

mo

Figure 1 Trouble Report

When did it occur? Was anybody hurt? Was there any property damage? Was there a work stoppage? Since insurance claims, worker's compensation awards, and, in some instances, lawsuits may hinge on the information contained in a trouble report, be sure to include precise times, dates, locations, treatment of injuries, names of any witnesses, and any other crucial information. Give a detailed analysis of what caused the problem. Be thorough and accurate in your analysis, and support any judgments or conclusions with facts. Be careful about your **tone;** avoid any condemnation or blame. If you speculate about the cause of the problem, make it clear to your **reader** that you are speculating. In your **conclusion,** state what has been done, what is being done, or what will be done to correct the conditions that led to the problem. This may include training in safety practices, better or improved equipment, protective clothing (for example, shoes or goggles), and so on.

The report shown in Figure 1, opposite, about an accident involving personal injury, was written by the foreman of a group of punch press operators for the plant's safety officer, at the safety officer's request. Since there were no witnesses, the foreman obtained the information for the report by talking to the plant nurse, hospital personnel, and the victim—and by inspecting the equipment used by the victim at the time of the accident.

try and

The **phrase** *try and* is colloquial for *try to.* Unless you are writing a casual personal letter, it is better to use *try to.*

CHANGE Please *try and* finish the report on time.
TO Please *try to* finish the report on time.

U

unity

Unity is singleness of purpose and treatment, the cohesive element that holds a piece of writing together; it means that everything in an article or paper is essentially about one thing or idea.

To achieve unity, the writer must select one **topic** and then treat it with singleness of purpose—without digressing into unrelated paths. The prime contributors to unity are a good outline and effective **transition.** After you have completed your outline, check it to see that each part relates to your subject. Be certain that your transitional terms make clear the relationship of each part to what precedes it.

Transition dovetails sentences and **paragraphs** like the joints of a well-made drawer. Notice, for example, how neatly the sentences in the following paragraph are made to fit together by the italicized words and **phrases** of transition:

EXAMPLE Any company that operates internationally today faces a host of difficulties. Inflation is worldwide. Most countries are struggling with other economic problems *as well. In addition,* many monetary uncertainties and growing economic nationalism are working against multinational companies. *Yet* ample business is available in most developed countries if you have the right products, services, and marketing organization. To maintain the growth Data Corporation has achieved overseas, we recently restructured our international operations into four major trading areas. *This* reorganization will improve the services and support that the corporation can provide to its subsidiaries around the world. *At the same time,* the reorganization establishes firm management control, ensuring consistent policies around the world. *So* you might say the problems of doing business abroad will be more difficult this year, but we are better organized to meet those problems.

The logical sequence provided by a good outline is essential to achieving unity. An outline enables the writer to lay out the most direct route from **introduction** to **conclusion** without digressing into side issues that are not related, or that are only loosely related, to the subject. Without establishing and following such a direct route, a writer cannot achieve unity.

up

Adding the word *up* to **verbs** often creates a redundant **phrase.**

CHANGE Next open *up* the exhaust valve.
TO Next open the exhaust valve.

CHANGE He wrote *up* the report.
TO He wrote the report.

(See also **conciseness.**)

usage

Usage describes the choices we make among the various words and constructions available in our language. The line between standard English and nonstandard English, or between formal and informal English, is determined by these choices. Your guideline in any situation requiring such choices should be appropriateness: Is the word or expression you use appropriate to your **reader** and subject? When it is, you are practicing good usage.

This book has been designed to help you sort out the appropriate from the inappropriate: Just look up the item in question in the index. A good **dictionary** is also an invaluable aid in your selection of the right word.

utilize

Utilize should not be used as a **long variant** of *use,* which is the general word for "employ for some purpose." When you are tempted to use this term, try to substitute *use.* It will almost always prove a clearer and less pretentious word.

CHANGE You can *utilize* the fourth elevator to reach the fiftieth floor.
TO You can *use* the fourth elevator to reach the fiftieth floor.

V

vague words

A vague word is one that is imprecise in the context in which it is used. Some words encompass such a broad range of meanings that there is no focus for their definition. Words such as *real, nice, important, good, bad, contact, thing,* and *fine* are often called "omnibus words" because they can mean everything to everybody. In speech we sometimes use words that are less than precise, but our vocal inflections and the context of our conversation make their meanings clear. Since writing cannot rely on vocal inflections, avoid using vague words. Be concrete and specific. (See also **abstract words/concrete words.**)

CHANGE It was a *meaningful* meeting, and we got *a lot* done.
TO The meeting resolved three questions: pay scales, fringe benefits, and work loads.

verb phrases

A verb phrase is a group of words that functions as a single **verb.** It consists of a main verb preceded by one or more **helping verbs.**

EXAMPLE He *is* (helping verb) *working* (main verb) hard this summer.

The main verb is always the last verb in a verb phrase.

EXAMPLES You *will file* your tax return on time if you begin early.
You *will have filed* your tax return on time if you begin early.
You *are* not *filing* your tax return too early if you begin now.

Questions often begin with a verb phrase.

EXAMPLE *Will* he *audit* their account soon?

Words can appear between the helping verb and the main verb of a verb phrase.

EXAMPLE He *is* always *working.*

The **adverb** *not* may be appended to a helping verb in a verb phrase.

EXAMPLES He *cannot work* today.
He *did not work* today.

verbals

Verbals, which are derived from **verbs,** function as **nouns, adjectives**, and **adverbs.** There are three types of verbals: **gerunds, infinitives,** and **participles.**

When the *-ing* form of a verb is used as a noun, it is called a **gerund.**

EXAMPLE *Seeing* is *believing.*

An **infinitive** is the root form of a verb, usually preceded by *to* (*to analyze, to build*). Infinitives can be used as nouns, adjectives, or adverbs.

EXAMPLES He likes *to jog.* (noun)
That is the main problem *to be solved.* (adjective)
The valve acts *to control* the flow of liquid. (adverb)

When a verb is used as an adjective, it is called a **participle;** present participles commonly end in *-ing,* past participles in *-ed, -en, -d,* or *-t.*

EXAMPLES A *burning* candle consumes stale tobacco smoke.
The *published* report contains up-to-date information.

verbs

A verb is a word, or a group of words, that describes an action (The antelope *bolted* at the sight of the hunters), states the way in which something or someone is affected by an action (He *was saddened* by the death of his friend), or affirms a state of existence (He *is* a wealthy man now).

TYPES OF VERBS

Verbs may be described as being either transitive verbs or intransitive verbs.

Transitive Verbs. A transitive verb is a verb that requires a direct object to complete its meaning.

EXAMPLES They *laid* the foundation on October 24. (*foundation* is the direct object of the transitive verb *laid*)
George Anderson *wrote* the treasurer a letter. (*letter* is the direct object of the transitive verb *wrote*)

Intransitive Verbs. An intransitive verb is a verb that does not require an object to complete its meaning. It is able to make a full assertion about the **subject** without assistance (although it may have modifiers).

EXAMPLES The water *boiled.*
The water *boiled* rapidly.
The engine *ran.*
The engine *ran* smoothly and quietly.

Although intransitive verbs do not have an **object,** certain intransitive verbs may take a **complement.** These verbs are called **linking verbs** because they link the complement to the subject. When the complement is a **noun** (or **pronoun**), it refers to the same person or thing as the noun (or pronoun) that is the subject.

EXAMPLES The winch *is* rusted. (*Rusted* is an **adjective** modifying *winch.*)
A calculator *remains* a useful tool. (*A useful tool* is a **subjective complement** renaming *calculator.*)

When the complement is an **adjective,** it *modifies* the subject.

EXAMPLES The study *was* thorough.
The report *seems* complete.

Such intransitive verbs as *be, become, seem,* and *appear* are almost always linking verbs. A number of others, such as *look, sound, taste, smell,* and *feel*, may function as either linking verbs or simple intransitive verbs. If you are unsure about whether one of them is a linking verb, try substituting *seem;* if the sentence still makes sense, the verb is probably a linking verb.

EXAMPLES Their antennae *feel* delicately. (simple intransitive verb)
Their antennae *feel* delicate. (linking verb; you could substitute *seem*)

Some verbs, such as *lie* and *sit,* are inherently intransitive.

EXAMPLE The patient should *lie* on his back during treatment. (*On his back* is an **adverb** phrase modifying the verb *lie.*)

FORMS OF VERBS

By form, verbs may be described as being either finite or nonfinite.

Finite Verbs. A finite verb is the main verb of a **clause** or sentence. It makes an assertion about its subject and can serve as the only verb in its clause or sentence. Finite verbs may be either transitive or intransitive (including linking) verbs. They are subject to changes in form to reflect **person** (I *see,* he *sees*), **tense** (I *go,* I *went*), and **number** (he *writes,* they *write*).

EXAMPLE The telephone *rang,* and the secretary *answered* it.

A **helping verb** (sometimes called an *auxiliary verb*) is used in a **verb phrase** to help indicate **mood,** tense, and **voice.**

EXAMPLES The work *had* begun.
I *am* going.
I *was* going.
I *will* go.
I *should have* gone.
I *must* go.

The most commonly used helping verbs are the various forms of *have* (*has, had*), *be* (*is, are, was,* and so on), *do,* (*did, does*), and *can (may, might, must, shall, will, would, should* and *could*). **Phrases** that function as helping verbs are often made up of combinations with the sign of the infinitive, *to*: for example, *am going to* and *is about to* (compare *will*), *has to* (compare *must*), and *ought to* (compare *should*).

The helping verb always precedes the main verb, although other words may intervene.

EXAMPLE Machines *will* never completely *replace* people.

Nonfinite Verbs. Nonfinite verbs are **verbals,** which, although they are derived from verbs, actually function as nouns, adjectives, or **adverbs.**

When the *-ing* form of a verb is used as a noun, it is called a **gerund.**

EXAMPLE *Seeing* is *believing.*

An **infinitive,** which is the root form of a verb (usually preceded by *to*), can be used as a noun, an adverb, or an adjective.

EXAMPLES He hates *to complain.* (noun, direct object of *hates*)
The valve closes *to stop* the flow. (adverb, modifies *closes*)
This is the proposal *to select.* (adjective, modifies *proposal*)

A **participle** is a verb form used as an adjective.

EXAMPLES His *closing* statement was very *convincing.*
The *rejected* proposal was ours.

PROPERTIES OF VERBS

Person is the grammatical term for the form of a **personal pronoun** that indicates whether the pronoun refers to the speaker, the person spoken to, or the person (or thing) spoken about. Verbs change their forms to agree in person with their subjects.

EXAMPLES I *see* (first person) a yellow tint, but he *sees* (third person) a yellow-green hue.
I *am* (first person) convinced, and you *are* (third person) not convinced.

Voice refers to the two forms of a verb that indicate whether the subject of the verb acts or receives the action. If the subject of the verb acts, the verb is in the active voice; if it receives the action, the verb is in the passive voice.

EXAMPLES The aerosol bomb *propels* the liquid as a mist. (active)
The liquid *is propelled* as a mist by the aerosol bomb. (passive)

In your writing, the active voice provides force and momentum, whereas the passive voice lacks these qualities. The reason is not difficult to find. In the active voice, the verb plays the key role in identifying what the subject is doing. The passive voice, on the other hand, consists of a form of the verb *to be* and a past participle of another verb. In the passive voice, the **emphasis** is on what is being done to the subject.

EXAMPLES Things are seen by the normal human eye in three dimensions: length, width, and depth. (*Things* takes precedence over the eye's function.)
The normal human eye sees things in three dimensions: length, width, and depth. (Here the eye's function—which is what the sentence is about—receives the emphasis.)

Number refers to the two forms of a verb that indicate whether the subject of a verb is singular or plural.

EXAMPLES The machine *was* in good operating condition. (singular)
The machines *were* in good operating condition. (plural)

Tense refers to verb forms that indicate time distinctions. There are six tenses: present, past, future, present perfect, past perfect, and future perfect. All verb tenses are derived from the three principal parts of the verb: the infinitive (without *to,* which normally precedes it); the past form; and the past participle. The following are the six tenses of *write* (infinitive), whose other two principal parts are *wrote* (past form) and *written* (past participle):

EXAMPLES I *write* (present, based on infinitive)
I *wrote* (past, based on past form)
I *will write* (future, based on infinitive)
I *have written* (present perfect, based on past participle)
I *had written* (past perfect, based on past participle)
I *will have written* (future perfect, based on past participle)

Note that *write* is an irregular verb; that is, its past and past participle forms are produced by internal changes: *write, wrote, written.* Regular verbs, on the other hand, form the past and past participle by the addition of the suffixes *-ed, -d,* or *-t: talk, talked; believe, believed; spend, spent.* The tenses of regular verbs, like those of irregular verbs, are derived from the three principal parts.

EXAMPLES I *believe* (present, based on infinitive)
I *believed* (past, based on past form)
I *will believe* (future, based on infinitive)
I *have believed* (present perfect, based on past participle)
I *had believed* (past perfect, based on past participle)
I *will have believed* (future perfect, based on past participle)

Each of the six tenses also has a progressive form. The progressive form is created by adding the appropriate tense of *be* to the present participle (*-ing*) form of the verb.

EXAMPLES I *am writing.* (present progressive)
I *was writing.* (past progressive)
I *will be writing.* (future progressive)
I *have been writing.* (present perfect progressive)
I *had been writing.* (past perfect progressive)
I *will have been writing.* (future perfect progressive)

(See also **tense.**)

CONJUGATION OF VERBS

The conjugation of a verb arranges all forms of the verb so that the differences caused by the changing of the tense, number, person, and voice are readily apparent. At the bottom of this page, and continued on the next page, is a conjugation of the verb *drive*.

very

The temptation to overuse **intensifiers** like *very* is great. Evaluate your use of them carefully. When you do use them, clarify their meaning.

EXAMPLE Bicycle manufacturers had a *very* good year: sales across the country were up 43 percent over the previous year.

In many sentences, however, the word can simply be deleted.

CHANGE The board was *very* angry about the newspaper report.
TO The board was angry about the newspaper report.

via

Via is Latin for "by way of." *(Please turn to page 704.)*

EXAMPLE The equipment is being shipped to Los Angeles *via* Chicago.

Tense	*Number*	*Person*	*Active Voice*	*Passive Voice*
Present		1st	I drive	I am driven
	Singular	2nd	You drive	You are driven
		3rd	He drives	He is driven
		1st	We drive	We are driven
	Plural	2nd	You drive	You are driven
		3rd	They drive	They are driven
Progressive Present		1st	I am driving	I am being driven
	Singular	2nd	You are driving	You are being driven
		3rd	He is driving	He is being driven
		1st	We are driving	We are being driven
	Plural	2nd	You are driving	You are being driven
		3rd	They are driving	They are being driven
Past		1st	I drove	I was driven
	Singular	2nd	You drove	You were driven
		3rd	He drove	He was driven
		1st	We drove	We were driven
	Plural	2nd	You drove	You were driven
		3rd	They drove	They were driven

Tense	*Number*	*Person*	*Active Voice*	*Passive Voice*
Progressive Past	Singular	1st	I was driving	I was being driven
		2nd	You were driving	You were being driven
		3rd	He was driving	He was being driven
	Plural	1st	We were driving	We were being driven
		2nd	You were driving	You were being driven
		3rd	They were driving	They were being driven
Future	Singular	1st	I will drive	I will be driven
		2nd	You will drive	You will be driven
		3rd	He will drive	He will be driven
	Plural	1st	We will drive	We will be driven
		2nd	You will drive	You will be driven
		3rd	They will drive	They will be driven
Progressive Future	Singular	1st	I will be driving	I will have been driven
		2nd	You will be driving	You will have been driven
		3rd	He will be driving	He will have been driven
	Plural	1st	We will be driving	We will have been driven
		2nd	You will be driving	You will have been driven
		3rd	They will be driving	They will have been driven
Present Perfect	Singular	1st	I have driven	I have been driven
		2nd	You have driven	You have been driven
		3rd	He has driven	He has been driven
	Plural	1st	We have driven	We have been driven
		2nd	You have driven	You have been driven
		3rd	They have driven	They have been driven
Past Perfect	Singular	1st	I had driven	I had been driven
		2nd	You had driven	You had been driven
		3rd	He had driven	He had been driven
	Plural	1st	We had driven	We had been driven
		2nd	You had driven	You had been driven
		3rd	They had driven	They had been driven
Future Perfect	Singular	1st	I will have driven	I will have been driven
		2nd	You will have driven	You will have been driven
		3rd	He will have driven	He will have been driven
	Plural	1st	We will have driven	We will have been driven
		2nd	You will have driven	You will have been driven
		3rd	They will have driven	They will have been driven

The term should be used only in routing instructions.

CHANGE His project was funded *via* the recent legislation.
TO His project was funded *through* the recent legislation.
OR His project was funded *as the result of* the recent legislation.

vogue words

Vogue words are words that suddenly become popular and, because of an intense period of overuse, lose their freshness and preciseness. They may become popular through their association with science, technology, or even sports. We include them in our vocabulary because they seem to give force and vitality to our language. Ordinarily, this language sounds pretentious in our day-to-day writing.

EXAMPLES *super, interface* (as a verb), *bottom line, input, mode, variable, state of the art, impact* (as a verb), *cutting edge, parameter, communication, feedback,* and many words ending in *-wise*

Obviously, some of these terms are appropriate in the right context. It is when they are used outside that context that imprecision becomes a problem.

EXAMPLE An *interface* was provided between the two companies. (appropriate)
CHANGE We must *interface* with the Purchasing Department. (inappropriate)
TO We must *cooperate* with the Purchasing Department.

This book discusses some vogue terms individually; use the index to find the word in question. For those not included, check a current **dictionary.** (See also **gobbledygook** and **trite language**.)

voice

In grammar, voice indicates the relation of the subject to the action of the verb. When the **verb** is in the active voice, the **subject** acts; when it is in the passive voice, the subject is acted upon.

EXAMPLES David Cohen *wrote* the advertising copy. (active)
The advertising copy *was written* by David Cohen. (passive)

Both of the sentences say the same thing, but each has a different **emphasis:** In the first sentence emphasis is on the subject, *David Cohen,* whereas in the second sentence the focus is on the object, *the ad-*

vertising copy. Notice how much stronger and more forceful the active sentence is.

One of the rules of good writing is always to use the active voice unless there is good reason to use the passive. Because they are wordy and indirect, passive sentences are hard for the **reader** to understand. Notice how much clearer the following passive sentence becomes when it is rewritten in the active voice:

> CHANGE Things *are seen* by the normal human eye in three dimensions: length, width, and depth.
>
> TO The human eye *sees* things in three dimensions: length, width, and depth.

Passive-voice sentences are wordy because they always use a **helping verb** in addition to the main verb, and an extra preposition if they identify the doer of the action specified by the main verb.

> ACTIVE Employees *resent* changes in policy.
>
> PASSIVE Changes in policy *are resented by* employees.

The active-voice version takes one verb (*resent*) and one preposition (*in*); the passive-voice version takes two verbs (*are resented*) and two prepositions (*in* and *by*). The passive-voice version is also indirect because it puts the doer of the action behind the verb instead of in front of it.

One difficulty with passive sentences is that they can bury the subject, or performer of the action, in **expletives** and prepositional phrases.

> CHANGE It *was reported* by Engineering that the new relay is defective.
>
> TO Engineering *reported* that the new relay is defective.

Sometimes writers using the passive voice fail to name the performer—information that might be missed.

> CHANGE The problem *was discovered* yesterday.
>
> TO The Engineering Department *discovered* the problem yesterday.

Very often the passive voice can be just plain confusing, especially when used in **instructions.**

> CHANGE Plates B and C should be marked for revision. (Are they already marked?)
>
> TO You *should mark* plates B and C for revision.
>
> OR *Mark* plates B and C for revision.

Another problem that sometimes occurs with the passive voice is **dangling modifiers.**

CHANGE Hurrying to complete the work, the wires *were connected* improperly. (Who was hurrying, the wires?)

TO Hurrying to complete the work, the technician improperly *connected* the wires. (Here, *hurrying to complete the work* properly modifies *technician.*)

There are, however, certain instances when the passive voice is effective or even necessary. Indeed, for reasons of tact and diplomacy, you might need to use the passive voice to *avoid* identifying the doer of the action.

CHANGE Your sales force didn't meet the quota last month. (active)

TO The quota wasn't met last month. (passive)

When the performer of the action is either unknown or unimportant, use the passive voice.

EXAMPLES The copper mine *was discovered* in 1929.
Fifty-six barrels *were processed* in two hours.

When the performer of the action is less important than the receiver of that action, the passive voice is sometimes more appropriate.

EXAMPLE Ann Bryant *was presented* with an award by the president.

When you are explaining an operation in which the reader is not actively involved (but not when you are giving the reader **instructions**) or when you are explaining a process or a procedure, the passive voice may be more appropriate.

EXAMPLE Area strip mining *is used* in regions of flat-to-gently rolling terrain, like that found in the Midwest and West. Depending on applicable reclamation laws, the topsoil may *be removed* from the area to be mined, *stored,* and later *reapplied* as surface material during reclamation of the mined land. After the removal of the topsoil, a trench *is cut* through the overburden to expose the upper surface of the coal to be mined. The length of the cut generally corresponds to the length of the property or of the deposit. The overburden from the first cut *is placed* on the unmined land adjacent to the cut. After the first cut *has been completed,* the coal *is removed,* and a second cut *is made* parallel to the first.

Anyone could be the doer of the action in this example; it really doesn't matter who does it. Therefore, it would be pointless for the writer to put this passage in the active voice. Do not, however, simply assume that any such explanation should be in the passive voice. Ask yourself, "Would it be of any advantage to the reader to know the doer of the action?" If you can answer no, then use the passive voice. But if your honest answer is yes, then use the active voice. The following example uses the active voice to explain an operation:

EXAMPLE In the operation of an internal combustion engine, an explosion in the combustion chamber *forces* the pistons down in the cylinders. The movement of the pistons in the cylinders *turns* the crankshaft. The crankshaft, in turn, *operates* the differential. The differential *turns* the rear axles, which *turn* the wheels and *move* the automobile.

Whether you use the passive or the active voice, however, be careful not to *shift* voices in a sentence.

CHANGE Ms. McDonald *corrected* the malfunction as soon as it *was identified* by the technician.

TO Ms. McDonald *corrected* the malfunction as soon as the technician *identified* it.

W

wait for/wait on

Wait on should be restricted in writing to the activities of waitresses and waiters—otherwise, use *wait for.* (See also **idioms.**)

EXAMPLES Be sure to *wait for* Ms. Sturgess to finish reading the report before asking her to make a decision.
Joe's feet ached from *waiting on* tables at the diner.

when and if

When and if (or *if and when*) is a colloquial expression that should not be used in writing.

CHANGE *When and if* your new position is approved, I will see that you receive adequate staff help.
TO *If* your new position is approved, I will see that you receive adequate staff help.
OR *When* your new position is approved, I will see that you receive adequate staff help.

where . . . at

In **phrases** using the *where . . . at* construction, *at* is unnecessary and should be omitted.

CHANGE *Where* is his office *at*?
TO *Where* is his office?

where/that

Do not substitute *where* for *that* to anticipate an idea or fact to follow.

CHANGE I read in the IEEE journal *where* molecules will be used in our process.
TO I read in the IEEE journal *that* molecules will be used in our process.

whether or not

When *whether or not* is used to indicate a choice between alternatives, omit *or not*; it is redundant, since *whether* itself communicates the notion of a choice.

> CHANGE The project director asked *whether or not* the request for proposals had been issued.
>
> TO The project director asked *whether* the request for proposals had been issued.

(See also **as to whether.**)

while

While, meaning "during an interval of time," is sometimes substituted for connectives like *and, but, although,* and *whereas*. Used as a connective in this way, *while* often causes **ambiguity.**

> CHANGE John Evans is sales manager, *while* Joan Thomas is in charge of research.
>
> TO John Evans is sales manager, *and* Joan Thomas is in charge of research.

Do not use *while* to mean *although* or *whereas*.

> CHANGE *While* Ryan Patterson wants the job of chief engineer, he has not yet asked for it.
>
> TO *Although* Ryan Patterson wants the job of chief engineer, he has not yet asked for it.

Restrict *while* to its meaning of "during the time that."

> EXAMPLE I'll have to catch up on my reading *while* I am on vacation.

who/whom

Writers often have difficulty with the choice of *who* or *whom*. *Who* is the subjective **case** form, *whom* is the objective case form, and *whose* is the possessive case form. When in doubt about which form to use, try substituting a **personal pronoun** to see which one fits. If *he* or *they* fits, use *who*.

> EXAMPLES *Who* is the congressman from the tenth district?
> *He* is the congressman from the tenth district.

If *him* or *them* fits, use *whom.*

EXAMPLES It depended on *them.*
It depended on *whom?*

It is becoming common to use *who* for the objective case when it begins a sentence, although some still object to such an ungrammatical construction. The best advice is to know your **reader.**

The choice between *who* and *whom* is especially complex in sentences with interjected **clauses.**

CHANGE These are the men *whom* you thought were the architects.
TO These are the men *who* you thought were the architects.

The parenthetical clause *you thought* tends to attract the form *whom,* a form that appears logical as the object of the verb form *thought.* In fact, *who* is the correct form, since it is the subject of the verb form *were.* Note that the interjected clause *you thought* can be omitted without disturbing the construction of the clause.

EXAMPLE These are the men *who* were the architects.

whose/of which

Whose should normally be used with persons; *of which* should normally be used with inanimate objects.

EXAMPLES The man *whose* car had been towed away was angry.
The mantle clock, the parts *of which* work perfectly, is over one hundred years old.

If these uses cause a sentence to sound awkward, however, *whose* may be used with inanimate objects.

EXAMPLE There are added fields, for example, *whose* totals should never be zero.

who's/whose

Who's is a **contraction** of *who is.*

EXAMPLE *Who's* scheduled to attend the productivity seminar next month?

Whose is the possessive for *who* or *of which.*

EXAMPLE *Whose* department will be affected by the budget cuts?

Who's and *whose* are not interchangeable.

-wise

Although the **suffix** *-wise* often seems to provide a tempting shortcut, it leads more often to inept than to economical expression. It is better to rephrase the sentence.

CHANGE Our department rates high efficiencywise.
TO Our department has a high efficiency rating.

The *-wise* suffix is appropriate, however, to **instructions** that indicate certain space or directional requirements.

EXAMPLES Fold the paper *lengthwise.*
Turn the adjustment screw half a turn *clockwise.*

word choice

As Mark Twain once said, "The difference between the right word and almost the right word is the difference between 'lightning' and 'lightning bug.'" The most important goal in choosing the right word in technical writing is the preciseness implied by Twain's comment. **Vague words** and **abstract words** defeat preciseness because they do not convey the writer's meaning directly and clearly. Vague words are imprecise because they can mean many different things.

CHANGE It was a *meaningful* meeting.
TO The meeting helped both sides understand each other's position.

In the first sentence, *meaningful* ironically conveys no meaning at all. See how the revised sentence says specifically what made the meeting meaningful. Although abstract words may at timcs be appropriate to your **topic,** their unnecessary use creates dry and lifeless writing.

ABSTRACT work, fast, food
CONCRETE sawing, 110 m.p.h., steak

Being aware of the **connotation** and denotation of words will help you anticipate the **reader's** reaction to the words you choose. Connotation is the suggested or implied meaning of a word beyond its dictionary definition. Denotation is the literal, or primary, dictionary meaning of a word.

Understanding **antonyms** and **synonyms** will increase your ability to choose the proper word. Antonyms are words with nearly the opposite meaning (*fresh/stale*), and synonyms are words with nearly the same meaning (*notorious/infamous*).

Malapropisms and **trite language** are likely to irritate your readers, by either confusing or boring them. Malapropisms are words that sound like the intended word but have different, sometimes even humorous, meanings.

> CHANGE The printer and the card reader are in *convention* with each other for the processor's attention.
>
> TO The printer and the card reader are in *contention* with each other for the processor's attention.

Trite language consists of stale and worn phrases, frequently **clichés.**

> CHANGE We will finish the project *quick as a flash*.
>
> TO We will finish the project *quickly*.

Avoid using technical **jargon** unless you are certain that all of your readers understand the terms. Avoid choosing words with the objective of impressing your reader, which is called **affectation;** also avoid using **long variants,** which are elongated forms of words used only to impress (*utilize* for *use, telephonic communication* for *telephone call, analyzation* for *analysis*). Using a **euphemism** (an inoffensive substitute for a word that is distasteful or offensive) may help you avoid embarrassment, but the overuse of euphemisms becomes affectation. Using more words than necessary (*in the neighborhood of* for *about, for the reason that* for *because, in the event that* for *if*) is certain to interfere with **clarity.**

Understanding the use of **clipped forms of words, compound words, blend words, idioms,** and **foreign words in English** will help you find the appropriate word for a specific use. Clipped forms of words are created by dropping the beginning or ending of a word (*phone* for *telephone, ad* for *advertisement*). Compound words are made from two or more words that are either hyphenated or written as one word (*mother-in-law, nevertheless*). Blend words are created by using two words to produce a third word (*smoke* + *fog* = *smog*).

A key to choosing the correct and precise word is to keep current in your reading and to be aware of new words in your profession and in the language. Be aware also, in your quest for the right word, that there is no substitute for a good **dictionary.**

word processing

Word processing enables you to type documents, such as **memorandums** and **reports,** into a computer, edit your text on a video screen, print out a well-formatted paper copy of your document, and then store it electronically for future **revision.** If this technology is new to you, it helps to think of a word processor as a sophisticated electric typewriter connected to a computer, television screen, and printer. The main advantage of word processing is that once you have typed your text into the computer, you can easily revise it, deleting errors and inserting corrections and creating a perfect final draft without having to retype the entire document.

Nearly all word-processing systems have the ability to create and manipulate texts in the following three ways: cursor-controlled editing, document formatting and printing, and file storage and retrieval.

Cursor-controlled or "full-screen" editing enables you to move the "cursor" (a solid or flashing light that marks your position on the video screen) through your text and to perform editing tasks, such as the following:

- inserting, deleting, or replacing text anywhere in a document;
- deleting, copying, or moving blocks of text, such as phrases, sentences, paragraphs, or whole pages;
- searching for and replacing individual letters, words, or phrases;
- returning the carriage automatically at the end of each line and automatically readjusting margins and page "breaks" as you edit.

Document formatting commands enable you to create individual or standard **formats** for your documents, including such features as:

- fixed margins and automatic **indentation**;
- automatic page numbering and **headers and footers**;

- automatic notes (end notes or footnotes);
- multiple copies of your documents using a variety of type styles and sizes, including boldface and italics.

File storage and retrieval enables you to store documents on a permanent magnetic medium, such as disk or tape, so that you may:

- save all original and revised copies of your documents in an electronic filing system and print a list of your files at any time;
- retrieve the documents at a later date for revision and reprinting;
- merge, copy, and delete documents at will.

Writers find the word processor efficient for recording ideas quickly when writing first drafts and for extensive rewriting during the revision process. In the business world, the computer is rapidly becoming a primary communication tool for all professionals. Because **correspondence,** reports, and **technical manuals** can be quickly produced and revised, many companies give writers and technical personnel word processors for everyday use. Businesses are beginning to use computers as management tools by linking word-processing technology to other computer applications, such as automatic document preparation, computer graphics, data base retrieval, electronic mail, and "mail-merge" programs that merge large lists of mailing addresses with form letters.

HARDWARE

The term *hardware* refers to the physical parts, or electronic and mechanical components, of a computer system. Most word processors use the same basic components. Figure 1 shows the five basic parts of a typical microcomputer system.

1. The computer *keyboard* allows you to type data and commands into the computer. It looks like a regular typewriter keyboard but contains several extra keys, including "function keys" for special operations, such as saving documents, listing files, moving blocks of text, and issuing search and replace commands.
2. The *main system unit* includes three segments: the central processing unit (CPU), internal memory, and disk drive units. The central processing unit is the "brain" that manipulates data and carries out programmed instructions. The computer's internal memory refers to the temporary storage used by the central processing unit for performing programming tasks and creating

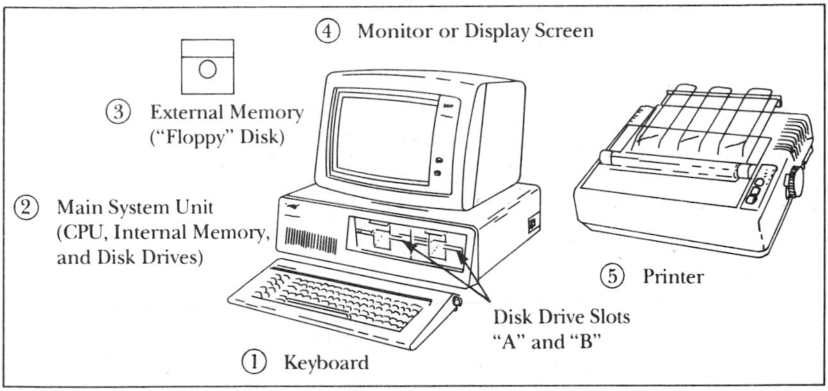

Figure 1 Typical Microcomputer System Used for Word Processing

temporary working files. The disk drive units, shown as slots "A" and "B" in Figure 1, are standard input/output (I/O) devices for reading and writing stored information back and forth between floppy disks and the computer's internal memory.

3. *External memory* refers to any permanent storage medium, such as the flexible or "floppy" disks, on which you can save all your documents. The advantage of floppy disks is that, like cassette tapes, they are portable. You can remove them from the computer for safe storage and reinsert them later to retrieve individual documents. A single floppy disk stores approximately 100 to 200 pages of text. Other forms of external memory include "hard disks" and magnetic tapes capable of storing much larger quantities of data than floppy disks can.
4. The *monitor* resembles a common television screen. It displays everything you type into the computer, the computer's responses, and any documents you retrieve from storage.
5. The *printer* is similar to a typewriter. When you give the appropriate command, the printer prints your text on paper. Microcomputers use three different kinds of printers: dot matrix, letter quality, and laser, each of which produces different typefaces of varying quality, as shown in Figure 2.

Dot matrix printers are fast, reliable printers whose print characters are composed of tiny dots. Although some writers feel that the "computer-generated" look of the dot image is suitable only for draft copies, this typeface is gaining greater acceptance as more

This is an example of dot matrix printing in draft mode.

This is an example of dot matrix near-letter quality printing.

This is an example of letter quality printing.

This is an example of laser printing with **bold**, *italic*, and **enlarged** typefaces.

Figure 2 Varying Qualities of Typeface Produced by Different Printers

writers use the double-strike or "overstrike" mode (shown in the second example of Figure 2) to produce "near letter quality" dot matrix print for memos and informal reports. *Letter quality printers* are slower than dot matrix printers and use a daisy wheel or thimble to print a clear, dark typeface similar to that of an office typewriter. *Laser printers* work on the same principle as do photocopy machines, using a jet-spraying device to deposit carbon images directly on paper. Laser printers can produce a great variety of high-quality typefaces comparable to professionally printed material.

SOFTWARE AND COMPUTER SYSTEMS

The term *software* refers to the precoded instructions or programs that tell the computer how to execute the commands you type into the computer. A microcomputer uses software, such as word processing, graphics, and electronic spreadsheet programs, stored on floppy disks. The advantage of the microcomputer is that all software and hardware "stand alone" in an easy-to-use, self-contained system. In a mini- or mainframe computer system, your individual terminal (keyboard and screen) connects with a large central computer, but you share the CPU, memory, printer, and word-processing software (usually consisting of "text editor" and "file formatter" programs) with other users. Although mainframe systems do not give individual writers exclusive use of the central computer, they provide a huge storage capacity and the ability to communicate with other users through electronic mail messages.

Distributed processing systems link microcomputers through a central mainframe computer, combining all the advantages of stand-alone microcomputing with large mainframe processing and communications. By linking or "networking" microcomputers through larger computer systems in remote locations, you can transmit documents electronically from one microcomputer to another microcomputer in the same building, across town, or thousands of miles away, where the documents can be viewed, printed out, revised, and transmitted back to you.

WORD PROCESSING AND THE WRITING PROCESS

The word processor is a powerful tool that can help you record your ideas quickly, improve your writing and revising skills, and create professionally designed documents. At the same time, however, good writing is still the result of careful planning, constant practice, and thoughtful revision. In some cases, word-processing technology can initially intrude on the writing process and impose certain limitations that many beginners overlook. The ease of making minor, sentence-level changes and the limitation of a twenty-four-line viewing screen, for example, may focus your attention too narrowly on the text's surface features so that you lose sight of the larger problems of **scope** and **organization.** Or the fluid and rapid movement of the text on the screen together with last-minute editing changes may allow typographical errors and mispellings to creep into the text. Also, as you master the machine and become enamored of its powerful revising capabilities, you may begin to "overwrite" your documents. Inserting phrases and rewriting sentences become so easy that you may find yourself writing more text and rewriting more extensively, but saying less. The following tips will help you avoid these pitfalls and develop writing strategies that take advantage of the benefits offered by word-processing technology.

TIPS ON USING THE WORD PROCESSOR TO IMPROVE YOUR WRITING SKILLS

1. Avoid writing first drafts on the computer without any planning or outlining. Plan your document by identifying your **objective, readers,** and scope and by completing your **research** before you begin **writing the draft** on the computer.
2. Use the word processor to brainstorm and organize an initial outline for your topic.
3. When you're ready to begin writing, you can overcome "writer's block" by practicing "free writing" on the computer. Free writing means typing your thoughts as quickly as possible without stopping to correct mistakes or complete your sentences.

Concentrate on what you are writing rather than on how you are writing it—concentrate on perfection later.

4. Use the search command to find and delete wordy phrases such as *that is, there are, the fact that, to be,* and unnecessary **helping verbs** such as *will.*
5. If the software is available, use spelling checkers and other specialized programs to identify and correct typographical errors, misspellings, and **grammar** and **diction.** Do not rely completely on computerized spelling checkers, however, since they cannot identify errors such as the misuse of *there* for *their.* Maintain a file of your most frequently misspelled or misused words, and use the search command to check them in your documents.
6. Avoid excessive editing and rewriting on the screen. Print out periodically a complete paper copy of your drafts for major revisions and reorganization.
7. Always proofread your final copy on paper, since the fluidity of the viewing screen makes it difficult to catch all of the errors in your manuscript. Print out an extra copy of your document for your peers to comment on before making final revisions.
8. When writing a single document for multiple readers, use the search command to find technical terms and other data that may need further explanation for secondary readers or inclusion in a **glossary.**
9. Use the computer for effective document design by emphasizing major headings and subheadings with boldface, by using the copy command to create and duplicate parallel headings throughout your text, and by inserting blank lines (carriage returns) and tab key spaces in your text to create extra white space around examples and **illustrations.**
10. Frequently "save" or store your text on disk during long writing sessions and always create an extra or "backup" copy of your documents for safekeeping.
11. Keep on file a standard version of certain documents, such as your **résumé** and **application letters,** so you can revise them to meet the needs of each new job opportunity.

writing a draft

You are well prepared to write a rough draft when you have established your **objective, readers'** needs, and **scope** and when you have done adequate **research** and **outlining.** Writing a rough draft is simply transcribing and expanding the notes from your outline into **paragraphs,** without worrying about **grammar,** refinements of language, or such mechanical aspects of writing as **spelling.** Refinement will come with **revision.**

Write a rough draft quickly, concentrating on converting your outline into sentences and paragraphs (see also **sentence construction**). Write as though you were explaining your subject to someone across the desk from you. Don't worry about a good opening. Just start. There is no need in a rough draft to be concerned about an **introduction** or **transitions** unless they come easily—concentrate on *ideas.* Don't try to polish or revise. Writing and revising are different activities. Keep writing quickly to achieve **unity** and proportion.

Even with good preparation, however, writing a draft remains a chore and an obstacle for many writers. Experienced writers use the following tactics to get started and keep moving. Discover which ones are the most helpful to you.

- Set up your writing area with the equipment and materials (paper, dictionary, source books, and so forth) you will need to keep going once you get started.
- Use whatever writing tools—separately or in combination—are most comfortable for you: pencil, felt-tip pen, typewriter, word processor, or whatever.
- Remind yourself that you are beginning a version of your writing project that no one else will read.
- Remember the writing projects you've finished in the past—you have completed something before, and you will this time.
- Start with the section that seems easiest to you—your reader will not know or care that you first wrote a section in the middle.
- Give yourself a time limit (10 or 15 minutes, for example) in which you will keep your pen or fingers on the typewriter moving, regardless of how good or bad your writing seems to you. The point is to keep moving.
- Don't let anything stop you when you are rolling along easily—if you stop and come back, you may not regain the momentum.
- But stop writing before you're completely exhausted; when you begin again, you may be able to regain your momentum.
- When you finish a section, give yourself a small reward—a short walk, a cup of coffee, a chat with a friend, an easy task, and so on.

- Reread what you've written when you resume writing. Often seeing what you've written will trigger the frame of mind that was productive.
- Switch the writing tools you were using when you stopped: go from ball-point pen to typewriter, for example.
- If all else fails, use a tape recorder to dictate a rough draft. (See the entry on **dictation**.)

The most effective way to start and keep going, however, is to use a good outline as a springboard and map for your writing. Your outline notes can become the topic sentences for paragraphs in your draft, as shown below. For example, notice that items III.A. and III.B. in the following topic outline become the topic sentences of the succeeding two paragraphs. And notice, too, that the subordinate items in the outline become sentences in the two paragraphs.

Outline

III. Advantages of Chicago as Location for New Plant
  A. *Outstanding Transport Facilities*
    1. Rail
    2. Air
    3. Truck
    4. Sea (except in winter)
  B. *Ample Labor Supply*
    1. Engineering and scientific personnel
      a. Many similar companies in the area
      b. Several major universities
    2. Technical and manufacturing personnel
      a. Existing programs in community colleges
      b. Possible special programs designed for us

Resulting Paragraphs

Probably the greatest advantages of Chicago as a location for our new plant is its excellent transport facilities. The city is served by three major railroads. Both domestic and international air cargo service is available at O'Hare International Airport. Chicago is a major hub of the trucking industry, and most of the nation's large freight carriers have terminals there. Finally, except in the winter months when the Great Lakes are frozen, Chicago is a seaport, accessible through the St. Lawrence Seaway.

> *A second advantage of Chicago is that it offers a large labor force.* An ample supply of engineering and scientific personnel is assured not only by the presence of many companies engaged in activities similar to ours but also by the presence of several major universities in the metropolitan area. Similarly, technicians and manufacturing personnel are in abundant supply. The seven colleges in the Chicago City College system, as well as half a dozen other two-year colleges in the outlying areas, produce graduates with associate degrees in a wide variety of technical specialties appropriate to our needs. Moreover, three of the outlying colleges have expressed an interest in establishing courses attuned specifically to our requirements.

Remember that as you write, your function as a writer is to communicate certain information to your readers. Do not try to impress them with a fancy writing **style.** Write in a plain and direct style that is comfortable and natural for both you and your reader—the first rule of good writing is to *help the reader*.

As you write your rough draft, keep in mind your reader's level of knowledge of the subject. Doing so will not only help you write directly to your reader, it will also tell you which terms you must define. (See also **defining terms.**)

When you are trying to write quickly and come to something difficult to explain, try to relate the new concept to something with which the reader is already familiar. Although **figures of speech** are not used extensively in technical writing, they can be very useful in explaining a complex process. In the rough draft, a figure of speech might be just the tool you need to keep moving when you encounter a complex concept that must be explained or described.

Above all, don't wait for inspiration to write your rough draft—you need to treat writing a draft in technical writing as you would any on-the-job task. Use the tips in this entry to help you complete the task. (See also the Checklist of the Writing Process.)

X

X-ray

X-ray, usually capitalized, is always hyphenated as an **adjective** and **verb** and usually hyphenated as a **noun**. If you hyphenate the noun, be sure to do so consistently.

EXAMPLES portable *X-ray* unit (adjective)
The technician *X-rayed* the ankle. (verb)
X-rays and gamma rays (noun)

Z

zeugma

A zeugma is a construction in which an adjective or verb is forced to apply to two words, even though it is grammatically or logically correct only for one of them. Poets, and occasionally public speakers, use zeugma deliberately for rhythmic effect; but it is imprecise and should therefore be avoided in technical writing.

> CHANGE The walls of the office *are* green, the rug, blue.
> TO The walls of the office *are* green; the rug *is* blue.

In this case *are* (a plural **verb**) is used for both subjects, *walls* (a plural **noun**) and *rug* (a singular noun), which is grammatically and logically incorrect.

Index

Index terms followed by page numbers refer to entries in the handbook. Terms without page numbers refer either to synonyms for entries or to topics discussed within an entry.